W9-CRL-519

NUCLEIC ACIDS: THE VECTORS OF LIFE

THE JERUSALEM SYMPOSIA ON
QUANTUM CHEMISTRY AND BIOCHEMISTRY

*Published by the Israel Academy of Sciences and Humanities,
distributed by Academic Press (N.Y.)*

1st JERUSALEM SYMPOSIUM:	*The Physicochemical Aspects of Carcinogenesis* (October 1968)
2nd JERUSALEM SYMPOSIUM:	*Quantum Aspects of Heterocyclic Compounds in Chemistry and Biochemistry* (April 1969)
3rd JERUSALEM SYMPOSIUM:	*Aromaticity, Pseudo-Aromaticity, Antiaromaticity* (April 1970)
4th JERUSALEM SYMPOSIUM:	*The Purines: Theory and Experiment* (April 1971)
5th JERUSALEM SYMPOSIUM:	*The Conformation of Biological Molecules and Polymers* (April 1972)

*Published by the Israel Academy of Sciences and Humanities,
distributed by D. Reidel Publishing Company (Dordrecht, Boston and Lancaster)*

6th JERUSALEM SYMPOSIUM:	*Chemical and Biochemical Reactivity* (April 1973)

*Published and distributed by D. Reidel Publishing Company
(Dordrecht, Boston and Lancaster)*

7th JERUSALEM SYMPOSIUM:	*Molecular and Quantum Pharmacology* (March/April 1974)
8th JERUSALEM SYMPOSIUM:	*Environmental Effects on Molecular Structure and Properties* (April 1975)
9th JERUSALEM SYMPOSIUM:	*Metal-Ligand Interactions in Organic Chemistry and Biochemistry* (April 1976)
10th JERUSALEM SYMPOSIUM:	*Excited States in Organic Chemistry and Biochemistry* (March 1977)
11th JERUSALEM SYMPOSIUM:	*Nuclear Magnetic Resonance Spectroscopy in Molecular Biology* (April 1978)
12th JERUSALEM SYMPOSIUM:	*Catalysis in Chemistry and Biochemistry Theory and Experiment* (April 1979)
13th JERUSALEM SYMPOSIUM:	*Carcinogenesis: Fundamental Mechanisms and Environmental Effects* (April/May 1980)
14th JERUSALEM SYMPOSIUM:	*Intermolecular Forces* (April 1981)
15th JERUSALEM SYMPOSIUM:	*Intermolecular Dynamics* (October 1982)

VOLUME 16

NUCLEIC ACIDS: THE VECTORS OF LIFE

PROCEEDINGS OF THE SIXTEENTH JERUSALEM SYMPOSIUM ON
QUANTUM CHEMISTRY AND BIOCHEMISTRY HELD IN
JERUSALEM, ISRAEL, 2–5 MAY, 1983

Edited by

BERNARD PULLMAN

Université Pierre et Marie Curie (Paris VI)
Institut de Biologie Physico-Chimique
(Fondation Edmond de Rothschild) Paris, France

and

JOSHUA JORTNER

Department of Chemistry
Tel-Aviv University, Tel-Aviv, Israël

D. REIDEL PUBLISHING COMPANY

A MEMBER OF THE KLUWER ACADEMIC PUBLISHERS GROUP

DORDRECHT / BOSTON / LANCASTER

7118-4855

CHEMISTRY

Library of Congress Cataloging in Publication Data

Jerusalem Symposium on Quantum Chemistry and Biochemistry
        (16th : 1983)
        Nucleic acids.

        (The Jerusalem symposia on quantum chemistry and biochemistry ;
v. 16)
        Includes index.
        1.   Nucleic acids–Congresses.   I.   Pullman, Bernard, 1919–
II.   Jortner, Joshua.   III.   Title.   IV.   Series.
QP620.J46   1983   574.87′328   83–16055
ISBN 90–277–1655–2

Published by D. Reidel Publishing Company,
P.O. Box 17, 3300 AA Dordrecht, Holland.

Sold and distributed in the U.S.A. and Canada
by Kluwer Academic Publishers,
190 Old Derby Street, Hingham, MA 02043, U.S.A.

In all other countries, sold and distributed
by Kluwer Academic Publishers Group,
P.O. Box 322, 3300 AH Dordrecht, Holland.

All Rights Reserved
© 1983 by D. Reidel Publishing Company, Dordrecht, Holland
No part of the material protected by this copyright notice may be reproduced or
utilized in any form or by any means, electronic or mechanical
including photocopying, recording or by any information storage and
retrieval system, without written permission from the copyright owner

Printed in The Netherlands

QP620
J46
1983
CHEM

TABLE OF CONTENTS

PREFACE

The 16th Jerusalem Symposium debated on one of the most important subjects of modern molecular biology : the nucleic acids. It continued the tradition of gathering the most distinguished experts in the field.

Placed under the auspices of the Israël Academy of Sciences and Humanities and the Hebrew University of Jerusalem it was sponsored by the Institut de Biologie Physico-Chimique (Fondation Edmond de Rothschild) of Paris and the National Foundation for Cancer Research of Bethesda, U.S.A. We wish to express our deep thanks to the Baron Edmond de Rothschild for his continuous generosity which guarantees the perenniality of these meetings and to the authorities of the National Foundation for Cancer Research, in particular Mrs Tamara Salisbury and Dr. Franklin Salisbury for having joined hands with us in this venture. We wish also to express our gratitude to the administrative staff of the Israël Academy and in particular to Mrs Avigail Hyam for the efficiency and excellency of the local arrangements.

Bernard Pullman
Joshua Jortner

BASE SEQUENCE, HELIX STRUCTURE AND INTRINSIC INFORMATION READOUT IN DNA

Richard E. Dickerson
Molecular Biology Institute
University of California, Los Angeles
Los Angeles, California 90024, U.S.A.

The binding of repressors and similar recognition proteins to specific DNA sequences involves intrinsic physical and chemical properties of both DNA and protein, most probably a specific pattern of hydrogen bonds. Single crystal analyses of double-helical DNA oligomers have revealed predictable sequence-dependent local variations in helix geometry, of a type that could contribute to the specificity of the intrinsic information readout process.

The readout of information from base sequence in DNA and its translation into polypeptide sequence is an extrinsic process, in the sense that it does not depend on an immediate physical or chemical similarity between the structure of the triplet codons of DNA and the amino acids of the growing protein chain. A quarter of a century of effort has failed to produce a plausible intrinsic connection between codons and polypeptide, and the translation system works only by virtue of a highly evolved battery of messengers, transfer RNA molecules, energy sources, and enzymes. In contrast, the readout of base sequence information by a repressor molecule when it binds selectively to its operator site is an intrinsic process, relying on a matching of size, shape, and (presumably) hydrogen-bonding properties of the DNA and protein.

What is the structural basis for this intrinsic information readout process? One obvious suggestion is a matching of hydrogen bond donors and acceptors on the edges of base pairs in major and minor grooves of the DNA, with hydrogen-bonding side chains on the protein (Seeman et al., 1976). As Figure 1 shows, the major groove edge of an A·T base pair exhibits the pattern: acceptor-donor-acceptor, and this same pattern with minor positional shifts is retained if the base pair is turned around, T·A. In contrast, the major groove edge of a G·C base pair has the asymmetric pattern: acceptor-acceptor-donor, and this pattern is reversed if the base pair is turned backward, C·G. Any recognition protein that interacted with the major groove would see a

1

B. Pullman and J. Jortner (eds.), Nucleic Acids. The Vectors of Life, 1–15.
© 1983 by D. Reidel Publishing Company.

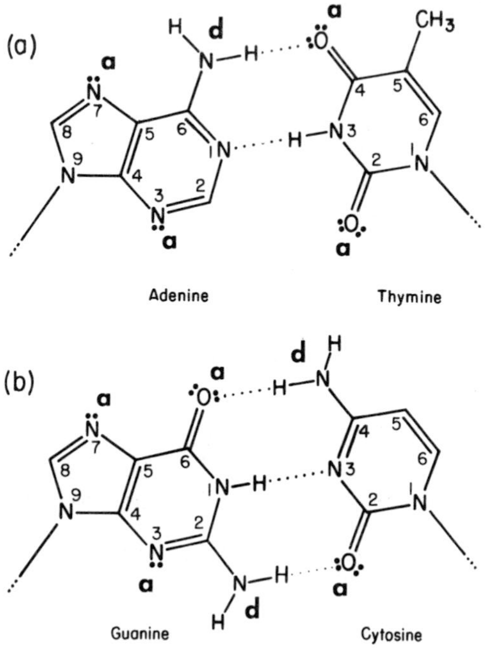

(a)

Adenine   Thymine

(b)

Guanine   Cytosine

Figure 1. Hydrogen - bonding patterns in A·T and G·C base pairs. $\underline{a}$ = Hydrogen bond acceptor, and $\underline{d}$ = Hydrogen bond donor. The major groove (upper edge of each drawing) exhibits the patterns:

A·T = $\underline{a}$ - $\underline{d}$ - $\underline{a}$

T·A = $\underline{a}$ - $\underline{d}$ - $\underline{a}$

G·C = $\underline{a}$ - $\underline{a}$ - $\underline{d}$

C·G = $\underline{d}$ - $\underline{a}$ - $\underline{a}$

The minor groove contains less information, only differentiating between G·C and A·T by presence or absence of the donor amine group.

characteristic pattern of hydrogen bond donors and acceptors on the floor of the groove, that would enable it to differentiate between base pairs, with a possible ambiguity between A·T and T·A. The minor groove is less informative, since A·T and G·C are both symmetric with respect to base reversal, and are distinguished only by the absence of the central -NH$_2$ donor in A·T and its presence in G·C.

Is this the end of the story? Is the DNA only a regular helix, whose role in the readout process is that of a passive template to be fitted by the protein molecule? Until a few years ago most observers would have answered in the affirmative, not out of conviction, but out of a lack of information about the effects, if any, of specific base sequence on local helix structure. Fiber diffraction methods can give only an average structure, and cannot reveal the details of local deviations from that average.[1] But single-crystal x-ray analyses of short DNA oligomers from all three helix families--A, B and Z--have begun to show us that, at least in the right-handed A and B forms, base sequence does indeed influence helix structure, in a manner that may contribute to the recognition process.

To date three A-DNA structures have been solved: CCGG (Conner et al., 1982), GGTATACC (Shakked et al., 1981, 1983) and GGCCGGCC (Wang et al., 1982a), and one A-helical RNA-DNA hybrid: r(GCG)d(TATACGC) (Wang et al., 1982b). B-DNA is represented by CGCGAATTCGCG, refined in several variants (Drew et al., 1981; Dickerson and Drew, 1981; Fratini et al., 1982). Z DNA has been studied as the tetramer CGCG (Drew et al.

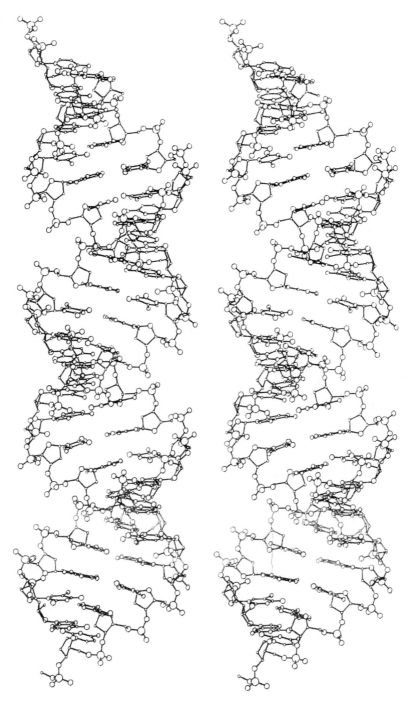

Figure 2.   Infinite A-DNA helix of sequence ..(GTATAC)$_n$..generated by repetition of the central six base pairs of the octamer GGTATACC.

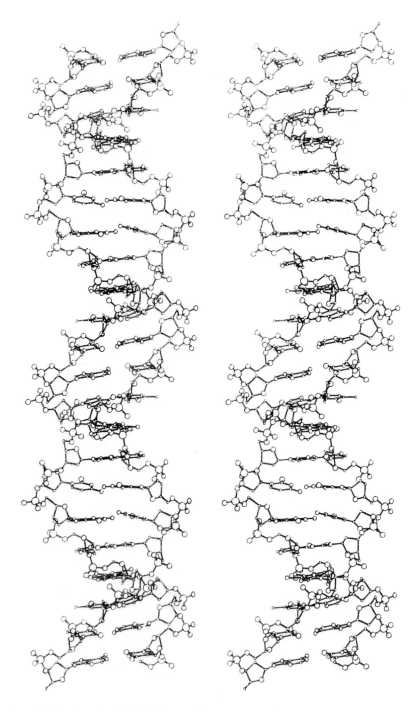

Figure 3. Infinite B-DNA helix ..(GCGAATTCGC)$_n$.., by repetition of the central ten base pairs of the dodecamer CGCGAATTCGCG.

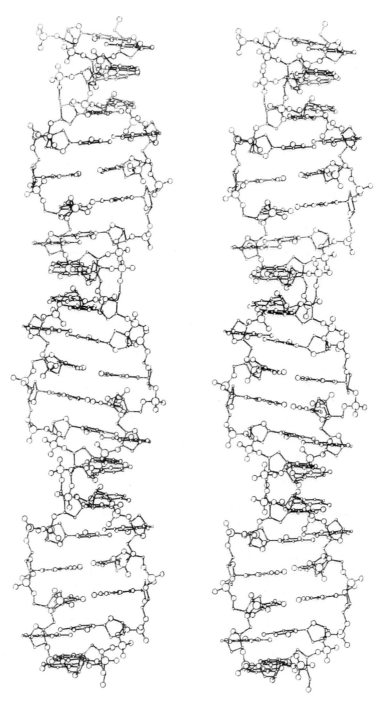

Figure 4.   Infinite Z-DNA helix ..(GCGC)$_n$.. generated by repetition of
the central four base pairs of the hexamer CGCGCG.

1978, 1980; Drew and Dickerson, 1981a), the hexamer CGCGCG (Wang et al.
1979), and the hexamer with 5-methylcytosine replacing cytosine (Fujii
et al., 1982a). Figures 2 - 4 are attempts to portray what extended
A, B and Z-DNA helices would look like, with preservation of the actual
sequence-induced variation as observed in the single-crystal analyses.
Each Figure has been produced by deleting the outermost base pairs from
an actual molecular helix, and then repeating the inner base pairs in a
way that generates an infinite helix. As expected, the A-DNA helix is
short and fat, with a deep major groove and shallow minor groove, the B
helix is taller and slimmer, with grooves of comparable depth, and the
Z-helix is elongated and thin, with a deep minor groove and extremely
shallow major groove. Z-DNA also is distinguished by its left-handed
helix sense, and the fact that the idealized helix repeat unit is two
successive base pairs, G·C and C·G, rather than a single base pair as
with A and B.

These three Figures give a truer picture of the variation along
real DNA double helices than do the idealized models derived from fiber
analysis. The two bases of a pair are usually not coplanar in A and B;
they exhibit a propeller twist along their mutual long axis. They also
can be rolled and tilted relative to the helix axis, and the local
helix rotation angle from one base pair to the next can vary considerably from the average value. Furthermore, a base pair may be shifted
along or perpendicular to its long axis, relative to the rest of the
helix stack. Careful examination of the B helix in Figure 3 reveals
that its minor groove is narrower in A·T regions than in G·C regions.
All of these are sequence-dependent effects, as we shall see.

What regions of these helices are sensed, and read, by repressors
during intrinsic information readout? The requirement of the Z helix
for alternating purine-pyrimidine sequence means that it is unlikely
to be used in coding, although it could be very important in control
of the expression of genetic information, either by recognition of the
left-handedness of the helix itself, or by the effect of helix reversal
on the degree of supercoiling of the DNA in which it is found. If, as
is generally assumed, the high-humidity B form is that which is present
when interactions with control proteins occur, then the wide major
groove is the most attractive candidate, both because of its accessibility and because of the richness of its information content (Figure 1).
The minor groove is less accessible, less informative, and is otherwise
occupied, being the site of the spine of hydration that is believed to
contribute significantly to the stabilization of the B form relative to
the A under high-humidity conditions (Drew and Dickerson, 1981b; Kopka
et al., 1983).

The considerable variation in local helix parameters was initially
a source of puzzlement, until C. R. Calladine (1982) brought order into
the field by a theory, aesthetically pleasing in its simplicity, based
on an application of principles of elastic beam mechanics to the DNA
double helix. Calladine's theory has been quantified and extended by
Dickerson (1983) into a set of algorithms for predicting helix varia-

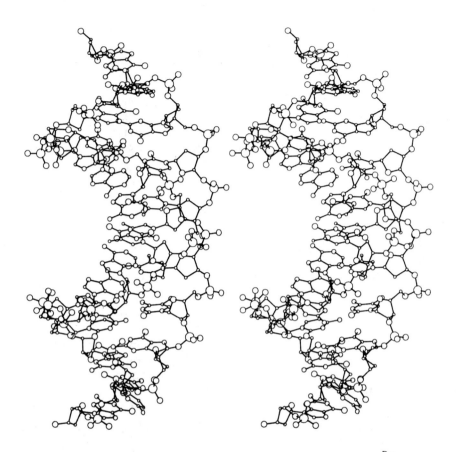

Figure 5.   Inclined view of the B-DNA dodecamer CGCGAATT^{Br}CGCG, with
the major groove seen in profile at the left.   Strand 1 begins with
base C1 at top front and ends with G12 at bottom front.   Strand 2 then
begins with C13 at bottom rear and ends with G24 at top rear.   Note how
the bases of each strand are tilted so that they stack efficiently atop
one another, almost as though the other strand were not present.   The
tilt necessary to bring about this stacking leads to positive propeller
twist at each base pair.

tion, and these predictions and their applicability to sequence recognition are the subject of this paper.

     The starting point for the analysis is the observation that base
pairs in both right-handed DNA helices, A and B, have a positive propeller twist because this tends to stabilize the helix by increasing
the stacking overlap between bases along each individual strand of the
helix (Figure 5). But as Figure 6 illustrates schematically, positive
propeller twist brings with it the disadvantage of too-close contact

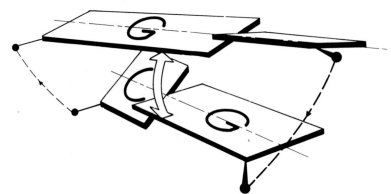

Figure 6. Positive propeller twist is a contrarotation of bases about the long axis of the pair, with the nearer base rotated clockwise as shown here. Black dots and lines connecting them to bases represent C1' atoms and glycosyl bonds. Phosphate backbones are schematized by dashed lines with arrows in the 5'-3' direction. Double-headed arrow represents minor groove steric clash between purines in a Pyr-Pur step.

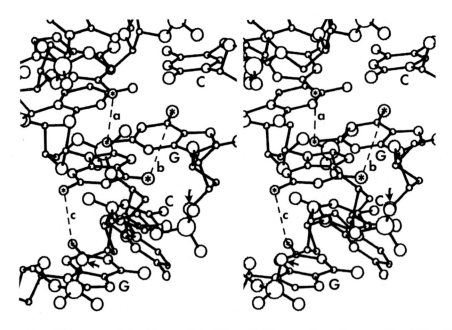

Figure 7. Closeup of bottom of helix of Figure 5. Bases C9, G10, C11, G12 are labeled from top to bottom by C, G, C, G. Stars and dashed lines indicate purine-purine cross-chain steric clash: a and c = minor groove clash at Pyr-Pur step; b = major groove clash at Pur-Pyr. The propeller twist at G10·C15 is flattened by dual steric clashes at a and b. Base C13 is tilted steeply by an intermolecular hydrogen bond.

between purine rings when they occur on opposite backbone strands at adjacent base pairs. If R = any purine (A or G) and Y = any pyrimidine (T or C), then at a Y-R step as in Figure 6 the close contact will be found in the minor groove, and at an R-Y step it will occur in the major groove. Both types are illustrated in B-helical CGCGAATT^{Br}CGCG in Figure 7. In this view into the minor groove, base G16 of base pair C9·G16 makes too close contact with base G10 of base pair G10·C15, as indicated by the dashed line labeled a. That same G10 purine ring also has a steric clash with base G14 of base pair C11·G14 just below it (dashed line b), and the double clash is partially relieved by flattening the propeller twist of base pair G10·C15. The symmetrically disposed base pair C3·G22 at the other end of the helix is similarly flattened, and this may be a general property of alternating Pur-Pyr sequences. Base pair C11·G14 is not flattened as much by clashes b and c because clash c is with a terminal base pair, which is unconstrained and free to adapt its own orientation rather than forcing the clash on its neighbor above. In a long, continuous stretch of alternating purines and pyrimidines, one would expect to find a uniform decrease in propeller twist at all base pairs.

Calladine (1982) proposed that the cross-chain steric hindrance between purines produced by propeller twist could be relieved by one or more of four strategies:

1. Flatten the propeller twist in one or both base pairs.
2. Open up the roll angle between successive base pairs on the side on which the clash occurs.
3. Shift one or both base pairs along their long axis in a direction that pulls the purine out of the base stack.
4. Decrease the helix rotation angle between the offending base pairs.

Strategy 1 is illustrated in Figure 7, and the double-headed arrow in Figure 6 illustrates the use of roll angle strategy 2. The roll angle θ_R is defined as the angle between best mean planes through two successive base pairs with respect to rotation around their long axes (Dickerson and Drew, 1981; Fratini et al., 1982). θ_R is positive if the angle between base pairs opens toward the minor groove, and negative if toward the major groove. Hence strategy 2 is implemented by increasing roll θ_R at Y-R steps as in Figure 6, and decreasing θ_R at R-Y steps. Strategy 3 involves concerted changes in main chain torsion angle δ (C5'-C4'-C3'-O3') at the two ends of a base pair (Figure 8), and offers a simple explanation for two phenomena noted early in the structure analysis of the CGCGAATTCGCG dodecamer: the preference of purines for larger δ angles than pyrimidines, and the principle of anticorrelation (Drew et al., 1981; Dickerson and Drew, 1981).

Three actual examples of cross-chain purine clash are illustrated in Figure 9 for the B-DNA dodecamer CGCGAATTCGCG, in a view directly down the helix axis. This view displays best the fact that clash can be lessened, no matter in which groove it occurs, by decreasing the helix rotation angle between base pairs (Strategy 4).

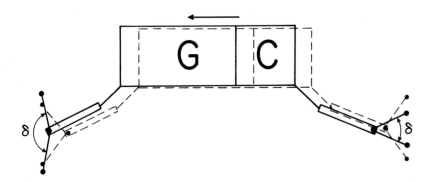

ANTICOMPLEMENTARITY OF TORSION ANGLES, δ

Figure 8.   Sliding a base pair along its long axis in B-DNA so that the purine is partially removed from the base pair stack increases main chain torsion angle δ at the purine (left) and decreases δ by a similar amount at the pyrimidine (right).   This explains both the observed preference of purines for larger δ angles than pyrimidines, and the principle of anticorrelation, or the fact that the pyrimidine and purine of a given base pair tend to exhibit δ angles that are equidistant to either side of a mean of δ = 123°.   Deoxyribose rings are shown only by flat slabs seen on edge at left and right, with their C3'-C4' bonds perpendicular to the plane of the diagram.

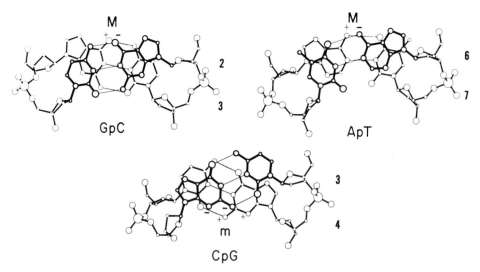

Figure 9.   Observed purine cross-chain clash in CGCGAATTCGCG.   Numbers at right are base numbers in strand 1-- e.g. --G2 and C3 in the GpC diagram.   + and - indicate that propeller twist causes the labeled atom to rise or fall out of the horizontal plane, and are shown only at positions of steric clash.   M and m denote major and minor groove clash at R-Y and Y-R steps, respectively.

From his analysis of strategy 3 and torsion angle variation, Calladine deduced that minor groove Y-R clashes were effectively twice as severe as major groove R-Y clashes. Following this, Dickerson has defined four sum functions, Σ_1 - Σ_4, by means of which the base sequence can be used to predict local variations in helix rotation angle, base plane roll, the difference between torsion angles δ at the two ends of a base pair, and propeller twist, respectively. The procedure for generating helix sum function Σ_1 is given below; for the others, see Dickerson (1983):

1. At every R-Y step, assume that steric clash is relieved by decreasing the local helix rotation angle by -2 arbitrary units. This motion of base pairs will increase the flanking rotation angles by +1 unit each. Hence the perturbation at the R-Y step and its neighbors will be: +1, -2, +1.
2. Let the sum function contributions at every Y-R step with its minor groove clash be twice as great: +2, -4, +2.
3. Build up sum function Σ_1 by carrying out the above analysis at every step of the sequence. Disregard homopolymer Y-Y and R-R steps, as they make no contribution to cross-chain repulsion.

Sum function Σ_1 is compared with the experimentally observed local variation in helix rotation angle in Figure 10, and the agreement is seen

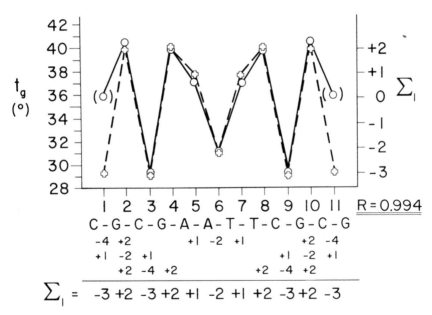

Figure 10. Comparison of observed helical twist angles, t_g, with sum function Σ_1 for CGCGAATT^{Br}CGCG. Solid lines = t_g; dashed lines = Σ_1. The way in which Σ_1 is generated from base sequence is shown below. The linear correlation coefficient between t_g and Σ_1 is R = 0.994.

to be excellent. Cross-chain clash of purines arising from propeller twist is an entirely adequate explanation for the observed local variation in helix rotation angles. The mean value of 36.0° was expected from fiber studies, but the wide variation between 29° and 40° was not. Now, however, it is seen to follow naturally from base sequence.

The algorithm for generating base roll sum function Σ_2 is similar, except that the Y-R contributions must be reversed in sign: -2, +4, -2, since one must open the roll angle toward the minor groove at Y-R steps and open it toward the major groove at R-Y steps. The results for the B dodecamer are shown in Figure 11. Again the agreement between observation and calculation is excellent. Roll angles were never considered in fiber studies, because they could not be observed. But Figure 11 indicates a variation of as much as 12° in the orientation of base plane edges, which means a considerable variation in the orientation of the potentially hydrogen-bonding groups on the base edges.

Sum functions Σ_3 and Σ_4 agree with observed base pair displacements along their long axes (measured as differences in δ) and with observed propeller twist angles, with correlations of 0.777 and 0.680, respectively. Plots equivalent to Figures 10 and 11 are to be found in Dickerson (1983).

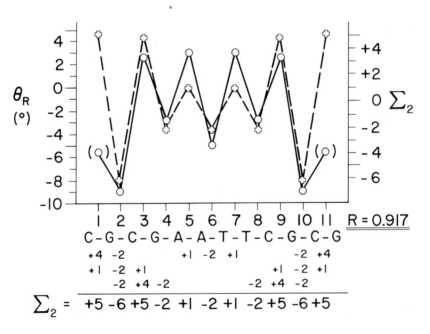

Figure 11. Comparison of observed base plane roll angles, θ_R, with sum function Σ_2 for CGCGAATT^{Br}CGCG. Solid lines = θ_R; dashed lines = Σ_2. The generation of Σ_2 from sequence is shown below. The correlation between observation and calculation is R = 0.917.

It is pleasing, of course, to have four such prominent local helix variations predicted by one hypothesis, and a simple one at that. But what is even more pleasing, and not a little surprising, is to find that the first two sum functions also hold for A-DNA. Table 1 shows the results of testing the sum functions against the four known A-helix structures. Roll angle behavior is explained successfully in all the known helices: B-DNA, A-DNA, and the RNA/DNA hybrid.   Helix rotation

Table 1. Correlation Between Observed and Predicted Helix Variation
         Using Sum Functions Σ_1 Through Σ_4

Helix	Variable	Mean	Range	R	Significant?
CGCGAATT[Br]CGCG	t_g vs. Σ_1	36.0°	29° - 40°	0.994	yes
(B-DNA)	θ_R vs. Σ_2	-1.9°	-9° - +3°	0.917	yes
	$\Delta\delta$ vs. Σ_3	0.0°	±74°	0.777	yes
	PT vs. Σ_4	17.7°	6° - 23°	0.680	yes
CCGG/CCGG	t_g vs. Σ_1	34.1°	31° - 37°	0.977	yes
(A-DNA)	θ_R vs. Σ_2	6.3°	3° - 13°	0.995	yes
	$\Delta\delta$ vs. Σ_3	0.0°	±32°	0.062	no
	PT vs. Σ_4	16.2°	16° - 21°	-0.134	no
GGTATACC	t_g vs. Σ_1	32.1°	30° - 34°	0.915	yes
(A-DNA)	θ_R vs. Σ_2	6.6°	1° - 15°	0.984	yes
	$\Delta\delta$ vs. Σ_3	0.0°	±2°	-0.611	no
	PT vs. Σ_4	12.4°	10° - 17°	-0.135	no
GGCCGGCC	t_g vs. Σ_1	33.2°	16° - 44°	0.991	yes
(A-DNA)	θ_R vs. Σ_2	6.3°	2° - 12°	0.896	yes
	$\Delta\delta$ vs. Σ_3	0.0°	±50°	0.442	marginal
	PT vs. Σ_4	8.6°	5° - 15°	-0.383	no
(GCG)TATACGC	t_g vs. Σ_1	33.3°	30° - 37°	0.180	no
(A-RNA/DNA	DNA only	31.5°	30° - 33°	0.986	yes
hybrid)	θ_R vs. Σ_2	9.7°	5° - 18°	0.835	yes
	$\Delta\delta$ vs. Σ_3	0.0°	±12°	0.365	no
	PT vs. Σ_4	11.0°	6° - 14°	-0.943	no

Data for CGCGAATT[Br]CGCG and CCGG/CCGG from Dickerson (1983), for GGTATA CC from Shakked et al. (1983), and for GGCCGGCC and (GCG)TATACGC from Wang et al. (1982a,b) and Fujii et al. (1982b). R is the linear cor-relation coefficient between the sum function and the variable.

behavior is predicted in all DNA, and begins to fail only in the RNA portion of the hybrid. Only the torsion angle and propeller twist functions are restricted to B-DNA. Torsion angles δ in A-DNA are confined nearer the value expected from fiber studies than in B-DNA, and sugar conformations are closer to the expected C3'-endo puckering. Sliding a base pair along its long axis in A-DNA is not an appropriate response to cross-chain purine clash, probably because the base pairs are steeply inclined away from perpendicularity to the helix axis. For possibly the same reason, the observed propeller twist appears to depend on factors other than steric clash of purines.

The mean values of helix rotation angle for B-DNA and A-DNA in Table 1 are as expected from fiber studies: 36° for B-DNA, and 32°-34° for A-DNA. The range of variation, however, is surprising: 11° for B-DNA, and as little as 3° (GGTATACC) to as much as 28° (GGCCGGCC) for A-DNA. The mean roll angle for B-DNA is -2°, reflecting the average perpendicularity of base pairs to the helix axis. For A-DNA the mean roll angle is a little over 6°, and this is easily explained from an inspection of Figure 2. To a first approximation, B-DNA is constructed from bases stacked atop one another along the helix axis, with sugar-phosphate backbone to lace the stack together. But A-DNA is more like a double-stranded ribbon, wound around an imaginary center pole along the helix axis, with the minor groove outside. This cylindrical winding forces the base pairs to open up accordion-like in the minor groove in a way that produces a mean θ_R of 6°. Superimposed on these mean values is a range of variation of roughly 12° for both helix types.

Are these sequence-dependent helix changes used in the intrinsic recognition process by repressors and other control proteins? A definitive answer will have to wait on x-ray analyses of complexes of DNA and protein. But variation of ±14° in rotational orientation of the potentially hydrogen-bonding groups about the helix axis cannot help but have a profound effect on the geometry of protein interaction. It would be strange indeed if these local variations in rotation and roll were not used in the recognition process. Single-crystal x-ray analysis has revealed that double-helical DNA has a richness and complexity of secondary structure that was entirely unsuspected from prior studies. Its significance to the biological functioning of the DNA is a blueprint for future research.

NOTES

[1]Arnott et al. (1983) have offered a "wrinkled" B-DNA model for poly d(GC):poly d(GC) with an alternating main chain conformation and with two base pairs as the helix repeating unit, as an example of the deduction of sequence-dependent helix variation from fiber analysis. But as the authors themselves point out, the stacking of base pairs is the same in both the wrinkled model, and the classical idealized B-DNA with one base pair as the helix repeat. The two models differ only in main chain conformation. Hence the wrinkled B-DNA model cannot represent a

sequence effect, since it is identical to the standard model in exactly those regions that contain sequence information. The lower residual error after refinement for the wrinkled model presumably could be a consequence of doubling the number of independent variables when two base pairs are chosen as the asymmetric helical unit.

REFERENCES

Arnott, S., Chandrasekaran, R., Puigjaner, L.C., Walker, J.K., Hall, I.H. and Birdsall, D.L.:1983, Nucl. Acids Res. 11, pp. 1457-1474.
Calladine, C.R.:1982, J. Mol. Biol. 161, pp. 343-352.
Conner, B.N., Takano, T., Tanaka, S., Itakura, I. and Dickerson, R.E.: 1982, Nature 295, pp. 294-299.
Dickerson, R.E.:1983, J.Mol. Biol. 166, in press.
Dickerson, R.E. and Drew, H.R.:1981, J. Mol. Biol. 149, pp. 761-786.
Drew, H.R. and Dickerson, R.E.:1981a, J. Mol. Biol. 152, pp. 723-736.
Drew, H.R. and Dickerson, R.E.:1981b, J. Mol. Biol. 151, pp. 535-556.
Drew, H.R., Dickerson, R.E. and Itakura, K.:1978, J. Mol. Biol. 125, pp. 535-543.
Drew, H.R., Takano, T., Tanaka, S., Itakura, K. and Dickerson, R.E.: 1980, Nature 286, pp. 567-573.
Drew, H.R., Wing, R.M., Takano, T., Broka, C., Tanaka, S., Itakura, I. and Dickerson, R.E.:1981, Proc. Natl. Acad. Sci.USA 78, pp. 2179-2183.
Fratini, A.V., Kopka, M.L., Drew, H.R. and Dickerson, R.E.:1982, J. Biol. Chem. 257, pp. 14686-14707.
Fujii, S., Wang, A.H.-J., van der Marel, G., van Boom, J.H. and Rich, A.:1982a, Nucl. Acids Res. 10, pp. 7879-7892.
Fujii, S., Wang, A.H.-J., van Boom, J.H. and Rich, A.:1982b, Nucl. Acids Res. Symp. Series 11, pp. 109-112.
Kopka, M.L., Fratini, A.V., Drew, H.R. and Dickerson, R.E.:1983, J. Mol. Biol. 163, pp. 129-146.
Seeman, N.C., Rosenberg, J.M. and Rich, A.:1976, Proc. Natl.Acad. Sci. USA 73, pp. 804-808.
Shakked, Z., Rabinovich, D., Cruse, W.B.T., Egert, E., Kennard, O., Sala, G., Salisbury, S.A. and Viswamitra, M.A.:1981, Proc. Royal Soc. Lond. B213, pp. 479-487.
Shakked, Z., Rabinovich, D., Kennard, O., Cruse, W.B.T., Salisbury, S.A. and Viswamitra, M.A.:1983, J. Mol. Biol., in press.
Wang, A.H.-J., Quigley, G.J., Kolpak, F.J., Crawford, J.L., van Boom, J.H., van der Marel, G. and Rich, A.:1979, Nature 282, pp. 680-686.
Wang, A.H.-J., Fujii, S., van Boom, J.H. and Rich, A.:1982a, Proc. Natl. Acad. Sci. USA 79, pp. 3968-3972.
Wang, A.H.-J., Fujii, S., van Boom, J.H., van der Marel, G.A., van Boeckel, S.A.A. and Rich, A.:1982b, Nature 299, pp. 601-604.

NEW WRINKLES ON DNA

Struther Arnott, R. Chandrasekaran, L. C. Puigjaner, and
J. K. Walker
Department of Biological Sciences, Purdue University,
West Lafayette, IN   47907

The B allomorph of poly d(GC):poly d(GC) at first sight appears to be quite similar to that of calf thymus DNA whose identical chains are described adequately as 10_1 helices of (average) mononucleotides. Nevertheless the chains of poly d(GC):poly d(GC) are significantly better described as 5_1 helices of dinucleotides in which GpC has more or less the same conformation as common B DNA but CpG has its C3'-03' conformation <u>gauche</u> <u>minus</u> rather than <u>trans</u>. Consequently the phosphate orientations of alternate nucleotides differ dramatically justifying describing the conformation of this molecule as a wrinkled form of B DNA.

The D forms of poly d(AT):poly d(AT) and poly d(IC):poly d(IC) are wrinkled similarly suggesting that such wrinkles are intrinsic to DNA sequences that have purine and pyrimidine nucleotides alternating.

The only known allomorph of poly d(A):poly d(T) is superficially similar to B DNA in having a 10_1 helix of pitch 3.2 nm. This turns out to have a heteronomous structure in which the poly d(T) chains are indeed B-DNA-like but the poly d(A) chains are A-DNA-like.

The idiosyncratic secondary structures of DNAs of peculiar sequence will affect their physical and biological structures profoundly.

INTRODUCTION

Polynucleotide duplexes in fibers are quite polymorphic (1): DNA of general sequence is at least trimorphic (A, B, C) (Fig. 1); so also are poly d(AT):poly d(AT) (A, B, and D) and poly d(GC):poly d(GC) (A, B, and Z) (2) (Figs. 1, 2 and 3). By contrast poly d(A):poly d(T) has been trapped in only one form (originally called B') (3) which, however, has two packing arrangements. Until recently only the Z allomorphs of poly d(GC):poly d(GC) (2,4) and the Z-like forms of poly d(AC):poly d(GT) (2), poly d(As^{4}T):poly d(s^{4}T) (2) and the

17

B. Pullman and J. Jortner (eds.), Nucleic Acids: The Vectors of Life, 17–31.
© *1983 by D. Reidel Publishing Company.*

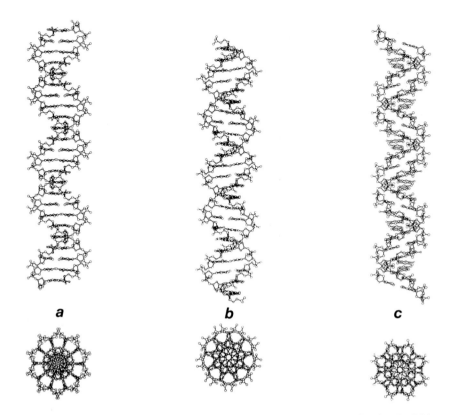

Figure 1: Mutually perpendicular views of 3 members of the B DNA
family with projected nucleotide lengths (a) 0.34 nm, (b) 0.32 nm,
(c) 0.30 nm, and rotations between adjacent residues (a) + 36°,
(b) + 40°, (c) + 45°. The changes in helical parameters and molecular
morphology is achieved without any <u>qualitative</u> changes in conformation
angles. In particular all their furanose rings are puckered C2'-<u>endo</u>.

C form of poly d(AG):poly d(CT) (1) were known to reflect their primary
dinucleotide structures in their secondary structures -- by having a
repeat unit 0.6-0.7 nm long (2) and therefore different conformations
in alternate nucleotides.

    There is now direct evidence that the B form of poly d(GC):poly d(GC)
also has a structural unit that is a dinucleotide with a projected
length of 0.68 nm. This is the meridional reflexion on the 5th layer
line of its X-ray fiber pattern (e.g. Fig. 4a) from oriented but poorly
crystalline samples. Fully crystalline, orthorhombic (P2$_1$2$_1$2$_1$) samples
do not show the diagnostic (005) reflexion (Fig. 4b) since this is absent
as a result of the symmetry of packing. On the other hand, the unit
cell of this orthorhombic form has different dimensions (<u>a</u> = 3.79 nm,
<u>b</u> = 3.61 nm, <u>c</u> = 3.36 nm) from those of calf thymus DNA (which are

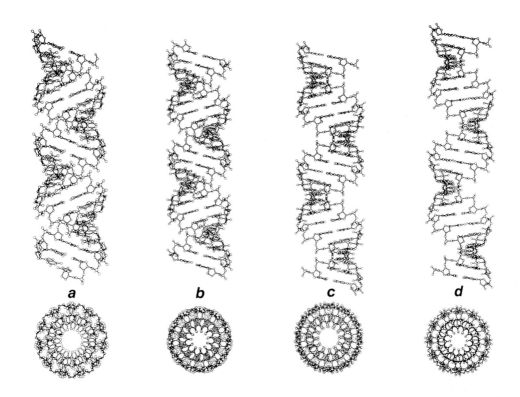

Figure 2: Mutually perpendicular views of 4 members of the A family
with projected nucleotide lengths (a) 0.26 nm, (b) 0.28 nm,
(c) 0.30 nm, (d) 0.33 nm and rotations between successive nucleotides
(a), (b) + 32.7°, (c), (d) + 30.0°. They all belong to the same
conformational genus and have C3'-endo-puckered furanose rings.

a = 3.08 nm, b = 2.24 nm, c = 3.37 nm) and contains 4 molecules rather
than 2. Clearly the surface of B form poly d(GC):poly d(GC) is
"wrinkled" in some way which distinguishes it from typical or "smooth"
B DNA.

     If such wrinkles are intrinsic to DNAs containing sequences of
alternating purine and pryimidine nucleotides we would expect to find
them also with poly d(AT):poly d(AT) and even on allomorphs which are
not B. In advance of detailed analyses X-ray fiber patterns
(Figs. 4c,d) suggest that this is indeed the case. The 8_1 D helices
of poly d(AAT):poly d(AAT) are packed in a pseudo-hexagonal
(a = b = 1.98 nm, c = 2.41 nm, γ = 120°), screw-disordered fashion as
if they were smooth helices (5) but the D helices of poly d(AT):poly d(AT)
(6) and poly d(IC):poly d(IC) pack in apparently isomorphous, tetragonal
crystal forms with a = b = 1.69 nm, c = 2.42 nm, γ = 90° and

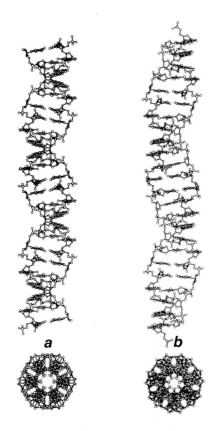

Figure 3: Mutually perpendicular
views of 2 members of the Z family
with projected nucleotide lengths
(a) 0.37 nm and (b) 0.39 nm and
rotations per dinucleotide of
(a) -60.0°, (b) -51.4°.

a = b = 1.70 nm, c = 2.45 nm, γ = 90° respectively (7). This is what
would be expected if the molecular structures in these two cases
strictly were 4_1 helices of dinucleotides i.e. wrinkled versions of
D DNA.

    In wrinkled B or D DNA duplexes the two antiparallel chains would,
of course, be identical as in all successful models of polynucleotide
duplexes until now. This is to say that the structures have diad axes
perpendicular to their helix axes. If the difference between the so-
called B' forms of poly d(A):poly d(T) (Fig. 4e,f) and other DNA
duplexes results from the chemical difference of its two strands then
we would expect them to be structurally different and the duplex not to
contain any diad axes. There are precedents for Watson-Crick-paired
polynucleotide chains having different structures e.g. in models
proposed for the triple-stranded complexes of poly d(T):poly d(A):poly d(T)
(3) and poly (U):poly (A):poly (U) (8) and for the DNA-RNA hybrid
poly r(A):poly d(T) (27). Unfortunately none of these 3 fibrous
structures is polycrystalline and therefore none provides sufficient
Bragg intensities to allow a detailed structural analysis. On the
other hand, the well-oriented, polycrystalline (Fig. 4e), orthorhombic
(a = 1.87 nm, b = 3.55 nm, c = 3.23 nm), β form of poly d(A):poly d(T)
with 263 reflexions accessible for measurement provides a data base

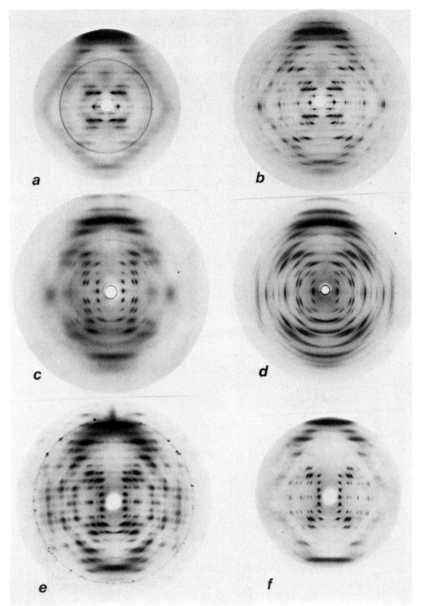

Figure 4: Various X-ray fiber diffraction patterns whose intepretation
is discussed in the text: (a) wrinkled B form poly d(GC):poly d(GC),
oriented but not fully crystalline and, therefore, causing an (005)
reflexion; (b) wrinkled B form poly d(GC):poly d(GC), oriented and
polycrystalline; (c) smooth D form poly d(AAT):poly d(ATT), oriented but
screw-disordered; (d) wrinkled D form poly d(IC):poly d(IC), oriented
and polycrystalline; (e) heteronomous β poly d(A):poly d(T), oriented
and polycrystalline; (f) heteronomous (?) α poly d(A):poly d(T),
oriented but screw-disordered.

rich enough to explore whether and to what extent the unique structure
of poly d(A):poly d(T) differs from those of other DNAs (9).

DNAs WITH ALTERNATING PURINE AND PYRIMIDINE SEQUENCES ARE WRINKLED

The strategy adopted by Arnott et al (7,9) to explore the
structures of poly d(GC):poly d(GC), poly d(AT):poly d(AT) and
poly d(A):poly d(T) was to use the Linked Atom Least Squares approach
(10,11) to obtain the best models of the more symmetric kind for each
structure. These have molecular symmetries respectively 2 2 10_1,
2 2 8_1, 2 2 10_1 and an average mononucleotide as the motif repeated
by the symmetry elements. These structures were then compared with the
best 2 5_1, 2 2 4_1 and 10_1 models with respectively dGpdCp or
dApdTp or dAp:dTp as the appropriate (larger) units of structure.
Invariably these larger asymmetric units, with their increased numbers
of degrees of freedom, permitted a better fit both with the diffraction
data and with the steric requirements. The question of whether the
improvements were significant was settled by applying Hamilton's test
(12). In every case the lower symmetry model was found to be
resoundingly superior at a level of significance better than 99.5%.

Left-handed versions of the successful, lower symmetry models were
also tested as alternative solutions and also found to be rejectable at
a significance level better than 99.5%.

Wrinkled B DNA

When the diad axis projections (Fig. 5a,b) of the wrinkled and
smooth forms of B DNA are compared they are seen to be very similar.
The subtle difference between them becomes emphatic when one views the
structures down their helix axes (Fig. 5a,b). These latter projections
make it immediately obvious why the crystal packings of smooth and
wrinkled B DNA are different.

Close-up views of the structures of PupPy and PypPu (Fig. 6)
reveal another interesting feature: the PupPy dinucleoside
monophosphate is essentially the same in wrinkled and smooth DNA, but
PypPu is different. This is most noticeable in the orientations of the
phosphate groups. In smooth B DNA the normals to the O1-P-O2 planes
make an angle of 49° to the helix axis. In wrinkled B DNA this angle
is similar (39°) in GpC but much larger (75°) in CpG.

It cannot be emphasized too strongly that the wrinkles in
poly d(PuPy):poly d(PuPy) DNAs are characterized by a significant
change in the shape of one set of nucleotides (PypPu) from the average
shape observed in smooth DNA. In wrinkled B DNA the root-mean-squared
difference in the main conformation from those in smooth B DNA is only
9.5° for PupPy but 36° in PypPu. Much of this difference is concentrated
in the conformation at C3'-O3' which in wrinkled DNA is gauche minus in
PypPu but trans in PupPy as in all nucleotides of smooth DNA.

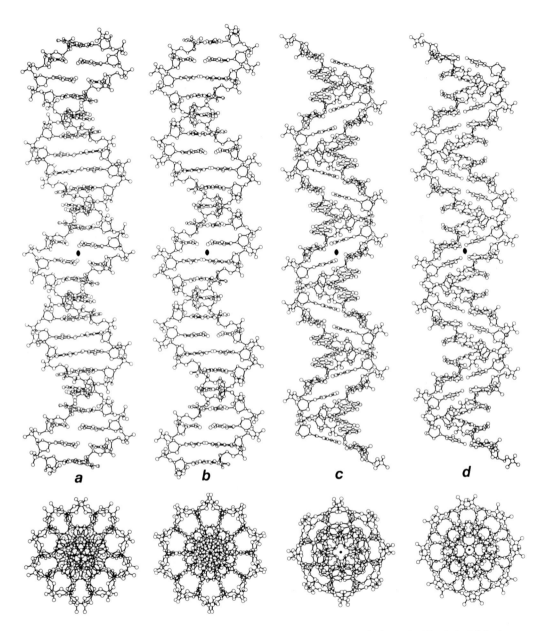

Figure 5: Mutually perpendicular projections of (a) wrinkled B DNA,
(b) smooth B DNA, (c) wrinkled D DNA, (d) smooth D DNA.

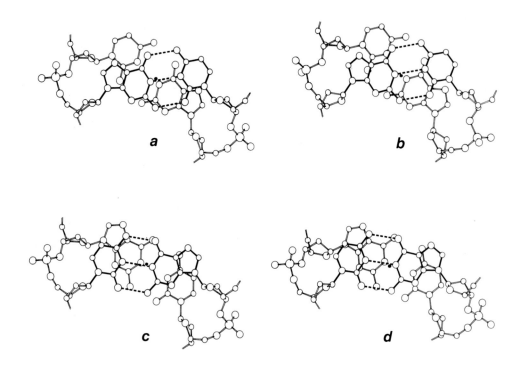

Figure 6: Detailed views of dinucleoside phosphate duplex fragments
in smooth and wrinkled B DNA. CpG in (a) smooth B DNA has a quite
different phosphate orientation from that in (b) wrinkled B DNA. By
contrast, GpC in (c) smooth and (d) wrinkled B DNA have quite similar
shapes. The small filled circles show the positions of the helix axes.

Wrinkled D DNA

     In the wrinkled D forms of poly d(IC):poly d(IC) and
poly d(AT):poly d(AT), a damped version of the B DNA wrinkle is
observed. Once again PuPy in the wrinkled allmorph has a shape like
the nucleotides of the smooth form but PypPu is different. As before
the differences between smooth and wrinkled DNA is most evident in the
orientations of the phosphate groups (Fig. 7). These distinctive
peripheral differences are emphasized in the views down the helix axes
(Fig. 5c,d).

     This finding that the wrinkles found in B form poly d(GC):poly d(GC)
survive in poly d(AT):poly d(AT) even in its D form suggests that such
wrinkles are intrinsic to all poly d(PuPy):poly d(PuPy).

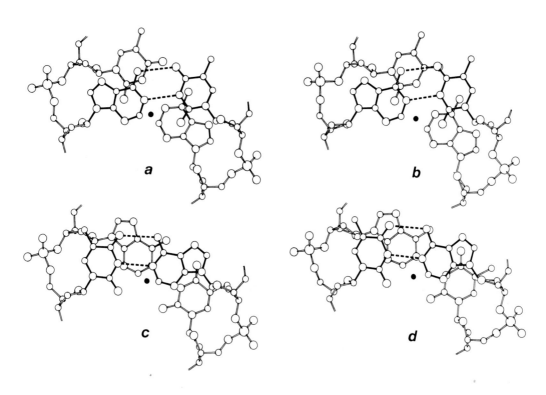

Figure 7: Detailed views of dinucleoside phosphate duplex fragments in
smooth and wrinkled D DNA. TpA in (a) smooth D DNA has a somewhat
different phosphate orientation from that in (b) wrinkled D DNA. By
contrast, ApT in (c) smooth and (d) wrinkled D DNA have quite similar
shapes. The small filled circles show the positions of the helix axes.

POLY D(A):POLY D(T) HAS A HETERONOMOUS STRUCTURE

When Arnott and Selsing (3) first studied poly d(A):poly d(T) (in
its α packing form) they assumed that the molecular structure was a
trivial variant of B DNA. The new study by Arnott et al (9) using the
richer data set provided by the polycrystalline β form of
poly d(A):poly d(T) revealed not only that the two chains had to be
different but that they were very different. The poly d(A) chain has
conformations corresponding to an A-type nucleic acid with C3'-endo
furanose rings. The poly d(T) chain, however, has B-type C2-endo-
puckered sugar rings. The credibility of this structure appears to be
beyond cavil since it emerged in a least-squares refinement of a model
in which both chains started off with similar B-like structures. In
addition, the result is in keeping with the conclusions of an
independent Raman Spectral study that poly d(A):poly d(T) contains both

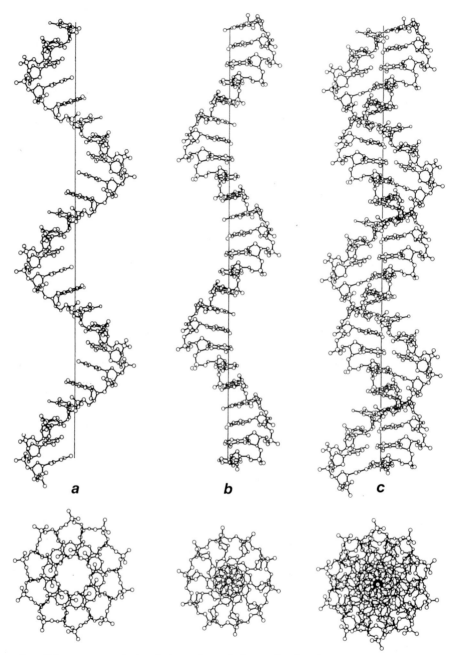

Figure 8: Views normal to (above) and down (below) the vertical helix axes of (a) the B-like poly d(T) strand which mates with (b) the A-like poly d(A) strand to produce (c) the heteronomous duplex.

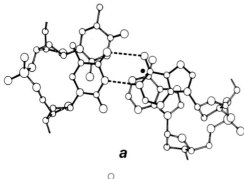

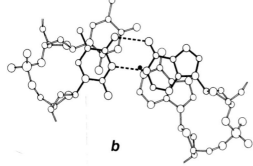

a

b

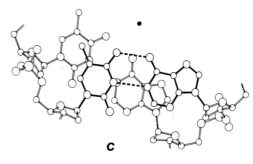

c

Figure 9: Helix axis projections
of dinucleoside phosphate
fragments of (a) the heteronomous
poly d(A):poly d(T) duplex,
(b) B DNA, and (c) an A DNA
allomorph. The base stacking in
the heteronomous model is quite
similar to that in B DNA although
the shapes of the TpT and ApA
nucleotides resemble their
conformational relatives in B and
A DNA respectively.

C2'-endo and C3'-endo furanose rings at ordinary temperatures (13).

Just how different the two chains of poly d(A):poly d(T) are is
emphasized in Fig. 8. On the other hand, it is evident (Table 1) that
each chain is a rather conventional member of its respective
conformational genus. The base-stacking (Fig. 9), however, is more
reminiscent of B DNA than A DNA.

Since the structure contains no diad axes the orientations of the
two bases need not be the same. In fact the adenine bases are more
nearly perpendicular to the helix axis than the thymine bases. The
dihedral angle between paired bases is quite high (29°) in keeping with
the trend observed for AT pairs in oligomers (14) and polymers (7).
With the former the base-twist of AT pair reaches 27°; with the latter
the twist in poly d(AT):poly d(AT) is 21° compared with 14° in
poly d(GC):poly d(GC).

IMPLICATIONS

Arnott et al (7) have defined wrinkles explicitly only for the
B form of poly d(GC):poly d(GC) and the D forms of poly d(IC):poly d(IC)
and poly d(AT):poly d(AT). There can be little doubt, however, that
the B forms of poly d(AT):poly d(AT) and poly d(AC):poly d(GT) are at
least as wrinkled as poly d(GC):poly d(GC). It follows that we would

Table 1: The values (and estimated standard deviations) of the molecular parameters for heteronomous poly d(A):poly d(T) and for the wrinkled forms of poly d(GC):poly d(GC) and poly d(AT):poly d(AT). Those of the smooth A, B and D DNA are given for comparison.

Parameters	Smooth A DNA	Heteronomous DNA		Smooth B DNA	Wrinkled B DNA		Smooth D DNA	Wrinkled D DNA	
		ApA	TpT		GpC	CpG		ApT	TpA
Backbone Conformations (°)									
θ(C4'-C3'-O3'-P)	-145	-170(4)	171(4)	-141	-156(4)	-85(4)	-162	-157(5)	-133(7)
θ(C3'-O3'-P-O5')	-78	-59(4)	-121(5)	-157	-158(5)	-169(4)	-121	-145(6)	-153(5)
θ(O3'-P-O5'-C5')	-50	-60(5)	-41(5)	-33	-30(5)	-66(5)	-73	-45(6)	-72(6)
θ(P-O5'-C5'-C4')	172	172(4)	-174(5)	138	126(4)	145(3)	157	142(5)	143(4)
θ(O5'-C5'-C4'-C3')	42	54(3)	43(4)	33	47(4)	22(3)	77	55(5)	69(5)
θ(C5'-C4'-C3'-O3')	79	81(1)	152(3)	142	143(2)	148(2)	139	142(3)	154(3)
Glycosylic Conformations (°)									
θ(C2'-C1'-N9-C4)	83	89(2)	-	141	143(3)	-	135	128(4)	-
θ(C2'-C1'-N1-C2)	83	-	147(3)	141	-	164(3)	135	-	138(4)
Furanose Conformations (°)									
θ(C4'-O4'-C1'-C2')	8	9(1)	-25(2)	-36	-35(2)	-16(3)	-20	-36(3)	-17(4)
θ(O4'-C1'-C2'-C3')	-32	-35(2)	42(1)	46	53(2)	37(2)	37	45(3)	37(3)
θ(C1'-C2'-C3'-C4')	43	43(1)	-42(2)	-38	-48(2)	-42(2)	-38	-38(3)	-43(3)
θ(C2'-C3'-C4'-O4')	-39	-40(1)	29(3)	19	30(2)	34(2)	28	19(3)	35(3)
θ(C3'-C4'-O4'-C1')	19	21(1)	-3(3)	10	3(3)	-12(3)	-6	10(4)	-11(4)
Furanose Endocyclic Bond Angles (°)									
τ(C4'-O4'-C1')	110	107(2)	109(1)	107	105(1)	108(1)	108	109(2)	110(2)
τ(O4'-C1'-C2')	106	106(2)	106(2)	104	105(1)	107(1)	106	104(2)	108(2)
τ(C1'-C2'-C3')	100	103(2)	98(1)	99	95(1)	100(1)	102	97(2)	96(2)
τ(C2'-C3'-C4')	101	96(2)	103(1)	103	100(1)	100(2)	100	105(2)	105(2)
τ(C3'-C4'-O4')	103	108(2)	104(1)	106	106(2)	107(2)	107	104(3)	101(3)
Dihedral Angle between Planes in a Base-pair (°)									
Δγ	-6	-29(2)		-13	-14(1)		-34	-21(2)	

expect enzymatic cleavage of poly d(AT):poly d(AT) to occur
preferentially at the P-O3' bonds in the ApT sequences (which we would
expect to have the same conformations as in smooth DNA) rather than the
TpA sequences (which have a different shape). This is exactly what has
been observed (15,16). We would expect also that the two nuclear
magnetic resonances of phosphorus atoms in alternating purine/pyrimidine
DNAs would indeed be different from one another. This has been observed
with poly d(AT):poly d(AT) (17).

It is also worth recalling that the lac repressor of E. coli
binds to poly d(AT):poly d(AT) nearly 10^3 times more strongly than to
a DNA of general sequence (18). Presumably this is partly because of
sequence elements common to the lac operator and poly d(AT):poly d(AT).
It is likely also that the operator has wrinkles in common with the
operator which has a high proportion of alternating purine/pyrimidine
sequences.

As far as poly d(A):poly d(T) is concerned, certainly its
persistently anomalous behaviour in solution is compatible with its
having usually a rather unconventional structure.

For example, it is possible to predict the ultraviolet circular
dichroism spectrum of a DNA of complex sequence by making an appropriate
linear combination of CD spectra of DNAs of repeated simple sequences
that are fragments of the complex sequence (19). A prerequisite for
success is that all the DNAs involved are isomorphous. Interestingly,
the measured CD spectrum of poly d(A):poly d(T) (20) has been shown to
be quite anomalous in this connection (21). The predicted spectrum
which would form a self-consistent set with those of other DNAs is
dramatically different from the measured version (21).

Another quite different experiment, involving incomplete scission
of poly d(A):poly d(T) molecules laid down on mica or calcite surfaces,
indicates that the pitch of this polymer is significantly different
from that of typical DNAs (22).

Further, it has been shown (23) that nucleosomes will not
form over a sufficiently (e.g. 80 nucleotide) long segment of
poly d(A):poly d(T) in a recombinant DNA molecule. This implies not
only that the segment has a structure substantially different from
general sequence DNA but also that the special structure is
sufficiently robust to resist the homogenising effect of flanking
sequences. Interestingly, however, smaller (e.g. 20 nucleotide)
segments can be accommodated in nucleosomes (23) although
poly d(A):poly d(T) tracts even of this size appear to retain their
idiosyncratic secondary structure (24,25). This may imply that such
tracts contrive to lie between the sites of closest histone-DNA
interaction (22,26) and could therefore play a role in phasing
nucleosomes on DNA. Obviously any unconventional DNA secondary
structure could produce the phenomena just discussed but it would be
surprising if they did not result from heteronomous duplexes resembling

(if not identical to) the ones we have been describing.

Monotonous $d(A)_n:d(T)_n$ tracts need not, of course, be the only sequences with these properties: clearly it is not unthinkable that any oligo d(Pu):oligo d(Py) segment, or any fragment that had approximately such a sequence, might serve equally well.

Heteronomous duplexes are not confined to polynucleotides with only A:T base pairs. Patterns similar to those of poly d(A):poly d(T) have been observed also with poly d(I):poly d(C) and with poly d(AI):poly d(CT) (1).

DNA-RNA hybrid duplexes are also prime candidates to form heteronomous structures. Although many such hybrids are known to be essentially isomorphous with well-characterized RNA-RNA duplexes which do have diad axes, some have been thought to have heteronomous structures: poly d(I):poly (C) which has a 10-fold helix of pitch P = 3.1 nm (28); poly (A):poly d(T) which has a pitch P = 3.4 nm and is thought by some (27) to be a 10-fold helix but by us to be an 11-fold helix isomorphous with poly d(A):poly d(U) (8) in which case the structure is unlikely to be heteronomous. There is a need, therefore, for more intensive investigations of the polymorphic range of DNA-RNA hybrids to determine how commonly heteronomous structures occur.

ACKNOWLEDGEMENTS

We thank Carol Jacobson for typing the text, Bill Boyle and Les Booth for photography and the NIH for a grant (GM17371).

REFERENCES

1. Leslie, A.G.W., Arnott, S., Chandrasekaran, R. and Ratliff, R. L.: 1980, J. Mol. Biol. 143, pp. 49-72.
2. Arnott, S., Chandrasekaran, R., Birdsall, D. L., Leslie, A.G.W. and Ratliff, R. L.: 1980, Nature 283, pp. 743-745.
3. Arnott, S. and Selsing, E.: 1974, J. Mol. Biol. 88, pp. 509-521.
4. Wang, A.H.-J., Quigley,, G. J., Kolpak, F. J., Crawford, J. L., van Boom, J. H., van der Marel, G. and Rich, A.: 1979, Nature 282, pp. 680-686.
5. Selsing, E., Arnott, S. and Ratliff, R. L.: 1975, J. Mol. Biol. 98, pp. 243-248.
6. Arnott, S., Chandrasekaran, R., Hukins, D.W.L., Smith, P.J.C. and Watts, L.: 1974, J. Mol. Biol. 88, pp. 523-533.
7. Arnott, S., Chandrasekaran, R., Puigjaner, L. C., Walker, J. K., Hall, I. H., Birdsall, D. L. and Ratliff, R. L.: 1983, Nucl. Acids. Res. 11, pp. 1457-1474.
8. Arnott, S. and Bond, P. J.: 1973, Nature New Biology 244, pp. 99-101.
9. Arnott, S., Chandrasekaran, R., Hall, I. H., and Puigjaner, L. C.: 1983, Nucl. Acids Res. 11, pp. xxx-xxx.

10. Arnott, S. and Wonacott, A. J.: 1966, Polymer 7, pp. 157-166.
11. Smith, P.J.C. and Arnott, S.: 1978, Acta Cryst. A34, pp. 3-11.
12. Hamilton, W. C.: 1965, Acta Crystallogr. 18, pp. 502-510.
13. Thomas, G. A. and Peticolas, W. L.: 1983, J. Am. Chem. Soc. 105, pp. 993-996.
14. Fratini, A. V., Kopka, M. L., Drew, H. R. and Dickerson, R. E.: 1982, J. Biol. Chem. 257, pp. 14686-14707.
15. Scheffler, I. E., Elson, E. L. and Baldwin, R. L.: 1968, J. Mol. Biol. 36, pp. 291-304.
16. Klug, A., Jack, A., Viswamitra, A., Kennard, O., Shakked, Z. and Steitz, T. A.: 1979, J. Mol. Biol. 131, pp. 669-680.
17. Shindo, H., Simpson, R. T. and Cohen, J. S.: 1979, J. Biol. Chem. 254, pp. 8125-8128.
18. Riggs, A. D., Lin, S.-Y. and Wells, R. D.: 1972, Proc. Nat. Sci., U.S.A. 69, pp. 761-764.
19. Gray, D. M. and Tinoco, I., Jr.: 1970, Biopolymers 9, pp. 223-244.
20. Wells, R. D., Larson, J. E., Grant, R. C., Shortle, B. E. and Cantor, C. R.: 1970, J. Mol. Biol. 54, pp. 465-497.
21. Arnott, S. (1975) Nucl. Acids Res. 2, pp. 1493-1502.
22. Klug, A., Rhodes, D., Smith, J., Finch, J. T. and Thomas, J. O.: 1980, Nature 287, pp. 509-516.
23. Kunkel, G. R. and Martinson, H. G.: 1981, Nucl. Acids Res. 9, pp. 6869-6888.
24. Strauss, F., Gaillard, C. and Prunell, A.: 1981, Eur. J. Biochem. 118, pp. 215-222.
25. Peck, L. J. and Wang, J. C.: 1981, Nature 292, pp. 375-378.
26. Finch, J. T., Brown, R. S., Rhodes, D., Richmond, T., Rushton, B., Lutter, L. C. and Klug, A.: 1981, J. Mol. Biol. 145, pp. 757-769.
27. Zimmerman, S. B. and Pheiffer, B. H.: 1981, Proc. Natl. Acad. Sci. U.S.A. 78, pp. 78-82.
28. Arnott, S., Chandrasekaran, R., Hall, I. H., Puigjaner, L. C., Walker, J. K. and Wang, Manlin.: 1982, "in Symposia on Quantitative Biology: Structures of DNA (Proceedings of the Forty-Seventh Cold Spring Harbor Laboratory Publications)", Vol. 47, pp. 53-65.

X-RAY ANALYSIS OF OLIGONUCLEOTIDES AND THE CHANGING CONCEPT OF DNA
STRUCTURE

Olga Kennard
University Chemical Laboratory, Lensfield Rd., Cambridge, U.K.

INTRODUCTION

X-ray fibre diffraction studies played a crucial part in the discovery
of the double helix and the role of DNA in the transfer of genetic
information [1]. During the past thirty years the parameters of the
various helical conformations were substantially refined [2] but fibre
diffraction, by its very nature, can only give information about
averaged structures. To determine finer details such as the influence of
specific base sequence on the local conformation, other techniques,
particularly single crystal X-ray analysis, have to be employed.

Single crystal methods have indeed been used with outstanding success to
analyse the structure of proteins, transfer RNA's and other
macromolecules at near-atomic resolution. The technique could not,
however, be applied to DNA fragments until pure deoxyoligonucleotides
became available in sufficient quantity for crystallisation experiments.
Developments in chemical synthesis by the triester method of
oligonucleotide assembly [3,4] and, more recently, rapid and high yielding
solid phase methods [5] made it possible to synthesise a variety of
oligonucleotides with different base sequences for X-ray analysis.

The rate determining step to-day is not so much the synthesis but the
preparation of single crystals of oligonucleotides which diffract
sufficiently well for X-ray analysis. The solution of the phase problem
for these structures also presents special difficulties and so far only
some seven different DNA fragments, ranging from tetranucleotides to a
dodecanucleotide and one RNA-DNA decamer have been fully analysed. These
studies have changed our concept of the DNA double helix from a
monotonous, uniform structure generated from a single base-pair unit to
a much more complex structure which is subtly influenced by base
sequence. The correlation between structure and sequence is likely to be
of fundamental importance in understanding how regulatory proteins
recognise and interact with specific base sequences.

33

B. Pullman and J. Jortner (eds.), Nucleic Acids: The Vectors of Life, 33–48.
© *1983 by D. Reidel Publishing Company.*

This paper summarises the published oligonucleotide structures and focuses on those which contain the right handed A type helix. It looks at the question of how far base sequence influences the local and global conformation of the helix, at crystal packing and at the possible biological implications of the structural results for protein nucleic acid interactions. A more extended analysis of this problem is in course of publication [6].

OLIGONUCLEOTIDE CRYSTAL STRUCTURES 1978-1983

All oligonucleotide structures published to date relate to self-complementary sequences. With one exception they crystallise in structures where the repeat unit contains at least two oligomers. These form antiparallel double helices by Watson-Crick hydrogen bonding between corresponding bases. Table 1 lists the different base sequences analysed and gives crystallographic data derived from one structure determination for each. Many of these sequences were analysed as different heavy atom derivatives, at different temperatures, different habits, or grown from solvents with different counter ions. No changes in helix type were observed in these multiple studies but small variations were found in the conformational parameters which give an indication of the relative influence of base sequence, crystal packing and other external forces.

OLIGONUCLEOTIDE X-RAY STRUCTURES 1978 - 1982

Compound	Resolution $\overset{o}{A}$	Space Group	Z	R value	Structure Type	Reference
d(pATAT)	1.04	$P2_1$	1	15.3	B	[7,8]
d(CGCGAATTCGCG)	1.90	$P2_1 2_1 2_1$	2	17.8	B	[18,19]
d(CGCGCG)	0.90	$P2_1 2_1 2_1$	2	13.0	Z	[14,15]
d(CGCG)	1.50	$P6_5$	3	19.0	Z	[17]
d(CGCG)	1.50	$C222_1$	2	19.9	Z	[16]
d(GGTATACC)	1.80	$P6_1$	2	19.8	A	[21,22]
d(GGBrUABrUACC)	2.25	$P6_1$	2	13.5	A	[21,22]
d(^{I}CCGG)	2.10	$P4_3 2_1 2$	2	20.5	A	[26]
d(GGCCGGCC)	2.25	$P4_3 2_1 2_1$	1	15.8	A	[27]
r(GCG)d(TATACGC)	1.90	$P2_1 2_1 2_1$	2	16.0	A	[25]

Table 1

The first oligonucleotide analysed d(pATAT) [7,8] is the exception in that it contains only a single oligomer in the asymmmetric unit of a cell with a two fold screw symmetry. A two base-pair long right handed double helical fragment, illustrated in Fig. 1 is, however, formed by hydrogen bonding between symmetry related molecules.

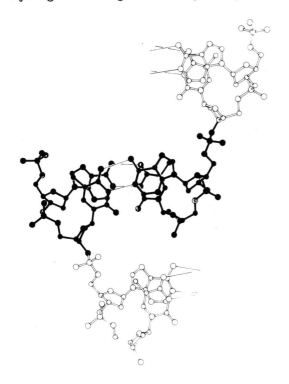

Figure 1 - Two base pair mini-helix with alternating sugar conformation observed in the high-resolution study of d(pATAT)

This helix has sequence dependent features. The phosphodiester conformations of the A-T and T-A sequences is different. The deoxyribose-pucker and the sugar-base orientation alternate along the helix depending on the nature of the base (3'-endo, typical of A-DNA for purine, and 2'endo, typical of B-DNA for pyrimidine). The assignment of the sugar pucker is unequivocal since the analysis was carried out at atomic resolution.

The crystal structure of d(pATAT) suggested a dinucleotide repeat for the alternating copolymer poly d(AT) in which every second phosphodiester linkage and sugar residue has a conformation different from that of the normal B form [9]. This "alternating B" model provided an explanation for some puzzling spectroscopic and enzymatic observations and suggested the existence of DNA helices with more than one base pair in the repeat unit [10]. The hypothesis was supported by subsequent spectroscopic [11] and fibre diffraction studies [11,12,13]. Values of various parameters [8] characterising the "alternating B" are given in Table 2.

COMPARISON OF B-DNA MODELS

	"Alternating" B		B-DNA
Sugar conformation	C3'endo	A	C2'endo
	C2'endo	B	
Sugar base orientation	0.0°	A	81°
	70.0°	T	
Backbone conformation α	−67.0°	A-T	−90°
	−125.0°	T-A	
Backbone conformation ξ	67.0°	A-T	−60°
	−70.0°	T-A	
Helical turn	31.2°	A.T	36°
	40.7°	T.A	

Table 2

Perhaps the greatest change in our concept of DNA structure was brought about by the discovery of a hitherto unconsidered left handed helix, with a two base pair repeat unit, in the crystal structure of the hexamer d(CGCGCG) [14,15].

The conformation and biological properties of this Z helix are described by Rich elsewhere in this volume. The structure is included in this brief survey to emphasise how much the X-ray analysis of oligonucleotide structures is dependent on having either high resolution X-ray data or the correct structural model. Experimental data on two forms of the tetramer d(CGCG) [16,17] was available for some years to a resolution of about 2Å. The crystal structures could, however, not be solved until the correct model was deduced from the high resolution study of the hexamer d(CGCGCG). Both tetramer structures contain variants of the Z helix.

A full turn of the B-type helix was found in the dodecamer d(CGCGAATTCGCG) [18,19] and is discussed by Dickerson in this volume. The remaining four structures listed in Table 2 are all of the A type, an unexpected frequency suggesting that either the crystallisation conditions or the base sequences stabilise the A form. It is interesting that one of these structures when examined in solution by proton n.m.r. was found to be in the B conformation [20]. The four structures are compared in some detail in the next section.

CRYSTAL DATA FOR A-TYPE STRUCTURES

Compound	Cell A			Volume A	Volume/base pair A
	a	b	c		
d(GGTATACC)	45.01	45.01	41.55	72899	1519
d(GGBrUABrUACC)	45.08	45.08	41.72	73425	1530
d(ICCGG)	41.10	41.10	26.70	45102	1409
d(GGCCGGCC) -8°C	42.06	42.06	25.17	44527	1392
d(GGCCGGCC) -18°C	40.51	40.51	24.67	40485	1265
r(GCG)d(TATACGC)	24.20	43.46	49.40	51956	1299

Table 3

CRYSTAL STRUCTURES WITH A-TYPE HELICES

The A-type helix was found in four oligomers listed in Table 3. The octanucleotide d(GGTATACC) [21,22] contains a central four base pair fragment which is part of a concensus sequence in the promoter regions of both prokaryotic and eukaryotic genes and is involved in RNA-polymerase binding, prior to unwinding of the DNA double helix and subsequennt transcription [23,24]. The hybrid r(GCG)d(TATACGC) [25] contains the same fragment and a comparison of the two structures allows us to examine the same base sequence in two different crystal lattices.

The two other oligomers d([I]CCGG) [26] and d(GGCCGGCC) [27] analysed contain the recognition sites (CCGG) and (GGCC) of the two restriction endonucleases Hpa II and Hae III repectively. Both enzymes make symmetrical double stranded cuts. Hpa I cleaves (CCGG) between the two cytosine residues on each strand while the cleavage of (GGCC) occurs between the central guanine and cytosine residues. The two structures are isomorphous as indicated by the cell constants in Table 3. In the crystal structure of d([I]CCGG) two double helical tetramers are stacked one upon another via a crystallographic two fold axis thus simulating a continuous octamer duplex d([I]CCGG/[I]CCGG), but with the connecting phosphate group absent. The octamer d(GGCCGGCC) has a crystallographic two fold symmetry. The packing of the duplexes in the crystal lattice of the two structures is identical and we are thus able to compare the characteristics of the same base sequence in two identical crystal environments. Such comparative studies open up the possibility of differentiating between the influence of specific base sequence and crystal forces on conformation.

The Structure of the Octamer d(GGTATACC)

Two derivatives of this octanucleoside heptaphosphate were analysed; the native crystals and the isomorphous bromouracil containing analogue d(GGBrUABrUACC) [21,22]. Both crystallised in the hexagonal space group P6$_1$ with one double helix in the asymmetric unit. The structures were solved by a new multi-dimensional search approach combining packing criteria and R factor calculations for low resolution X-ray data [28]. Both A and B helices were used in the initial trials which rapidly discriminated in favour of the A conformation. This search method is an extremely promising one for the analysis of oligonucleotide structures which often do not diffract to a high resolution. It is, however, critically dependent on having a near correct model for the initial trials and would not be suitable for structures containing oligonucleotides with novel conformations.

The uniform A-DNA helix used for the initial searches was refined by treating the bases and phosphates as rigid groups but allowing the sugar conformations and the various torsion angles, including those of the sugar phosphate backbone, to vary. The starting model and the refined octamer are compared in the space filling drawings of fig. 2.

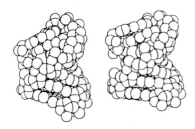

Figure 2 - Space filling drawings of the A-DNA viewed perpendicular to the helix axis. (a) the starting model based on fibre data and used in the initial stages of the solution of the single crystal structure of d(GGTATACC) (b) refined model

a   b

The most striking change in the model is the opening up of the major groove as a result of the symmetrical bending of the octamer helix about its pseudo two-fold axis. The bend is to the right in the plane of the paper in Fig.3 which is a stereoscopic view of the octamer perpendicular to the helix and pseudo 2-fold axes.

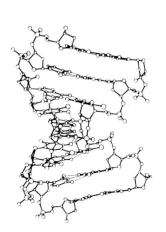
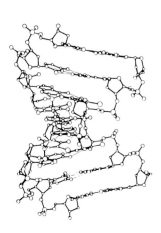

Figure 3 -
Stereoscopic view
of the octamer
duplex d(GGTATACC)
perpendicular to
the helix axis

The conformation of the sugar moieties is retained as 3'endo, but considerable variations are observed in the tilt of the base pairs with respect to the helix axis. The tilt values are much smaller for the end base pairs than for the central ones.

The difference in the depth of the major and minor grooves and the variations in the base overlap can best be observed in the stereoscopic view of the octamer down the helix axis in Fig. 4.

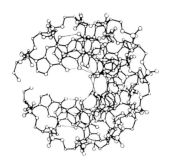
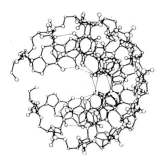

Figure 4 -
Stereoscopic view
of the octamer
duplex d(GGTATACC)
along the helix
axis

There is a remarkable similarity in the stacking patterns of the bases at equivalent steps and a well defined pseudo two-fold axis reflects the symmetry of the base sequence d(GGTATACC). Other characteristics of the base stacking such as the discontinuity at the T-A junction and the interstand stacking of the adenines can also be seen clearly.

The base overlaps in the first four steps of the refined octamer are
shown in fig.5 and compared with an A-DNA fibre model. The overlap of
the G-C base pairs (5a) is similar to the fibre pattern. The T-A steps
(5c) show the interstrand purine-purine stacking refered to above,
typical of the A-type double helix. The largest variations are observed
for the purine-pyrimidine steps G-T where the G.C base pair is shifted
relative to the T.A base pair by nearly 1Å. The shift is towards the
major groove and results in an energetically favourable parallel
arrangement of the keto groups.

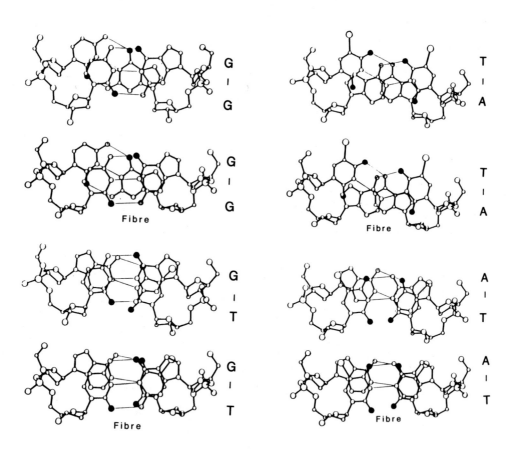

Figure 5 - Projected views from the first four steps of the octemer
d(GGTATACC) compared with equivalent steps of the A-DNA fibre model.
The view is perpendicular to the mean plane through the upper base
pair.

The sequence dependent symmetry of the octamer is also reflected in the variations in the local helical parameters which display a wide spectrum of A-type helical conformation ranging from nearly 10 to 12 fold structures. The local twist values are illustrated in fig.6. The largest sequence dependent variations occur at the central T-A-T-A fragment where the angles alternate between low values (av.29.9°) for T-A and high values (av.34.1°) for A-T. Similar sequence dependent variations are found in other local helical parameters and even in the thermal vibration of the individual bases. The backbone torsion angles on the other hand do not appear to show such variations and will be discused later in this paper.

Figure 6 - Helix angles (t) calculated from the angle between adjacent interstrand C1'-C1' vectors, projected down the global helix for d(GGTATACC)

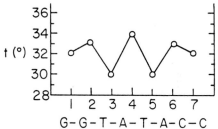

The packing of the molecules in the crystal lattice is achieved by stacking of the end base pairs parallel to and in van-der-Waals contact with the sugar-phosphate backbone of neighbouring molecules. The octamer duplexes pack in infinite spirals about the 6_1 axis. Adjacent spirals related by 2_1 screw axes interleave forming continous cylinders (fig.7). Most of the solvent molecules are situated in these central channels - some 70 of which were located in refinement. Neither the spermine molecules nor the counter ions could be assigned with any certainty but it is hoped that they will be located when the data set collected on a synchrotron source to 1.65Å resolution is refined.

Figure 7 - A portion of the extended crystal structure of d(GGTATACC) projected down the axis, illustrating the packing of adjacent helices and the channels where the solvent molecules and counter ions are located

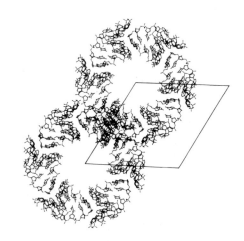

The structure of the octanucleotide was described in some detail to illustrate the conformational features which can be identified with some confidence from an X-ray analysis carried to a resolution of around 2Å. Sequence dependent changes are clearly observed which appear to arise from the stacking interactions between adjacent bases. The tendency for maximising base stacking results in changes in local helical parameters and in a global change of the helix which affects particularly the dimensions of the major and minor grooves.

Comparison of the A-type Helices

The three other structures containing A-type helices, d(GGCCGGCC) [27], d(^{I}CCGG) [26] and r(GCG)d(TATACGC) [25] were all analysed to about the same resolution as d(GGTATACC). The first two structures, as already mentioned, were isomorphous; the tetramer in d(^{I}CCGG) simulating a continuous eight base pair helix by stacking of one double stranded molecule on top of another. The two halves are related by a crystallographic two fold axis and the octamer d(GGCCGGCC) also has a two-fold symmetry axis imposed by the crystal structure. It is notable that a similar two fold axis was found in d(GGTATACC) as a consequence of symmetry of the base sequence. The two helices d(^{I}CCGG/^{I}CCGG) and d(GGCCGGCC) are straight but the RNA-DNA hybrid appears bent.

The global helical conformation of the four duplexes are significantly different as illustrated in Figure 8. The RNA-DNA decamer (III) approximates most closely to the 11 fold fibre A-DNA form with a 20° base tilt and 2.6Å rise per base pair. In the other three structures the bases have significantly smaller tilt values (12 -14°) and the rise per residue is increased (2.8-3.0Å). Large variations are observed in the groove width, particularly the major groove which becomes wider (4.7-7.9Å) as the base tilt decreases.

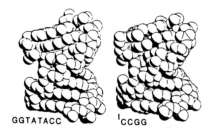 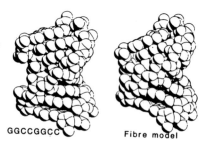

GGTATACC ^{I}CCGG   GGCCGGCC   Fibre model

Figure 8 - Comparison of space filling drawings of the three A-DNA structures with the fibre model

The stacking patterns of succesive base pairs reflect most closely the influence of base sequence on the local helical conformation. With data which was kindly made available for d(^{I}CCGG) by Professor Dickerson and for d(GGCCGGCC) by Professor Rich it was possible to compare in detail the stacking patterns in these structures with those in d(GGTATACC). Full details are given in another publication [6] but the salient features are summarised here.

In all three structures large variations are observed for each type of dinucleotide sequence. In d(GGCCGGCC) the stacking patterns of the two crystallographically non-equivalent homopolymer steps are similar though their corresponding helix twist and roll angles are appreciably different. In contrast the two non-equivalent homopolymer steps in the d(^{I}CCGG/^{I}CCGG) duplex are different; one step shows considerable overlap between the two cytosines, whereas in the other step the overlap is between the guanines. A tendency to interstrand stacking between cytosine and guanine bases is observed in both d(GGTATACC) and d(GGCCGGCC). The two different purine-pyrimidine steps in G-T (=A-C) and A-T in the d(GGTATACC) octamer have different stacking patterns, as have the two G-C steps in d(GGCCGGCC), although the corresponding twist and roll angles are similar. This suggests that the relative orientation of the bases are determined primarily by electostatic and van der Waals interaction between the bases.

The pyrimidine-purine steps in all three studies show the interstrand purine-purine partial stacking typical of the A-helix. The stacking modes of T-A in d(GGTATACC) and C-G in d(^{I}CCGG) are similar whereas that of C-G in d(GGCCGGCC) is markedly different. Since the latter two structures are isomorphous this indicates that the relative orientation of adjacent base pairs may be modulated by neighbouring base pairs.

Although the stacking pattern and local parameters of the RNA-DNA decamer have not yet been analysed in full detail there appear to be close similarities in the central TATA segment with the corresponding region in d(GGTATACC) octamer.

In contrast to the variability of the base stacking and of the related local helical parameter the backbone conformation is remarkably similar in all four structures. Values are given in Table 4 which also lists the values for the A-DNA helix deduced from fibre diffraction studies. The differences in base stacking appear to be accomodated by small concerted changes in the various torsion angles, without significant conformational changes in the sugar-phosphate backbone.

AVERAGE TORSION ANGLES

$$O3'-P-\alpha-O5'-\beta-C5'-\gamma-C4'-\delta-C3'-\varepsilon-O3'-\zeta-P-O5' \quad X$$

	α	β	γ	δ	ε	ζ	X
d(GGTATACC)	-62	173	52	88	-152	-78	-160
d(ᴵCCGG)	-73	180	64	80	-161	-67	-161
d(GGCCGGCC)	-75	185	56	91	-166	-75	-149
RNA-DNA decamer	-69	175	55	82	-151	-75	-162
A-DNA	-50	172	41	79	-146	-78	-154
B-dodecamer	-63	171	54	123	-169	-108	-177

Table 4

The sugar conformation in d(GGCCGGCC) is 3'endo except for the central four base pairs in which the alternate residues have conformations closer to B-DNA. The terminal guanosine sugar of d(ᴵCCGG) is also in this conformation. In the RNA-DNA decamer both the ribose and deoxyribose sugars are in the 3'endo conformation.

Strong correlations were observed between various torsion angles, particularly those about the C5'-O5' and C4'-C5' bonds and about the phosphodiester linkage O3'-P-O5'.

Crystal Packing and Intermolecular Interaction

The major intermolecular interaction in all four crystal structures is a stacking interaction between the flat base pairs at the ends of one molecule with the sugar phosphate backbone in the broad, shallow groove of an adjacent molecule. In d(GGTATACC), as already illustrated, the molecules are organised about a sixfold screw axis whereas in d(ᴵCCGG) and d(GGCCGGCC) the four-fold screw axis is utilised.

The same stacking arrangement is observed in the RNA-DNA decamer where the duplexes are organised around the two fold screw axes. It has been suggested by Rich and co-workers [27] that this stacking mode is utilised by certain non intercalative drugs, such as benzo[a]pyrene which have dimensions similar to a G-C base pair, when binding to DNA. In d(GGCCGGCC) and d(ᴵCCGG) only stacking interactions were found, whereas in in DNA-RNA decamer and the d(GGTATACC) the duplexes were also linked by hydrogen bonds.

The intermolecular stacking observed in these structures would not be possible if the duplexes were in the B-type conformation, since the minor groove of the B helix is deep and narrow. In the B-type structure

[18,19] a very different intermolecular geometry is found. It may well be that the common intermolecular interaction observed in these structures helps to stabilise the A-conformation.

The DNA content of the structures discussed here is only around 40% to 50%. The water, spermine and counter ions are located in channels, running through the structures. These channels are cylindrical in d(GGTATACC) with a diameter of 30Å and elliptical in the other two DNA structures with cross-sections 10Å by 20Å. The weak diffracting power of the crystals is very likely related to the disordered species in these channels. In none of the structures have the spermine molecules or the counter ions been located.

SUMMARY

The seven different oligonucleotide structures which have been analysed at atomic or near atomic resolution during the past five years have brought about a profound change in our view of DNA. The variability of the double helix was already suspected from spectroscopic and biochemical experiments but new forms of DNA, such as the Z-form, the multiplicity in the number of base pairs forming the repeat unit of the helix and local change in conformation were established as a result of single crystal X-ray investigations.

A comparison of the A-type helix found in four different crystal structures gives some insight into the correlation between conformation and base sequence.

The present indication is that specific sequence influences both global and local conformation. Base stacking seems to provide the driving force for the local variations. The tendency is to maximise the stacking overlap by concerted changes in the backbone torsion angle and the sugar-base orientation. Crystal structures containing A helices are characterised by intermolecular stacking arrangements which bring the flat end base pair of one duplex in close proximity to the sugar phosphate backbone in the shallow minor groove of another duplex. This kind of stacking might, itself, have a biological significance.

Single crystal structure analysis have so far been confined to comparatively short oligonucleotides and could rarely be extended to atomic resolution. They do, however, hold out great promise of achieving a better understanding of the mechanism of protein-nucleic acid interaction. The next decade will undoubtedly see the analysis of a wide variety of oligonucleotides and their complexes with drugs and proteins. Nature still has many surprises in store for us.

ACKNOWLEDGEMENTS

I thank my collegues Dr. W.B.T. Cruse, Professor D. Rabinovich, Dr. S. Salisbury, Dr. Z. Shakked and Professor M.A. Viswamitra for their collaboration during the period covered in this review; and Dr. S. Salisbury and Professor M.A. Viswamitra for their helpful comments on the manuscript. Long term financial support was provided by the Medical Research Council.

REFERENCES

1. Watson, J.D. and Crick, F.H.C. (1953) Nature (Lond) 171 pp.737-740

2. Arnott, S. and Hukins, D.W.L. (1972) Biochemical and Biophysical Research Communication 47 pp. 1504-1509

3. Stravinsky, J. and Hozumi, T., Narang, S.A., Bahl, C.P. and Wu, R. (1977) Nucl. Acid. Res. 4 pp. 353-371

4. a) Beaucage, S.L. and Caruthers, M.H. (1981) Tet. Lett. 22 pp. 1859-3661
   b) Letsinger, R.L. and Lunsford, W.B. (1976) J. Amer.Chem. Soc. 98 pp. 3655-3661

5. Gait, M.J., Mathes, H.W.D., Singh, M, Sproat, B.S. and Titmas, R.C (1982) Nucleic Acid Res.10 pp. 6243-6248

6. Shakked, Z. and Kennard, O. (1983) The A-Form of DNA in Structural Biology. Eds. McPherson A. and Jurnak F. John Wiley and Sons, N.Y. To be published.

7. Viswamitra, M.A., Kennard, O. Jones, P.G., Sheldrick, G.M., Salisbury, S.A. and Falvello, L. (1878) Nature (Lond) 273 pp. 697-688

8. Viswamitra, M.A., Shakked, Z., Jones, P.G., Sheldrick, G.M., Salisbury, S.A. and Kennard, O. (1982) Biopolymers 21 pp. 513-532

9. Viswamitra, M.A., Kennard, O., Shakked, Z., Jones, P.G., Sheldrick, G.M., Salisbury, S.A. and Falvello, L.(1978) Curr. Sci. 9 pp. 289-292

10. Klug, A., Jack, A., Viswamitra, M.A., Kennard, O.,Shakked, Z. and Steitz, T.A. (1979) J. Mol. Biol. 131 pp. 699-680

11. Patel, D.J. (1980) in Abstracts of Nucleic Acids : Interaction with Drugs and Carcinogens. British Biophysical Society Meeting, London

12. Shindo, H., Simpson, R.T.and Cohen, J.S.(1979) J. Biol. Chem. 245 pp. 8125-8128

13. Shindo, H. and Zimmerman, S.B. (1980) Nature (Lond) 283 pp. 690-691

14. Wang, A.H.J., Quigley, G.J., Kolpak, F.J., Crawford, J.L., van Boom, J.L., van der Marel, G. and Rich, A. (1979) Nature (Lond) 282 pp. 680-686

15. Wang, A.H.J., Quigley, G.J., Kolpak, F.J., van Boom, J.H. and Rich, A. (1981) Science 211 pp. 171-176

16. Drew, H., Takano, T., Tanaka, S., Itakura, K., Dickerson, R.E. (1980) Nature (Lond) 286 pp. 567-573

17. Crawford, J.L., Kolpak, F.J., Wang, A.H.J., Quigley, G.J., van Boom J.H., van der Marel, G. and Rich, A. (1980) Proc. Nat. Acad. Sci. U.S.A. 77 pp. 4016-4020

18. Wing, R, Drew, H., Takano, T., Broka, C., Tanaka, S., Itakura, K. and Dickerson, R.E. (1980) Nature (Lond) 287 pp. 755-758

19. Drew, H.R., Wing, R.M., Takano, T., Broka, C., Tanaka, S., Itakura, K. and Dickerson, R.E. (1981) Proc. Nat. Acad. Sci. U.S.A. 78 pp. 2179-2183

20. Reid, D.G, Salisbury, S.A., Bellard, S., Shakked, Z. and Williams, D.H. Biochemistry (1983) In press

21. Shakked, Z., Rabinovich, D., Kennard, O., Cruse, W.B.T., Salisbury, S.A. and Viswamitra, M.A. (1981) Proc. Roy. Soc. London 213 pp. 479-487

22. Shakked, Z. Rabinovich, D., Cruse, W.B.T., Egert, E., Kennard, O., Salisbury, S.A. and Viswamitra, M.A. (1983) J. Mol. Biol. In press

23. Pribnow, D. (1979) In Biological Regulation and Development. Goltberger, R.F. Ed. Phenum Press, N.Y. pp. 219

24. Siebelist, U., Simpson, R.B. and Gilbert, W. (1980) Cell 20 pp. 269-281

25. Wang, A.H.J., Fujii, S., van Boom, J.H., van der Marel, G.A., van Boeckel, S.A.A. and Rich,A. (1982) Nature (Lond) 299 pp. 601-604

26. Conner, B.N., Takano, T., Tanaka, S., Itakura, K. and Dickerson, R.E.
(1982) Nature (Lond) 295 pp. 294-299

27. Wang, A.H.J., Fujii, S., van Boom, J. and Ritch, A. (1982) Proc. Nat.
Acad. Sci. U.S.A. 79 pp.3968-3972

28. Rabinovich, D., Shakked, Z. (1983) Acta Cryst A. In press.

DYNAMIC STRUCTURE OF Z-DNA

M. Leng, B. Hartmann, B. Malfoy, J. Pilet and M. Ptak
Centre de Biophysique Moléculaire, C.N.R.S.,
1A, avenue de la Recherche Scientifique,
45045 Orléans cedex, France.

Abstract. - We first show that under topological constraints segments of natural DNA containing (A.T) and (G.C) base pairs can adopt the Z conformation. This is deduced from the titration of form V DNA by the antibodies to Z-DNA. We then report some results on exchange kinetics of the protons involved in hydrogen bonds between base pairs of poly(dG-m^{5}dC).poly(dG-m^{5}dC), in Z conformation. There are two classes of protons, a fast one the exchange half-time of which is 40 mn and a slow one the exchange half-time of which is 960 mn. These values are deduced from infrared absorption measurements in ^{2}H$_2$O and H$_2$O. These half-times are in good agreement with those determined for several alternating polynucleotides in Z-form and confirm the difference between the dynamic structure of B and Z-forms. Finally, we have studied the thermal stability of (m^{5}dC-dG)$_3$ by ^{31}P nuclear magnetic resonance, circular dichroism and ultra-violet absorption. In 3 M NaClO$_4$, (m^{5}dC-dG)$_3$ has the Z conformation. The melting curve of the hexanucleotide is biphasic. The first transition occurs over a large range of temperature. It corresponds to a conformational change due to an intramolecular non-cooperative process. This new conformation is not the B-form. The second transition is cooperative and corresponds to the melting of a double helix into single strands.

INTRODUCTION

Immunological methods have brought strong lines of evidence for the presence of segments having the Z-conformation in natural DNA. Z-DNA was discovered from a X-ray crystallographic study of oligonucleotide (dC-dG)$_n$ (1,2). Z-DNA is a left-handed double helix in which Watson-Crick base pairing is preserved. The guanine bases adopt the *syn* conformation and the cytosine the *anti* conformation. The dguanosine residues have the C3' endo conformation and the dcytidine residues the C2' endo conformation. The repetitive unit is a dinucleotide (general review 3). Numerous studies have shown that poly(dG-dC).poly(dG-dC) can adopt the Z conformation in solution and that this conformation is stabilized by several chemical modifications of base residues (general review 4). Z-DNA is a strong

B. Pullman and J. Jortner (eds.), Nucleic Acids: The Vectors of Life, 49–60.
© *1983 by D. Reidel Publishing Company.*

immunogen and the antibodies to Z-DNA are specific for Z-DNA (5,6). The antibodies bind to polytene chromosomes of *Drosophila melanogaster* and *Chironomus thummi* as visualized by fluorescent staining (7-9). Z-DNA immunoreactivity was also detected in *Stylonychia mytilus* (10) and in rat tissues (11). Most, but not all, nuclei of rat cerebellum, liver, kidney and testis were stained. In testis, nuclei of spermatogonia located at the periphery of the seminiferous tubules were heavily stained whereas the nuclei of spermatocytes, spermatides and sperm located closer to the lumen remained largely unstained.

The antibodies to Z-DNA do not bind to linear calf thymus DNA (42 % G+C) or *Micrococcus luteus* DNA (72 % G+C) even in high salt solution (5, 6). The staining of chromosomes and tissues has been performed in physiological salt conditions. This suggested that some other factors can be important to stabilize the Z-form. We have studied form V DNA prepared from pBR 322 DNA by annealing covalently closed complementary single strands according to the method of Stettler (12,13). The titration of form V DNA by the antibodies to Z-DNA indicates that about 40 ± 10 % of base residues have the Z conformation. Conformational constraints can stabilize segments of natural DNA containing (A.T) and (G.C) base pairs in the Z conformation.

The accessibility of some base atoms in Z-DNA is different from that in B and A-DNA (3,14). *In vivo*, many compounds as chemical carcinogens bind covalently to DNA. A question is to know the relative reactivity of some target atoms in these structures. We had shown that the chemical carcinogens N-hydroxy-2-aminofluorene binds to B-DNA and almost not to Z-DNA and proposed that the very weak binding of the carcinogen to Z-DNA could be due to the dynamic structure of Z-DNA (15). We here report some results on the exchange rates of protons involved in hydrogen bonds between base pairs in poly(dG-m^{5}dC).poly(dG-m^{5}dC) and we confirm that the dynamic structures of B-DNA and Z-DNA are different.

Numerous studies carried out on oligonucleotides have been very useful for the understanding of nucleic acids. The hexanucleotide pentaphosphate (m^{5}dC-dG)$_3$ has been crystallized. It forms a left-handed Z-DNA helix which is similar to that of the unmethylated Z-DNA structure with slight modifications due to the methyl groups (16). In high salt concentration, the hexamer (dC-dG)$_3$ has the Z conformation as judged by Raman spectroscopy and circular dichroism (17,18). In this paper, we report some results on the conformation of the hexanucleotide pentaphosphate (m^{5}dC-dG)$_3$ studied by circular dichroism and ^{31}P NMR. In 3 M NaClO$_4$ and at 5°C, the hexanucleotide adopts the Z-form or a Z-like form. As the temperature is increased, far below the Tm of the cooperative melting double helix-single strands, a conformational change due to a noncooperative intramolecular process occurs. The hexanucleotide adopts a new conformation which is not the B-form as observed in low salt conditions and at low temperature.

Titration of form V DNA by antibodies to Z-DNA

Form V DNA was prepared from pBR 322 DNA by annealing covalently
closed complementary strands (12). Antibodies to Z-DNA were elicited in
rabbits immunized with poly(dG-dC).poly(dG-dC) chemically modified by
the reaction with the monofunctional platinum derivative chlorodiethyle-
netriamino platinum(II) chloride (6). This polynucleotide, in which 12 %
of the base residues are modified, adopts the Z conformation in physiological conditions (19). The antibodies are specific for Z-DNA and do not
cross-react with native or denatured linear DNA, RNA, oligonucleotides
and nucleotides (20). On the other hand, the antibodies bind to form V
DNA as shown by radioimmunoassay and electron microscopy (13).

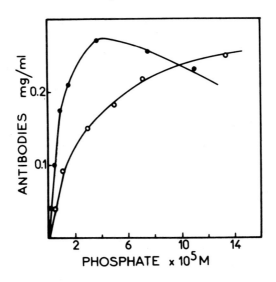

Figure 1. Titration of form V DNA by antibodies to Z-DNA. Amount of precipitated antibodies to Z-DNA (mg/ml antiserum) as a function of added
antigen. The concentration of the antigen is expressed in moles of nucleotide residues. Antigen, ($\bullet$) poly(dG-dC).poly(dG-dC), solvent 3 M
NaCl, (O) form V DNA, solvent 0.2 M NaCl.

The quantitative precipitation of form V DNA in 0.2 M NaCl and of
poly(dG-dC).poly(dG-dC) in 3 M NaCl (Z-form) was performed by addition
of increasing amount of antigen to a constant volume of antibodies. The
amount of precipitated antibodies was estimated by the Folin-Cocaltieu
method. The results are shown in figure 1. At the left of the equivalence point, there is an excess of antibodies and all the DNA is precipitated. In the precipitate, the molar ratio nucleotide residues over antibodies is equal to 8 for poly(dG-dC).poly(dG-dC)-antibody complexes and
to 20 for form V DNA-antibody complexes. The antibodies are mainly immunoglobulin G (20) and thus each antibody binding site covers about 4 nucleotide residues of poly(dG-dC).poly(dG-dC). Assuming that 4 nucleotide
residues are covered by each antibody binding site in form V DNA-antibody
complexes, one deduces that 40 % of the nucleotide residues adopt the Z

conformation in form V DNA. Because of the experimental error mainly due
to the presence of open circular DNA as shown by gel electrophoresis
(we recall that in our experimental conditions, the antibodies do not
bind to form I and form II DNA), we estimate that about 40 ± 10 % of the
nucleotide residues are in Z conformation. These results show that the
segments of DNA recognized by the antibodies must include sequences
other than stretches of alternating (G-C) residues (the sequence of pBR
322 DNA is known ; there is one $(GpC)_3$ sequence and 21 $(GpC)_2$ sequences
and even the number of (G-C) doublets, about 10 % of the total residues,
cannot account for the 40 % of base residues in Z conformation). In other
words, the segments recognized by the antibodies contain (A.T) and (G.C)
base pairs. Thus, under topological constraints, sequences of natural
DNA can adopt the Z conformation, conclusion in agreement with very recent works (21-23,38,39).

Exchange rate of protons in $poly(dG-m^5dC).poly(dG-m^5dC)$

Behe and Felsenfeld have shown that the midpoint of the B-form-Z-
form transition of $poly(dG-m^5dC).poly(dG-m^5dC)$ occurs at lower salt concentration than that of poly(dG-dC).poly(dG-dC), 0.7 M and 2.7 M NaCl
respectively (24,25). This transition occurs even at lower salt concentration in presence of traces of divalent or polyvalent ions (24).

In solution, double helical nucleic acids are known to be subjected
to thermal fluctuations resulting in transient conformations with opened
base pairs. A useful probe of the dynamic aspects of nucleic acid structure is the tritium exchange technique (26,27). We had used this technique to study the dynamic structure of poly(dG-dC).poly(dG-dC) and had
found that this varies depending on whether high or low salt conditions
are used.

The exchange rates of the five protons involved in hydrogen bonds
between base pairs in poly(dG-dC).poly(dG-dC) in B and Z-forms are different. In the B-form, there is one class of fast protons with an exchange half-time of about 6 minutes while in the Z-form there are two classes
of protons, one fast class (3 protons) with an exchange half-time of 20
minutes and a slow class (2 protons) with an exchange half-time of 420
minutes (28). These two slow protons are those of the amino group of cytosine residues as deduced from the study of $poly(dI-br^5dC).poly(dI-br^5dC)$
which has the B conformation in low salt conditions and the Z conformation in high salt conditions (29).

The exchange rates of the protons involved in hydrogen bonds between
base pairs can also be determined by infrared absorption measurements.
The infrared absorption spectra of oriented films of poly(dG-dC).poly(dG-
dC) present some significant differences in H_2O and ^2H_2O (30). The deute-
ration rates of exchangeable protons were deduced from the change in absorption near 1700 cm^{-1}. Similar differences in the H_2O and ^2H_2O spectra
of $poly(dG-m^5dC).poly(dG-m^5dC)$ were observed (results not shown).

Figure 2 is the variation of the absorbance at 1693 cm^{-1} as a func-

tion of time. As in the case of poly(dG-dC).poly(dG-dC), there are two
classes of protons. A quantitative analysis of this curve indicates that
the exchange half-time of the fast class is 40 minutes and that of the
slow class is 960 minutes.

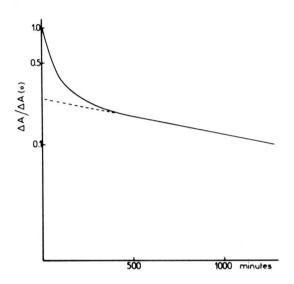

Figure 2. Hydrogen/deuterium exchange in poly(dG-m^{5}dC).poly(dG-m^{5}dC)
observed by infrared absorption spectroscopy at 1693 cm^{-1}. The figure
represents the variation of $\Delta A(t)/\Delta A(0)$ that is the normalized amplitude
in a semi-logarithmic scale versus time. $\Delta A(t)$ is the difference between
the infrared absorbance of the band maximum at time t and the absorbance
at the same frequency at the end of the exchange. Relative humidity 90 %,
temperature 6°C, salt content of the film $\cong$ 6 % w/w.

In order to related the measured exchange rate constants to the parameters governing the dynamic structure of the double helix, a reaction
scheme for the exchange process is required. If we postulate the exchange process can only occur during a transient open state, the most simple
reaction scheme is the one proposed by Teitelbaum and Englander (26,27)
which can be written

$$\text{closed} \underset{k_{cl}}{\overset{k_{op}}{\rightleftharpoons}} \text{open} \xrightarrow{k_{ch}} \text{exchanged}$$

where k_{op} and k_{cl} are respectively the opening and closing rate constants
and k_{ch} is the chemical rate constant for the exchange process of an exchangeable proton of a nucleotide in the open state with the solvent proton.

In the following table are given the exchange half-times and the
values of $K_{eq} = k_{op}/k_{cl}$ for the Z-form of several polynucleotides, values
deduced from infrared absorption experiments (this work and 29,30). The

results relative to the B-form of poly(dG-dC).poly(dG-dC) and poly(dI-br[5]dC).poly(dI-br[5]dC) are given for comparison.

		half-times		$K_{eq} = \dfrac{k_{op}}{k_{cl}}$
Poly(dG-dC).poly(dG-dC)		67	1440	$0.8 \ 10^{-5}$
Poly(dG-m[5]dC).poly(dG-m[5]dC)	Z-form	40	960	$1.2 \ 10^{-5}$
Poly(dG-br[5]dC).poly(dG-br[5]dC)		34	930	$1.2 \ 10^{-5}$
Poly(dI-br[5]dC).poly(dI-br[5]dC)		90	850	$1.3 \ 10^{-5}$
Poly(dG-dC).poly(dG-dC)	B-form		19	$62 \ 10^{-5}$
Poly(dI-br[5]dC).poly(dI-br[5]dC)			20	$56 \ 10^{-5}$

Table. Conformation dependence of proton exchange kinetic.

The exchange half-times of the fast class in Z-DNA are about 2 to 3 times larger than the half-times in B-DNA and about 10 to 30 times smaller than the half-times of the slow class. K_{eq} is about 50 times larger for the B-form than for the Z-form. As we have shown that the closing rate constants for the B and Z-forms are about the same (29), one can conclude that an open state is about 50 times less likely in Z-DNA than in B-DNA.

This large difference in the dynamic structure between left-handed and right-handed double helices might be important in biological process that require molecular recognition. It might explain some recent results on the reactivity of the chemical carcinogen N-hydroxy-2-aminofluorene (N-OHAF) versus B and Z-DNA (15,31). In given conditions, a large amount of aminofluorene residues are covalently bound to B-DNA whereas almost no residues are bound to Z-DNA. N-OHAF reacts mainly on the C8 of guanine residues *via* the formation of a nitrenium ion (32). The accessibility of C8 atom is larger in Z-DNA than in B-DNA but the electrostatic potential is smaller in Z-DNA than in B-DNA for crystal structures (1,14, 33). The larger electrostatic potential cannot explain our results since the carcinogen N-acetoxy-2-aminofluorene which reacts mainly on the C8 of guanine residues *via* a nitrenium ion (32) binds equally well to B-DNA and Z-DNA (15). Another point is that N-OHAF reacts only with native calf thymus DNA and not with oligonucleotides and nucleotides (34). Thus, there are some interactions between the double helix and the carcinogen. N-OHAF does not react with Z-DNA which might be due to the dynamic structure of Z-DNA. We assume that a necessary step in the reaction between the carcinogen and Z-DNA is an intercalation of the carcinogen between the base pairs (intercalation has to be taken in a large sense, i.e., interaction with nucleotide residues) which is much less likely in Z-DNA than in B-DNA.

Conformation of the hexanucleotide $(m^5dC-dG)_3$

The description of Z_I , Z_{II} and Z' conformations of oligonucleotides at atomic resolution has been derived from single-crystal structures of self-complementary hexanucleotide pentaphosphate $(dC-dG)_3$ and
tetranucleoside triphosphate $(dC-dG)_2$ (1,2). Very recently, the crystal
structure of the hexanucleotide pentaphosphate $(m^5dC-dG)_3$ has been determined showing the existence of a left-handed Z-DNA helix which is similar to the unmethylated Z structure (16). We report here a first investigation of the conformation of $(m^5dC-dG)_3$ by using ultra-violet absorption, circular dichroism and ^{31}P NMR techniques.

At low salt concentration $(0.2$ M $NaClO_4)$, a stable conformation was
found which can be assigned to the B-family, especially as judged by circular dichroism. The spectrum shown in figure 3 presents a first positive band centered at 280 nm and then a negative band centered at 255 nm.

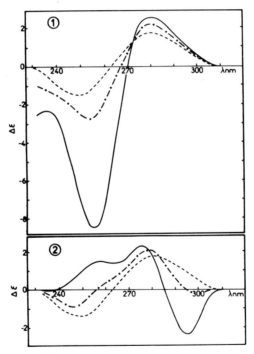

Figure 3.   Circular dichroism of $(m^5dC-dG)_3$. Variation of $\Delta\varepsilon$ $(M^{-1}$ $cm^{-1})$
as a function of wavelength, solvent 0.2 M $NaClO_4$, 5 mM Tris-HCl pH 7.5
(1), 3 M $NaClO_4$, 5 mM Tris-HCl pH 7.5 (2). Temperature in (1) (————)
0°C, (—·—) 55°C, (----) 80°C ; in (2) (————) - 10°C, (—·—) 43°C, (----)
80°C.

The melting of this B-form is a cooperative transition and the Tm depends
linearly on the oligonucleotide concentration (results not shown). The
enthalpy for the melting of the oligonucleotide is equal to 60 kcal
$mole^{-1}$, a value which is in the range characterizing a double helix-

single strands transition (35).

A conformational transition is induced by addition of salt. This transition is complete at 3 M NaClO$_4$. As shown in figure 3, the circular dichroism spectrum of this high salt form has a first negative band centered at 295 nm and then a large positive band. This spectrum looks like the spectra of poly(dG-dC).poly(dG-dC) and poly(dG-m^{5}dC).poly(dG-m^{5}dC) in the Z conformation (24,25).

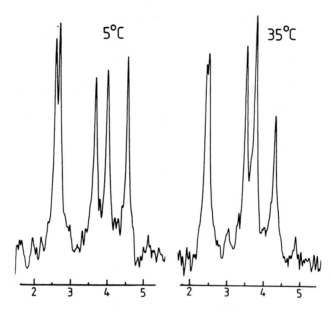

Figure 4. Proton noise decoupled 36.43 MHz ^{31}P NMR spectra of (m^{5}dC-dG)$_3$, c = 6 10^{-3} M in ^{2}H$_2$O : H$_2$O mixture (2:1), 10 mM Tris-HCl buffer pH 7.5, 3 M NaClO$_4$ as a function of temperature. The spectra are not normalized. The chemical shifts are relative to internal trimethylphosphate without correction for temperature dependence of the standard.

The ^{31}P NMR spectrum of (m^{5}dC-dG)$_3$ in the Z conformation contains five narrow lines spread over a 2.14 ppm chemical shift range with a characteristic low field doublet which can be assigned to the two (dGpm^{5}dC) phosphate groups (40), the three other (m^{5}dCpdG) phosphate groups giving a triplet which is high field shifted relatively to the previous doublet (figure 4). The centers of these two groups of resonances separated by ≅ 1.5 ppm coincide with those of the two broad lines characterizing the ^{31}P resonance of poly(dG-dC).poly(dG-dC) and poly(dI-br^{5}dC).poly(dI-br^{5}dC) in the Z conformation (36,29). The exchange frequency between B and Z-forms is slow on the NMR scale (< 5 Hz lower limit) since at intermediate salt concentrations, the ^{31}P NMR spectra result from the sum of B-form and Z-form characteristic spectra (results not shown). The complete assignment of the spectrum in 3 M NaClO$_4$ is not achieved at present and one cannot associate this spectrum with a well-defined Z conformation. More likely, the conformation in solution has to

be considered as an average conformation because of fast intramolecular
motions involving bases, sugar rings and phosphodiester backbones. It
has to be pointed out that ^{31}P NMR is not sensitive to concentration
effects. As shown later, at high hexanucleotide concentration, the formation of end-to-end aggregates stabilized by base stacking as in crystal
structures is likely to occur. The ^{31}P NMR spectrum of DNA fragments of
various lengths is almost independent of the length of the rod-like molecules (37).

   The aggregation phenomenon interferes with the conformational behavior of the Z-form in the premelting range. At low (m^{5}dC-dG)$_3$ concentration (up to 2 10^{-4} M), equilibrium ultracentrifuge experiments clearly
prove that there is no aggregation (manuscript in preparation). As shown
in figure 5, in this concentration range, the circular dichroism signal
at 295 nm decreases as the temperature is raised from -8°C to about 30°C
(at higher temperature the cooperative double helix-single strands transition occurs as usually observed for oligonucleotide duplexes). The melting is reversible. This variation of $\Delta\varepsilon_{295}$ corresponds to an intramolecular change which acts more on the high wavelength transition since
$\Delta\varepsilon_{255}$ does not significantly vary in the same range of temperature. At
higher oligonucleotide concentration, aggregation occurs and both values
$\Delta\varepsilon_{295}$ and $\Delta\varepsilon_{255}$ depend upon concentration and temperature (not shown).

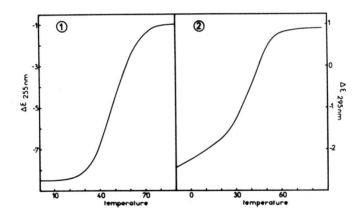

Figure 5. Variation of $\Delta\varepsilon$ as a function of temperature. $\Delta\varepsilon_{255}$ for the
B-form (1) and $\Delta\varepsilon_{295}$ for the Z-form (2) of (m^{5}dC-dG) as a function of
temperature. Same experimental conditions as in figure 3.

   The ^{31}P NMR spectra, recorded at oligonucleotide concentrations
ranging from 1 to 6 10^{-3} M are modified by an increase of temperature
in the premelting range. The relative intensity of the low field doublet
characterizing the Z-form decreases whereas one of the component of the
high field triplet increases (figure 4). This is assumed to be due to a
conformational change, exchange frequency of which is slow on the NMR
scale (<< 50 Hz). The percentage of the Z-form decreases as that of this
new conformation increases. This new conformation is not a B-form or a

single strand. A transition from the Z-form to the B-form can be excluded from the ultra-violet absorption results (manuscript in preparation).
The characteristic lines of a B-form were never observed in the ^{31}P NMR
spectrum of the Z-form of $(m^5dC-dG)_3$ as the temperature was raised from
5° to 40°C. This new conformation cannot yet be ascribed to a precise
form in the present state of our investigation. As in the Z-form the
stereochemical unit is a dinucleotide, the conformational change should
affect the two terminal $(m^5dCpdGp)$ dinucleotides leading to this new
conformation which then dissociates into single strands at higher temperature.

CONCLUSION

      In conclusion, the presence of segments of natural DNA having the
Z conformation has been revealed by the antibodies to Z-DNA (7-10). The
biological role of Z-DNA is not yet known. It has been suggested by Rich
that Z-DNA might be involved in the mechanism of regulation (3). The results here reported on oligonucleotides and polynucleotides confirm that
the dynamic structure of B-DNA differs strongly from that of Z-DNA. One
expects that these differences can be important in the reactions between
DNA and several ligands. For example, they might explain the very weak
binding of the chemical carcinogen N-hydroxy-2-aminofluorene to Z-DNA
whereas this carcinogen binds strongly to B-DNA. In a more general way,
it is tempting to speculate that the amount of Z-DNA in chromosomes depends upon cell cycle and thus the reactivity of several ligands with
some sequences in B or Z-form, will depend upon the cell cycle.

REFERENCES

1. Wang, A.H.J., Quigley, G.J., Kolpak, F.J., Crawford, J.L., van Boom,
    J.H., van der Marel, G. and Rich, A.: 1979, Nature 282, pp 680-686.
2. Drew, H., Takano, T., Tanaka, S., Itakura, K. and Dickerson, R.E. :
    1980, Nature 286, pp 567-573.
3. Rich, A. : 1983, in Structure, Dynamics, Interactions and Evolution
    of Biological Macromolecules, Hélène, C., ed., D. Reidel Publishing
    Company, Dordrecht, pp 3-21.
4. Leng, M. : 1983, in Structure, Dynamics, Interactions and Evolution
    of Biological Macromolecules, Hélène, C., ed., D. Reidel Publishing
    Company, Dordrecht, pp 45-53.
5. Lafer, E.M., Möller, A., Nordheim, A., Stollar, D.B. and Rich, A. :
    1981, Proc. Natl. Acad. Sci. USA 78, pp. 3546-3550.
6. Malfoy, B. and Leng, M. : 1981, FEBS Letters 132, pp 45-48.
7. Nordheim, A., Pardue, M.L., Lafer, E.M., Möller, A., Stollar, D.B.
    and Rich, A : 1981, Nature 294, pp 417-422.
8. Leng, M., Hartmann, B., Malfoy, B., Pilet, J., Ramstein, J. and Sage,
    E. : 1983, in 47th Symposium on Quantitative Biology, Cold Spring
    Harbor, in press.
9. Lemeunier, F., Derbin, C., Malfoy, B., Leng, M. and Taillandier, E.:
    1982, Exp. Cell. Res. 141, pp 508-513.

10. Lipps, H.J., Nordheim, A., Lafer, E.M., Ammermann, D., Stollar, D.B. and Rich, A. : 1983, Cell 32, pp 435-441.
11. Morgenegg, G., Celio, M.R., Malfoy, B., Leng, M. and Kuenzle, C.C. : 1983, Nature, in press.
12. Stettler, U.H., Weber, H., Koller, Th. and Weissmann, C. : 1979, J. Mol. Biol. 131, pp 21-40.
13. Lang, M.C., Malfoy, B., Freund, A.M., Daune, M. and Leng, M. : 1982, The Embo J. 1, pp 1149-1153.
14. Pullman, B., Pullman, A. and Lavery, R. : 1983, in Structure, Dynamics, Interactions and Evolution of Biological Macromolecules, Hélène, C., ed., Dordrecht, D. Reidel Publishing Company, pp 23-44.
15. Leng, M., Freund, A.M., Malfoy, B., Malinge, J.M., Pilet, J. and Rio, P. : 1983, in Cellular Responses to DNA Damage. UCLA Symposia on Molecular and Cellular Biology, New Series, Friedberg, E.C. and Bridges, B.R., eds, Alan R. Liss, Inc. New York, vol. 11, in press.
16. Fujii, S., Wang, A.H.J., van der Marel, G., van Boom, J.H. and Rich, A. : 1982, Nucleic Acids Res. 10, pp 7879-7892.
17. Thamann, T.J., Lord, R.C., Wang, A.H.J. and Rich, A. : 1981, Nucleic Acids Res. 9, pp 5443-5457.
18. Quadrifoglio, F., Manzini, G., Vasser, M., Dinkelspiel, K. and Crea, R. : 1981, Nucleic Acids Res. 9, pp 2195-2206.
19. Malfoy, B., Hartmann, B. and Leng, M. : 1981, Nucleic Acids Res. 9, pp 5659-5669.
20. Malfoy, B., Rousseau, N. and Leng, M. : 1982, Biochemistry 21, pp 5463-5467.
21. Nordheim, A., Lafer, E.M., Peck, L.J., Wang, J.C., Stollar, D.B. and Rich, A. : 1982, Cell 31, pp 309-318.
22. Peck, L.J., Nordheim, A., Rich, A. and Wang, J.C. : 1982, Proc. Natl. Acad. Sci. USA 79, pp 4560-4564.
23. Singleton, C.K., Klysik, J., Stirdivant, S.M. and Wells, R.P. : 1982, Nature 299, pp 312-316.
24. Behe, M. and Felsenfeld, G. : 1981, Proc. Natl. Acad. Sci. USA 78, pp 1619-1623.
25. Pohl, F.M. and Jovin, T.M. : 1972, J. Mol. Biol. 67, pp 375-396.
26. Teitelbaum, H. and Englander, S.W. : 1975, J. Mol. Biol. 92, pp 55-78.
27. Teitelbaum, H. and Englander, S.W. : 1975, J. Mol. Biol. 92, pp 79-92.
28. Ramstein, J. and Leng, M. : 1980, Nature 288, pp 413-414.
29. Hartmann, B., Pilet, J., Ptak, M., Ramstein, J., Malfoy, B. and Leng, M. : 1982, Nucleic Acids Res. 10, pp 3261-3277.
30. Pilet, J. and Leng, M. : 1982, Proc. Natl. Acad. Sci. USA 79, pp 26-30.
31. Rio, P., Loukakou, P. and Leng, M., 1983, Nucleic Acids Res., submitted.
32. Miller, J.A. and Miller, E.C. : 1979, in Environmental Carcinogenesis, Occurrence, Risk, Evaluation and Mechanism, Emmelot, P. and Kriek, E., eds, Elsevier/North Holland Biomedical Press, p 25-50.
33. Lavery, R. and Pullman, B. : 1981, Nucleic Acids Res. 9, pp 4677-4688.
34. Spodheim-Maurizot, M., Saint-Ruf, G. and Leng, M. : 1979, Nucleic Acids Res. 6, PP 1683-1694.

35. Pohl, F.M. : 1974, Eur. J. Biochem. 42, pp 495-504.
36. Patel, D.J., Canuel, L.L. and Pohl, F.M. : 1979, Proc. Natl. Acad. Sci. USA 76; pp 2508-2511.
37. Hogan, M.E. and Jardetzky, O. : 1980, Biochemistry 19, pp 3460-3465.
38. Brahms, S., Vergne, J., Brahms, J.G., Dicapua, E., Bucher, P.H. and Koller, Th. : 1982, J. Mol. Biol. 162, pp 473-493.
39. Pohl, F.M., Thomae, R. and Dicapua, E. : 1982, Nature 300, pp 545-546.
40. Jovin, T.M., vand de Sande, J.H., Zarling, D.A., Arndt-Jovin, D.J., Eckstein, F., Fuldner, H.H., Greider, C., Grieger, I., Hamori, E., Kalisch, B., McIntosh, L.P. and Robert-Nicoud, M. : 1983, 47th Symposium on Quantitative Biology, Cold Spring Harbor, in press.

RELATIVE STABILITY OF B AND Z STRUCTURES IN OLIGODEOXYNUCLEOTIDES WITH DIFFERENT ALTERNATING BASE SEQUENCE AND LENGTH

F.Quadrifoglio, G.Manzini, N.Yathindra[+] and R.Crea[++]
Laboratorio di Chimica delle Macromolecole, Istituto di Chimica, Università di Trieste, 34127 Trieste, Italy
+ On leave of absence from Dept.of Biophysics, Madras University, India
++ Creative Biomolecules, Inc., Brisbane, Ca 94005, U.S.A.

ABSTRACT

The conformation of several oligodeoxynucleotides with alternating purine-pyrimidine sequences in aqueous salt solution has been studied by means of CD technique. Although the solution conditions should favor the Z-conformation, only C-G sequences are found in this structure. $(G-C)_n$ sequences with n up to 5 do not assume the left-handed conformation. The concatamer obtained with $d(AT)_3(CG)_3$ in high salt solution was found in B conformation whereas the concatamer obtained with $d(TA)_3(CG)_3$ in the same conditions showed alternating blocks of B and Z structures. The effect of base-stacking in this behavior is discussed.

INTRODUCTION

Since the discovery of left-handed double-stranded Z-DNA in single crystals of oligodeoxynucleotides of alternate dC-dG sequences[1-3] a number of experimental results have shown that this conformation is stable also in aqueous solution of high ionic strength[4-7] and it is the same as observed earlier by Pohl and Jovin in poly(dG-dC).poly (dG-dC).[8] The presence of chemically modified bases such as m^5C, Br^8G, m^7G and the like in poly(dG-dC).poly(dG-dC) was also found to promote the B → Z transition at ionic concentrations near or even below physiological levels.[9-12]

Possible biological implications of this novel structure have been demonstrated by the use of immunological techniques which have shown that Z-DNA is indeed present in nuclear material.[13-14] The biological role of Z-DNA has still to be elucidated but hypothesis on its regulatory function in genetic activity has been advanced.[13] However, there are many intriguing stereochemical problems to be understood. Foremost among them are the base sequence combination and the environmental factors that would promote Z-like structures.

61

B. Pullman and J. Jortner (eds.), Nucleic Acids: The Vectors of Life, 61–74.
© *1983 by D. Reidel Publishing Company.*

The presence of alternating <u>syn</u> and <u>anti</u> conformations is an important requirement for the left-handed Z-helix. As the <u>syn</u> is relatively a high energy conformation for pyrimidines compared to purines, it follows that any alternating purine-pyrimidine sequence seems a natural choice to fulfill this requirement. However, only dC-dG sequences in linear polymers have been found to be able to assume the Z-structure in solution. Sequences like dG-dT, with its complementary dA-dC, appear condidate in view of recent work[15],[16] but more persuasive evidence is necessary.

Another important aspect concerns the length of alternate sequences necessary to induce Z-structure in linear DNA fragments under appropriate conditions. In natural DNA the alternate sequences are rarely long and it has to be understood whether short (up to 10 base pairs) sequences can be transformed to left-handed conformations in the presence of B DNA chains and what are the stereochemical or environmental conditions to obtain this. It has already been shown that fairly long dC-dG sequences can assume a Z-form in appropriate salt conditions when enclosed in B-form chain[17],[18] but this evidence is lacking for short alternating sequences. Z-conformation has also been obtained in low salt conditions when a highly negative supercoiling is induced in plasmids containing inserts of alternate sequences and even in plasmids without such insertions[18-20]. Very extreme topological constraints, like those obtained with form V of plasmid DNA, force the polynucleotide chain to assume a Z conformation for about 40% of its length which cannot be accounted for by the actual extent of alternate sequences[21],[22] These data suggest that even natural sequences can be transformed to Z-structure in extreme conditions.

We have initiated a research program in our laboratory with the principal aim of finding if :
a) alternating sequences, different from dC-dG, can also assume a Z-conformation in linear DNA fragments and in what conditions;
b) short alternating sequences, potentially Z-helicogenic, can be induced to assume the Z-conformation when they are inserted in long B-form chains.

In this communication we report some preliminary results concerning the conformational stability in different salt conditions of several alternating oligodeoxynucleotides of different lengths and sequences including those which contain only dC-dG or dG-dC sequences.

MATERIALS AND METHODS

d(CG)$_2$, d(CG)$_3$, d(CG)$_4$, d(GpCpGpCpGpm^{5}C), d(CpGpCpApTpGpCpG), d(GpCpGpTpApCpGpC), d(GTAC)$_2$, d(AT)$_3$(CG)$_3$ and d(TA)$_3$(CG)$_3$ have been synthesized by the phosphotriester method according to the procedure reported previously.[23] Purification of the unblocked oligonucleotides was performed by high performance liquid chromatography on anionic exchange (Permaphase AAX, Dupont) and reverse-phase resins (Bondapak C$_{18}$, Waters Ass.).

d(GC)$_3$, d(GC)$_4$, d(GC)$_5$, d(TG)$_5$ and d(CA)$_5$ were obtained from P-L Biochemicals.

The concentration of the oligonucleotides was calculated by recording the absorption spectrum at 80°C and using the extinction coefficients calculated after enzymic hydrolysis with pancreatic DNAse (Sigma). All solutions contained 1 mM phosphate buffer, pH = 7.2. All the salts were of the highest purity available. Absorption spectra were run with a Cary 219 spectrophotometer equipped with a thermostattable cell holder. Temperature increase was obtained with a Haake F3 thermo-stat with a PG 10 programmer. The rate of temperature increase in the melting experiments was 0.1°/min. CD spectra were recorded with a Jasco J-500A with a thermostattable cell holder. Stoppered quartz cells of suitable pathlength were used throughout. All the melting profiles were fully reversible.

RESULTS

a) Oligodeoxynucleotides with dC-dG Sequence

We have examined the conformational state of (dC-dG)$_n$ with n = 2, 3, 4 in aqueous media of varying ionic strength. When n = 3, 4 the sample undergoes a transition from B to Z form as a function of ionic strength. The midpoint of the transition is not much different from that found with poly(dG-dC).poly(dG-dC)[8] The termal stability of the sample in both conformations depends on n and on its concentration. (dC-dG)$_2$ does not assume a complete B or Z conformation when C $\simeq 10^{-3}$M and the temperature is above 0°C.

b) Oligodeoxynucleotides with dG-dC Sequence

We have examined the conformational state of (dG-dC)$_n$ with n = 3, 4, 5 at low and high NaCl concentration. In all cases a conformation of the B-family has been observed. Fig.1 shows the CD spectrum of (dG-dC)$_4$ at different NaCl concentrations, which is representative of all the compounds of this series. The CD changes observed can be explained with the transformation from classical B form to C form.

We have also examined a sample of (dG-dC)$_3$ containg m^{5}C at the 3'end to see a possible stabilizing effect of methyl cytosine on the left-handed structure. Again no Z conformation could be observed at high ionic strength. All the samples were also checked in the presence of suitable amount of cations which are known to stabilize the Z-structure, such as Ca^{++}, Mg^{++}, Tb^{+3} but no effects different from those produced by NaCl were observed.

c) Oligodeoxynucleotides with GT and AC Sequences

Poly(d GdT).poly(d AdC) shows in high ionic media a CD spectrum which resembles in some aspects that of poly(dG-dC).poly(dG-dC) in Z

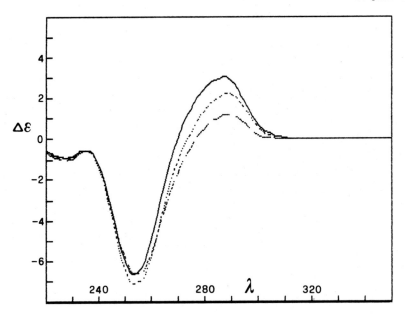

Fig.1 CD spectra of d(GC)$_4$, C = 5.3x10^{-5}M, at different NaCl concentra-
       tions: (——) 0.1M, (- - -) 2.6M, (— —) 5.2M. T = 10°C. Tris buffer
       0.1M, pH = 7.2.

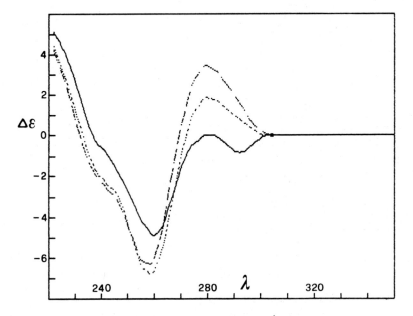

Fig.2 CD spectra of d(GTAC)$_2$, C = 1.8x10^{-4}M, at different salt
       concentrations: (— —) 0.01M NaCl, (- - -) 2M CsF, (——) 7M CsF.
       T = 10°C. Tris buffer 0.01M, pH = 7.2.

conformation.[15,16] We have studied in the same conditions the 1:1 mixture of the two decadeoxynucleotide $(dT-dG)_5:(dC-dA)_5$ and the self complementary octadeoxynucleotide $d(GTAC)_2$. In both cases the addition of different salts such as NaCl, CsF, $CaCl_2$ produced only moderate and non-cooperative changes on the CD spectrum. Fig.2 shows the data obtained on $d(GTAC)_2$ with CsF. The results cannot be explained as due to the induction of a Z-conformation, probably a C-type conformation is stabilized at high ionic strength.

d) Oligodeoxynucleotdies with Mixed Alternate Sequences

Two octadeoxynucleotides with reversed sequences were examined: d(CGCATGCG) and d(GCGTACGC). Both give duplexes at room temperature and at concentration around 10^{-5} (expressed in moles of chain). Both contain sequences Z-helicogenic (CpG) and sequences supposed to give in appropriate conditions Z-structure. The two samples were examined in different solvent media (0 ÷ 5 M NaCl, 0 ÷ 6 M CsF, 0 ÷ 2 M $CaCl_2$) and at different temperatures. No evidence for the presence of a Z-conformation was found according to CD results (Figs.3, 4). B to coil transitions were observed by increasing temperature.

e) Concatamers with Blocks of Alternating Sequences

Two dodecadeoxynucleotides of sequences $d(AT)_3(CG)_3$ and $d(TA)_3(CG)_3$ were investigated. Both oligomers are only 50% self-complementary and in solution form concatamers whose length and conformational stability is a function of concentration and temperature. Fig.5 and 6 show this behavior for $d(AT)_3(CG)_3$ and $d(TA)_3(CG)_3$, respectively. The melting profiles are biphasic as $(AT)_3$ and $(TA)_3$ blocks melt before $(CG)_3$ blocks. At low temperatures the two concatamers are stable and their conformation can be checked by CD measurements. Fig.7 shows the results obtained at three different temperatures and high ionic strength for $d(AT)_3(CG)_3$. The lowest temperature corresponds to the ordered concatamer, the intermediate one corresponds approximately to the dimer with $d(CG)_3$ blocks in duplex and the highest temperature to the single strands in coil conformation.

CD results show that for $d(AT)_3(CG)_3$ in concatameric form the conformation assumed is of the B type whereas at intermediate temperature the $d(CG)_3$ blocks are (at least partially) in Z conformation.

On the other hand, the corresponding results for $d(TA)_3(CG)_3$ show that even in concatameric form the $(CG)_3$ blocks assume a Z conformation, (Fig.8). In fact increasing temperature through the first melting no appreciable change in the negative 294 nm CD band is observed whereas a change in the negative 250 nm CD band parallels the absorbance change due to the $d(TA)_3$ blocks melting. The negative 294 nm CD band disappears in a parallel fashion with the second increase in absorbance (the concatamer obtained with $d(TA)_3(CG)_3$ is composed of short alternating

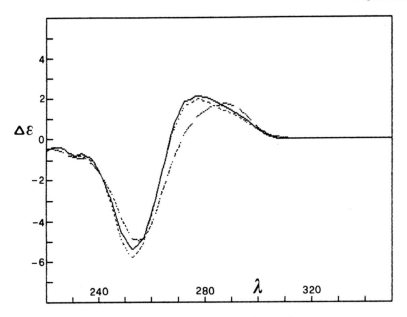

Fig.3 CD spectra of d(CGCATGCG), C = 6.0x10^{-5}M, at different NaCl
concentrations: (——) 0.15M, (- - -) 1.1M, (— —) 5M. T = 15°C.
Tris buffer 0.01M, pH = 7.2.

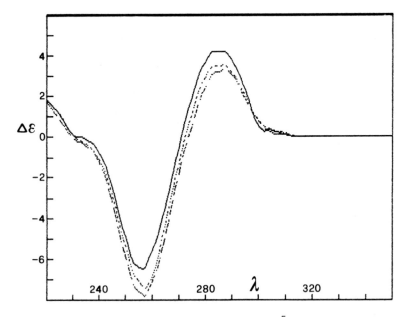

Fig.4 CD spectra of d(GCGTACGC), C = 5.5x10^{-5}M, at different NaCl
concentrations: (——) 0.15M, (- - -) 1.0M, (— —) 5M. T = 15°C.
Tris buffer 0.01M, pH = 7.2.

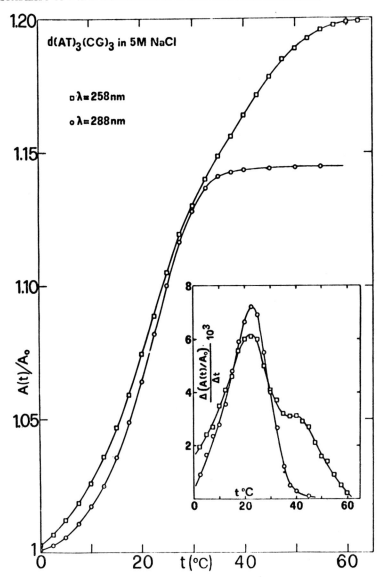

Fig.5 Melting profiles recorded at two different wavelengths for
d(AT)$_3$(CG)$_3$ in 5M NaCl. The insert shows the derivative curves.
d(AT)$_3$(CG)$_3$ = 8x10^{-5} monomol/l.

B and Z helices). These results show that the conformation of the
d(CG)$_3$ blocks depends on the sequence of the preceding and succeeding
alternating blocks when in ordered conformations. That the nagative
band at 294 nm in the concatamer is not due to some C-type negative
band can be demonstrated by the study of the NaCl concentration dependence
of the CD bands (Fig.9). All the bands change cooperatively with NaCl

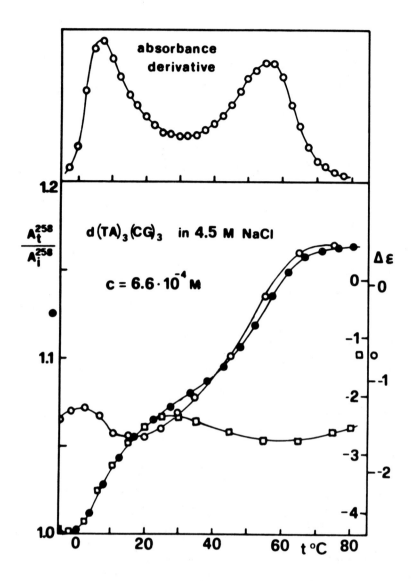

Fig.6 Temperature dependence of relative absorbance at 258 nm (O) and
       of ellipticity at 250 nm (□) and 293 nm (O) for d(TA)₃(CG)₃ in
       4.5 M NaCl, 1 mM phosphate buffer, pH = 7.2.

concentration as expected for a B to Z transition and contrary to what
would be found for a B to C transformation.

    If the concatamer is composed of short alternating B and Z mini-
helices its CD spectrum can be approximated as the sum of the contributions of the CD of d(TA)₃ in B form and that of d(CG)₃ in Z form.

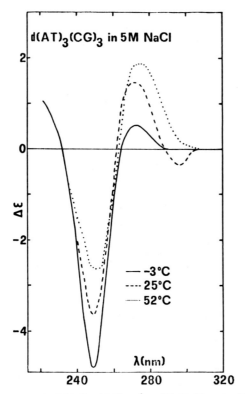

Fig.7   CD spectra of d(AT)$_3$(CG)$_3$ in 5M NaCl at different temperatures.
d(AT)$_3$(CG)$_3$ = 8x10^{-5} monomol/l.

Fig.10 shows the calculated and experimental CD curves. The agreement
can be considered fairly good.

DISCUSSION

     CD results obtained with several oligodeoxynucleotides of alternat-
ing purine-pyrimidine show that only (dC-dG)$_n$ sequences easily assume
the left-handed Z-structure in solution of high ionic strength. Although
the other oligodeoxynucleotides investigated conform to the alternating
sequence requirement, they show no tendency to exhibit Z-type CD spectra
in high salt solution. The finding that even the oligodeoxynucleotides
(dG-dC)$_n$, with n up to 5, do not assume the left-handed Z-conformation
in appropriate media of high ionic strength appears particularly
surprising. This result suggests that the difference in free energy of
the two conformations is the result of a delicate balance in which
stacking energies (not necessarily confined to base-base interaction)
play a fundamental role at least for short chains. It is known that two
stacking geometries exist in a Z structure[1] and it seems that only
stacking of C over G favors energetically the Z structure with respect

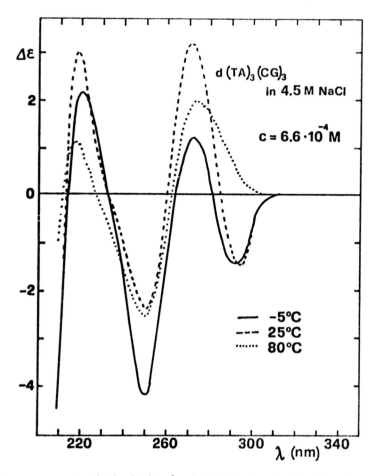

Fig.8   CD spectra of $d(TA)_3(CG)_3$ in 4.5 M NaCl, 1 mM phosphate buffer,
        pH = 7.2, at different temperatures.

to B structure, whereas the stacking of G over C opposes it. For
$(dC-dG)_n$ the number of C/G stackings is always greater than the number
of G/C stackings whereas the reverse is true for $(dG-dC)_n$. Published
results on $(dG-dC)_8$[24] (even if the sample used was a mixture of $(dG-dC)_8$
and $(dC-dG)_8$ according to the procedure used in the preparation) suggest
that this oligomer is in complete Z conformation at high ionic strength.
From this and from our results it follows that B and Z states have near
the same free energy at high salt concentration for $(dG-dC)_n$ when n is
6 or 7. It follows also that there exists a significant free energy
contribution different from stacking which stabilizes Z conformation
and, with the growing of the chain, overcomes the unfavorable G over C
stacking energy.

      The unfavorable base stacking energy may also be one of the reasons
why the oligodeoxynucleotides with mixed alternate sequences cannot be

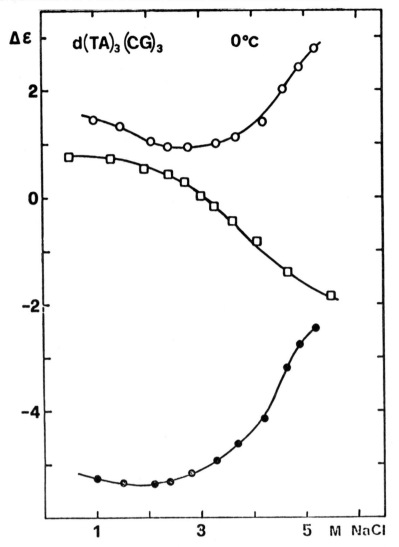

Fig.9   Dependence of ellipticity at 250 (●), 272 (O) , and 293 nm
(□) for d(TA)$_3$(CG)$_3$ at 0°C on NaCl concentration.

stabilized in Z conformation. This holds only for alternate sequences
in linear chains. In fact, it has been demonstrated that alternate
sequences can be induced to assume a left-handed structure when enclosed
in a circular DNA under topological constraints.[19,20] In this case the
free energy necessary for the flipping from more stable B to less stable
Z conformation is supplied through relaxation of the degree of supercoiling. Interaction of Z structure with specific protein has been
envisaged as further source of stabilization energy.[25] When the topological constraints are very severe, as in the case of form V of plasmid DNA,
it could happen that even non alternating sequences can assume a Z-

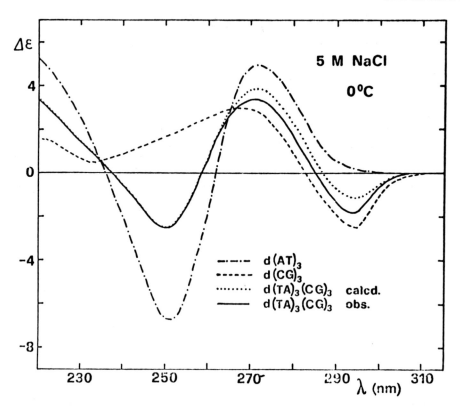

Fig.10   Comparison between observed (——) and calculated (···) CD
spectra of d(TA)$_3$(CG)$_3$ in 5M NaCl at 0°C. The CD spectra
of d(CG)$_3$ (---) and d(AT)$_3$ (-·-·-) used for calculations
are also shown.

conformation, as judged from the works of Pohl et al.[21] and Brahms at
al.[22]

The importance of stacking energies is apparent also from the CD
results obtained in high ionic strength with d(AT)$_3$(CG)$_3$ and d(TA)$_3$(CG)$_3$
concatamers.[26,27] The only significant difference between the two chains
is the junction between the two alternate blocks: a T/C junction in the
former concatamer and an A/C junction in the latter. Nevertheless in the
former concatamer the two blocks have the same B conformation whereas in
the latter both assume the conformation more typical of the solvent
condition: B for the (TA)$_3$ stretch and Z for the (CG)$_3$ stretch. It appears
that an A/C junction allows coexistence of B and Z conformation in a
concatamer, whereas T/C junction does not despite the flexibility allowed
by the nick. What is not achievable in a concatamer with a certain degree
of freedom should be much more difficult to obtain in an unnicked chain,
at least in the absence of topological or other constraints and for
short alternate sequences. In other words, it appears that the free
energy requirements necessary to induce a Z conformation in a stretch

of alternate sequence flanked by B type helices strongly depends on the stereochemistry of stacking in the junction regions. It is interesting to note that recent studies on two different plasmids have shown the highest free energy contribution for the B → Z transition arises from the junctions.[19]

ACKNOWLEDGEMENTS

This work was supported by Consiglio Nazionale delle Ricerche, Roma, Italy, Grant N.81.01714.03, by Ministero della Pubblica Istruzione and by Università degli Studi di Trieste. Sincere thanks are due to Mrs.G. Fabris for her excellent administrative and editorial assistance and to Mr.A.Vici for his skillful technical cooperation.

REFERENCES

1. Wang, A.H.J., Quigley, G.J., Kolpak, F.J., Crawford, J.L., van Boom, J.H., Van der Marel, G., and Rich, A.: 1979, Nature 282, pp.680-686.
2. Crawford, J.L., Kolpak, F.J., Wang, A.H.J., Quigley, G.J., van Boom, J.H., van der Marel, G., and Rich, A.: 1980, Proc.Natl.Acad.Sci.USA 77, pp.4016-4020.
3. Drew, H., Takano, T., Tanaka, S., Itakura, K., and Dickerson, R.E.: 1980, Nature 286, pp.567-573.
4. Mitra, C.K., Sarma, M.H., and Sarma, R.H.: 1981, Biochemistry 20, pp. 2036-2041.
5. Thamann, T.J., Lord, R.C., Wang, A.H.J., and Rich, A.: 1981, Nucleic Acids Res. 9, pp.5443-5457.
6. Wu, H.M., Dattagupta, N., and Crothers, D.M.: 1981, Proc.Natl.Acad. Sci.USA 78, pp.6806-6811.
7. Patel, D.J., Kozlowski, S.A., Nordheim, A., and Rich, A.: 1982, Proc. Natl.Acad.Sci.USA 79, pp.1413-1417.
8. Pohl, F.M. and Jovin, J.M.: 1972, J.Mol.Biol. 67, pp.375-396.
9. Behe, M. and Felsenfeld, G.: 1981, Proc.Natl.Acad.Sci.USA 78, pp. 1619-1623.
10. Lafer, E.M., Möller, A., Nordheim, A., Stollar, B.D., and Rich. A.: 1981, Proc.Natl.Acad.Sci.USA 78, pp.3546-3550.
11. Möller, A., Nordheim, A., Nichols, S.R., and Rich, A.: 1981, Proc. Natl.Acad.Sci.USA 78, pp.4777-4781.
12. Santella, R.M., Grunberger, D., Nordheim, A., and Rich, A.: 1982, Biochem.Biophys.Res.Comm. 106, pp.1226-1232.
13. Nordheim, A., Pardue, M.L., Lafer, E.M., Möller, A., Stollar, B.D., and Rich, A.: 1981, Nature 294, pp.417-422.
14. Lipps, H.J., Nordheim, A., Lafer, E.M., Ammermann, D., Stollar, B.D., and Rich, A.: 1983, Cell 32, pp.435-441.
15. Vorlickovà, M. Kypr, J., Stokrova, S. and Sponar, J.: 1982, Nucleic Acids Res. 10, pp.1071-1080.
16. Zimmer, Ch., Tymen, S., Marck, Ch., and Guschlbauer, W.: 1982, Nucleic Acids Res. 10, pp.1081-1091.

17. Klysik, J., Stirdivant, S.M., Larson, J.E., Hart, P.A., and Wells, R.D
    1981, Nature 290, pp.672-677.
18. Peck, L.J., Nordheim, A., Rich, A., and Wang, J.C.: 1982, Proc.Natl.
    Acad.Sci.USA 79, pp.4560-4564.
19. Nordheim, A., Lafer, E.M., Peck, L.J., Wang, J.C., Stollar, B.D., and
    Rich, A.: 1982, Cell 31, pp.309-318.
20. Singleton, C.K., Klysik, J., Stirdivant, S.M., and Wells, R.D.: 1982,
    Nature 299, pp.312-316.
21. Pohl, F.M., Thomae, R., and Di Capua, E.: 1982, Nature 300, pp.545-546
22. Brahms, S., Vergne, J., Brahms, J.G., Di Capua, E., Bucher, Ph., and
    Koller, Th.: 1982, J.Mol.Biol. 162, pp.473-493.
23. Crea, R., Kraszewski, A., Hirose, T., and Itakura, K.: 1978, Proc.
    Natl.Acad.Sci.USA 75, pp. 5765-5769.
24. Patel, D.J., Canuel, L.L., and Pohl, F.M.: 1979, Proc.Natl.Acad.Sci.
    USA 76, pp.2508-2511.
25. Nordheim, A., Tesser, P., Azorin, F., Kwon, Y.H., Moller, A., and Rich
    A.: 1982, Proc.Natl.Acad.Sci.USA 79, pp.7729-7733.
26. Quadrifoglio, F., Manzini, G., Vasser, M., Dinkelspiel, K., and Crea,
    R.: 1981, Nucleic Acids Res. 9, pp.2195-2206.
27. Quadrifoglio, F., Manzini, G., Dinkelspiel, K., and Crea, R.: 1982,
    Nucleic Acids Res. 10, pp.3759-3768.

ELECTROSTATIC MOLECULAR POTENTIAL VERSUS FIELD IN NUCLEIC ACIDS AND THEIR CONSTITUENTS.

A. Pullman, B. Pullman and R. Lavery
Institut de Biologie Physico-Chimique,
13 rue Pierre et Marie Curie, 75005 Paris.

ABSTRACT

The molecular electrostatic potential (MEP) and the molecular electrostatic field (MEF) are associated with significantly different patterns of distribution in the nucleic acids and their constituents and are influenced differently by such environmental factors as counterion screening. They also govern the electrostatics of interaction of DNA with different types of species, the MEP being of particular significance in this respect for interaction with cations and the MEF for the association with neutral dipolar molecules. A number of exemples are given to illustrate this situation.

INTRODUCTION

The polyanionic character of the nucleic acids and the fact that these macromolecules often interact with species which are positively charged in their active form suggest that the electrostatic properties of the nucleic acids can play an important role in their reactive behaviour. We have been able to confirm this situation in a series of publications dealing with the molecular electrostatic potential (MEP) of these fundamental biopolymers and their constituents (for reviews see ref. 1, 2). These potentials, when combined with a measure of the steric accessibility of the sites under attack, have notably enabled the explanation of many experimental data concerning the reactivity of the nucleic acids towards charged electrophiles.

Recently a significant new development in these studies has been achieved by the extension of the computations to the evaluation of the related molecular electrostatic field (MEF) of the same macromolecules [3,4]. This has the important consequence of enlarging the possibility of investigating their biochemical reactivity towards a different type of reagent, namely, neutral, dipolar molecules, of which the most important representative is water. This advantage springs from the laws of electrostatics which tell us that while the energy of a point charge at a chosen point in space, due to a potential generated by a charge distri-

75

B. Pullman and J. Jortner (eds.), Nucleic Acids: The Vectors of Life, 75–88.
© *1983 by D. Reidel Publishing Company.*

bution, is simply the product of the potential at this point multiplied by the magnitude of the charge, the electrostatic interaction energy between a point dipole placed at the same point is equal to the scalar product of the dipole moment with the field generated at this point by the same charge distribution. The two concepts appear thus as complementary in their applicability to fundamental types of biochemical reactivity.

We present some essential results obtained in this respect for DNA and its constituents. We recall that the potential (V) and the field ($\underline{E}$) of a molecule are defined by :

$$V(P) = \sum_{\alpha} \frac{Z_\alpha}{|\underline{r}_{\alpha P}|} - \int \frac{\rho(i)}{|\underline{r}_{iP}|} d\tau_i$$

$$\underline{E}(P) = \sum_{\alpha} \frac{Z_\alpha \underline{r}_{\alpha P}}{|\underline{r}_{\alpha P}|^3} - \int \frac{\rho(i) \underline{r}_{iP}}{|\underline{r}_{iP}|^3} d\tau_i$$

where Z_α is the charge of nucleus α, distant by $r_{\alpha P}$ from point P and $\rho(i)$ is the electronic distribution whose volume element $d\tau_i$ is at a distance r_{iP}. Although the field is thus simply the derivative of the potential with respect to distance, the distribution of these two electrostatic characteristics may be very different for a given system, because of their very different dependence on distance. The aim of this paper is to present the extent, nature and significance of this difference for the nucleic acids and their constituents.

METHODS

The methods of evaluating the potentials and the fields in biomolecules and biopolymers have been described in references [1-4] and will not be repeated here. May we just recall that for simple biomolecules, of limited dimensions, such as the fundamental constituents of the nucleic acids, phosphates, sugars, purine and pyrimidine bases, they are computed from the electronic distributions of these subunits, obtained from ab initio SCF calculations. The potential and the field of the macromolecules are constructed by the superposition of the potentials and fields of all the subunits forming the nucleic acid, appropriately positioned in space. In order to facilitate the calculation of the macromolecular electrostatic properties, the electronic distribution of each subunit is replaced by a multicenter multipole expansion which is capable of reproducing the exact electrostatic properties of the subunit down to a distance of 2 Å from its constituent atoms. Below this distance the exact electronic distribution must, however, be employed.

Numerous representations are available for describing the molecular potentials and fields. Among the most significant ones, to which we shall refer in this paper, are :

a) Plane potentials or fields (potential or field intensity maps, whose minima or maxima represent, respectively, the main site potentials or fields at reactive centers).
b) Surface envelope potentials or fields (potentials or fields on envelopes formed by the intersection of spheres centered on each atom of DNA, with radii equal to the van der Waals radius of the atoms concerned, multiplied, if desired, by a factor F ; they are generally presented in the form of their projection onto a two-dimensional "window"). For technical details on this last particularly significant representation see ref. [5].

RESULTS AND DISCUSSION

We shall carry out the comparison of MEP with MEF on two levels, first with respect to the constituents of the nucleic acids and then with respect to these biopolymers themselves.

1) Representative constituents of DNA.

We consider first, in some detail, the examples of cytosine and of the phosphate group [6]. Fig. 1 contains a graphical representation of the MEP (Fig. 1a) and MEF (Fig. 1b) in the plane of cytosine. Fig. 2 represents the same quantities in the plane containing the phosphorus and the anionic oxygens of the phosphate group. For the sake of the discussion note that the potentials are deeper and the fields are greater the darker the shadings on these figures (as well as on the following figures in this paper).

From the point of view of the comparison of the distribution of MEP and MEF it may be noted that both on fig. 1 and 2 the zones of strong field are less extended than those of potential and that while deep potentials are distributed over relatively a large zone, the maxima of the field are concentrated closer to certain atoms. Thus in fig. 1 cytosine exhibits a zone of deep potential located between its carbonyl oxygen O_2 and ring nitrogen N_3 (the potentials surrounding the rest of the base being only weakly negative or positive) whereas the strong field is split into two separate zones around each of these two atoms. The same features are clearly visible in fig. 2 for the phosphate group, where we observe a continuous zone of deep potential between the two anionic oxygens, with the minimum occuring almost at the midpoint between them, whereas the fields are separated into two zones concentrated around these oxygens.

A similar situation occurs in the remaining constituents of the nucleic acids with the interesting overall result that, as illustrated in table I, no parallelism is observed, for instance, between the minimal potential and the maximal field of the four bases of DNA : while the potentials are the deepest in guanine followed by cytosine, the fields are the strongest in the two purines, adenine followed by guanine. Among the isolated subunits, the deepest potential (-6.2 volt) and the strongest field (2.5 volt) are, however, both due to the phosphate group.

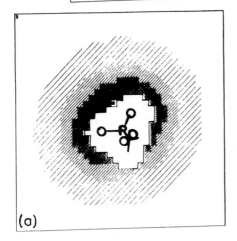

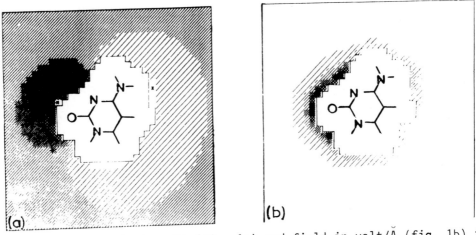

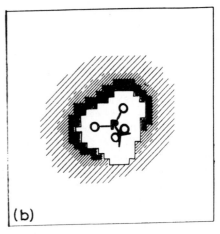

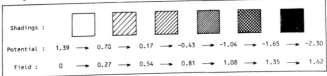

Figure 1. Potential, in volts(fig. 1a) and field in volt/Å (fig. 1b) in the plane of cytosine. Significance of the shadings :

Shadings :	□	▨	▨	▨	▨	■
Potential :	1.39 →	0.73 →	0.17 →	-0.43 →	-1.04 →	-1.65 → -2.30
Field :	0 →	0.27 →	0.54 →	0.81 →	1.08 →	1.35 → 1.62

Figure 2. Potential, in volts (fig. 2a) and field in volt/Å (fig. 2b) in the plane of the P-anionic oxygens of the phosphate group. Significance of the shadings :

Shadings :	□	▨	▨	▨	▨	■
Potential :	-1.08 →	-1.95 →	-2.82 →	-3.69 →	-4.55 →	-5.42 → -6.33
Field :	0 →	0.43 →	0.87 →	1.30 →	1.73 →	2.17 → 2.60

TABLE I

Minimal potentials (volts) and maximal fields (volts/Å) on the surface envelopes of the nucleic acid bases.

Base	Potential	Field
G	-2.3	1.5
A	-1.3	1.6
C	-2.0	1.4
T	-1.2	1.3

In conformity with the predicted utilization of these two electrostatic indices, the MEPs have been shown to be good guides to the reactivity of the bases toward electrophilic agents [1, 7] and the MEFs for the evaluation of their hydration shells [6].

2) Nucleic acids.

The differences between the distributions of the MEP and MEF are very significantly increased when we go over from their constituents to the nucleic acids themselves. These differences have several aspects.

A) A fundamental difference appears in the variation of the magnitude of the minimal potentials and the maximal fields on going over from the constituents to the double helical macromolecules. This phenomenon is illustrated in table II in which the results concerning the macromolecular system refer to one complete turn of B-DNA [1, 7]. It is observed that while the values of the minimal potentials at the reactive sites of the bases or of the phosphates increase manyfold when going from the subunit to the double helix (with intermediate variations for units of intermediate complexity such as mononucleotides or single stranded polynucleotides), the values of the fields undergo only very small changes. This is a most striking demonstration of the different distance dependence of these two electrostatic properties : the relatively weak dependence of the potentials permits large superposition effects, leading to a continuous increase of their maximal values, while the much more rapid decrease of the fields preserves the more local character of their maxima.

We have been able to demonstrate abundantly the significance of the deepening of the potential wells when going over from the constituent units of DNA to the double helix (with an intermediate situation in single stranded polynucleotides) by relating it to a parallel increase of the reactivity of these species towards electrophilic reactants, a phenomenon particularly striking in view of the parallel decrease in the accessibility of the involved sites [1, 8, 9]. The most significant examples were found in the series of chemical carcinogens, whose reactive

TABLE II

Minimal potentials (volts) and maximal fields (volts/Å) at the bases
and the phosphate group (P⋆) in B-DNA compared to the corresponding
values in the isolated subunits. The values in B-DNA are taken from
 studies of a complete helical turn, with G-C or A-T sequences [4].

	Minimal potential		Maximal field	
	Isolated subunit	B-DNA	Isolated subunit	B-DNA
G	-2.3	-27.5	1.5	1.5
A	-1.3	-27.1	1.6	1.3
C	-2.0	-26.2	1.4	1.2
T	-1.2	-27.3	1.3	1.2
P⋆	-6.2	-26.6	2.5	2.6

forms are well-known to be electrophilic cations produced with or without metabolic activation. Details of the examples may be found in the
above references. The only exception to this common behaviour appears to
be N-acetoxy-N-2-acetylaminofluorene, which reacts less readily with
double-stranded helical polynucleotides than with single-stranded ones
or with mononucleotides [10], [11]. The exceptional behaviour of this carcinogen may be attributed to the particular mechanism of its action upon
DNA which involves intercalation into the double helix and implies thus
a strong deformation of the substrate. As indicated already in ref. [8]
and [9], the examples concerning the relation between the reactivity of
electrophiles and the electrostatic potential in nucleotide systems are
in no way limited to carcinogens. A similar pattern of reactivity with
nucleic acids and their constituents may be demonstrated in a group of
antitumor agents, the antibiotics of the pyrrole(1, 4)benzodiazepine
series, such as anthramycin, sibromycin, tomaymycin and neothramycin
[12-16]. Quite generally, these compounds form a covalent bond with the
NH_2 group of the guanine moiety of double stranded DNA (in striking
analogy to the adduct formed between DNA and the carcinogenic metabolic
product of benzo [a] pyrene) ; they interact much less with denatured
DNA and are unreactive towards synthetic polynucleotides or nucleic acid
bases.

      B) A second fundamental difference between the MEP and the MEF
in nucleic acids concerns the location of their most significant values.
Thus it is shown [3], [4] that while the deepest potentials are located in
the grooves of the double helix, the greatest fields are concentrated
on the phosphates of the backbone. This is illustrated in fig. 3 for a
model of B-DNA, consisting of one turn of its double helix with a homo-
genous A-T sequence (ref. [3]). The model is presented diagrammatically in
the central figure (3b), from which it can be recognized that the minor
groove is situated in its upper part and the major groove below. The
distributions of the potential (fig. 3a) and field (fig. 3c) are pre-

sented on the surface envelope of the helix, following the technique described in ref. [5]. (Very similar results are obtained for the G-C sequence, with the only difference that in this polymorph of DNA the deepest potentials are in the minor groove for A-T sequences and in the major groove for G-C sequences). The situation presented in fig. 3 is a most impressive manifestation of the different distance dependence of the two electrostatic properties studied. Thus, as we have seen before, when the constituents of the nucleic acids are considered separately, the phosphates are characterized by both the deepest potential and the greatest field. In the double helix, because of the long range of the potentials, their superposition creates the deepest minima between the two strands, in the grooves ; the shorter range of the electrostatic field attenuates these superposition effects and the strongest fields remain close to the charged phosphates. When the computations of the potentials in DNA are carried out not at the surface envelopes but at the sites of their different minima, which are located much closer to the phosphate group or the bases, the two types of minima become competitive, with the deepest minimum,which is associated with N7 of guanine, being still slightly (3 kcal/mole) deeper than the one associated with the phosphate groups. Further, the potential minimum associated with the phosphate oxygen which is oriented towards the inside of the double helix is appreciably deeper ($\sim$ 50 kcal/mole) than the one associated with the phosphate oxygen pointing toward the exterior of the helix.

Altogether we may thus expect the grooves to constitute particularly reactive centers towards electrophilic reagents and the phosphates to be the main sites of water fixation. We have already indicated on the examples of the electrophilic carcinogens or antitumor drugs the reality of the affinity of the base atoms for these reactants. A particularly striking recent demonstration of the preferential affinity of cations for the grooves of the B-DNA double helix is, however, provided by the studies of Skuratovskii et al. [17, 18] who used synchrotron radiation diffraction data for phage T2 CsDNA fibers to determine the coordinates of the caesium ions in crystalline form of B-DNA. Cs^+ is a particularly interesting test for the theoretical computations because no similar precise information can be obtained for the "light" DNA salts (Li^+, Na^+, Mg^{2+}) and also because Cs^+ is almost unhydrated and represents thus a "simple" type of a metal cation whose interaction with DNA is devoid of the "perturbing" effects of hydrogen bonding of the hydrated species. In Skuratovskii's et al. experiments, the Cs^+ ions are found to be located close to the dyad axes lying between the planes of adjacent bases pairs, in the grooves of the double helix "instead of near the phosphate groups". In particular, the Cs^+ ions are in direct contact with the base atoms in the narrow groove, in which they manifest a preference for AT sequences (an aspect of the interaction to which we shall come back later). The cations in the wide groove are the closest to the phosphates from which they are nevertheless separated by a hydration layer 1-2 water molecules thick.

The preferential hydration of the anionic oxygen atoms of the phosphate groups of DNA is well substantiated since the classical work

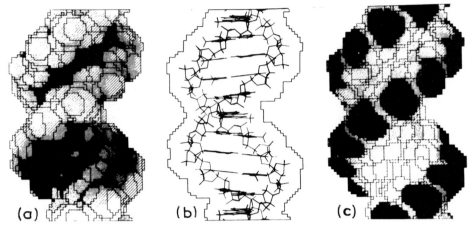

Figure 3.   Potential in volts (fig. 3a), a diagrammatic representation (fig. 3b) and the field in volts/Å (fig. 3c) in a turn of B-DNA double helix, with A-T base sequence. Significance of shadings :

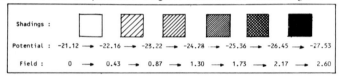

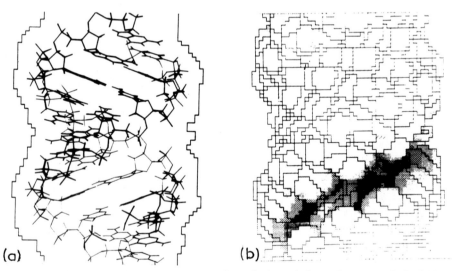

Figure 4.   Diagrammatic representation (fig. 4a) and potential (fig. 4b), in A-DNA (G-C sequence). The deepest potential, indicated by M, is on a phosphate anionic oxygen at the edge of the major groove. Significance of shadings :

of Falk et al. [19-21]. It has received recently a remarkable confirmation through an X-ray study of a single crystal oligomeric B-DNA docecamer (CGCGAATTCGCG) by Dickerson et al.[22] in which, contrary to indications from a room temperature and insufficiently resolved spectrum [23], the refinement of the resolution (through the study of the same sequence at 16°K and of the 9-bromo variant (CGCGAATTBrCGCG) at 7°C in 60% 2-methyl-2,4-pentanediol) indicated the phosphates as the sites of the strongest hydration (for a more detailed analysis see[24]). This feature was invisible in the primitive experiment because of high temperature factors, indicating thermal or static disorder of the backbone atoms.

The phosphates are not, of course, the only hydration sites in oligo- and polynucleotides. The results of the dodecamer crystal, in particular, indicate that, in the major groove, the majority of the N and O atoms on the edges of the base pairs bind water molecules. This correlates with our evaluation of relatively strong field associated with the atoms N7, O6 of guanine, N7 of adenine and O4 of thymine[24]. In the minor groove, the dominant feature of the crystal is a regular spine of water molecules binding between successive adenine N3 and thymine O2 atoms. a result again in agreement with our calculations indicating that these latter two atoms are associated with relatively strong fields. The extension and regularity of this spine is probably due to a particularly favourable geometrical arrangements of the water receptor sites in this groove of A-T sequences and is considered by Dickerson et al. [22,23] as a possible source of the stability of the B form of DNA. In G-C sequences the formation of a similar minor groove spine is prevented by the perturbing effect of the NH_2 group [22,25]. This geometrically favoured water bridging of the O2 atoms of thymine and N3 atoms of adenine in the minor groove of A-T sequences of B-DNA and the resulting stabilizing effect on this form are reminiscent of the same effect obtained, with the Cs^+ ions, as mentioned above, and also of similar effects observed upon the binding of netropsin, distamycin A and related compounds to DNA (for references and general discussion see [26,27]).

In relation with the problem of hydration it may be worthwile adding that explicit computations on the hydration scheme of the A-T and G-C base pairs, carried out in our laboratory [28,29] indicate that the first pair binds 1 or 2 more water molecules than the second. A reverse prediction is made by Clementi et al.[30] who claim that the hydration of the G-C pairs should exceed that of the AT pairs. The available experimental results from a number of different sources [31-33] confirm our estimates and contradict those of ref. [30]. To complete the picture it must also be added that, although experimental data are scarce, the information available, essentially concerning the series of uracil and its derivatives [34-37], confirms our numerical evaluation of the energies of interaction of water molecules with the bases and the similarity from this point of view of uracil and thymine. On the contrary, they invalidate completely the very different predictions based on a Monte Carlo treatment of the same subject in ref. [30]. Quoting from a recent work of Wierzchowski et al. [37] which presents the results of a thermochemical determination of the enthalpies of hydration of uracil and 5-alkylura-

cils : "From a quantitative point of view the results of recent Monte
Carlo calculations (ref. [30]) of binding energies at 300° of 40-water
molecule clusters by uracil and thymine do not find support in our data.
Neither can the large difference between thymine and uracil in the
average interaction energy of the water molecules with the base (-154.77
and - 322.92 KJ mol^{-1}, respectively) nor that in the net balance ($\Delta\bar{U}$)
of the energies of base-water and water-water interactions (196.6 and
8.4 KJ mol^{-1}, respectively) be understood in the light of our findings
... The difference in $\Delta\bar{U}$ by as much as about 23-fold (!) between thymine
and uracil remains in open contradiction with the contribution of only
about 10% of the CH_3-group to ΔH°_{int} of thymine". Obviously, the Monte-
Carlo computations of ref. [30] cannot be trusted.

We have concentrated our attention till now essentially on B-DNA
because it is for this form that most of the rather scarce, available
experimental support of our theoretical results may be found. The computations on the distribution of the potential and the field have,
however, been performed for all the major types of DNA conformers [2, 4].
Broadly speaking, the general results are similar in all the allomorphic
forms of DNA, in the sense that, for instance, in all of them the zones
of deepest potentials are generally located in the grooves of the surface envelopes while the strongest fields are connected with the phosphate groups. Each form, presents, however, its own specific features.
Thus, in particular, the balance of potential between the grooves of
DNAs is largely a function of the conformation of the allomorphs and,
specifically, of the position of the base pairs with respect to the
helical axis [2], being very similar when the base pairs are centrally
located (e.g. B-DNA) [5] and much more negative in the deeper of the two
grooves for those conformations where the base pairs are notably displaced (e.g. A or Z-DNA) [38, 39]. It is important, however, to also pay
attention to fine details. Thus when describing the distribution of the
potential in A-DNA [38] we observed that "the most negative potentials are
seen to be concentrated in the major groove of the double helix" which
is dominant in this form. We added, however, immediately that : "the
 minimum potential of - 790 kcal/mole lies on the surface of a phosphate
anionic oxygen on the edge of the groove". The situation is illustrated
in fig. 4 for the poly (dG) . poly (dC) model of A-DNA, where the minimum is designated by the letter M. (A similar result is obtained for the
poly (dA) . poly (dT) model of A-DNA). Now, our remark on the displacement of the deepest minimum of potential towards the phosphate anionic
oxygens acquires a particular significance in view of the remarkable
finding of Skuratowskii et al. [40] that, in distinction to the situation
described previously for B-DNA, a similar study of the A form of CsDNA
fibers indicates that all the cations are, in this case, located on the
periphery of the double helix, near the phosphates and have no close
contacts with base atoms.

As to the fields, the strongest ones are always associated, as
stated above, in all forms of DNA, with the phosphate groups but this,
of course, does not imply a uniform field distribution around the double
helix, for the very reason that in some conformations the phosphate

groups approach one another closer across one groove. In such cases the stronger fields occur on the same side of the double helix as the stronger potentials, namely, on the side containing the narrower and deeper groove. Such is the case e.g. for A-DNA, where as we have seen above the major groove is dominant, and Z-DNA, where the minor groove is dominant.

The differences between the distributions of the potential and the field have also been investigated for a molecule of RNA, namely tRNA[Phe] [41, 42] and references indicated therein. In this highly convoluted molecule, a general correlation between structure and potential may be noted in the sense that phosphates or bases situated at the extremities of the L-shaped molecule have generally weaker potentials than those located towards its center ("core"). This is due, evidently to the concentration of the phosphates in the center of tRNA[Phe] and to the resulting important superposition effect of their long range potentials. On the contrary, the strongest fields are found all along the phospho-diester backbone of the system, due to their much shorter range of action. Moreover, the fields in the grooves of the double helical sections of tRNA[Phe] are generally much weaker than those on the bordering backbone, while the reverse caracterizes the distribution of the potential, a situation similar to that observed in DNA and explainable by the same reasons.

C) The third aspect of the difference between MEP and MEF concerns the influence of counterion binding upon these properties. This effect was investigated using a model saturated screening of the nucleic acids consisting of binding a sodium cation in a bridged position between the two anionic oxygens of each phosphate group, 2.15 Å from each oxygen in the plane containing these atoms and the phosphorus atom. We have verified, however, that the exact nature of the saturated counterion distribution, or the type of cation employed, have relatively little effect on the resulting, general image of the electrostatic properties of the screened acids.

The most spectacular difference in the influence of the screening on MEP and MEF concerns its effect on the magnitudes of these two electrostatic indices. While the deepest potentials are still observed in the grooves of the screened acids and the strongest fields on the backbone the absolute values of the potentials are profoundly reduced, while the fields are increased with respect to those of the unscreened acids with the maxima occuring close to the phosphate-sodium ion pairs 3, 4. This last result, surprising perhaps at first sight to some, is in fact easily understood when one realizes that this zwitterionic pair is a zone where the gradient of the potential will be very large. The phenomenon is presented numerically for a few representative cases in table III. Concerning the particular case of the MEPs one may note that their reduction in the screened systems leaves unperturbed the sequence-dependant relative order of the two grooves in the nucleic acids which have "symmetrically" positioned base pairs (with respect to the helix axis, e.g. B-DNA) but may reverse this order in the nucleic acids with "unsymmetrically" positioned base pairs (e.g. Z-DNA), in which the

preference for one of the grooves is, as seen previously, sequence
independent.

TABLE III

Surface minimal potentials (volts) and surface maximal field (volts/Å)
for B-, A- and Z-DNA in the naked and screened states.

Model	Unscreened			Screened		
	Min. Potential		Max.	Min. Potential		Max.
	Minor groove	major groove	field backbone	minor groove	major groove	field backbone
B-DNA poly(dG).poly(dC)	-26.3	-27.5	2.6	-4.2	-4.8	2.9
B-DNA poly(dA).poly(dT)	-27.3	-26.1	2.6	-5.0	-3.2	2.9
B-DNA poly(dG).poly(dC)	-30.1	-33.7	2.9	-3.2	-4.7	3.5
Z-DNA poly(dG-dC). poly(dG-dC)	-31.9	-29.7	2.8	-2.8	-3.5	3.3

The influence of counterion binding on the electrostatic properties
of tRNA[Phe] follows a similar pattern [39] :while it produces a weakening
of the potential over the entire surface of the macromolecule, it leaves
the field distribution almost unchanged, except in limited zones around
the ion binding positions where it is strongly enhanced.

CONCLUSION

It is obvious that in spite of their relationship the MEPs and MEFs
are associated with significantly different patterns of distribution in
the nucleic acids and evolve differently under the effect of such environmental factors as counterion screening. They also govern the electrostatics of interaction of DNA with different types of species, the MEP
being of particular significance in this respect for interactions with
cations and the MEF for the association with neutral dipolar molecules.
The two concepts are thus complementary in their applicability to the
study of the fundamental types of biochemical reactivity. The few examples to which we were able to apply so far our theoretical results confirm this distinction.

ACKNOWLEDGMENT

This work was supported by the National Foundation for Cancer Research and the authors wish to thank Professeur Albert Szent-Gyorgyi for inspiring discussions on the role of water in biological systems. They also wish to thank Dr. I. Ya Skuratovskii (Institute of Molecular Genetics, USSR Academy of Sciences, Moscow) and R.E. Dickerson (Molecular Biology Institute, University of California at Los Angeles) for the communication of manuscripts prior to publication.

REFERENCES

1. Pullman, A. and Pullman, B. : 1981, Quart. Rev. Biophys. 14, pp. 289-380.
2. Pullman, B., Lavery, R. and Pullman, A. : 1982, Eur. J. Biochem. 124, pp. 229-238.
3. Lavery, R., Pullman, A. and Pullman, B. : 1982, Theor. Chim. Acta 62, pp. 93-106.
4. Lavery, R. and Pullman, B. : 1982, Nucl. Acids Res. 10, pp. 4383-4395.
5. Lavery, R. and Pullman, B. : 1981, Int. J. Quantum Chem. 20, pp. 259-272.
6. Lavery, R., Pullman, A. and Pullman, B. : 1983, Biophysical Chem. 17, pp. 75-86.
7. Pullman, A. and Pullman, B. : 1981, in Chemical Applications of Atomic and Molecular Electrostatic Potential, P. Politzer and D.G. Truhlar ed., Plenum Publish. Corp. N.Y., pp. 381-405.
8. Pullman, A. and Pullman, B. : 1980, Int. J. Quantum Chem, Quantum Biol. Symp. 7, pp. 245-259.
9. Pullman, B. and Pullman, A. : 1980, in Carcinogenesis : Fundamental Aspects and Environmental Effects. Proceedings of the 13th Jerusalem Symposium in Quantum Chemistry and Biochemistry. Pullman, B., Ts'o, P.O.P. and Gelboin, H. eds., Reidel Publishing Co. Dordrecht, Holland, pp. 55-66.
10. Kapulser, A.M. and Michelson, A.M. : 1971, Biochim. Biophys. Acta 232, pp. 436-450.
11. Kriek, E. : 1974, Biochim. Biophys. Acta 355, pp. 177-187.
12. Maruyama, I.N., Tanaka, N., Kondo, S. and Umazawa, H. : 1981, Biochem. Biophys. Res. Comm. 98, pp. 970-975.
13. Kohn, K.W., Glaubiger, D. and Spears, C.L. : 1974, Biochim. Biophys. Acta 361, pp. 288-302.
14. Hurley, L.H. and Petrusek, R. : 1979, Nature 282, pp. 529-531.
15. Hurley, L.H., Gairola, Ch. and Zmijewski, M. : 1977, Biochim. Biophys. Acta 475, pp. 521-535.
16. Glaubiger, D., Kohn, K.W. and Charney, E. : 1974, Biochim. Biophys. Acta 361, pp. 303-311.
17. Skuratovskii, I. Ya, Volkova, L.I., Kapitonova, K.A. and Bartenev, V.N. : 1979, J. Mol. Biol. 134, pp. 369-374.
18. Bartenev, V.N., Volkova, L.I., Golovanov, E.I., Kapitonova, K.A., Mokulskii, M.A. and Skuratovskii, I. Ya. : J. Mol. Biol., in press.
19. Falk, M., Hartman, K.A. Jr. and Lord, R.C. : 1962, J. Am. Chem.

Soc. 84, pp. 3843-3848.

20. Falk, M., Hartman, K.A. Jr. and Lord, R.C. : 1963, J. Am. Chem. Soc. 85, pp. 3843-3848.

21. Falk, M., Hartman, K.A. Jr. and Lord, R.C. : 1963, J. Am. Chem. Soc. 85, pp. 391-395.

22. Kopka, M.L., Fratini, A.V., Drew, H.R. and Dickerson, R.E.: 1983, J. Mol. Biol. 163, 129-146.

23. Drew, N.R. and Dickerson, R.E. : 1981, J. Mol. Biol. 151, pp. 535-556.

24. Lavery, R. and Pullman, B., Studia Biophys., in press.

25. Lavery, R. and Pullman, B. : 1981, Nucl. Acids Res. 9 , pp. 7141-7051.

26. Pullman, B. and Pullman, A. : 1981, Studia Biophys. 86, pp. 95-102.

27. Pullman, B., Pullman, A. and Lavery, R. : in Structure, Dynamics, Interactions and Evolution of Biological Macromolecules, Helene, C. ed., Reidel Publish. Co., Dordrecht, Holland, p. 23.

28. Goldblum, A., Perahia, D. and Pullman, A. : 1978, FEBS Letters 91, pp. 213-215.

29. Pullman, B., Miertus, S. and Perahia, D. : 1979, Theoret. Chim. Acta 50, pp. 317-325.

30. Clementi, E. and Corongiu, G. : 1980, J. Chem. Phys. 72, pp. 3979-3992.

31. Tunis, M.J.B. and Hearst, J.E. : 1968, Biopolymers 6, pp. 1325-1344.

32. Tunis, M.J.B. and Hearst, J.E. : 1968, Biopolymers 6, pp. 1345-1355.

33. G.M. Mrevlishvili, G.M., Dzhaparidze, G. Sh., Sokhadze, V.M., Tatishvili, D.A. and Orvelashvili, L.V. : 1980, Molecular Biol. (U.R.S.S.) English Ed. 15, pp. 265-271.

34. Yanson, I.K., Teplitsky, A.B. and Sukhodub, L.F. : 1979, Biopolymers 18, pp. 1149-1162.

35. Sukhodub, L.F., Yanson, I.K. Shelkovski, V.S. and Wierzchowski, K.L. : 1982, Studia Biophysica 87, pp. 223-224.

36. Sukhodub, L.F., Yanson, I.K., Shelkovski, V.S. and Wierzchowski, K.L. : 1982, Biophys. Chem. 15, pp. 145-155.

37. Teplitsky, A.B., Glukhove, O.T., Sukhodub, L.F., Yanson, I.K., Zielenkiewicz, Z., Zielenkiewicz, W., Kosinski, J. and Wierzchowski, K.L. : 1982, Biophys. Chem. 15, pp. 139-147.

38. Lavery, R. and Pullman, B. : 1981, Nucl. Acids Res. 9, pp. 4677-4688.

39. Zakrzewska, K., Lavery, R. and Pullman, B. : 1981, in Biomolecular Stereodynamics, R.H. Sarma Ed., Adenine Press, N.Y. pp. 163-183.

40. Alexeev, D.G., Bertenev, V.N., Kapitonova, X.A., Volkova, L.I. and Skuratovskii, I. Ya : 1981, in Symposium on Biophysics of Nucleic Acids and Nucleoproteins, Tallinn, Abstract F11.

41. Lavery, R. and Pullman, A. : 1982 (June), in New Horizons in Quantum Chemistry Proc. of the 4th International Congress of Quantum Chemistry, Uppsala, Löwdin, P.O. and Pullman, B. Eds, Reidel Publishing Co., Dordrecht, Holland, p. 439.

42. Pullman, A. : Studia Biophysica, in press.

PROBING FOR AND WITH LEFT-HANDED DNA: POLY[d(A-br^{5}C)·d(G-T)], A MEMBER OF A NEW FAMILY OF Z-FORMING DNAs.

Thomas M. Jovin, Lawrence P. McIntosh, David A. Zarling,
Donna J. Arndt-Jovin
Max Planck Institut für biophysikalische Chemie, Postfach 968,
D-3400 Göttingen, FRG

Michel Robert-Nicoud
Lehrstuhl für Entwicklungsphysiologie, Universität Göttingen,
D-3400 Göttingen, FRG

Johan H. van de Sande
Department of Medical Biochemistry, University of Calgary,
Alberta, Canada T2N 4N1

ABSTRACT

Left-handed helical conformations can be adopted by the poly[d(A-C)·d(G-U)] family of DNAs bearing at least one methyl or halogen (bromine, iodine) substitution at the pyrimidine C5 heterocyclic position. A representative polynucleotide is poly[d(A-br^{5}C)·d(G-T)]. Elevation of salt concentration and temperature shifts the equilibrium in favor of the left-handed state. The spectral changes (u.v. absorption, c.d., NMR, Raman) accompanying the highly co-operative and reversible R(right)-L(left) transitions are similar to those observed with poly[d(G-C)]. Some, but not all, antibodies raised against the left-handed members of the poly[d(G-C)] family also recognize the new class of polymers, thus establishing the existence of sequence-specific determinants in the binding of certain anti-Z DNA immunoglobulins. The latter have been used to establish the presence of left-handed regions in natural plasmid and viral DNAs and in the giant salivary gland polytene chromosomes of Chironomus thummi thummi. The Z-DNA specific immunofluorescence in fixed polytene chromosomes has been quantitated by laser scanning and photon counting and is particularly intense in certain regions, such as the telomeres.

B. Pullman and J. Jortner (eds.), Nucleic Acids: The Vectors of Life, 89–99.
© *1983 by D. Reidel Publishing Company.*

INTRODUCTION

Synthetic polynucleotides with defined sequences have provided the classical model structures for the elucidation and classification of DNA conformation. It has recently become apparent that double helical DNA can be highly polymorphic, as best exemplified by the B and Z forms of poly[d(G-C)] which are right- and left-handed, respectively (1,2). In order to study the factors determining the expression of DNA polymorphism, we have employed synthetic polynucleotides with systematic base (3,5) and backbone (4) substitutions. The resultant polymers differ greatly in the propensity for adopting a left-handed conformation (3-5) and are distinguishable immunochemically (6).

RESULTS AND DISCUSSION

Spectroscopic properties of poly[d(A-br^{5}C)·d(G-T)]

The ultraviolet absorption and c.d. spectra of poly[d(A-br^{5}C)·d(G-T)] are shown in Figure 1. The spectra obtained at

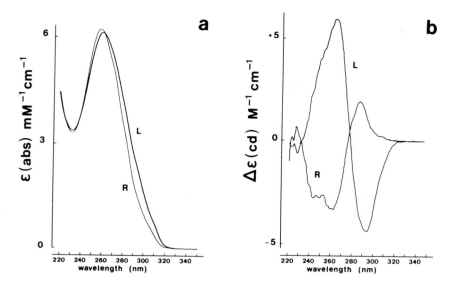

Figure 1. Absorption (a) and c.d. (b) spectra of poly[d(A-br^{5}C)·d(G-T)] in the right(R)- and left(L)-handed conformations. R) 0.12 mM DNA in 11 mM Tris-HCl, pH 7.6, 11 mM NaCl, 0.1 mM EDTA, at 22 °C. L) conditions as in (R) with the exception of the NaCl concentration (4.4 M) and temperature (53 °C). At the high salt concentration but at 22 °C, the right-handed conformation is maintained; however, a reduction in the intensity of the positive peak of the c.d. spectrum is noted (data not shown; 3). The synthesis and characterization of poly[d(A-br^{5}C)·d(G-T)] are given in reference 3.

low salt concentration (designated R) are characteristic for B DNA. Increasing the salt concentration and temperature leads to a co-operative reversible transition evidenced by a red shift in the absorption spectrum and a near inversion in the c.d. spectrum (curves designated L). The spectral changes are similar to those observed with other members of the [d(A-C)·d(G-U)] family (3,5) as well as with poly[d(G-C)] (7), a finding which provides the basis for equating the R and L states with the B and Z conformations, respectively. Additional evidence for this assignment is shown in Figure 2, which depicts the ^{31}P NMR and laser Raman spectra of the R and L forms of poly[d(A-br[5]C)·d(G-T)]. The NMR spectrum at low salt concentration

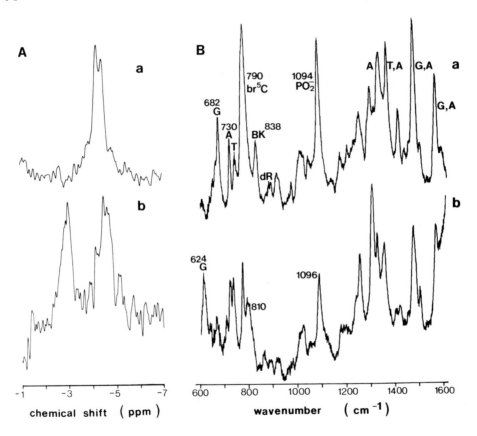

Figure 2. ^{31}P-NMR (A) and laser Raman (B) spectra of poly[d(A-br[5]C)·d(G-T)] in the R (a) and L (b) conformations. The NMR spectra were obtained at 30 °C with 2.7 mM DNA in 10 mM Tris-HCl, pH 7.6, 1 mM EDTA, and either (a) 10 mM NaCl, or (b) 5.4 M NaCl. Chemical shifts are relative to trimethylphosphate. The Raman spectra were recorded at 25 °C with ca. 20 mM DNA in (a) 2 mM Tris-HCl, pH 7.6, 0.2 mM EDTA, and 0.1 M NaCl, or (b) 8 mM Tris-HCl, pH 7.6, 0.2 mM EDTA, and 5.4 M NaCl. The frequencies of certain known resonances are indicated (see text).

shows two closely spaced resonances, a characteristic finding with polynucleotides having an alternating dinucleotide structure (3,5,8,9). In the case of the L conformation at high salt concentration, one of the resonances is markedly shifted downfield, leading to a separation of 1.5 ppm, a value virtually identical to that observed with the Z form of poly[d(G-C)] (8). Similarities are also observed in the corresponding Raman spectra (Figure 2B), notably the shifts in a) the ring breathing resonance of G (682->624 cm^{-1}); b) the PO$_2$$^-$ symmetrical stretch frequency (1094->1096 cm^{-1}); and c) the backbone stretch frequency (838->810 cm^{-1}). These data suggest the existence of structural elements characteristic of Z DNA (10,11), i.e. the <u>syn</u> configuration of the purine (G) glycosidic bond, and the C3'- <u>endo</u> sugar pucker. Other changes are present in frequencies corresponding to the four bases; thus, a direct comparison with the spectrum of poly[d(G-C)] presents obvious difficulties.

The R-L transitions of poly[d(A-br^{5}C)·d(G-T)] are highly co-operative and reversible with respect to both salt and temperature

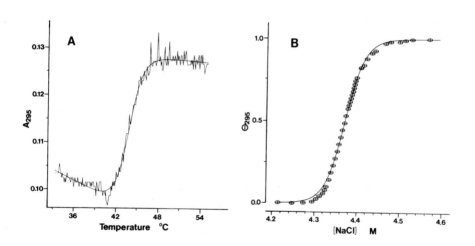

Figure 3. Thermally (A) and salt (B) induced R-L transitions of poly[d(A-br^{5}C)·d(G-T)]. Absorbance changes at 295 nm were monitored. A) The thermal R->L transition was measured with a 50 μM DNA solution in 12 mM Tris-HCl, pH 7.6, 0.1 mM EDTA, and 4.39 M NaCl, which was heated from 30 °C to 70 °C at a rate of 0.03 °C min^{-1}. An analysis according to a two-state model yielded the thermodynamic parameters: T$_t$ (transition midpoint) = 43.7 °C, ΔH= 0.81 MJ mole^{-1}, ΔS= 2.6 kJ mole^{-1} deg.$^{-1}$. B) A similar DNA solution as described in (A) but initially in the L form (4.8 M NaCl, 40.4 °C, 716 μl) was titrated continuously (symbols) with a 10 mM Tris-HCl, pH 7.6, 0.1 mM EDTA buffer at a rate of 0.203 μl min^{-1}. The smooth curve represents a two-state analysis of the degree of transition Θ$_{295}$ = [L]/([L]+[R]) which yielded the values: C$_t$ (transition midpoint) = 4.37 M NaCl, Hill coefficient = 190.

perturbations (Figure 3). The reactions are complete within a few
seconds and are characterized by high values of the Hill coefficient
(170-190) and of the apparent transition enthalpy change (0.8-1.1 MJ
mole^{-1}). The transition midpoints (salt concentrations and transition
temperatures) are inversely related, a property shared by all members of
the [d(A-C)·d(G-U)] family (3,5).

Immunochemical properties of poly[d(A-br^{5}C)·d(G-T)]

Left-handed Z DNA is highly immunogenic as evidenced by the
numerous antibodies which have been raised to poly[d(G-C)] and its
derivatives (6,13-16). The presently available members of the
[d(A-C)·d(G-U)] DNA family cannot be maintained in the L form under
physiological conditions (unless exposed to topological stress in the
form of negative supercoiling; 17,18). Nonetheless, they can be used as
substrates for testing the binding specificities of polyclonal and
monoclonal anti-Z DNA antibodies (6,12). Figure 4 compares the
properties of two antibodies by radioimmunoassay. A rabbit polyclonal
IgG (T-4), raised against chemically brominated poly[d(G-C)] (6),

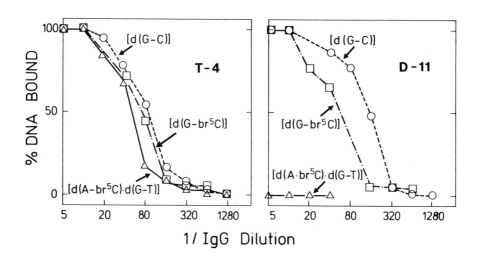

Figure 4. Specificities of anti-Z DNA antibodies measured by
radioimmunoassay. (T-4): rabbit polyclonal IgG (6). Serial two-fold
dilutions were made of a solution with a total IgG concentration of 2.8
mg ml^{-1}, of which approximately 2% was anti-Z DNA specific (6). (D-11):
monoclonal IgG (15). The serially diluted antibody solution was 0.4 mg
ml^{-1}. The radioactive DNA probes were used at 0.5 μg ml^{-1}.
Poly[d(G-C)], poly[d(G-br^{5}C)] and poly[d(A-br^5)·d(G-T)] were assayed in
40 mM Tris-HCl, pH 7.2, 4 mM EDTA containing 4.0 M, 0.2 M, and 5.0 M
NaCl, respectively. The reactions were carried out at room temperature
for 1 hr and either anti-rabbit or anti-mouse IgG was used as the second
antibody. Additional details are given in reference 6.

interacts strongly with poly[d(G-C)] and poly[d(G-br^{5}C)] and only slightly less well with poly[d(A-br^{5}C)·d(G-T)]. In contrast, a monoclonal murine IgG (D-11; reference 15) induced by the same immunogen binds the pyrimidine substituted polymer poly[d(G-br^{5}C)] poorly relative to poly[d(G-C)], and poly[d(A-br^{5}C)·d(G-T)] not at all. We attribute these contrasting anti-Z DNA specificities to immunoglobulin recognition of different surface features on left-handed DNA (bases, backbone) which are either common to or distinct in the d(G-C) and [d(A-C)·d(G-T)] families of sequences. The antibodies do not bind the respective right-handed B forms of the various polynucleotides.

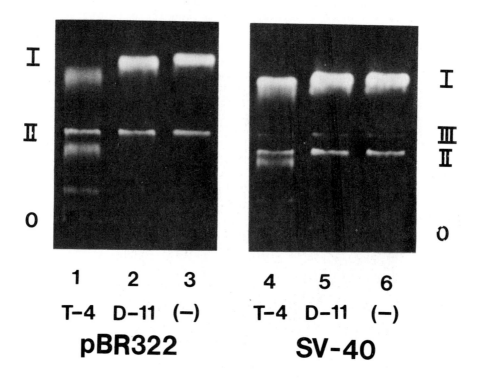

Figure 5. Electrophoretic retardation of plasmid and viral DNAs by bound anti-Z DNA antibodies. DNAs (0.4 μg) were incubated with 39 μg or 2 μg of T-4 and D-11 IgGs, respectively, for 1 hr at 24 °C followed by 3 hrs at 10 °C. The final ionic strength was approximately 0.04 and 0.16 M in the case of T-4 and D-11, respectively. Electrophoresis was in 0.7% agarose gels in 0.1 M Tris, 76 mM borate, 25 mM EDTA at 10 °C. (0= origin with electrophoretic migration upwards). The positions of DNA Forms I (supercoiled), II (relaxed), and III (linear) are indicated for the control lanes without antibody. Additional details are given in reference 6.

Demonstration of left-handed regions in natural DNA and chromosomes

Topological stress induced by negative supercoiling of plasmid and viral closed circular double-stranded DNA molecules can potentiate the B-Z transition of $d(G-C)_n$ and $d(A-C)_n$ inserts (17-21). This property has been demonstrated by alterations in the molecular hydrodynamic properties (sedimentation constant and electrophoretic mobility) which reflect the relaxation of writhe as a result of changes in helical sense. In addition, the binding of antibodies to supercoiled DNA can be used to score the existence of left-handed regions (6,15,21,22). This phenomenon is demonstrated in Figure 5 in which the binding of T-4 IgG to pBR322 and to SV-40 DNAs lacking synthetic oligonucleotide inserts is accompanied by electrophoretic retardation of the DNA-IgG complex in agarose gels (6,15). Only the mobilities of the negatively supercoiled Form I molecules are altered and the appearance of new slower bands

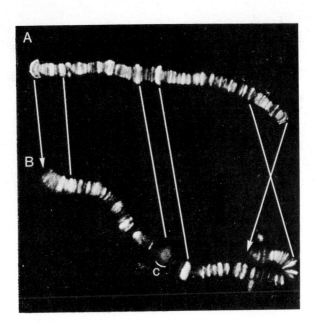

Figure 6. Comparison of indirect immmunofluorescence patterns of Chironomus thummi thummi polytene chromosome III obtained with different anti-Z DNA antibodies. A) polyclonal Z-6 rabbit IgG with broad anti-Z DNA specificity (6). B) monoclonal D-11 IgG (15) with a bias for $d(G-C)_n$ sequences. The lines connect corresponding regions on the two chromosomes. The arrows point to the two non-fluorescent telomeres of the chromosomes stained with D-11 IgG. The symbol c designates the centromere. The explanted salivary glands were fixed for 1 min in 3:1 EtOH/HOAc followed by 5 min in 45% HOAc. They were then squashed and frozen on dry ice. Staining was as described in reference 12.

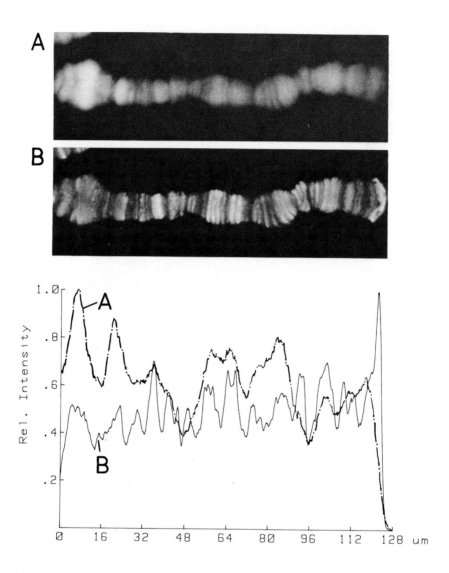

Figure 7. Scanning photometry of a fixed <u>Chironomus</u> <u>thummi</u> <u>thummi</u> polytene chromosome. Scan is from the centromere to the <u>right-hand end</u> of chromosome I. A) fluorescence from bound DNA dye Hoechst 33342; scanning spot of 1.5 μ imaged from Hg lamp. B) indirect immunofluorescence from anti-Z polyclonal T-4 rabbit IgG; 0.7 μ scanning spot using a He-Cd laser. Single pass scans were performed with a step resolution of 0.25 μ and a frequency of 100 Hz. Additional details are given in reference 12.

suggests the formation of oligomeric species crosslinked by bivalent IgGs. Under the conditions shown, the polyclonal T-4 IgG is much more effective than the monoclonal D-11, possibly reflecting the relative sequence specificities indicated above.

The same IgG probes have been used to locate regions of left-handed DNA in highly polytenized Dipteran chromosomes. Indirect (4,12,23,24) and direct (12) immunofluorescence measurements readily demonstrate binding patterns on the chromosomes of Drosophila melanogaster (4,12,23) and Chironomus thummi thummi (4,12,24). Our results indicate that the fluorescence is concentrated in the DNA-rich bands of acid-fixed chromosomes (4,6,12,24; Figure 6). The Z-6 (a preparation similar to T-4, reference 6) and D-11 antibodies show differential binding patterns. Some bands demonstrate comparable staining intensities with both antibody preparations but other areas, including the centromere and telomere of chromosome III (shown in Figure 6), are bright in the case of T-4 but very weak upon staining with D-11 IgG. It is tempting to attribute these differences to the relative sequence specificities of the immunoglobulins, particularly in view of the reported presence of d(A-C) stretches in telomeric clones (25,26). We are pursuing this question using two approaches: a) in situ hybridization with specific probes from the [d(A-C)·d(G-U)] family, and b) quantitative estimation of bound antibody and its distribution in cytological preparations. Figure 7 depicts high resolution scans of the fluorescence images of the right-hand end of a Chironomus number I chromosome. A comparison of the curves derived from the fluorescent DNA bisbenzimidazole dye, Hoechst-33342, and the immunofluorescence elicited by bound T-4 anti-Z DNA IgG clearly show local differences in absolute (12) and relative intensities. These arise from the intrinsic modulations of DNA density associated with the underlying band-interband structure, but also reflect the exceptionally high extent of antibody binding at certain loci, i.e. the telomere at the end of the chromosome (6,12).

Photon counting measurements of preparations similar to those in Figures 6 and 7 yield a suprisingly high estimate (0.02-0.1%) for the frequency of left-handed regions in the DNA of acid-fixed polytene choromosomes (12). Current experiments on unfixed preparations (24; manuscript in preparation) are directed toward establishing whether such conformations exist in the presence of histone and other chromosomal proteins and toward specifying the influences of local or global topological states, as well as of the ionic environment. We propose that Z DNA may fulfill a structural role in the condensation and association of chromatin and chromosomes (4,6,12,24). Noteworthy in this connection is the pronounced tendency for self-assocation exhibited by Z DNA under many ionic conditions, particularly those involving divalent cations (4,27,28). The "Z* DNA" which is produced shows extensive intermolecular interactions, as evidenced by sedimentability in low centrifugal fields.

The relaxed sequence requirements for the R-L (B-Z) transition outlined in this report and related publications would suggest that left-handed DNA may be relatively abundant in nature, a circumstance which renders the elucidation of its biological role(s) even more compelling than before.

ACKNOWLEDGEMENTS

We acknowledge the support of the Alexander von Humboldt Foundation, the Volkswagen Foundation, the Alberta Heritage Foundation for Medical Research, the Medical Research Council of Canada, and NATO. We are indebted to Drs. E. Weimer, F. Eckstein, and D. Pörschke for assistance with the scanning microscopy, NMR spectroscopy, and thermal transitions, respectively. C. Lalande provided expert technical support.

REFERENCES

1. Zimmerman,S.B.: 1982, Ann. Rev. Biochem., 51, pp. 395-427.
2. Cold Spring Harbor Symp. Quant. Biol.: 1982, 47, in press.
3. McIntosh,L.P., Zarling,D.A., van de Sande,J.H., and Jovin,T.M.: 1983, J. Mol. Biol., submitted.
4. Jovin,T.M., van de Sande,J.H., Zarling,D.A., Arndt-Jovin,D.J., Eckstein, F., Füldner,H.H., Greider,C., Grieger,I., Hamori,E., Kalisch.B, McIntosh, L.P., Robert-Nicoud,M.: 1983, Cold Spring Harbor Symp. Quant. Biol., 47, in press.
5. McIntosh,L.P., Grieger,I., Eckstein,F., Zarling,D.A., van de Sande,J.H., and Jovin,T.M.: 1983, J. Mol. Biol. submitted.
6. Zarling,D.A., Arndt-Jovin,D.J., Robert-Nicoud,M., McIntosh,L.P., Thomae,R., and Jovin,T.M.: 1983, J. Mol. Biol., submitted.
7. Pohl,F.M., and Jovin,T.M.: 1972, J. Mol. Biol., 67, pp. 375-396.
8. Cohen,J.S., Wooten,J.B., and Chatterjee,C.L.: 1981, Biochemistry, 20, pp. 3049-3055.
9. Eckstein,F. and Jovin,T.M.: 1983, Biochemistry, in press.
10. Brahms,S., Vergne,J., Brahms,J.G., Di Capua,E., Bucher,Ph., and Koller, Th.: 1982, J. Mol. Biol., 162, pp. 473-493.
11. Wang,A.H., Quigley,G.J., Kolpak,F.J., van der Boom,J.H., van der Marel,G., and Rich,A.: 1981, Science (Wash), 211, pp. 171-176.
12. Arndt-Jovin,D.J., Robert-Nicoud,M., Zarling,D.A., Greider,C., Weimer,E., and Jovin,T.M.: 1983, Proc. Natl. Acad. Sci. USA, in press.
13. Malfoy,B. and Leng,M.: 1981, FEBS Letters, 132, pp. 45-48.
14. Lafer,E.M., Möller,A., Nordheim,A., Stollar,B.D., and Rich,A.: 1981 Proc. Natl. Acad. Sci. USA, 78, pp. 3546-3550.
15. Pohl,F.M., DiCapua,E., and Thomae,R: 1982, Nature, 300, pp. 545-546.
16. Möller,A., Gabriel,J.E., Lafer,E.M., Nordheim,A., Rich,A., and Stollar, B.D.: 1982, J. Biol. Chem., 257, pp. 12081-12085.
17. Haniford,D.B. and Pulleyblank,D.E.: 1983, Nature, 302, pp. 632-634.

18. Nordheim,A., and Rich,A.: 1983, Proc. Natl. Acad. Sci., 80, pp. 1821-1825.
19. Klysik,J., Stirdivant,S.M., Larson,J.E., Hart,P.A., and Wells,R.D.: 1981, Nature, 290, pp. 672-677.
20. Peck,L.J., Nordheim,A., Rich,A., and Wang,J.C: 1982, Proc. Natl. Acad. Sci. USA, 79, pp. 4560-4564.
21. Nordheim,A., Lafer,E.M., Peck,L.J., Wang,J.C., Stollar,B.D., and Rich,A. 1982, Cell, 31, pp. 309-318.
22. Lang,M.C., Malfoy,B., Freund,A.M., Daune,M. and Leng,M.: 1982, The EMBO J., 1, pp. 1149-1153.
23. Nordheim,A., Pardue,M.L., Lafer,E.M., Möller,A., Stollar,B.D., and Rich, A.: 1981, Nature, 294, pp. 417-422.
24. Robert-Nicoud,M., Arndt-Jovin,D.J., Zarling,D. and Jovin,T.M.: 1982, in "Mobility and Recognition in Cell Biology". eds. H. Sund and C. Veezer eds, (Gruyter Berlin), pp. 281-290.
25. Appels,R., Dennis,E.S., Smyth,D.R., and Peacock,W.J.: 1981, Chromosoma, 84, pp. 265-277.
26. Walmsley,R.M., Szostak,J.W., and Petes,T.D.: 1983, Nature, 302, pp. 84-86.
27. van de Sande,J.J., and Jovin,T.M.: 1982, The EMBO J., 1, pp. 115-120.
28. van de Sande,J.H., McIntosh,L.P., and Jovin,T.M.: 1982, The EMBO J., 1, pp. 777-782.

STRUCTURES OF DNA :
A CASE STUDY OF RIGHT AND LEFT HANDED DUPLEX IN THE B-FORM

V.Sasisekharan,M.Bansal & G.Gupta†
Molecular Biophysics Unit
Indian Institute of Science
Bangalore 560 012
INDIA.

ABSTRACT

A detailed analysis of the conformational space of a polynucleotide chain with mononucleotide as a repeating unit led to eight possible combinations of the six torsion angles of the backbone as possible for both right and left-handed helical duplexes for DNA. Out of these combinations, a total of five models, three right-handed and two left-handed double helices are seen to be compatible with X-ray data of B-DNA. A comparison of these models with the available infra-red data for B-DNA however indicated that only two models, one right and the other left-handed are in good agreement. Both these models could be refined to low values of the R-factor, thus reaffirming our hypothesis that both right and left-helical structures are plausible models for DNA.

INTRODUCTION

It is well established that a base-paired dinucleoside monophosphate fragment is the smallest unit embodying all the major sources of flexibility as well as the essential stabilizing forces present in a nucleic acid duplex. The backbone of the nucleic acid chain has six bonds about which rotations are possible. In addition there is considerable flexibility possible about the glycosidic bond. If the repeating unit is designated as O5'-C5'-C4'-C3'-O3'-P-O5' (viz. a 3'-nucleotide) then according to the latest IUPAC-IUB nomenclature[1] the backbone torsion is completely defined by the sequence of torsion angles (β , γ , δ , ϵ , ζ , α). The ranges for these six torsion angles, obtained from X-ray diffraction data and from theoretical energy calculations are listed in Table 1. In general the theoretically obtained ranges encompass the experimentally observed values. The torsion angles α , γ and ζ, denoting the conformation about the bonds P-O5', C5'-C4', and O3'-P can take up all the three possible staggerred orientations viz. g^+,t,g^-. The conformation about the O5'-C5' bond, denoted by β is, however, strictly confined to the trans orientation while the torsion angle δ which describes the sugar pucker can be broadly classified

101

B. Pullman and J. Jortner (eds.), Nucleic Acids: The Vectors of Life, 101–111.
© 1983 by D. Reidel Publishing Company.

Table 1. Theoretically allowed ranges (from conformational
energy calculations) and the observed ranges (from single-
crystal x-ray diffraction studies) for the six backbone
torsion angles which define the polynucleotide conformation.

Torsion Angle[a]	Theoretical range[b]	Experimental observation
α (P-05')	40 - 80 (g$^+$)	47 - 92
	160 - 200 (t)	146 - 223
	260 - 320 (g-)	271 - 302
β (05'-C5')	150 - 210 (t)	160 - 221
γ (C5'-C4')	40 - 80 (g$^+$)	41 - 68
	160 - 200 (t)	157 - 191
	260 - 320 (g-)	-
δ (C4'-C3')	70 - 100 (g$^+$)	74 - 100^c
	130 - 160 (t)	134 - 158^d
ε (C3'-03')	150 - 300 (t,g-)	195 - 266
ζ (03'- P)	40 - 80 (g$^+$)	55 - 81
	160 - 200 (t)	163 - 168
	260 - 320 (g-)	283 - 294

[a]
  The torsion angles are given following the new
  IUPAC-IUB nomenclature[1].

[b] g$^+$, t, g- denote gauche$^+$, trans, gauche$^-$
  conformations respectively.

[c] C3'-endo sugar

[d] C2'-endo sugar

as E^3 if the sugar has C3'-endo conformation and as E^2 for
the C2'-endo sugar conformation[2]. One striking feature to
be noted is that the large allowed range predicated by
theory for the torsion angle ε is in contrast to the
general concept of a rigid near trans conformation about
the C3'-03' bond.

In principle therefore, the three allowed ranges of α, γ
and ζ , two ranges each of δ and ε along with the trans
conformation for β lead to (3x3x3x2x2x1) = 108 possible

allowed conformations for a nucleotide . However, not all of these lead to viable base-paired double-helical structures for nucleic acids[3,4]. In this paper we briefly describe the combinations of torsion angles which lead to stereochemically acceptable models for the various double helical forms of DNA and then give the details of the possible models for B-DNA and their agreement with the available X-ray pattern as well as infra-red data.

STEREOCHEMICAL GUIDELINES FOR BUILDING DOUBLE-HELICAL DUPLEXES

The possible combinations of the six torsion angles which can lead to duplex structures were first examined by Sasisekharan et al[5,6], and by Sundaralingam's group[7,8] for the trans conformation of ε . Recent detailed investigations carried out by our group reveal that only eight distinctly different classes of conformational combinations lead to the formation of double-helical structures of both right and left handedness, and these are listed in Table 2. Of these the first six were obtained directly by calculations on the helical parameters of polynucleotides with mononucleotide as a repeating unit and incorporating the trans conformation about the C3'-O3' (ε) and O5'-C5' (β) bonds. In five of the six cases, small variations in the ζ - α space; i.e. in the phosphodiester conformations lead to stereochemically acceptable right and left uniform helices (henceforth denoted as RU and LU helices), provided some stereochemical correlations are maintained between the P-O torsions and the sugar pucker[3], as well as the glycosidic torsion angle χ [4].

The remaining two combinations correspond to the torsion angle ε about the C3'-O3' bond being in the g⁻ conformation, with the sugar pucker being confined to the C2'-endo region. This then gives the two combinations (t,t,E2,g⁻,t,g⁺) and (t,g⁺,E2,g⁻,t,t) which also can lead to both RU and LU helices.

In all the seven sterically allowed helical domains, the glycosyl torsion angle χ falls in the anti region ($170° < \chi < 260°$) and hence both purines, pyrimidines can be accommodated in the RU and LU helices of these domains. However, while stereochemically acceptable duplex models can be built with the mononucleotide conformation in any of the above seven domains not all these structures give good agreement with the observed fibre patterns of the various forms of DNA. As mentioned earlier, here we have only addressed ourself to the various uniform helical models for B-DNA and judged their acceptability on the basis of their agreement with the crystalline X-ray fibre data for the B-form[9] and the available infra-red data[10].

B-DNA MODELS AND THEIR COMPARISON WITH THE X-RAY DATA

Initially we had confined our attention to the near trans conformation for ε and arrived at three models that are compatible with the B-DNA X-ray data, both on qualitative (molecular transform) and quantitative

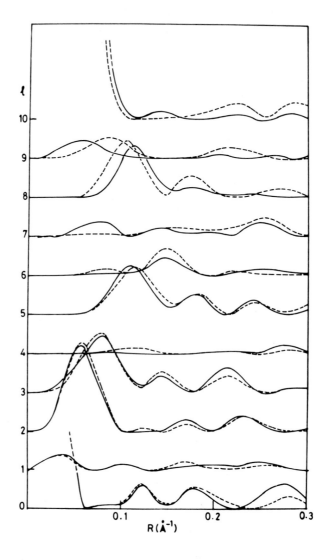

Figure 1. Fourier transform of the RU model (———) in the (t,g^+,E^2,t,tg^-) domain and LU model (- - - - -) in the (t,t,E^2,g^-,tg^+) domain. It is seen that the two molecular transforms are nearly identical, particularly for reciprocal space radius $\rho < 0.22$ Å^{-1} where most of the crystalline reflections are observed[9].

Table 2 : The eight possible combinations of backbone torsion
angles (termed as domains) which lead to both right
and left-handed double-helical models[+].

(O5'-C5')	(C5'-C4')	(C4'-C3')	(C3'-O3')	(O3'-P)	(P-O5')
β	γ	δ	ε	ζ	α
t	g^+	E^3	t	g^-	g^-
t	t	E^3	t	g^-	t
t*	g^-	E^3	t	g^-	g^+
t	g^+	E^2	t	t	g^-
t	t	E^2	t	t	t
t	g^-	E^2	t	t	g^+
t	g^+	E^2	g^-	t	t
t	t	E^2	g^-	t	g^+

+ For purposes of easy identification the sugar conformation is
denoted as E^2 or E^3 (for δ being t and g^+ respectively) and
the phosphodiester torsions are given together without a comma
when describing the domains.

* This nucleotide conformation is sterically disallowed for a
polynucleotide chain.

(R-factor) basis[11,12]. Two of the three are right-handed models while
the third is a left-handed model. Of the two right handed models one
has a C2'-endo sugar pucker while the other has a C3'-endo sugar pucker.
The left handed model has a C2'-endo sugar pucker. A left-handed model
with C3'-endo sugar pucker is not compatible with the X-ray data of
B-DNA[9]. The conformational parameters for the right handed models
correspond to (t,g^+,E^2,t,tg^-) and (t,g^+,E^3,t,g^-g^-), while the conformation of the left-handed model corresponds to (t,g^+,E^2,t,tg^-). It
may be noted that these models follow the stereochemical guideline
postulated earlier by us, namely the C2'-endo sugar pucker is accompanied
by tg^- phosphodiester torsions while for the C3'-endo sugar pucker, the
phosphodiester torsion angles are in the g^-g^- conformation. The details
of these models have been published elsewhere [11,12]. For reasons

stated in the next section further refinement with the X-ray data was
carried out only for the right handed B-DNA structure in the domain
(t,g^+,E^2,t,tg^-). The refined model gave an improved fit to the data
with the R-factor going down from 0.36 to 0.30. The conformational
parameters and other relevant parameters for all the three models are
given in Table 3 and the coordinates of the refined right-handed
structure are given in Table 4.

Table 3: The conformational and other parameters for the five
models which are in agreement with the x-ray data.

Para-meters	RU helices			LU helices	
	$(t,g^+,E^3,$ $t,g^-g^-)$	$(t,g^+,E^2,$ $t,tg^-)*$	$(t,t,E^2$ $g^-,tg^+)$	$(t,g^+,E^2,$ $t,tg^-)$	$(t,t,E^2$ $g^-,tg^+)*$
β	179	141	238	135	228
γ	75	40	176	36	160
δ	97	141	140	137	136
ϵ	184	225	285	241	297
ζ	292	208	199	204	178
α	269	313	72	270	51
χ^+	215	250	228	186	189
θ_x	−7	−3	+2	−3	+2
θ_y	9	−5	−5	0	0
D	0.3	−0.6	−0.4	−1.2	−1.0
r_p	9.2	9.2	9.5	8.9	9.1
s_p	14.6	14.0	13.0	11.2	13.2
R-factor	0.36	0.30	0.40	0.37	0.32
τ_1	70	46	17	39	54
τ_2	49	79	79	58	72

* Helical structures in these two domains could be refined to even
lower values of R-factor, but the final models given are the ones which
give good agreement with the x-ray data as well as being free of both
intra and inter-molecular short contacts. The packing parameters θ and
d (described in Ref.12) are $2°$ and 0.328 for the RU model and $4°$ and
0.325 for the LU helix. The scale factor (K) and attenuation parameter
(B) have values 116.1 and −11.6 A^{-2} for RU and 105.6 and −20.0 A^{-2} for
the LU model.

+ The glycosyl torsion χ is given according to the IUPAC-IUB nomenclature and corresponds to the torsion angle $(O4'-C1'-N9-C4)$.

We have recently investigated B-DNA models with ε in the g⁻ conformation[3]. Stereochemically satisfactory B-DNA models could be built only for the C2'-endo sugar pucker and not for C3'-endo sugar pucker. Comparison of these models with the X-ray data led to (t,t,E^2,g^-,tg^+) conformation of the nucleotide unit for both RU and LU helices for B-DNA. The gross features of these two models are quite similar; so also their molecular transforms[3]. On refining the models further it was found that only the LU model in this domain could be improved to fit better with the X-ray data. The R-factor being 0.32, a value comparable to the value of 0.30 obtained for the RU model in the ε = t region. The conformational parameters for both the models are given in Table 3 but only the co-ordinates of the LU model are given in Table 4.

Table 4: Cylindrical polar coordinates of the final RU and LU models which are in good agreement with the available x-ray data[9] as well the infra-red data[10].

(a) RU duplex in (t,g^+,E^2,t,tg^-) domain			
Atom	r(Å)	ϕ (°)	z(Å)
O3'	8.02	95.77	2.85
P	9.20	93.67	1.83
O1P	9.42	99.50	0.81
O2P	10.46	92.37	2.53
O5'	8.80	84.83	1.16
C5'	8.57	77.28	1.99
C4'	7.68	70.10	1.33
C3'	7.87	69.53	-0.17
C2'	6.67	75.01	-0.77
C1'	5.60	71.18	0.21
O4'	6.27	72.23	1.49
N9G	4.50	81.93	0.25
C8G	4.89	98.00	0.28
N7G	4.14	111.70	0.30
C6G	1.72	128.87	0.29
C5G	2.81	104.85	0.28
C4G	3.19	79.24	0.25
N3G	2.87	54.37	0.22
C2G	1.68	37.28	0.23
N1G	0.53	84.61	0.26
O6G	2.40	158.47	0.32
N2G	2.32	2.47	0.20
N1C	4.50	81.93	0.25
C6C	5.03	97.08	0.28
C5C	4.53	112.32	0.31
C4C	3.15	118.19	0.30
N3C	2.30	95.98	0.26
C2C	3.22	75.45	0.23

O2C	3.38	54.10	0.19
N4C	3.20	142.23	0.32

(b) LU duplex in (t,t,E^2,g^-,tg^+) domain.

Atom	$r(Å)$	$\phi(°)$	$z(Å)$
O3'	7.66	93.80	-2.86
P	9.13	94.07	-2.25
O1P	9.59	102.14	-1.81
O2P	10.14	91.36	-3.21
O5'	9.05	87.72	-1.02
C5'	8.73	78.80	-1.24
C4'	7.80	74.88	-0.19
C3'	7.36	64.28	-0.60
C2'	5.88	65.11	-1.00
C1'	5.51	75.00	0.08
O4'	6.72	82.21	0.20
N9G	4.47	86.14	-0.14
C8G	4.93	101.83	-0.16
N7G	4.29	115.72	-0.20
C6G	1.96	136.33	-0.19
C5G	2.92	111.03	-0.18
C4G	3.10	85.28	-0.15
N3G	2.69	60.25	-0.12
C2G	1.42	45.76	-0.13
N1G	0.61	118.91	-0.16
O6G	2.73	160.68	-0.22
N2G	2.00	2.70	-0.10
N1C	4.47	86.14	-0.14
C6C	5.09	100.71	-0.16
C5C	4.68	115.90	-0.20
C4C	3.33	123.17	-0.20
N3C	2.37	104.13	-0.16
C2C	3.16	81.23	-0.13
O2C	3.19	59.03	-0.10
N4C	3.50	145.54	-0.22

* Both G.C and A.T base pairs can be accommodated in the RU and LU models given here, though we have listed only the coordinates for a G.C base pair.

The coordinates of successive units in the helix can be generated by applying the following transformations:

$\phi = \phi + 36°$ and $z = z + 3.4$ Å for the RU helix;

$\phi = \phi - 36°$ and $z = z + 3.4$ Å for the LU helix.

The antiparallel chain in the duplex can be generated by putting $\phi = -\phi$ and $z = -z$ for both RU and LU helices.

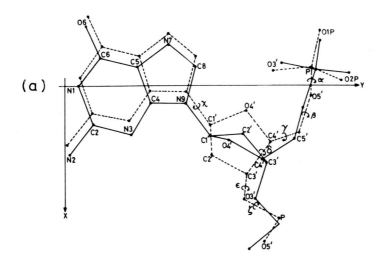

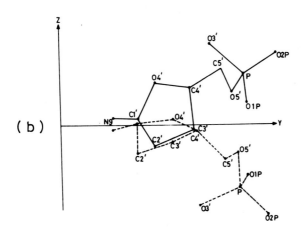

Figure 2.(a)   A projection down the helix axis for the RU model
(———) and LU model (----) clearly showing that most of the atoms
are within   0.5 A in the two models.   In the case of the furanose
ring atoms the C2' and C3' atoms in one structure are respectively
equivalent to the O4' and C4' atoms in the other.   Taking this
equivalence into consideration the sugar atoms are also within
0.8 A in the two models.   (b) A projection perpendicular to the
dyad x-axis, showing the near mirror symmetry about the y-axis, for
the polynucleotide backbone of the two models.

DISCUSSION

It is known that infra red measurements of oriented fibres of DNA could give information regarding the orientation of the phosphate groups in terms of two angles τ_1 and τ_2. The angle τ_1 describes the angle the vector joining the two pendent oxygens O1P and O2P makes with the helix axis while τ_2 gives the angle the bisector of O1P-P-O2P makes with the helix axis. For the B DNA fibres the experimentally observed values of τ_1 and τ_2 are $56°$ and $70°$ respectively[10]. Table 3 lists the τ_1 and τ_2 values calculated for the B-DNA models described earlier. It may be noted that the τ_1 and τ_2 values for these models are widely different. Interestingly, it turns out that for a given handedness the calculated τ_1 and τ_2 values of various helices in a particular helical domain do not change significantly. On the other hand, structures of both handedness in a given helical domain could be refined to fit better with the X-ray data. Thus, for example, although the right handed structures in domain I — of Table 3 could perhaps be refined against X-ray data, the values of τ_1 and τ_2 did not change significantly and these values are far removed from the observed values of τ_1 and τ_2. Similarly, the LU helices in the (t,g^+,E^2,t,tg^-) domain and RU helices in (t,t,E^2,g^-,tg^+) domain have τ_1 and τ_2 values not in agreement with the observed data. Thus, only RU helices in the (t,g^+,E^2,t,tg^-) domain and the LU helices in the (t,t,E^2,g^-,tg^+) domain have τ_1 and τ_2 values comparable to the observed values. Hence, only these two types of helices were taken up for further refinement and the parameters for the final model arrived at are given in Table 3. A comparison of these models for agreement with both X-ray data (R-factor) and IR data (observed τ_1 and τ_2 values) therefore led to the conclusion that the above two models are the only acceptable models for B-DNA. The salient features of these models are discussed further. Both the models have very low base tilts (θ_x) and the gross features of these models viz. the phosphate radius ($\overset{\cdot\cdot}{r}_p$) and the chain separation (Sp) are also very similar. Hence, the molecular transforms of both the models shown in Fig.1 are nearly identical. In fact the two models are extremely alike when seen in projection perpendicular to the helix axis as shown in Fig.2a. The near mirror symmetry of the atoms of the backbone of the two models can be readily seen in Fig.2b. The major difference between these two structures lies in the glycosidic torsion which links the base to the sugar of the backbone. Therefore, it is not surprising that both the models lead to R value within a difference of 0.02 only. It may be noted that the RU model has a slightly lower R value than the LU model. On the other hand, the LU model gives a significantly better fit than the RU model as far as the τ_1 and τ_2 values are concerned. Thus, if we have to select a 'best-model' for B-DNA, taking into consideration the allowed stereochemistry of the backbone, favourable stacking arrangement, agreement with available X-ray and IR data, it must be concluded that it is impossible to choose between the RU helix with (t,g^+,E^2,t,tg^-) conformation and the _LU helix with (t,t,E^2,g^-,tg^+) conformation.

†Present address : Chemistry Department
SUNYA, 1400 Washington Avenue
New York 12222, USA.

REFERENCES

1. IUPAC-IUB Nomenclature for polynucleotide chains, 1983,Eur.J.
Biochem. 131,pp. 9-13
2. Rao,S.N. and Sasisekharan,V. in 'Conformation in Biology',
(Eds. R.Srinivasan and R.H.Sarma) Adenine Press, New York, 1982,
pp.323-333
3. Gupta,G., Rao,S.N. and Sasisekharan, V. 1982,FEBS Letts. 150,
pp. 424-428
4. Rao,S.N. and Sasisekharan, V. 1983, Int. J. Biol. Macromol.
5, pp.83-88
5. Sasisekharan, V., Sato, T. and Langridge, R. 1971, Fed.Proc.
Abs. 30, pp.1219
6. Sasisekharan,V. in 'Jerusalem Symposium on Quantum Chemistry
and Biochemistry' (Eds. E.D. Bergman and B.Pullman) Academic
Press, New York, 1973,Vol. 5, pp. 247-260
7. Sakurai, T and Sundarlingam, M. 1971, Am.Cryst.Assocn.Meeting
Abs., pp. 91
8. Yathindra and Sundarlingam, M. 1976,Nucl. Acid Res. 3 pp.729-747
9. Arnott,S. and Hukins, D.W.L. 1973,J.Mol.Biol. 81, pp.93-105
10. Pilet, J. and Brahms,J. 1973,Biopolymers 12, pp.387-407
11. Gupta, G., Bansal, M. and Sasisekharan, V. 1980, Proc. Natl. Acad.
Sci. U.S.A. 11 , pp.6486-6490
12. Gupta, G., Bansal, M. and Sasisekharan, V. 1980 , Int.J.Biol.
Macromol. 2,pp. 368-380.

TWO DIMENSIONAL NMR INVESTIGATION OF THE STRUCTURAL PROPERTIES OF DNAS

David R. Kearns, Peter A. Mirau, Nuria Assa-Munt and
Ronald W. Behling
Department of Chemistry, B-014
University of California-San Diego
La Jolla, CA 92093

ABSTRACT: We describe the use of 2D NOE and truncated driven NOE (TNOE) spectroscopy to investigate structural features of three alternating homopolymer systems; poly(dA-dT), poly(dG-dC) (B- and Z-forms), and poly(dI-dC). Poly(dA-dT) and poly(dI-dC) have Watson-Crick base pairing, not Hoogsteen pairing as recently proposed for fibers. The B-form of poly(dG-dC), poly(dA-dT) and poly(dI-dC) exhibit qualitative structural features expected for B-family DNA. Specifically, all have anti-glycosidic torsional angles. This is contrasted with results from Z-form poly(dG-dC) which exhibits a syn-torsional angle as well as other differences in the 2D NOE spectrum. 2D NOE measurements on the decamer d(ATATCGATAT)$_2$ in H$_2$O were used to delineate the various spin lattice relaxation pathways (dipolar, exchange) of the imino protons in this molecule.

INTRODUCTION

After a number of years of quiet progress in the field, there has been a dramatic renewal of interest in the structural and dynamic properties of DNA. Much of this is due to x-ray diffraction studies on single crystals of short DNA duplexes which have provided us with atomic resolution structures of several different DNA molecules (Wang et al., 1979; Wing et al., 1980; Drew et al., 1980). These studies have included the discovery of left-handed Z-DNA (Wang et al., 1979), and of large sequence effects on the conformations of right-handed helices (Drew et al., 1981). Supercoiled DNAs have been found to have a number of interesting structural and dynamic properties (Wang, 1980; Bendel et al., 1983), particularly with regard to the effect of supercoiling density on stabilization of Z-DNA (Klysik et al., 1981; Nordheim & Rich, 1983; Haniford & Pulleyblank, 1983), which may play a role in biological control and regulation of genomic information. Parallel to the x-ray diffraction studies, there has been important progress in the application of nuclear magnetic resonance (NMR) spectroscopy in solution

113

B. Pullman and J. Jortner (eds.), Nucleic Acids: The Vectors of Life, 113–125.
© 1983 by D. Reidel Publishing Company.

state studies on DNA (Feigon et al., 1982; Kearns, 1983). In this
article we describe the application of two NMR techniques, two
dimensional nuclear Overhauser spectroscopy (2D NOE) (Jeener, 1971;
Jeener et al., 1979) and truncated driven nuclear Overhauser
enchancements (TNOE) (Wagner & Wuthrich,1979), to an examination of the
structural features of some synthetic DNAs. The molecules studied
include poly(dA-dT), poly(dG-dC), poly(dI-dC) and d(ATATCGATAT)[2].
Poly(dA-dT) was chosen for study because it is probably the most
thoroughly studied synthetic DNA. Poly(dG-dC) has attracted
considerable attention since it can be converted from a right-handed to
the left-handed (Pohl & Jovin, 1972) duplex, and since d(CGCGCG)[2] was
the first DNA duplex found to crystallize in the Z-form (Wang et al.,
1979). Poly (dI-dC) is of interest because studies carried out prior to
the discovery of Z-DNA suggested that it may exist in a left-handed
structure (Mitsui et al., 1970). There is the additional interest in
both the poly(dA-dT) and the poly(dI-dC) systems as a result of
suggestions that these molecules may have non-Watson-Crick base pairing
schemes (Drew & Dickerson, 1982), and may possibly be left-handed
(Ramaswamy et al., 1982). With poly(dI-dC) and poly(dA-dT) we used TNOE
experiments to demonstrate that the base pairing structure is Watson-
Crick. The 2D NOE method was then used to examine qualitative
structural features in these molecules and to compare the structure of
poly(dI-dC) and poly(dA-dT) with each other and with poly(dG-dC) in both
the right- and left-handed forms. In the case of d(ATATCGATAT)[2] we have
used the 2D NOE technique in 100% H_2O to study relaxation of the
exchangeable imino protons in this molecule and the dynamics of base
pair opening as manifested by exchange with water.

MATERIALS AND METHODS

     Poly(dI-dC), poly(dA-dT) and poly(dG-dC) were obtained from P-L
Biochemicals and buffer reagents were obtained from Sigma Chemical Co.
The molecular weight of the DNA polymers was reduced by ~5-6 hrs of
sonication using a W-375 Ultrasonics sonicator with a 50% duty cycle at
0°C in 0.25 M NaCl, 0.01 M cacodylate, and 0.01 M EDTA at pH 7.0 (Granot
et al., 1982). The size of the polymers was determined by
electrophoresis on 7% polyacrylamide gels vs. DNA restriction fragments.
After sonication the DNA was ~50-70 base pairs in length. The samples
were ethanol precipitated twice, dried, and dissolved in the appropriate
buffer. For the NMR measurements the samples typically contained 25 A_{260}
of the DNA polymer. The DNA decamer, d(ATATCGATAT)[2], was synthesized in
this laboratory, as described elsewhere (Denny et al., 1982). Samples,
dissolved in 10 mM sodium phosphate, pH 7.0, 0.1 M NaCl, were contained
in either a Wilmad 508 cp microcell (3 mM duplex, 120 μl) or a thin wall
tube.

     ^{1}H NMR spectra were obtained at 360 MHz on a Fourier transform NMR
spectrometer equipped with an Oxford Instruments magnet and a Nicolet
1180-E computer. Pure absorption phase 2D NOE spectra were acquired
using the $(90^{\circ}-t_1-90^{\circ}-t_m-90^{\circ}-t_2)_n$ pulse sequence with alternate block

accumulation.

In order to conduct 2D NOE measurements in 100% water, it was necessary to suppress the very large (100 M) solvent peak. This was accomplished using the pulse scheme shown in Fig. 1.

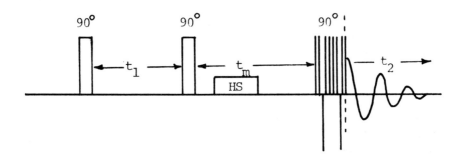

Fig. 1. A pulse scheme used to obtain 2D NOE spectra in water.

The first two pulses are the usual 90° "hard" pulses, however the third 90° pulse is a time-shared Redfield 214 pulse which is adjusted to minimize excitation of the water peak (Wright et al., 1981). To gain additional suppression of the water signal, alternate delayed acquisition (ADA) was used (Roth et al., 1980). In this method, a free induction delay is first collected in the normal fashion. A second free induction decay is then collected, after a delay of $t=1/(2\Delta\upsilon)$, where $\Delta\upsilon$ is the difference between the carrier frequency and the location of the water resonance, and the two signals are added together. Additional features include application of a homospoil pulse (HS) during the mixing period and the usual phase cycling to suppress axial ridges due to recovery of equilibrium magnetization during the mixing time, τ_m (Jeener et al., 1979).

Truncated driven NOE (TNOE) spectra (Wagner & Wüthrich, 1979) in water were obtained by irradiating the lowfield imino resonances for various lengths of time followed immediately by a time-shared Redfield observation pulse. A control spectrum with an off-resonance irradiation was obtained using the same irradiation time and subtracted from the on-resonance irradiation to give the difference spectra.

RESULTS AND DISCUSSION

1. ·Base Pairing Structure of Poly(dA-dT) and Poly(dI-dC)

In view of the recent suggestions, based on x-ray diffraction data that both poly(dA-dT) and poly(dI-dC) may have non-Watson-Crick base

pairing structures in fibers (Drew & Dickerson, 1982), it was important
to firmly establish the base pairing structures of these two molecules
in solution. This is accomplished by the TNOE experiments shown in Fig.
2.

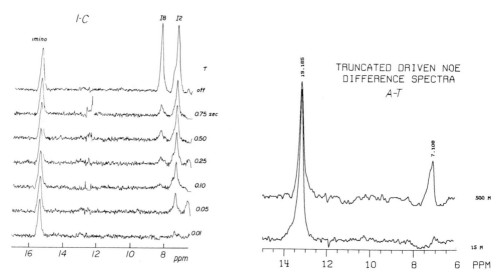

Fig. 2. The TOE difference spectra for poly(dI-dC) (left) and poly(dA-
dT) (right) in H₂O solution. In each case a control spectrum was
obtained with off-resonance irradiation and subtracted from the spectrum
where the lowfield resonance from the imino protons irradiated for the
length of time shown to the right of the spectra.

In both cases the strongest interaction is between the imino proton and
the purine H2 proton (either AH2 or IH2). This result is only compatible
with Watson-Crick base pairing (see Fig. 3) since the imino proton in
Hoogsteen or reverse Hoogsteen pairing is closest to the purine H8
proton and very distant from the H2 proton.

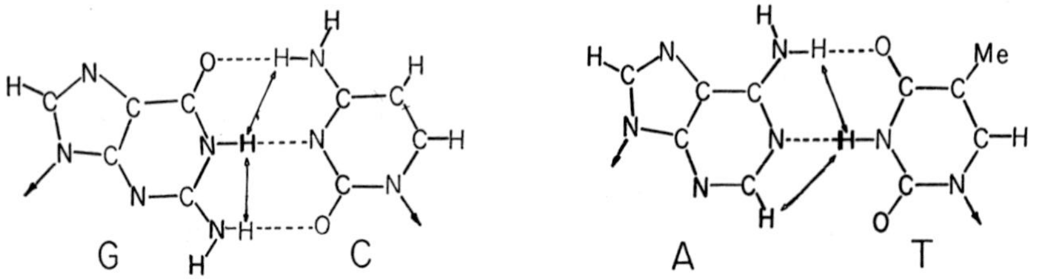

Fig. 3. The standard Watson-Crick base pairs showing the base protons
in close proximity to the imino proton.

While these results cannot rule out the possibility that different base pairing structures exist in the fiber, the TNOE experiments confirm the widely accepted notion that these two molecules have normal Watson-Crick pairing in solution. Having settled this issue, we now turn to an examination of the other structural features in these simple sequence alternating polymers using 2D NOE methods.

2. 2D NOE Investigation of Polymer Structures

The 2D NMR techniques have recently been applied to DNA decamer duplexes (Feigon et al., 1982), but the results presented here represent the first time these methods have been used on much higher molecular weight (50-70 base pair) material. This is an important extension since there is evidence for substantial "end effects" with short DNA duplexes.
    Poly(dA-dT). A stacked plot of a 2D NOE experiment on poly(dA-dT) at 20°C is shown in Fig. 4.

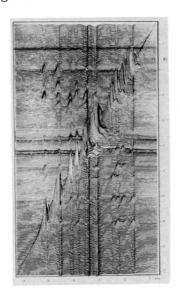

Fig. 4. A pure absorption phase 2D NOE spectrum of poly(dA-dT) in 0.1 M NaCl, pH=7, and T=20°C. A mixing time of 20 ms was used.

The largest peaks in the 2D NOE spectrum appear along the diagonal which bisects the two frequency axes. This diagonal spectrum, which looks similar to the usual one-dimensional spectrum, arises from resonances of protons which did not (cross) relax during the mixing time (Wagner et al., 1981). The relatively small peaks which are situated symmetrically off the diagonal arise from dipole-dipole induced cross relaxation during the mixing time, τ_m. Since cross relaxation rates vary with the inverse sixth power of the separation between two interaction protons, a strong cross peak located at frequencies (f_1, f_2) implies a short internuclear separation (probably less than 3.5 Å) between the two interacting protons. For purposes of discussion, it is easier to consider contour plots of the 2D NOE spectra, and one such plot for poly(dA-dT) is shown in Fig. 5.

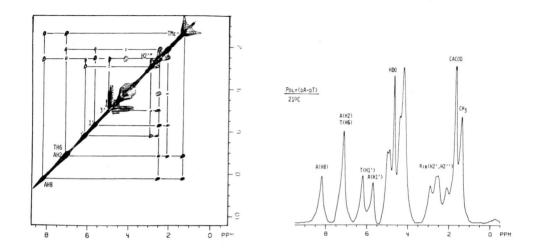

Fig. 5. A contour plot of the 2D NOE spectrum of poly(dA-dT) obtained at 21°C using a mixing time of 50 ms (left). The regular spectrum is shown at the right.

The AH8 resonance (8.2 ppm) exhibits strong cross peaks with sugar peaks at 2.5 ppm (from the H2' sugar proton on the _same_ nucleotide unit) and a very prominent cross peak with the thymine methyl resonance located at 1.3 ppm. The AH8 cross peak with the H2' sugar proton is to be expected since the separation between these two protons is between 1.9-2.5 Å for most DNA models. The cross peak between AH8 and T-methyl protons is especially interesting since it arises from an _inter_-base interaction between A and T bases in the _same_ strand (5'A$_p$T3'). Since the intensities of the cross peaks between the TH6 and the methyl protons on the same base and the AH8-TMe peak are comparable, we conclude that the interproton distances are comparable (average interproton distance of ~3.0 Å). The TMe resonance at 1.3 ppm also exhibits weak cross peaks (not evident in the contour plot shown) with sugar resonances located at 2.9 and 2.5 ppm. Simple geometrical consideration show that the H2' and H2'' protons of the thymine nucleotide cannot be close to the methyl protons. Therefore, the resonances at 2.9 and 2.5 ppm must be assigned to the H2' and H2'' protons of the adenine nucleotide group. In the _anti_-conformation the AH8 proton is located closer to the H2' than to the H2'' proton in the same nucleotide unit, and the strong cross peak with AH8 and the 2.5 resonance indicates that the 2.9 resonance must be due to H2''. The sugar resonance at 2.1 ppm can be assigned to the H2' resonance of the T nucleotide, since it has a large cross peak with TH6 (impossible if it were the H2'' resonance) and a somewhat weaker interaction with the AH8 resonance. Qualitatively we find that the H2' proton on the thymine sugar is closer to the TH6 proton than it is to the AH8, but not significantly closer. The H1' resonances at 6.22 and 5.7 (from AH1' and TH1', respectively) each exhibit two strong cross peaks with H2' and H2'' sugar protons on the <u>same</u> sugar residue, with the H2'' interaction being the stronger one. This provides a second means of identifying these two resonances. The fact that the H1' protons exhibit only very weak (second order?) interactions with the AH8 or the TH6 resonances is indicative of an <u>anti</u>-conformation of the nucleotide units. These preliminary 2D data are qualitatively consistent with a right-handed B-family conformation. We note, however, some small differences in the relative intensities of the AH8-sugar and the TH6-sugar cross peaks, suggesting that the conformations of the sugar groups are not the same for the two nucleotide units, (e.g. AH8 exhibits only a single strong cross peak with H2',H2'' resonances, whereas TH6 exhibits two). An alternation in the structure indicated by previous ^{31}P NMR (Shindo et al., 1979) and ^{13}C NMR (Shindo, 1981) studies is, therefore, also evident in the 2D NOE spectrum, but selective and biselective spin lattice relaxation rate measurements will be required to quantitate the differences. A left-handed, Z-DNA structure is incompatible with the 2D-NOE data. Many other peaks were evident in the 2D spectrum, but these will be discussed elsewhere.

Poly(<u>dG-dC</u>). Figure 6 shows the 2D NOE spectrum of the low salt (B-form) of poly(dG-dC) obtained at 50°C with a 50 ms mixing time.

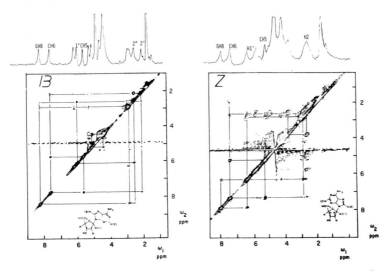

Fig. 6. (A) The pure absorption phase 2D NOE spectrum of B-form (0.1 M NaCl) of poly(dG-dC) obtained at 50°C with a 50 ms mixing time (left) and of Z form (4.5 M NaCl) poly(dG-dC) obtained at 20°C with a 50 ms mixing time (right).

Many of the cross peaks in this spectrum arise from protons which are attached to adjacent atoms (i.e. CH5-CH6, H1'-H2'', etc.), with the strongest of these (located at 7.5, 5.6 ppm), arising from the interaction of the cytosine CH5 and CH6 protons that are separated by a distance of 2.4 Å. Strong cross peaks are also observed between the CH6 and GH8 and the H2',H2'' sugar protons. The H1' protons at 6.0 and 6.2 ppm show strong cross peaks with the H2'' protons at 2.8 and 2.5 ppm and the H3' protons at 4.7 and 4.9 ppm show strong cross peaks with the H2' protons located at 2.7 and 2.0 ppm as well as to the H4' protons located in the broad envelope centered at 4.3 ppm. Since the broad peak at 4.3 ppm shows cross peaks only to the H3', this peak must contain the H4' and H5' protons. Because the H4' and H5' resonances overlap, no H5'-H4' cross peaks are observed. Rather, the peak appears to be broadened in the f_1 domain. The H2' and H2'' protons at 2.7 and 2.8 ppm are unresolved and show no cross peak while a strong cross peak is observed between the H2' at 2.5 ppm and the H2'' at 2.0 ppm. The higher field position of the cytosine H2' and H2'' protons is expected on the basis of model studies (Tran-Dinh et al., 1982). Assuming this is also true for poly(dG-dC), it is possible to assign the lowest field H1', H3', and H2' and H2'' protons to the guanine base. Information about the conformation of poly(dG-dC) is manifested in the cross peaks which arise from base-sugar or base-base interactions. For the low salt form of poly(dG-dC), cross peaks are observed between GH8-GH2'', GH8-CH2'', and CH6-CH2'. From the intensity of the cross peaks, it appears as if GH8 interacts more strongly with an H2'' proton on the same nucleotide. The CH6 shows only a small cross peak with its own sugar H2' proton, because its relaxation is dominated by the CH5-CH6 interaction. In this regard, we note that these features of the 2D spectra of low salt poly(dG-dC) are very similar to those found in poly(dA-dT). These strong interactions of the base protons (GH8,CH6) with H2',H2'' sugar protons, but not the H1' sugar protons, clearly indicate G and C have anti-glycosidic torsional angles. While the 2D NOE spectrum indicates that there are small differences in the strength of interaction of the GH8 and CH6 protons with the sugar protons, the various features seen in the 2D spectrum of poly(dG-dC) are consistent with those expected from a right-handed, B-like DNA.

The 2D NOE spectrum of the high salt (Z) form of poly(dG-dC) is different from that obtained in low salt in several important ways. The CH6 proton still exhibits a very strong interaction with CH5, although the interaction with an H2' H2'' proton is somewhat weaker. This indicates that the C nucleotide has retained an anti-conformation. The GH8 proton, on the other hand, shows a very strong cross peak with the GH1' proton but none with the H2' and H2'' proton resonances which have coalesced into a broad envelope centered at 2.5 ppm. The strong cross peak between GH8 and the H1' sugar resonance is due to the syn-glycosidic torsional angle for G in Z-DNA. A syn-conformation for the G nucleotide was first indicated by x-ray diffraction studies on crystals (Wang et al., 1979) and subsequently established in solution for poly(dG-dC) by Patel et al. (1982) using transient NOE techniques. The conversion from the anti- to the syn-conformation decreases the GH8-CH1' interproton distance from 3.7 to 2.7 Å and this interaction dominates

the GH8 cross relaxation. Only small cross peaks are observed to the
H2' and H2'' region. The strongest cross peak in the spectrum is again
due to the CH5-CH6 interaction and only a small cross peak is observed
between the CH5 and the H2'' protons. The cross peaks due to protons on
adjacent atom of the base and sugar also appear, but they are less well
resolved than in the B-form. The 2D NOE spectrum shown in Fig. 6 also
indicates other new cross peaks have developed in the Z-DNA between
various sugar protons which were not so evident in the B-form. This
suggests that other conformational features may be discernible, but
identification of these weaker interactions will require examination of
deeper contours and/or cross sectional slices.

    Poly(dI-dC). A number of studies have shown that poly(dI-dC) has
unusual properties, including an inverted long wavelength CD spectrum
and a "bizarre" fiber diffraction pattern (Mitsui et al., 1970). An
inverted CD spectra is often taken as evidence for a left-handed helix,
and there have been several proposals that poly(dI-dC) might have a
left-handed helical conformation (Mitsui et al., 1970; Drew & Dickerson,
1982). The results of the TNOE experiment, shown in Fig. 2, clearly
establish that the imino proton in poly(dI-dC) is located closer to the
IH2 proton than to the IH8 proton and, therefore, that the base pairing
scheme is Watson-Crick. The possibility of Hoogsteen pairing in solution
is eliminated. Additional information about structural features in
poly(dI-dC) can be extracted from the 2D NOE results presented in Fig.
7.

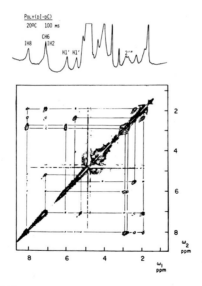

Fig. 7. The 2D NOE spectrum of poly(dI-dC) at 20°C obtained with a 100
ms mixing time.

The 2D NOE spectrum of poly(dI-dC) is quite similar to that observed for poly(dA-dT). Like AH8, IH8 shows strong cross peaks to the H2' and H2'' protons at 2.9 and 2.5 ppm. Cross peaks with the H1' resonances are absent or weak. The overlapping IH2-CH6 peak show cross peaks similar to those observed for TH6 in poly(dA-dT). In the 2D NOE spectrum of poly(dI-dBr⁵C) (not shown), the IH2 and CH6 resonances are clearly resolved and these same cross peaks are observed to the CH6 and no cross peaks with the IH2 resonance are observed (data not shown). We, therefore, believe that the cross peaks to the 7.1 ppm resonance are due solely to CH6. The pattern of the H2' and H2'' protons is almost identical to that observed for the B form of poly(dG-dC) and poly(dA-dT). These results strongly indicate that the conformation of poly(dI-dC) is similar to that of B form poly(dG-dC). Our results, therefore, conclusively rule out a Z conformation for poly(dI-dC). A left-handed helix with Watson-Crick pairs is likely to have a different nucleotide conformation and therefore, when the coordinates of other proposed left-handed helices are examined, we suspect that these too may be eliminated as possible models for the structure of poly(dI-dC).

3.   2D-NOE Studies of the Exchangeable Imino Protons in (ATATCGATAT)₂

As a final example of the application of 2D NMR techniques to DNA we present the results obtained on the decamer d(ATATCGATAT)₂. Using the pulse scheme depicted in Fig. 1 we have been able to successfully measure the 2D NOE spectrum of a DNA duplex in 100% H₂O with the results shown in Figs. 8 and 9.

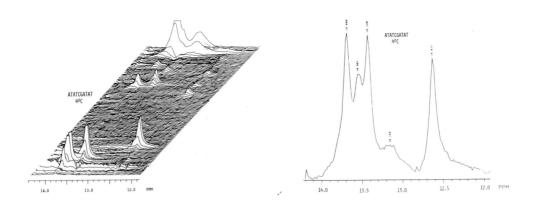

Fig. 8. A stacked plot of the 2D NOE spectrum of d(ATATCGATAT)₂ obtained in H₂O at 4°C using a mixing time of 120 ms (left). Only the lowfield quadrant of the spectrum is shown. The regular lowfield spectrum at 4°C is shown at the right.

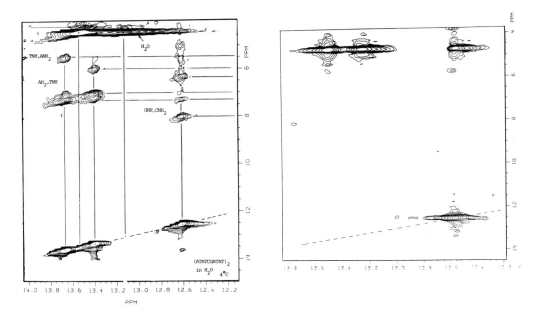

Fig. 9. Contour plots of the 2D NOE spectra of d(ATATCGATAT)₂ obtained at 4°C (left) and 33°C, τ_m=10 ms (right) in water. The diagonal is indicated by a dashed line.

The effect of temperature on the 2D NOE spectrum is illustrated in Fig. 9. In the contour plot of the 4°C data shown in Fig. 9, the key features are as follows. The thymine imino protons (resonating between 14-13 ppm) exhibit strong cross peaks with the AH2 resonances (~8 ppm) and with the water resonance located at 4.4 ppm. Much weaker cross peaks with the A-amino protons are also evident. From a comparison of the relative intensities of the cross peaks with water, one immediately obtains a qualitative evaluation of the importance of exchange of the imino protons with water to the relaxation of the imino protons. Thus, the peak at ~13.65 ppm is exchanging more slowly with water than is the resonance at ~13.5 ppm. Interestingly, there is a very strong water cross peak with a resonance located at ~13.1 ppm, even though the diagonal peak is absent. This demonstrates that the 2D spectrum is able to detect exchange of magnetization between two protons, even though the diagonal peak from one of those protons has disappeared. This memory of the origin of the magnetization will be useful in establishing exchange pathways. We note that the cross peaks arising from the interaction of the T-imino and the A-amino protons is quite weak. This is probably due to the fact that two amino protons are involved and that their resonances are quite broad, making it difficult to see cross peaks with them.

The single lowfield resonance from the GC base pair in this

molecule exhibits a number of cross peaks at $4°C$, including interactions with both the C- and G-amino protons, and to a much lesser extent, with the water protons. In the $33°C$ 2D NOE spectrum all diagonal peaks are greatly reduced in intensity except that due to the G-NH resonance, which now exhibits a strong cross peak with water. Exchange with water protons is, therefore, the major pathway for relaxation of the G-NH resonance at this temperature. It would be very useful for assignment purposes if cross peaks between imino protons in adjacent base pairs could be observed in these 2D NOE spectra, however these are generally expected to be quite small and, therefore, difficult to observe. Nevertheless, with extensive signal averaging it should be possible to observe these. At this point, however, it is evident that the 2D NOE spectra obtained in H_2O can be extremely valuable in qualitatively determining the order of exchange rates for individual resonances in lowfield resonances in a molecule and in studies of sequence effects on exchange rates in DNA.

CONCLUDING REMARKS

The results presented in this paper represent some of the first applications of 2D NOE techniques to relatively high molecular weight DNA duplexes. The encouraging results obtained here and in other recent studies (Feigon et al., 1982; Hore et al., 1982) indicate that much can now be learned about the interesting structural and dynamic properties of DNA molecules in solution using these techniques.

ACKNOWLEDGMENTS

This work was supported by grants from the American Cancer Society (grant CH32), and the National Science Foundation (PCM-7911571). We especially thank Dr. John M. Wright for his efforts in programming the computer to carry out the 2D experiments and to Drs. C.A.G. Haasnoot and C.W. Hilbers for their suggestion of incorporating ADA to further suppress the water peak in H_2O, and a preprint of their work. We gratefully acknowledge the efforts of Drs. W.A. Denny and W. Leupin in synthesizing the decamers.

REFERENCES

Bendel, P., Laub, O., and James, T.L.: 1983, J. Am. Chem. Soc. 104, p. 6748.
Denny, W.A., Leupin, W., and Kearns, D.R.: 1982, Helv. Chim. Acta. 65, p. 2372.
Drew, H., Takano, T., Tanaka, S., Itakura, K., and Dickerson, R.E.: 1980, Nature 286, p. 567.
Drew, H.R., Wing, R.M., Takano, T., Broka, C., Tanaka, S., Itakura, K., and Dickerson, R.E.: 1981, Proc. Natl. Acad. Sci. 78, p. 2179.
Drew, H.R., and Dickerson, R.E.: 1982, The EMBO J. 1, p. 663.

Feigon, J., Wright, J.M., Denny, W.A., and Kearns, D.R.: 1982, 47th Cold Springs Harbor Symposium on DNA Structure.

Feigon, J., Wright, J.M., Leupin, W., Denny, W.A., and Kearns, D.R.: 1982, J. Am. Chem. Soc. 104, p. 5540.

Granot, J., Assa-Munt, N., and Kearns, D.R.: 1982, Biopolymers 21, p. 873.

Haniford, D.B., and Pulleyblank, D.E.: 1983, Nature 302, p. 632.

Hore, P.J., Scheek, R.M., Volbeda, A., Kaptein, R., and van Boom, J.H.: 1982, J. Magn. Reson. 50, p. 328.

Jeener, J.: 1971, Ampere International Summer School, Basko Polji, Yugoslavia.

Jeener, J., Meier, B.H., Bachman, P., and Ernst, R.R.: 1979, J. Chem. Phys. 71, p. 4546.

Kearns, D.R.: 1983, "NMR Studies of Conformational States and Dynamics of DNA," (G. Fasman, Editor), in CRC Critical Reviews in Biochemistry, in press.

Klysik, J., Stirdivant, S.M., Larson, J.E., Hart, P.A., and Wells, R.D.: 1981, Nature 290, p. 672.

Mitsui, Y., Langridge, Shortle, B.E., Cantor, C.R., Grant, R.C., Kodama, M., and Wells, R.D.: 1970, Nature 228, p. 1166.

Nordheim, A., and Rich, A.: 1983, Proc. Natl. Acad. Sci. USA 80, p. 1821.

Patel, D.J., Kozlowski, S.A., Nordheim, A., and Rich, A.: 1982, Proc. Natl. Acad. Sci. USA 79, p. 1413.

Pohl, F.M., and Jovin, T.M.: 1972, J. Mol. Biol. 67, p. 375.

Ramaswamy, N., Bansal, M., Gupta, G., and Sasisekharan, V.: 1982, Proc. Natl. Acad. Sci. USA 79, p. 6109.

Roth, K., Kimber, B.J., and Feeney, J.: 1980, J. Magn. Reson. 41, p. 302.

Shindo, H., Simpson, R.T., and Cohen, J.S.: 1979, J. Biol. Chem. 254, p. 8125.

Shindo, H.: 1981, Eur. J. Biochem. 120, p. 309.

Tran-Dinh, S., Neumann, J.M., Huynh-Dinh, T., Igolen, J., and Kan, S.K.: 1982, Org. Magn. Reson. 18, p. 148.

Wagner, G., and Wuthrich, K.: 1979, J. Magn. Reson. 33, p. 675.

Wagner, G., Kumar, A., and Wuthrich, K.: 1981, Eur. J. Biochem. 114, p. 375.

Wang, A.H.-J., Quigley, G. J., Kolpak, F.J., Crawford, J.L., van Boom, J.H., van der Merel, G., and Rich, A.: 1979, Nature 282, p. 680.

Wang, J.C.: 1980, Trends in Biochem. Sci. 5, p. 219.

Wing, R., Drew, H., Takano, T., Broka, C., Tanaka, S., Itakura, K., and Dickerson, R.E.: 1980, Nature 287, p. 755.

Wright, J.M., Feigon, J., Denny, W.A., Leupin, W., and Kearns, D.R.: 1981, J. Magn. Reson. 45, p. 514.

SECONDARY STRUCTURE OF DNA IN SOLUTION BY NMR METHODS

Jack S. Cohen, Chi-wan Chen and Richard H. Knop[+]
Laboratory of Theoretical and Physical Biology, National
Institute of Child Health and Human Development and
[+]National Cancer Institute-Navy Oncology Branch, NIH,
Bethesda, MD 20205

Abstract: The conformations and transitions of polydeoxynucleotides
(ca. 100 bp) in solution have been investigated by ^{31}P NMR and CD
spectroscopy.   Several alternating purine-pyrimidine base sequences,
including poly-d(AT), -d(Gm^{5}C), and -d(a^{2}AT), exhibit two resolved ^{31}P
signals in low salt concentration.   This indicates the presence of
alternating (dinucleotide repeat) phosphodiester conformations (g⁻g⁻,
tg⁻) for these presumed B-forms.   The addition of several salts to
each of these sequences results in a somewhat different response.   The
continuous increase in the separation of the ^{31}P peaks for d(AT) and
d(a^{2}AT) is indicative of a fast exchange process, presumed to involve
gradually tighter helical winding for d(AT).   While this transition is
linear and non-cooperative for d(AT), it is cooperative for d(a^{2}AT),
as indicated by comparative fits to both the NMR and CD data with an
equation including a cooperativity (Hill) coefficient.   The ^{31}P
spectra for d(Gm^{5}C) shows a slow exchange, cooperative transition on
addition of salts, in which a characteristic signal appears at -3 ppm,
presumably corresponding to the tg$^+$ conformation of d(Gpm^{5}C) in the Z-
form.   These results highlight the selective secondary structures of
different base sequences in different solution conditions.   We have
also studied the effects of adriamycin, an anti-cancer drug known to
bind to DNA, on the B to Z transition of d(Gm^{5}C).   The presence of
adriamycin inhibits the Mg^{2+}-induced B to Z transition and reduces the
degree of cooperativity of the transition.   Additionally, adriamycin
alone converts the Z-form to the B-form in a cooperative manner.
These results indicate that adriamycin binds preferentially to the B-
form, which could be relevant to its mode of action.

1. INTRODUCTION

     Nuclear   magnetic   resonance   (NMR)   spectroscopy   provides
information on the conformations of polydeoxynucleotides in solution
(1,2).   For example, ^{31}P NMR spectra of several alternating purine-
pyrimidine sequences exhibit two resolved signals of equal area (3,4),
indicating a dinucleotide repeat geometry, and the phosphodiester
conformations (gg or tg) can generally be assigned.   In carrying out

127

B. Pullman and J. Jortner (eds.), Nucleic Acids: The Vectors of Life, 127–139.
© 1983 by D. Reidel Publishing Company.

A. Short oligonucleotides (< 15 bp)

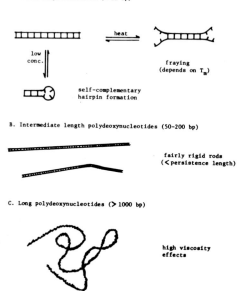

B. Intermediate length polydeoxynucleotides (50-200 bp)

C. Long polydeoxynucleotides (> 1000 bp)

Fig. 1. Effects of length on nucleotide structures. In A, as well as hairpin formation, concatamers can form. To avoid these problems and the broadening effect of high viscosity on NMR signals in C, we have used materials of intermediate length, corresponding to B.

these studies it is necessary to use materials which are not too long to avoid broad NMR linewidths; on the other hand, very short oligonucleotides may not be completely double-stranded (Fig. 1). Consequently we have used synthetic polymers which have been sonicated under controlled conditions to be of intermediate length (5).

In the work reported here we have examined poly-d(AT), -d(a^{2}AT) and -d(Gm^{5}C) duplexes. In each case the ^{31}P NMR and circular dichroism (CD) results enabled us to characterize their conformations and conformational transitions as a function of salt concentration. We have also studied the effects of adriamycin, a potent anti-cancer drug, on the B to Z transition of d(Gm^{5}C)

2. SALT-INDUCED CONFORMATIONAL TRANSITION OF POLY-d(AT)

We have shown (2,3) that poly-d(AT) exhibits two resolved ^{31}P NMR signals of equal area in solution in low salt concentration (Fig. 2). Upon the addition of various salts, both peaks shift. The nature of the shifts depends on the salt used, and we have recently described these shifts in detail compared to an oligonucleotide control (6). Generally the peak separation increases, as shown for CsF in Figure 2. By contrast, when ethanol is added the doublet collapses into a singlet resonance (Fig. 3). We have interpreted the latter as a

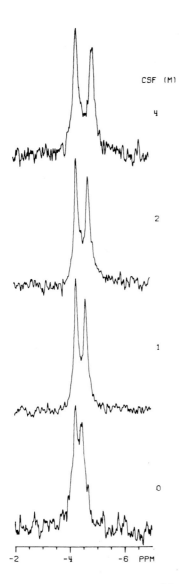

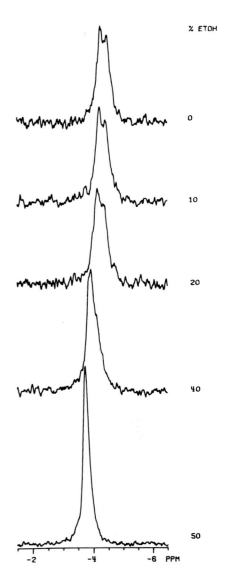

Fig. 2. Dependence of the ^{31}P NMR
spectra of poly-d(AT) at 109 MHz and
37°C on CsF concentration. The
^{1}H-decoupled. spectra were obtained
with 5-10,000 scans accumulated
using a 2 sec accumulation time, 4K
data points, and a 4 kHz sweepwidth.
Samples were 1-3 mg in 1 ml of 50 mM
Tris, 100 mM NaCl, 1 mM EDTA at pH 7.0,
containing 20-50% D_2O.

Fig. 3. Dependence of ^{31}P
NMR spectra of poly-d(AT)
on ethanol concentration
(% v/v) at 37°C.

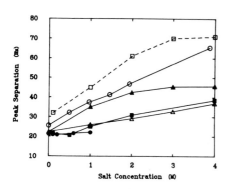

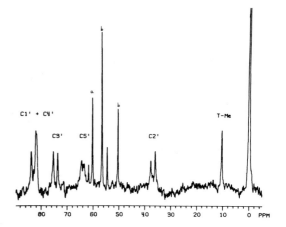

Fig. 4. Dependence of the peak
separation on salt concentration
for poly-d(AT); O, CsF; ●, MgCl$_2$;
▲, CsCl; ■ , NaCl; △, LiCl; data
for poly-d(a^{2}AT) in NaCl, □ .

Fig. 5. ^{13}C NMR spectrum at 125
MHz of poly-d(AT) at 37°C, with
^{1}H-decoupling, 64,000 scans,
using a 2.5 sec repetition time,
32 K data points, and a 30 kHz
sweepwidth. The sample contained
5 mg poly-d(AT) in 0.3 ml 33 mM
phosphate buffer, 100 mM NaCl,
1 mM EDTA at pH 7.0. The peaks
marked a and b derive from
residual Tris and EDTA.

change from an alternating B-form with a dinucleotide repeat structure
in low salt (2,3), to an A-form with a mononucleotide repeat unit upon
the addition of ethanol (8) involving an unwinding of the double helix
(7,8). In four monovalent salts studied (6) the doublet separation
increased (Fig. 4), indicating a persistence of the dinucleotide
repeat unit. That this alternation extends to the deoxyribose rings
is shown by the presence of a doublet for C2' and C3' resonances in
the ^{13}C NMR spectrum of poly-d(AT) (9,10) (Fig. 5), presumably
corresponding to ^{2}E and ^{3}E conformations.

Since a downfield shift of ^{31}P phosphate resonances has been
correlated with increased trans (t) proportion in the phosphodiester
geometry (1), we have assigned the downfield (left) component to a tg$^-$
conformation, and the upfield component to the standard Watson-Crick
g$^-$g$^-$ conformation (4,6). By comparison with the X-ray structure of
d(AT)$_2$, which is not helical but which does have a dinucleotide repeat
unit (11), we have assigned the downfield component to d(TpA) and the
upfield to d(ApT) (4,6). These assignments have recently been
confirmed by the elegant use of phosphorothioate substitution (12).
We do not, however, know if this alternating structure is similar to
that speculated on by Klug <u>et al</u>. (13) or others (14).

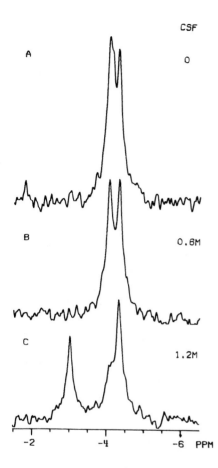

CSF

A   O

B   0.6M

C   1.2M

-2   -4   -6 PPM

Fig. 6. Effect of CsF concentration on the ^{31}P NMR spectrum of
poly-d(Gm^5C) obtained from PL
Biochemical and sonicated to ca
100 bp. The spectra were recorded
at 50°C, the shoulder at -4.2 ppm
in C derives from hybrid-template
formation (10). The solution was
5 mM Tris at pH 8.0 with 0.1 mM EDTA.

$[\theta] \times 10^{-4}$

1.0
0.0
-1.0
-2.0
-3.0
-4.0

0.0   0.3   0.6   0.9   1.2   1.5

[MgCl$_2$] (mM)

Fig. 7 The B to Z transition of
poly-d(Gm^5C) upon the addition
of $MgCl_2$ from the CD spectra in
molar ellipticity [θ]: ●, 293 nm;
O, 252 nm. The polymer was in 5
mM Tris, 50 mM NaCl, and 0.1 mM
EDTA at pH 8.0. After each
addition of $MgCl_2$ the mixture was
heated at 55-65°C for 3 min to
facilitate the transition, and
then allowed to cool. The CD
measurements were at room
temperature.

The salt-induced transition of poly-d(AT) can be considered to
involve tighter winding of the double helix involving an increase in
the winding angle (15-17). The gradual change in the chemical shifts
observed (Fig. 2) represents a fast equilibrium process (6,18), with a
rate limit from the linewidth (in Hz x 2π) > 180 sec^{-1}. However, from
both the CD (19) and ^{31}P NMR data the transition is clearly non-
cooperative for poly-d(AT) (Fig. 4).

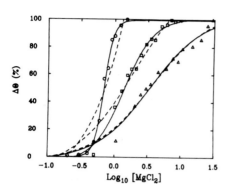

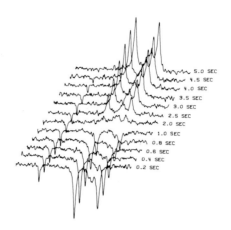

Fig. 8. Fit of CD data from figure 7 (treated as described in text) to equation 1, O. Dashed line, fit with n = 1; solid line, fit with n allowed to vary. Similar fits for Mg^{2+}-induced transition in the presence of adriamycin (see section 5); □ , 1 adriamycin/33 nucleotides; and Δ, 1 adriamycin/16 nucleotides. The results of the fitting procedures are given in Table 1.

Fig. 9. Determination of ^{31}P T$_1$ value for the Z-form of poly-d (Gm^{5}C) using the standard inversion recovery sequence, with 2000 scans and a 14 sec delay time. The T$_1$ values for the down-field (-3.15 ppm) peak and the up-field (-5.15 ppm) peak were 3.18 ± 0.13 and 3.10 ± 0.19 sec. respectively. The value for both peaks in the B-form were 3.31 ± 0.32 sec.

3. THE B TO Z TRANSITION OF POLY-d(Gm^{5}C)

Using CD spectroscopy it has been shown that the left-handed Z-form of poly-d(Gm^{5}C) is formed in lower salt concentrations than that of the unmethylated copolymer (20). We have also investigated this transition for the methylated copolymer with ^{31}P NMR spectroscopy using several salts (10). A peak with a characteristic chemical shift of -3 ppm appears (Fig. 6), as for poly-d(GC) (4,21) and its oligomers (22), but at lower salt concentrations. We assigned this resonance to d(Gpm^{5}C) in the tg$^+$ conformation of the Z-form (23), and this assignment has also been confirmed using phosphorothioate substitution (12). The nature of the appearance of this peak indicates a slow interconversion from B to Z, and from the separation of the signals (in Hz x 2π) we can place a lower limit on the rate of < 900 sec^{-1}.

From the raw CD data for this transition (Fig. 7) it is clear that this transition is cooperative. However, in order to quantitate this effect, we have fitted this and subsequent data with a modified

Hill equation for an equilibrium between two states of the form (24),

$$y = y_\infty + (y_0 - y_\infty)/(1 + (x/K)^n)$$ (1)

where y is the spectroscopic response, x is the concentration of added reagent, y_0 the response when $x = 0$, y_∞ the maximal response when $x = \infty$, K is the concentration of x at the midpoint (equivalent to the equilibrium constant) and n is the degree of cooperativity that determines the slope of the curve.

Table 1 - Fit of Data to Equation (1) for poly-d(Gm^{5}C)[a]

Agent for B-Z transition	Adriamycin/ nucleotide	n	pK	K (mM)	Root mean square error
MgCl$_2$	0	1.0	10±1.2		15.7
		6.1±0.6	-0.16±0.08	0.7	3.3
	1:33	1.0	0.3±0.1		7.0
		1.9±0.2	0.14±0.03	1.4	3.9
	1:16	1.0	0.56±0.07		4.7
		1.1±0.2	0.54±0.08	3.6	4.8
Adriamycin		1.0	-1.14±0.09		4.8
		2.0±0.19	-1.47±0.02	0.03	2.7

[a]Two fits were carried out to each set of CD data: with $n = 1$ (non-cooperative), and with n allowed to vary. The pK value is the fitted $\log_{10} K$ value.

To apply this equation the CD data at two wavelengths was converted to percent change and averaged, and the concentration was expressed as $\log_{10} x$. Fits were made with $n = 1$, i.e. non-cooperative, and n allowed to take any value (Fig. 8). No other constraints were used in these fits, and either three or four fitted parameters were used. The fitted values determined are shown in Table 1, and the Mg^{2+}-induced B-Z transition of poly-d(Gm^{5}C) was found to have a cooperativity coefficient >6, indicating a highly cooperative process.

In carrying out these studies we also determined the spin-lattice relaxation times of the two forms in solution (Fig. 9), in order to test for any major differences in relaxation rates. These in turn would reflect differences in phosphodiester mobility (25). However, the values showed no such differences. Thus, we have not yet carried out detailed correlation time calculations.

4. THE SALT-INDUCED CONFORMATIONAL TRANSITION OF POLY-d(a^{2}AT)

Upon addition of salts to poly-d(a^{2}AT), a transition has been observed using CD spectroscopy (26). In order to further characterize

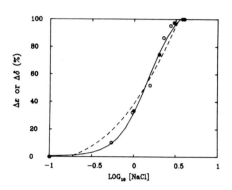

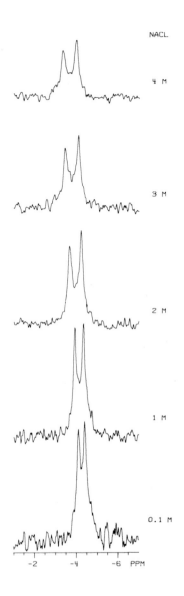

Fig. 11. Fit of equation (1) to CD data ($\varepsilon_L - \varepsilon_R$ or $\Delta\varepsilon$), 0, and ^{31}P NMR chemical shift separation $\Delta\delta$, ●, for poly-d(a^2AT). Values of parameters were, n = 2.4 ± 0.3, pK = 0.17 ± 0.03 (K = 1.48 M) and RMS error = 4.0. The RMS error for the non-cooperative fit (n = 1) was 7.6.

Fig. 10. Dependence of ^{31}P NMR spectra of poly-d(a^2AT) on NaCl concentration at 37°C. The solution was 2 mM cacodylate buffer at pH 7.0, with 100 mM NaCl and 1 mM EDTA.

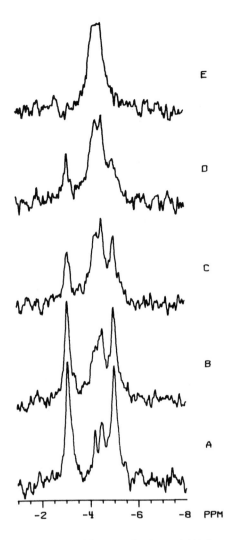

Fig. 12. Effect of the addition of adriamycin on the ^{31}P NMR spectrum of poly-d(Gm^{5}C) at 37°C. The Z form was prepared by the addition of 6 mM MgCl$_2$ in the presence of 50 mM NaCl and was converted back to the B form by adriamycin.

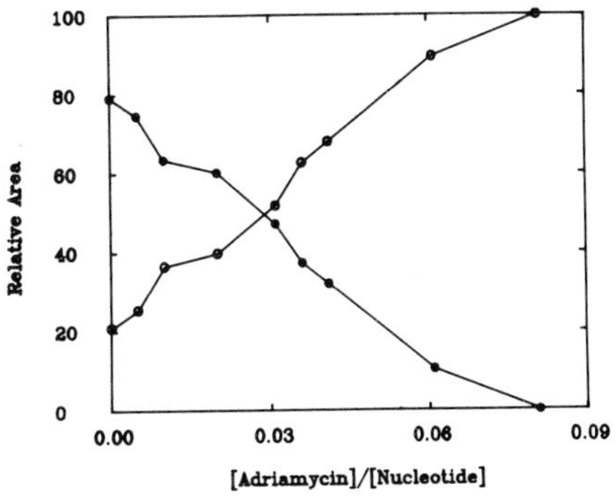

Fig. 13. Plots of areas of fitted peaks from Figure 12 of the Z form (●) and the B form (O).

this transition we carried out a ^{31}P NMR study, and observed a very similar phenomenon (Fig. 10) to that found for poly-d(AT) (Fig. 2), namely a fast exchange process with a rate limit similar to that for poly-d(AT) but different from the slow B to Z transition observed for poly-d(Gm^{5}C) (Fig. 6). The doublet resonances represent two different

phosphodiester conformations of a dinucleotide repeat structure. The response of the separation of the two resonances to salt is however different in this case from that of d(AT) (Fig. 4), and the final chemical shift value (-3.3 ppm) for the downfield component of d(a^2AT) is very close to that found at -3 ppm for the Z form (see above). These spectra (Fig. 10) could represent a transition to a Z form, but at a faster rate than that observed for the d(GC) copolymers.
Combining the NMR data with the CD data and fitting with equation (1) clearly showed that the transition for poly-d(a^2AT) was cooperative (Fig. 11).

5. EFFECT OF ADRIAMYCIN ON THE B TO Z TRANSITION OF POLY-d(Gm^5C)

Adriamycin is a potent anti-cancer drug, the structure of which is,

Adriamycin binds strongly to DNA, and is known to intercalate (27). Addition of adriamycin to poly-d(Gm^5C) in the B-form in low salt lead to a gradual broadening of the ^{31}P NMR signals. However, similar addition to poly-d(Gm^5C) in the Z-form resulted in a conversion to the B-form (Fig. 12). The areas of the peaks corresponding to the two forms are plotted in Fig. 13. Similarly, CD spectra showed that adriamycin converted the Z to the B form (Fig. 14). This data as a function of adriamycin/nucleotide concentration (Fig. 15) was fitted with equation (1), and it was found to be somewhat cooperative (Fig. 16), with n = 2 (Table 1).

In order to further investigate the effect of adriamycin on this B to Z transition, the Mg^{2+}-induced change was monitored by CD spectroscopy in the presence of adriamycin. The resulting curves showed a marked increase in the concentration of Mg^{2+} required to bring about the transition (Fig. 17 compared to Fig. 7). Fitting equation (1) to this data (Fig. 8, Table 1) showed that while a better fit was obtained with a degree of cooperativity (n = 1.9) for a ratio of one adriamycin per 33 nucleotides, at twice this concentration there was in effect no difference between the fits with n set equal to one (non-cooperative) or allowed to vary. The values of n and K in this latter case are essentially indeterminate, but it can be concluded that the transition in this case is effectively non-cooperative. Given the smooth change of the fitted K value from 0.7

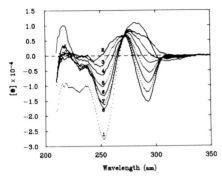

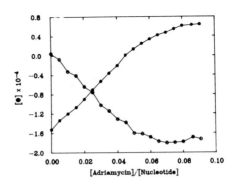

Fig. 14. CD spectra (molar ellipticity [θ] at room temperature showing the effect of adriamycin on poly-d(Gm⁵C) conformation. 1. Original B-form of poly-d(Gm⁵C) in the presence of 5 mM Tris/50 mM NaCl/0.1 mM EDTA/ pH 8.0; 2. Z-form after addition to 1 of MgCl₂ (final concentration 2 mM); 3-8, addition of adriamycin to 2: 3, 1.0 µM, [adriamycin]/[nucleotide] ([A]/[N]) = 0.005 or 1:200; 4, 2.0 µM, [A]/[N] = 0.02 or 1:50; 5, 3.0 µM, [A]/[N] = 0.03 or 1:33; 6, 4.0 µM, [A]/[N] = 0.04 or 1:25; 7, 4.9 µM, [A]/[N] = 0.05 or 1:20; 8, 8.8 µM, [A]/[N] = 0.091 or 1:11.

Fig. 15. Plot of CD molar ellipticity [θ] of poly-d (Gm⁵C) as a function of adriamycin/nucleotide concentration, showing the Z to B transition; ●, 293 nm; O, 252 nm.

to 1.4 and to 3.6 mM [Mg²⁺] in these experiments, it is clear that adriamycin inhibits the B to Z transition of poly-d(Gm⁵C) and causes the transition to change from a highly cooperative (n = 6) to a less cooperative process (Fig. 8, Table 1).

6. CONCLUSIONS

The three alternating purine-pyrimidine sequence polydeoxynucleotides reported here each exhibit a ³¹P NMR spectrum characteristic of a dinucleotide repeat structure in solution. While the salt-induced transition for poly-d(a²AT) appears essentially similar to that for poly-d(AT) as judged by the NMR spectra (Figs. 2 and 10), in the latter case it is a non-cooperative process (Fig. 4) while for d(a²AT) it is clearly cooperative (Fig. 11). In view of the similarity of the ³¹P NMR spectrum of d(a²AT) in the higher salt concentration to that of the Z-form of d(Gm⁵C) it is not possible to conclude whether the high salt form of d(a²AT) is either similar to that of poly-d(AT), or is a Z-form.

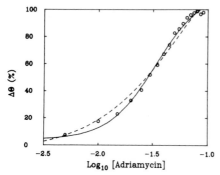

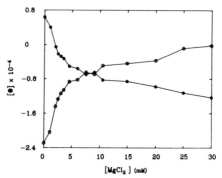

Fig. 16. Fit using equation (1) to CD data (from figure 15, treated as described in text) for the effect of adriamycin on the Z form of poly-d (Gm^{5}C). The fitted parameter values are given in Table 1.

Fig. 17. The B to Z transition of poly-d(Gm^{5}C) from CD spectra in molar ellipticity [θ] as a function of MgCl$_2$ concentration in the presence of adriamycin (7.1 µM or 1:6 adriamycin/ nucleotide); ●, 293, nm; O, 252 nm.

The highly cooperative Mg^{2+}-induced B to Z transition of poly-d(Gm^{5}C) is inhibited by adriamycin and changed into a less cooperative process (Fig. 8). Adriamycin also converts the Z-form into the B-form with some degree of cooperativity (Fig. 16). A large proportion of CG sequences in mammalian DNA are methylated at the C5 position (28) and this methylation may be correlated with the "turning off" of gene activity (29). It has also been speculated that B-DNA is present in active genes while the Z-form corresponds to inactive regions of DNA (30). Since adriamycin appears to bind preferentially to the B-form and to convert Z-DNA into that form, it is tempting to conclude that it could be exerting its influence by turning on inactive genes. However, the chemical processes involved in adriamycin interactions in the living organism are sufficiently complex that this simple explanation must be considered very tentative.

Acknowledgement: We thank David Rodbard for helpful discussions.

References

1. Gorenstein, D.G.: 1981, Ann. Rev. Biophys. Bioeng. 57, 355-386.
2. Chen, C., and Cohen, J.S.: 1983, in "Phosphorus-31 NMR: Principles and Applications," Gorenstein, D.G. (ed.), Academic Press, New York.
3. Shindo, H., Simpson, R.T., and Cohen, J.S.: 1979, J. Biol. Chem. 254, 8125-8128.
4. Cohen, J.S., Wooten, J.B., and Chatterjee, C.L.: 1981, Biochemistry 20, 3049-3055.
5. Chen, C., Cohen, J.S., and Zador, A.: 1981, J. Biochem. Biophys.

Methods 5, 293-295.
6. Chen, C., and Cohen, J.S.: 1983, Biopolymers 22, 879-893.
7. Ivanov, V.J., Minchenkova, L.E., Scholykina, A.K., and Poletayev, A.I.: 1973, Biopolymers 12, 89-110.
8. Lee, C.H., Mizusawa, H., and Kakefuda, T.: 1981, Proc. Natl. Acad. Sci. USA 78, 2838-2842.
9. Shindo, H.: 1981, Europ. J. Biochem. 120, 309-313.
10. Chen, C., Cohen, J.S., and Behe, H.: 1983, Biochemistry in press.
11. Viswamitra, M.A., Shakked, Z., Jones, P.G., Sheldrick, G.M., Salisbury, S.A., and Kennard, O.: 1982, Biopolymers 21, 513-533.
12. Eckstein, F., and Jovin, T.: 1983, unpublished results.
13. Klug, A., Viswamitra, M.A., Kennard, O., Shakked, Z., and Steitz, T.A.: 1979, J. Mol. Biol. 131, 669-680.
14. Gupta, G., Bansal, M., and Sasisekharan, V.: 1980, Int. J. Biol. Macromol. 2, 368-380.
15. Struddert, D.S., Patroni, M., and Davis, R.C.: 1972, Biopolymers 11, 761-779.
16. Zimmer, C., and Luck, G.: 1974, Biochim. Biophys. Acta 361, 11-32.
17. Chan, A., Kilkuskie, R., and Hanlon, S.: 1979, Biochemistry 18, 84-91.
18. Patel, D.J., Kozlowski, S.A., Suggs, J.W., and Cox, S.D.: 1981, Proc. Natl. Acad. Sci. USA 78, 6063-6067.
19. Vorlickova, M., Kypr, J., Kleinwachter, V., and Palacek, E.: 1980, Nuc. Acids Res. 8, 3965-3973.
20. Behe, M., and Felsenfeld, G.: 1981, Proc. Natl. Acad. Sci. USA
21. Simpson, R.T., and Shindo, H.: 1979, Nuc. Acids Res. 7, 481-492.
22. Patel, D.J., Canuel, L.L., and Pohl, F.M.: 1979, Proc. Natl. Acad. Sci. USA 76, 2508-2511.
23. Wang, A.H., Quigley, G.J., Kolpak, F.J., van der Marel, G., van Boom, J.H., and Rich, A.: 1981, Science 211, 171-176.
24. DeLean, A., Munson, P.J., and Rodbard, D.: 1978, Am. J. Physiol. 235, E97-E102.
25. Shindo, H.: 1980, Biopolymers 19, 509-522.
26. Miles, T., and Howard, F.: 1983, these proceedings.
27. Fritzsche, H., Triebel, H., Chaires, J.B., Dattagupta, N., and Crothers, D.M.: 1982, Biochemistry 21, 3940-3946.
28. Razin, A., and Riggs, A.D.: 1980, Science 210, 604-610.
29. McGhee, J.D., and Ginder, G.D.: 1979, Nature 280, 419.
30. Nordheim, A., Purdue, M.L., Lafer, E.M., Moller, A., Stollar, B.D., and Rich, A.: 1981, Nature, 294, 417-422.

DEPENDENCE OF OLIGONUCLEOTIDE CONFORMATIONAL PROPERTIES ON SUGAR RING AND BASE-SEQUENCE. ¹H AND ¹³C N.M.R. STUDIES.

David B. Davies
Department of Chemistry, Birkbeck College,
Malet Street, London WC1E 7HX, UK.

Interest has developed in the possibility that base sequence might affect the local conformational properties of nucleic acids and may lead to specific recognition sites for protein or drug binding. Part of the interest stems from the fact that variations in DNA structure with base sequence have been observed; for example, the recent crystal structure of a DNA dodecamer by Dickerson and Drew (1981) showed several departures from the ideal, regular helical structure of B-DNA that might depend on base sequence; Klug and co-workers (1979) proposed a non-classical "alternating B" configuration for poly(dA-dT) brought about by differential overlap of successive bases along the same chain; alternating C-G oligomers are known to form left-handed, zigzag, Z-DNA (Wang *et al.*, 1979; Arnott *et al.*, 1980; Drew *et al.*, 1980). On the other hand base sequence dependent perturbations of structure are not found in the A-form of DNA (Shakked *et al.*, 1981; Connor *et al.*, 1982). An attempt has been made to rationalise the sequence dependent stacking of bases in B-DNA in terms of simple steric repulsive forces between purine bases in consecutive base-pairs but on opposite backbones, and to show that the repulsive forces between base-pairs are resisted by stresses in the helical backbone (Calladine, 1982).

Many planar aromatic drugs, carcinogens and mutagens interact with DNA and RNA primarily by intercalating between the planes of the bases and results for oligonucleotide and polynucleotide-drug complexes in the crystalline state (Neidle, 1981) and in solution (Krugh, 1981) have recently been reviewed. There is considerable evidence that many intercalating molecules prefer py-pu sequences (Voet, 1977), particularly in RNAs. Using classical potential energy calculations, Broyde and Hingerty (1979) showed that there is a fundamentally different overlap of bases in py-pu v. pu-py base sequences in the A-RNA-11 conformation which occurs regardless of the specific base substituents and that the different stacking patterns, which result in a greater glycosidic bond mobility in py-pu sequences, permit the transition of preferred sequences to the intercalated conformation at lower energy cost.

In view of these X-ray crystallographic results it is pertinent to

B. Pullman and J. Jortner (eds.), Nucleic Acids: The Vectors of Life, 141–154.
© *1983 by D. Reidel Publishing Company.*

investigate which fundamental conformational properties of nucleic acids
in solution depend on base sequence, how base sequence alters the conformational properties in solution and which conformational properties
of nucleic acids respond to other perturbations such as drug binding.
A comparison of oligoribo- and oligodeoxyribonucleotides would be of
interest. Nuclear magnetic resonance spectroscopy (n.m.r.) has made
important contributions to our understanding of the conformations
(Davies 1979a, Sarma 1979, 1981; Altona 1982) and interactions (Krugh
1981; Sarma 1981, Altona 1982) of oligonucleotides in solution.
Adequate n.m.r. methods are available for determining sugar-ring
conformations (Altona and Sundaralingam, 1973, de Leeuw et $al.$, 1980) and
four out of the six bonds of the sugar-phosphate backbone utilising
$^1H-^1H$, $^{31}P-^1H$ or $^{31}P-^{13}C$ vicinal coupling constants (Davies 1979a) as
summarised for the nucleotide unit in Figure 1. The conformation of
the nucleotide unit is usually considered in terms of the sugar-
phosphate backbone, the sugar-ring and the glycosidic bond conformation
and the α - ζ notation used in Figure 1 follows recent IUPAC-IUB nomenclature recommendations.[1]

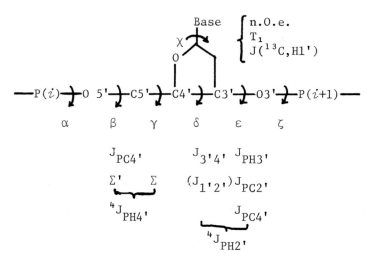

Figure 1. N.m.r. methods to determine conformational properties of
oligonucleotides in solution. N.B. $\Sigma = (J_{4'5'} + J_{4'5''})$ and $\Sigma' = (J_{P5'} + J_{P5''})$ Hz

     In this work the relation between the sugar-ring and backbone
conformational properties (β,γ and ϵ bonds) of oligo- and polynucleotides
will be considered. Most of the necessary ^1H n.m.r. information (Σ,Σ',
$J_{1'2'}$ and $J_{3'4'}$) is available in the literature from the work of Sarma
and co-workers (Lee et $al.$, 1976, Ezra et $al.$, 1977, Cheng and Sarma,
1977). The ^{13}C n.m.r. results of oligoribonucleotides ($J_{PC2'}$ and $J_{PC4'}$)
are taken mainly from the work of Alderfer and Ts'O (1977) supplemented
by a few results from other workers. In this work ^{13}C (and ^1H) n.m.r.
observations of a number of di-deoxyribonucleoside phosphates have been
observed and analysed in a manner similar to that for the oligoribonucleotides (Davies, 1983). It is shown that both the backbone C5'-C4'

(and O5'-C5') bonds and the C3'-O3' bond depend on the sugar ring
conformation but, whereas the backbone C5'-C4' bond depends on base
sequence, the C3'-O3' bond does not. These results are in general
agreement with the conformational analysis of nucleic acids by
Sundaralingram (1969, 1982).

RESULTS AND DISCUSSION

1. Relation between sugar ring and C5'-C4' bond conformation.

(i) Ribonucleotides. Hruska (1973) first demonstrated the
dependence of $J_{3'4'}$ and $\Sigma(=J_{4'5'} + J_{4'5''})$ magnitudes of pyrimidine
nucleosides and nucleotides indicating that an increase in sugar ring
N conformation (previously described as C3'-endo) is accompanied by an
increase in the C5'-C4' bond γ_+ conformation (60°, previously described
as gg). An approximate 1:1 relation between these conformational
properties was determined (Davies, 1979a) though paucity of data pre-
cluded determination of a meaningful correlation for purine derivatives
at the monomer level. Similarly, quantitative analysis of available Σ
and Σ' results for monomers and oligomers indicated that the interdependence of C5'-C4'(γ_+) and O5'-C5'(β_t) conformations are different
for purine (-pPur) and pyrimidine (-pPyr) nucleotides $i.e.$ the 5' side
of the backbone is influenced by the base (Davies, 1979b).

With the advent of a conformational model by Altona (1975) that
related the sugar ring conformations of oligoribonucleotides to the
proportion of base-stacked conformer, it was possible to explore the
variations in C5'-C4' and O5'-C5' bond conformations as shown in Fig. 2
$i.e.$ an increase in base-stacking is accompanied by an increase in the
C5'-C4' bond 60° conformation (γ_+) and concomitantly, the O5'-C5' bond
180° conformation, β_t (previously labelled g'g' or ϕ_a). Differences in
behaviour were observed for different base sequences of dinucleoside
phosphates (Davies, 1978).

$$i.e. \quad pu - pu \; > \; py - pu \; > \; pu - py \; \simeq \; py - py \qquad (1)$$

The effect of base-stacking on backbone conformation depends, to some
extent, on the amount of overlap of bases involved in the stacking
process, the distance apart of the bases and the angle between two
successive base planes. Although little of this information is available for such molecules in solution and the behaviour summarised in
equation (1) cannot yet be completely rationalised, the [1]H n.m.r.
results of oligoribonucleotides (XpY) definitely show that the nature
of the base -pY affects the O5'-C5' and C5'-C4' bond conformations.

(ii) Deoxyribonucleotides. N.m.r. studies of deoxyribonucleotides
(Altona $et\ al.$, 1976, Cheng and Sarma, 1977) showed no marked dependence of the sugar ring conformation with temperature which might be
interpreted in terms of the greater flexibility of the deoxyribose ring
compared to the ribose ring in oligonucleotides. The sugar rings of

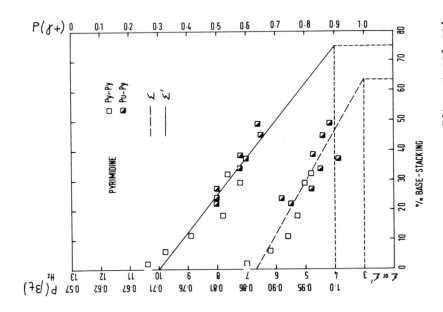

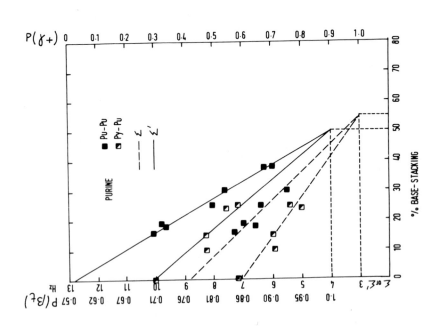

FIG. 2   Correlation between base stacking and backbone O5'-C5' bond, (plotted as Σ') and C5'-C4' bond, (plotted as Σ) conformers for purine and pyrimidine oligoribonucleoside phosphates (Davies 1978, reproduced by kind permission)

deoxyribonucleotides are found to exist predominantly in the S (C2'-endo) conformation and there is no simple relation between sugar ring conformation and base-stacking as found for oligoribonucleotides. One might expect that the C5'-C4' and O5'-C5' backbone conformational properties could be used to indicate base-stacking proclivities of oligodeoxyribonucleotides but that might preclude an investigation of the dependence of backbone conformational properties with base sequence. Indeed analysis of available ^{1}H n.m.r. data on deoxyribonucleotides shows no clear interdependence of Σ v Σ' magnitudes on base ring as found for oligoribonucleotides. At the present time no n.m.r. method is available to determine quantitatively the proportion of base-stacked conformers in oligoribonucleotides and, hence, investigate the sequence dependence of O5'-C5'-C4' backbone conformational properties.

2. Relation between sugar ring and C3'-O3' bond conformation.

(i) Ribonucleotides. The conformational properties of the C3'-O3' bond of nucleotides in solution may be determined approximately using $^3J_{PH3'}$ magnitudes and more accurately when combined with $J_{PC2'}$ and $J_{PC4'}$ magnitudes. Previous work (Davies, 1979a, Sarma, 1979) has shown that $J_{PH3'}$ magnitudes of oligoribonucleotides (8.0 ± 0.3 Hz) and oligodeoxyribonucleotides (6.9 ± 0.2 Hz) are approximately constant and can be rationalised in terms of a two state symmetrical model in which the phosphate group is considered to exist in an ε_t or ε_- conformation at an angle ±ϕ with respect to the C3'-H3' bond as shown in Figure 3. The magnitude of ϕ varies for ribonucleotides (ca 35°) and deoxyribonucleotides (ca 40°). For ribonucleotides the dependence of the sugar ring and the C3'-O3' bond conformations of the Xp-unit of oligonucleotides may be represented according to equation (2).

$$
\begin{array}{cc}
\text{N.}\varepsilon_t & \text{RIBO:} \\
\text{N.}\varepsilon_- \quad \text{S.}\varepsilon_t & \varepsilon_t \ ca.\,205 \\
\text{S.}\varepsilon_- & \varepsilon_-^t \ ca.\,275
\end{array}
\qquad (2)
$$

A simplified version of this conformational model has recently been verified quantitatively (Davies, 1983). The model is summarised by the dotted equilibrium in equation (2) which assumes a direct dependence of the two-state C3'-O3' bond model with a two-state sugar ring conformational model (Figure 3b). A plot of available ^{1}H and ^{13}C n.m.r. results for oligo- and polynucleotides exhibits a linear dependence between $(J_{3'4'} - J_{1'2'})$ and $J(_{PC4'} - J_{PC2'})$ magnitudes with slope (ca 1.6) that passes close to the origin as shown in Figure 4. The linear dependence can be predicted according to equation (3), where p_N and p_S are the relative populations of sugar ring N and S conformations p_t and p_- the relative populations of the ε_t and ε_- conformations, respectively, and coupling constants signified as $^*J$ indicate that different magnitudes are expected for deoxyribonucleotides compared to ribonucleotides to account for different substituent electronegativity effects at C2'. (see below).

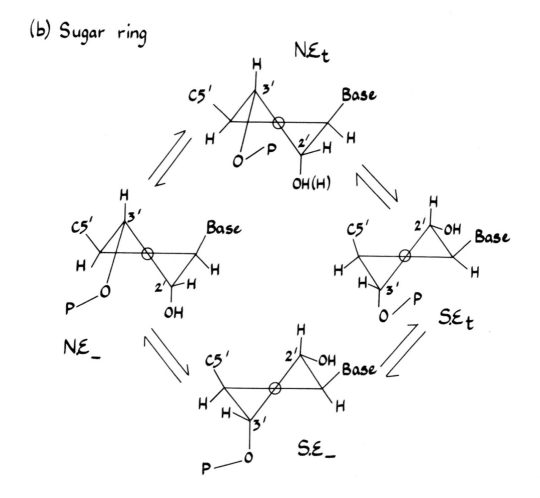

FIG. 3 The two-state symmetrical conformational models of
(a) C3'-03' bonds and (b) sugar rings of nucleotides.

$$\frac{(J_{3'4'} - {}^*J_{1'2'})}{(J_{PC4'} - {}^*J_{PC2'})} \simeq \frac{(P_N - P_S)}{(P_t - P_-)} \times \frac{({}^*J_{1'2} + J_{3'4})}{[{}^*J(120+\phi) - {}^*J(120-\phi)]} \qquad (3)$$

For ribonucleotides observed magnitudes of $(J_{1'2'} + J_{3'4'})$ and $(J_{PC2'} + J_{PC4'})$ are constant as demanded by the appropriate two-state conformational models, and the corresponding values of ϕ calculated from ${}^1H$ n.m.r. (ca 35°) and ${}^{13}C$ n.m.r. (ca 37 + 2°) are about the same; the latter value of ϕ was calculated according to equation (4) from an average sum of coupling constants (9.0 ± 0.2 Hz, Figure 4) and the appropriate Karplus parameters[2] for P-O-C-C vicinal coupling constants $i.e.$ A = 6.4, B = 1.3, C = 1.2 Hz (Davies and Sadikot, 1982).

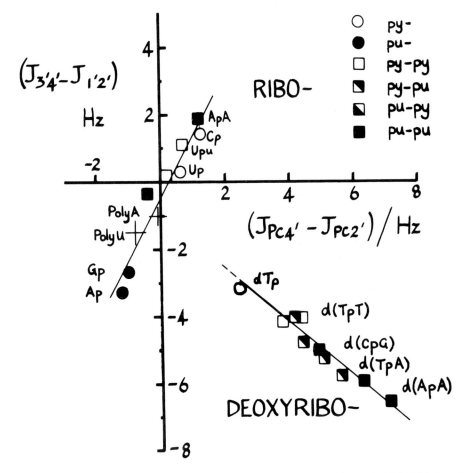

FIG. 4 Relation between sugar ring $(J_{3'4'} - J_{1'2'})$ and C3'-O3' bond $(J_{PC4'} - J_{PC2'})$ conformations of nucleotides. (i) Ribo- (data taken from Alderfer and Ts'o 1977, Davies 1983) (ii) Deoxyribo- (this work)

$$(J_{PC2'} + J_{PC4'}) = J(120 + \phi) + J(120 - \phi)$$
$$= -A\cos^2\phi + B\cos\phi + (1.5A + 2C) \tag{4}$$

An estimate can be made of the magnitude of the slope of the linear correlation for oligoribonucleotides in Fig. 4 using ϕ *ca* 35° and $(J_{1'2'} + J_{3'4'})$ *ca* 9.5 Hz

$$i.e. \qquad (p_N - p_S) \times 9.5 \,/\, (p_t - p_-) \times 6.8$$

Comparison of the observed (*ca* 1.6) and calculated slopes indicates that there is an approximate 1:1 dependence of the sugar ring and C3'-O3' bond conformational properties of the form $N.\varepsilon_t \rightleftharpoons S.\varepsilon_-$. Within the limitations of the present measurements, it seems that there is no sequence dependence of this conformational correlation.

(*ii*) Deoxyribonucleotides. The combination of ^1H and ^{13}C n.m.r. measurements may also be used to explore the sugar ring and C3'-O3' bond conformational properties of deoxyribonucleotides. The 50 MHz ^{13}C n.m.r. spectra of a number of di-deoxyribonucleoside phosphates have been measured as a function of temperature (Davies and Hardiman, to be published). A typical ^{13}C n.m.r. spectrum for d(ApA) in D_2O solution[3]

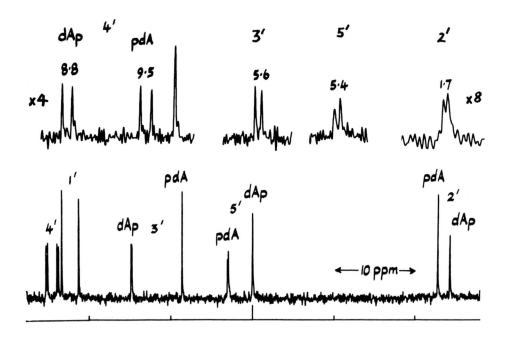

FIG. 5   50 MHz proton-noise decoupled ^{13}C n.m.r. spectrum (with expansions of phosphorus-coupled signals) of d(ApA) in D_2O solution at 300K (Data resolution 0.2 Hzpt.$^{-1}$)

is shown in Figure 5. Assignment of C1'-C5' signals is straightforward by comparison with results for ribonucleotides (Alderfer and Ts'O, 1977) and the expectation that the upfield signals belong to C2' as in the monomer (Niemczura and Hruska, 1980). Assignment of carbon atoms to Xp- and -pY units is made by consideration of which carbon atoms are expected to exhibit two and three bond $^{31}P-^{13}C$ couplings and the expected magnitudes of coupling constants $i.e.$ $^{2}J_{PC3'}$(Xp-) and $^{2}J_{PC5'}$(-pY) are constant (ca 5.4 Hz) at all temperatures, $J_{PC2'}$(Xp-) and $J_{PC4'}$(Xp-) are expected to vary with temperature and $J_{PC4'}$(-pY) is large (8-10) Hz resulting from an overwhelming preference for the O5'-C5' 180° (β_{t}) conformation. Variations in $J_{PC2'}$ and $J_{PC4'}$ magnitudes were observed for the deoxyribonucleotides studied, though the sum of ($J_{PC2'}$ + $J_{PC4'}$) was constant (Figure 6) $i.e.$ an average value of 10.2 ± 0.2 Hz. The ^{13}C n.m.r. results do not support the conclusions of recent molecular mechanics calculations (Dhingra and Saran, 1982) which predict ε_{t} (ca 180°) for both N and S conformations of deoxyribonucleotides as this would imply that not only is the sum ($J_{PC2'}$ + $J_{PC4'}$) constant but also that individual magnitudes of $J_{PC2'}$ and $J_{PC4'}$ do not vary. It has been suggested by Niemczura and Hruska (1980) that coupling constant magnitudes of mononucleotides may also be explained in terms of averaging within a single broad range of C3'-O3' bond conformations (ε $ca.$195-270) as found for the limited range of crystal structure results.

In this work it is assumed that the C3'-O3' bond of deoxyribonucleoside phosphates depends on the sugar ring conformation similar to that found for ribonucleotides according to equations (2) and (3), though with the $proviso$ that magnitudes of ϕ are different as determined from ^{1}H n.m.r. results. The dependence of these conformational properties may be explored by a plot of ($J_{3'4'}$-$J_{1'2'}$) $versus$ ($J_{PC4'}$ - $J_{PC2'}$) as shown in Figure 4. It should be noted that the $J_{1'2'}$ and $J_{3'4'}$ magnitudes were determined in the present work on the same solutions on which ^{13}C n.m.r. measurements were performed; small systematic variations in $J_{1'2'}$ and $J_{3'4'}$ magnitudes were observed that indicated an increase in the sugar ring S conformation of the Xp-unit at lower temperatures, where one might expect greater base-stacking to occur.

The plot of ($J_{3'4'}$-$J_{1'2'}$) against ($J_{PC4'}$-$J_{PC2'}$) magnitudes of di-deoxyribonucleoside phosphates shown in Figure 4 exhibits a linear correlation between sugar ring and C3'-O3' bond conformational properties. The negative slope ($ca.$ -0.65) shows that the correlation is opposite to that observed for ribonucleotides $i.e.$ the sugar ring S conformation is associated with C3'-O3' bond ε_{t} ($ca.$ 200°) and the sugar ring N conformation is associated with C3'-O3' bond ε_{-} ($c\acute{a}.$ 280°) according to equation (5).

$$N.\varepsilon_{t} \rightleftharpoons S.\varepsilon_{t}$$
$$N.\varepsilon_{-} \longrightarrow S.\varepsilon_{-}$$

DEOXYRIBO:

ε_{t} $ca.$200
ε_{-}^{t} $ca.$280

(5)

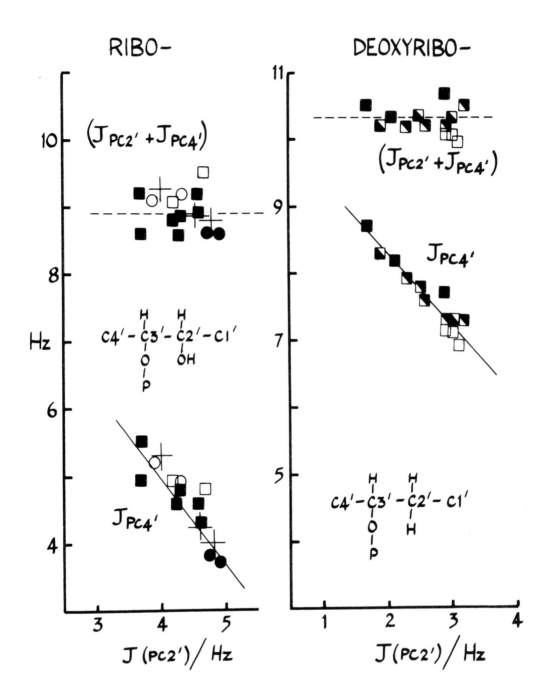

FIG. 6 Dependence of $J_{PC4'}$ and $(J_{PC2'} + J_{PC4'})$ on $J_{PC2'}$ magnitudes of oligo- and polynucleotides. (*i*) Ribo-(data taken from Alderfer and Ts'O 1977, Davies 1983) (*ii*) Deoxyribo- (this work)

The results also show that within experimental error the relation
between sugar ring and C3'-O3' bond conformational properties of deoxyribonucleotides is independent of base sequence as found recently for
oligo- and polyribonucleotides (Davies, 1983). The intercept (Figure 4)
may be explained in part by the effect on both $J_{1',2'}$ and $J_{PC2'}$ magnitudes
in equation (3) of the electropositive substituent at C2' in the deoxyribose ring compared to the hydroxyl groups of the ribose ring.

An estimate can be made of the magnitude of the slope of the linear
correlation in Figure 4 using ϕ (ca 40°), $(J_{1',2'} + J_{3',4'})$ ca 10.8 Hz for
deoxyribonucleotides (Cheng and Sarma, 1977) and an approximate electronegativity correction of 1 Hz for $J_{PC2'}$

$i.e.$ $(p_N - p_S) \times 10.8 \, / \, (p_t - p_-) \times 8$

Small variations in calculated slopes (1.3-1.4) occur depending on the
reliability of different Karplus relations used for coupling in H-C-O-P
molecular fragments (Altona, 1982), the particular Karplus relation
used for P-O-C-C coupling (Davies and Sadikot, 1982), the way in which
the substituent electronegativity correction is applied and even differ-
ent values of $J_{PH3'}$, ca 6 Hz, which corresponds to ϕ ca 45° ($i.e.$ ε_t ca
195, ε_- ca285)[4]. Even though these various modifications are taken into
account, it is apparent that there is a substantial difference in the
magnitudes of observed and calculated slopes for deoxyribonucleoside
phosphates in equation (3) indicating that the change in the sugar ring
conformation $(p_N - p_S)$ of the deoxyribose ring is about one half the
magnitude of the change in the C3'-O3' bond conformation $(p_t - p_-)$
compared to the approximate 1:1 dependence found for ribonucleotides.

Although the conformational model for deoxyribonucleotides in
equation (5) differs from those suggested previously (Cheng and Sarma,
1977; Dhingra and Saran, 1982), there is a broad area of agreement in
the results because the predominant S conformation observed for the
sugar ring indicates that the predominant conformation for the C3'-O3'
bond is ε_t (ca 200°) which is similar to that suggested previously from
[1]H n.m.r. measurements (Cheng and Sarma, 1977) and rationalised by
molecular mechanics calculations (Dhingra and Saran, 1982). The differences in the conformational models arises from observation of a
variation in the C3'-O3' bond conformation with sugar ring conformation
in the present work (Figure 4) compared to the suggestion that the
C3'-O3' bond conformation is the same for both N and S conformations of
the sugar ring in deoxyribonucleotides. The novel features of the
present analysis are to determine the quantitative relation between
sugar ring and C3'-O3' bond conformational properties of nucleotides,
to show that ribose and deoxyribose oligomers have opposite conformational dependencies, to show that the magnitude of the correlation
observed for deoxyribonucleotides is about one-half that observed for
ribonucleotides, and to show that the interrelation between sugar ring
and C3'-O3' backbone conformational properties of both ribo- and deoxyribonucleotides is independent of base sequence.

CONCLUSIONS

(*i*) N.m.r. studies show that there is an approximate 1:1 dependence of the sugar ring conformation of ribonucleotides with both the backbone C5'-C4' and C3'-O3' bonds such that an increase in sugar ring N conformation is accompanied by an increase in the C5'-C4' bond 60° conformation (γ_+) and the C3'-O3' bond ε_t conformation, *ca* 205°. These conformational changes are expected to occur with an increase in base-stacking of oligoribonucleotides in solution and confirm the behaviour found in the solid state by X-ray crystallography.

(*ii*) Using a two-state symmetrical conformational model for the C3'-O3' bond the present ^{1}H and ^{13}C n.m.r. results on di-deoxyribonucleoside phosphates show that the relation between the sugar ring and C3'-O3' backbone conformational properties differs from that of oligoribonucleotides in three respects: firstly, the available ε_t (and ε_-) conformations are slightly different (ε *ca* 205/275 ribo- ; ε *ca* 200/280 deoxyribo-); secondly, the conformational dependence observed for di-deoxyribonucleoside phosphates indicates that the sugar ring S conformation is associated with the C3'-O3' bond ε_t conformation in agreement with previous n.m.r. studies (Cheng and Sarma, 1977) and that the small population of sugar ring N conformation might be associated with the minor conformer ε_-; a third difference is that the dependence of the sugar ring and C3'-O3' bond conformation for oligodeoxyribonucleotides (*ca* 1:2) is much less that the 1:1 dependence found for oligoribonucleotides.

(*iii*) For oligoribonucleotides it was found that the C3'-O3' bond conformational properties are independent of base sequence whereas the O5'-C5'-C4' bond conformational properties depend not only on the nature of the -pY fragment but may also depend on base sequence. The results for oligoribonucleotides in solution confirm the summary of X-ray crystallographic results where it is found that "the sugar pucker has a greater influence on the conformation of the 3' side of the backbone while the 5' side of the backbone is (mostly) influenced by the base" (Sundaralingam 1969, 1982). A similar conclusion can be made for the 3' side of the backbone for oligodeoxyribonucleotides and further work is in progress to assess the behaviour for the 5' side of the backbone in these molecules.

ACKNOWLEDGEMENTS

I would like to thank Marian Hardiman and John Christofides for help in running the ^{1}H and ^{13}C n.m.r. spectra, and the M.R.C. for providing access to the Bruker WH500 spectrometer (N.I.M.R.) and for providing computing facilities at Birkbeck College. I would also like to thank Professor B. Pullman for the invitation and the organisers for their hospitality.

REFERENCES

Alderfer, J.L. and Ts'o, P.O.P. : 1977, Biochemistry, 16, pp.2410-2416

Altona, C. : 1975, in Structure and Conformation of Nucleic Acids and Protein-Nucleic Acids Interactions (Sundaralingam, M. and Rao, S.T. eds.) Univ. Park Press, Baltimore pp.613-629

Altona, C., van Boom, J.H. and Haasnoot, C.A.G. : 1976, Eur. J. Biochem. 71, pp.557-562

Altona, C. : 1982, Recl. Trav. Chim. Pays-Bas, 101, pp.413-434

Altona, C. and Sundarlingam, M. : 1972, J. Amer. Chem. Soc. 94, pp.8205-8212 ; ibid. 1973, 95, pp.2333-2344

Arnott, S., Chandrasekaran, R., Birdsall, D.L., Leslie, A.G.W. and Ratliff, R.L. : 1980, Nature (London) 283 pp.743-745

Broyde, S. and Hingerty, B. : 1979, Biopolymers, 18, pp.2905-2910

Calladine, C.R. : 1982, J. Mol. Biol. 161, pp.343-352

Cheng, D.M. and Sarma, R.H. : 1977, J. Amer. Chem. Soc., 99, pp.7333-7348

Connor, B.N., Takano, T., Tanaka, S., Itakura, K. and Dickerson, R.E. : 1982, Nature (London) 295, pp.294-299

Davies, D.B. : 1978, in Nuclear Magnetic Resonance Spectroscopy in Molecular Biology (Pullman, B. ed.) D. Reidel Pub. Co., Holland, pp.71-85.

Davies, D.B. : 1979a, Progress in NMR Spectroscopy, 12, pp.135-225

Davies, D.B. : 1979b, J. Chem. Soc. Perkin 2, pp.975-980

Davies, D.B. and Sadikot, H. : 1982, Org. Magn. Reson., 20, pp.180-183

Davies, D.B. : 1983, Biopolymers, in the press

Dickerson, R.E. and Drew, H.R. : 1981, J. Mol. Biol., 149, pp.761-786; Proc. Natl. Acad. Sci. U.S.A., 78, pp.7318-7322

Drew, H.R., Takano, T., Itakura, K. and Dickerson, R.E. : 1980, Nature (London) 286, pp.567-573

Ezra, F.S., Lee, C-H., Kondo, N.S., Danyluk, S.S. and Sarma, R.H. : 1977, Biochemistry, 16, pp.1977-1987.

Dhingra, M.M. and Saran, A.A. : 1982, Biopolymers, 21, pp. 859-872

Fratini, A.V., Kopka, M.L., Drew, H.R. and Dickerson, R.E. : 1982, J. Biol. Chem., 257, pp.14686-14707

Haasnoot, C.A.G., de Leeuw, F.A.A.M. and Altona, C. : 1980, Tetrahedron, 36, pp.2783-2792

Hruska, F.E. : 1973, in Conformations of Biological Molecules and Polymers (Bergmann, E.D. and Pullman, B. eds.) Pergamon Press, Oxford, pp.345-360

Klug, A., Jack, A., Viswamitra, M.A., Kennard, O., Shakked, Z. and Steitz, T.A. : 1979, J. Mol. Biol., 131, pp.669-680

Krugh, T. : 1981, Topics in Nucleic Acid Structure (Neidle, S. ed.) MacMillan Pub. Co., London, pp.197-217

Lee, C-H., Ezra, F.S., Kondo, N.S., Sarma, R.H. and Danyluk, S.S. : 1976, Biochemistry, 15, pp.3627-3638

de Leeuw, H.P.M., Haasnoot, C.A.G. and Altona, C. : 1980, Israel J. Chem., 20, pp.108-126

Neidle, S. : 1981, Topics in Nucleic Acid Structure (Neidle, S. ed.) MacMillan Pub. Co., London, pp. 177-196

Niemczura, W.P. and Hruska, F.E. : 1980, Can. J. Chem., 58, pp.472-478

Sarma, R.H. : 1979, in Nucleic Acid Geometry and Dynamics (Sarma, R.H.
    ed.) Pergamon Press, Oxford, pp.3.
Sarma, R.H. : 1981, in Topics in Nucleic Acid Structure (Neidle, S. ed.)
    MacMillan Pub. Co., London, pp.33-63
Shakked, Z., Rabinovich, D., Cruse, W.B.I., Egert, E., Kennard, O.,
    Sala, G., Salisbury, S.A. and Viswamitra, M.A. : 1981, Proc. Roy.
    Soc. ser. B., pp.479-487
Sundaralingam, M. : 1969, Biopolymers, 7, pp. 821-860
Sundaralingam, M. : 1973, in Conformations of Biological Molecules and
    Polymers (Bergmann, E.D. and Pullman, B. eds.) Pergamon Press,
    Oxford, pp.417-455
Sundaralingam, M. : 1975, in Structure and Conformation of Nucleic Acids
    and Protein- Nucleic Acid Interactions (Sundaralingam, M., and Rao,
    S.T. eds.) Univ. Park Press, Baltimore, pp.487-524
Sundaralingam, M. : 1982, in Conformations in Biology (Srinivasan, R.,
    and Sarma, R.H. eds.) Adenine Press, N.Y. pp.191-225
Viswamitra, M.A., Shakked, Z., Jones, P.G., Sheldrick, G.M., Salisbury,
    S.A and Kennard, O. : 1982, Biopolymers, 21, pp.513-533
Voet, D. : 1977, Nature (London) 269, pp.285-286
Wang, A.H-J., Quigley, G.J., Kolpak, F.J., Crawford, J.L., van Boom, J.H.,
    van der Marel, G. and Rich, A. : 1979, Nature (London) 282, pp.680-
    686

FOOTNOTES

1. Abbreviations and Symbols for the Description of Conformations
of Polynucleotide Chains : 1983, Eur. J. Biochem., 131, pp.9-15 :
this book, pp. 559-565.
2. The form of the Karplus relation used throughout this work is
$J = A\cos^2\theta - B\cos\theta + C$.
3. ^1H and ^{13}C n.m.r. measurements made on samples from Sigma Chem.
Co. : pu-pu (■d(ApA); pu-py (◥d(ApT); py-pu (◥d(TpA) and d(CpG)
a kind gift from S.A. Salisbury, Cambridge) and py-py(□d(TpT)).
4. Magnitudes of J_{PH3}, *ca* 6 Hz have been observed for some di-
deoxyribonucleoside phosphates in this work and quoted for some
molecules by Cheng and Sarma, 1977.

CONFORMATIONAL FLEXIBILITY OF NUCLEIC ACIDS

Thomas L. James, Peter Bendel, Jacqueline L. James, Joe W. Keepers, Peter A. Kollman, Aviva Lapidot, Joseph Murphy-Boesch, and John E. Taylor.
Department of Pharmaceutical Chemistry, University of California, San Francisco, CA 94143, U.S.A.

Abstract: In consideration of the importance of conformational flexibility to the functioning of nucleic acids, NMR studies have been carried out to elucidate aspects of dynamics with molecular mechanics calculations providing supplementary insight. Analysis of ^{31}P and ^{13}C NMR relaxation data in terms of plausible motions in a DNA helix indicate that winding and unwinding, base tilting and base pair propelling are not viable, but sugar repuckering and other limited bond rotations on the nanosecond time scale will account for the data. Molecular mechanics calculations imply that the internal motions are highly localized, not concerted along the length of the DNA helix. ^2H and ^{15}N NMR experiments on labeled nucleic acids reveal that the base moieties also experience some motions, limited relative to those of the backbone. ^{31}P and ^1H NMR studies comparing a closed duplex DNA, pIns36, with linear DNA reveal that segmental motion (apparently bending) occurs two orders of magnitude faster in the cdDNA. The imino proton resonances indicate that there are dynamically-averaged structural differences between linear DNA and supercoiled pIns36 in the A-T base pairs but not the G-C base pairs.

INTRODUCTION

In recent years, it has become evident that nucleic acids can exhibit a considerable degree of conformational flexibility. Most of the biological processes involving nucleic acids, including control processes, require recognition of unique sites on the nucleic acids, which may be short-lived. X-ray crystallography, theoretical calculations, and NMR studies have been primarily responsible for the new concept of double-stranded nucleic acids possessing a variety of conformations with a significant amount of motional freedom in the helix backbone. Conformational variants may exist for a long time or may exist transiently. Even those which exist for only a matter of nanoseconds may be functionally important structures. For example, a typical intercalating agent (drug or mutagen) can diffuse 5-10Å in a

155

B. Pullman and J. Jortner (eds.), Nucleic Acids: The Vectors of Life, 155–167.
© *1983 by D. Reidel Publishing Company.*

nanosecond.

We (1-6) and others (7-12) have been using NMR to provide information about dynamics in nucleic acids. NMR relaxation experiments
can reveal the frequencies of motion and, less definitively, the
amplitudes of the motion. However, NMR relaxation studies can only
obliquely indicate the nature of the motions; i.e., some plausible
motions do not effect a sufficiently large reorientation to account
for the relaxation, so one can eliminate certain items from a list of
feasible motions. It should be noted also that deuterium NMR line-
shape analysis has the potential for discriminating between diffusive
and jump motions if the motional frequency is in the correct range.
Theoretical calculations can permit examination of the limitations
imposed on any motions by structural and energetic considerations
(13). We have consequently used the constraints in frequency and
amplitude prescribed by analysis of our NMR relaxation results (5) to
examine the nature of the backbone motions in DNA via molecular mechanics calculations (14).

In this respect, we discuss some of our recent NMR results on
nucleic acids which serve to elucidate the backbone and base moieties'
motions. In particular, we will consider the extent of our knowledge
of backbone motions obtained from ^{13}P and ^{13}C NMR relaxation and
molecular mechanics calculations. The results from ^{2}H and ^{15}N NMR
yield some insight into base pair mobility. ^{31}P and ^{1}H NMR data
reveal some surprising differences between linear and supercoiled
DNA. Molecular mechanics calculations have been carried out to investigate some possible distinctions between linear and supercoiled DNA.

RELATIONSHIP OF NMR RELAXATION TO NUCLEIC ACID MOTIONS

NMR Relaxation

To characterize the motions as completely as possible, ideally
several NMR relaxation parameters should be measured as a function of
magnetic field strength for several nuclei on the nucleic acids. It
will be seen from the equations below that the various relaxation
parameters have a different functional dependence on molecular
motions. Of course, several nuclei will reflect the motions at their
various locations on the nucleic acid and should, as a whole, engender
a consistent picture of the molecular motions.

The spin-lattice relaxation time T_1, the spin-spin relaxation
time T_2, the nuclear Overhauser effect NOE, and the rotating frame
spin-lattice relaxation time in the presence of an off-resonance rf
field $T_{1\rho}^{off}$ are given in terms of the spectral densities $J_n(\omega)$ for
dipolar coupling to one hydrogen by (1,6):

$$\frac{1}{T_1} = K[J_0(\omega_H - \omega_I) + 3 J_1(\omega_I) + 6 J_2(\omega_H + \omega_I)] \tag{1}$$

$$\frac{1}{T_2} = \frac{1}{2T_1} + K [2 J_o(0) + 3 J_1(\omega_H)] \qquad (2)$$

$$NOE = 1 + \frac{\gamma_H}{\gamma_I} \frac{[6 J_2(\omega_H + \omega_I) - J_o(\omega_H - \omega_I)]}{[J_o(\omega_H - \omega_I) + 3 J_1(\omega_I) + 6 J_2(\omega_H + \omega_I)]} \qquad (3)$$

$$\frac{1}{T_{1\rho}^{off}} = K[\sin^2\theta_e \{2 J_o(\omega_e)\}] + \frac{1}{T_1} \qquad (4)$$

$$K = \frac{h^2 \gamma_I^2 \gamma_H^2}{20 r^6} \qquad (5)$$

where the subscript I is for the relaxing nucleus and subscript H is for hydrogen, γ_I and γ_H are the gyromagnetic ratios, ω_I and ω_H are the angular Larmor frequencies, and r is the internuclear distance. Similar expressions can be written for other relaxation mechanisms.

The spectral density terms $J_n(\omega)$ contain the motional information. Although it is possible to interpret NMR relaxation in terms of a model-free effective correlation time, which is a measure of the rate of motion, and a generalized order parameter (15), we have chosen to consider various potential motions of double-stranded nucleic acids within the framework of present knowledge of the structural features of double helical DNA. Having chosen a particular motion, an appropriate mathematical model representing the spectral densities can be constructed. In fact, fluctuations in internuclear distance, as well as orientation, with molecular motion can be explicitly taken into account (5).

Possible Molecular Motions in the Double Helix

The types of motions in double helical nucleic acids is not unlimited; the molecular geometry places constraints on the motions which may occur. Figure 1 shows many of the plausible motions for a double helix. The so-called "breathing" motion occurs in the time range of milliseconds to seconds and is too slow to affect relaxation of nuclei other than some protons. The end-for-end tumbling and "speedometer cable" motion of a long "worm-like" coil become, respectively, rotation around the short axis and around the long axis if the nucleic acid is a short, rigid rod. Two other possible segmental motions not depicted in Fig. 1 are in distinctly different time regimes. At the slow end (greater than milliseconds for sufficiently large amplitudes) are the Rouse-Zimm normal coordinate modes (16); these will be too slow to affect the relaxation. Librational motions, generally accorded to be on the order of 10^{11}–10^{12} sec^{-1} due to thermally induced displacements of groups of atoms, could influence NMR relaxation.

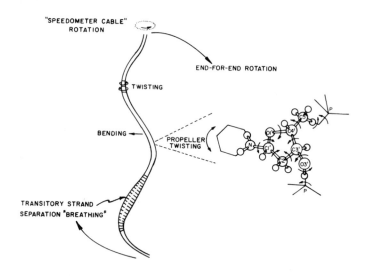

Figure 1. Plausible motions in a double helical nucleic acid.

MOLECULAR MOTIONS IN NUCLEIC ACIDS

Consideration of molecular motions will be made using both NMR data and molecular mechanics calculations. We will consider first data from backbone nuclei, then data from base nuclei. An emphasis will be placed on internal motions in this discussion, since our previous studies have shown that relaxation of backbone nuclei in nucleic acids of length greater than a couple of persistence lengths is most consistent with an overall isotropic bending motion on the order of microseconds (1,2,6).

Motions of the Backbone

We have previously examined the ^{31}P and ^{13}C NMR relaxation of DNA and RNA, both single- and double-stranded (1-3,5,6). Several of the feasible molecular motions illustrated in Fig. 1 have been tested to see if they could account for the observed NMR relaxation. The ^{31}P relaxation had contributions from chemical shift anisotropy as well as dipolar interactions; fluctuations in the proton–phosphorus internuclear distances were explicitly incorporated into the spectral densities. For each type of motion, a large variety of amplitudes and frequencies were examined. The results of the calculations are summarized in the following Table 1.

The motional models employed in the calculations presume that the six polynucleotide backbone torsion angles of each nucleotide unit are

Table 1. Plausible Motional Models for ^{13}C and ^{31}P Relaxation in the DNA Backbone

Internal or segmental motion model	Ribose ^{13}C, ^{31}P data fit?	1/jump rate (ns)
1. Twisting (jumps between 8 bp/turn and 11.3 bp/turn)	no	
2. Base tilting (jumps between $-10°$ and $+17°$)	no	
3. Base propeller twisting (jumps between $0°$ and $6°$)	no	
4. Combination base twist and tilt	no	
5. Sugar puckering ($\pm35°$ out of mean plane)	all ^{13}C except C5'	5
6. Sugar puckering ($\pm35°$) plus rotation about C4'-C5' bond (±15-$25°$)	all ^{13}C	5 (0.4-3)
7. Sugar puckering ($\pm35°$ plus rotation about C3'-03' ($\pm25°$)	all ^{13}C except C5' and ^{31}P	5 (1.5-2.5)

independent of any other nucleotide units in the polymer. The twisting model of Table 1 could not come close to fitting the experimental data since the amplitudes of reorientation of internuclear vectors and chemical shift anisotropy tensors were too small. However, if the twisting motions were concerted over the entire DNA, the amplitudes could conceivably become sufficiently large to contribute to NMR relaxation. A question is therefore raised: are the motions detected by NMR relaxation independent or concerted? The NMR experiments do not address this question, but molecular mechanics calculations can.

Olson (17) recently suggested that 1,3-crankshaft type motions of alternate bonds can occur with little dislocation of bases. We have carried out molecular mechanics calculations on [d(C-G-C-G-A-A-T-T-C-G-C-G)]$_2$ and [d(A)]$_{12}$·[d[T]]$_{12}$ (14). Each of the backbone torsion angles in a nucleotide unit was "forced" to values other than those in the normal B-DNA conformation and the remaining torsion angles were permitted to relax the double helix to its lowest energy. It was found that compensating torsion angle changes maintain the base stacking energy in the double helix. The compensations were highly localized with 1,3-crankshaft motions prominant, but with further sugar repuckering in the same nucleotide unit occurring as well. Base dislocations were negligible, and the effects were not propagated either along the polynucleotide chain or to the other nucleotide in the base pair. These results imply that from the viewpoint of molecular energetics, there is no incentive for concerted motions to occur along the length of a DNA double helix; rather any internal motions are highly localized.

We note that the calculated barrier of 3-5 kcal/mol between gauche and trans conformations in B-DNA (14) is consistent with the internal motions occurring on a nanosecond time scale.

Motions of the Bases

Our earlier NMR experiments indicated that the motions in double-stranded DNA and RNA were largely the same (1-3). The ribonucleic acid poly(I) was labeled with deuterium at the purine C8 position by exchange with deuterium oxide. Deuterium NMR experiments were carried out on the single-stranded species and double-stranded poly(I)·poly(C) for dry powders, hydrated powders and solutions. Representative spectra are shown in Fig. 2. The quadrupole splitting of 128-130 KHz observed in the dry fibers for both single- and double-stranded species implies no detectable motional averaging in the lineshapes. However, measured deuterium T_1 values of 320 ms and 630 ms for the single-stranded and double-stranded species, respectively, are inconsistent with completely rigid bases and suggest that low amplitude ($\lesssim 5^\circ$) anisotropic motions, which would not be evident in the lineshape, could occur even in the dry fiber.

As water is added to the samples, it is apparent in Fig. 2 that changes occur. For single-stranded poly(I), a signal from the hydrated fiber is not detectable in 48 hours of signal averaging (a small signal at 6% intensity due to some dry fibers is observed) but an isotropic, Lorentzian line is obtained in a few minutes in solution. For poly(I)·poly(C), a signal with full quadrupolar splitting is observed in the hydrated fiber but no signal could be detected from the solution. Woessner et al. (18) showed that T_{2e}, the time constant for the echo amplitude decay in a quadrupolar echo experiment, decreases substantially if an isotropic motion comparable to the quadrupole coupling constant occurs. In our case, a minimum value of $T_{2e} = 8.9\mu s$ can be calculated at a correlation time of 1.8 μs. In fact, values of T_{2e} will be less than our instrument receiver dead time for correlation times between 0.2 μs and 200 μs. The spectra of Fig. 2 indicate that the single-stranded poly(I) first passes through this regime before poly(I)·poly(C) as water is added to the dry fiber; indeed poly(I)·poly(C) does not emerge from that regime when the polymer is in solution, unlike poly(I).

In the first ^{15}N NMR study of DNA (4), the spectra suggested that the base moiety experienced internal motions with some restriction in amplitude or frequency compared with backbone motions. We have followed up those initial experiments with measurements of the ^{15}N NMR relaxation parameters at several magnetic field strengths. As seen in Fig. 3, nearly all the nitrogen resonances from the uniformly ^{15}N-labeled DNA bases are resolved. This provides us with a large data base for examining motions in the base moiety. Table 2 lists the experimental ^{15}N NMR relaxation parameters for DNA.

As in the case of backbone nuclei mentioned above, we have consi-

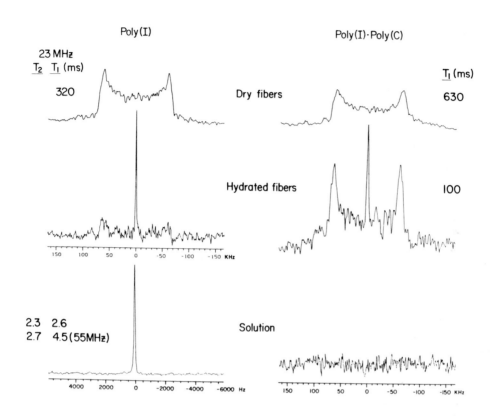

Figure 2. 23 MHz deuterium NMR spectra of C-8 labeled poly(I) and poly(I)·poly(C) at 23°C. The fiber spectra were obtained using the quadrupolar echo technique. The relatively narrow isotropic line in the spectra from the hydrated fibers is due to residual deuterons in the deuterium-depleted water. The water line in the solution spectrum of the duplex, which appears after hours of signal averaging, was removed to provide easy visual comparison with the isotropic line in the solution spectrum of poly(I) which was obtained in 10 minutes of signal averaging.

dered various possible molecular motions. The results of the calculations performed thus far for all adenine nuclei and the amino groups of guanine and cytosine are summarized in Table 3. Although we do not know the details of the base motions, it is evident that only relatively large amplitude base motions occurring about an order of magnitude slower than backbone motions will account for the data.

Table 2. ^{15}N-DNA† Relaxation Data

Peak Identification	T_1(sec)			Linewidth(Hz)[a]			NOE[b]		
	2.35T	5.64T	11.75T	2.35T	5.64T	11.75T	2.35T	5.64T	11.75T
G-NH$_2$	.06*	-	.8*	31	*	*	.19	.51	.50
A-NH$_2$	.05*	-	.7*	42	*	*	.31	.55	.60
C-NH$_2$	.04	-	.7	31	72	115	.10	.51	.60
TN1	*	1.75*	*	24*	*	*	.41	*	.81
GN1	*	.03*	*	39*	*	*	.0	*	.61
CN1	*	1.21*	*	25*	*	*	.26	*	.93
TN3	*	.50*	*	18*	*	*	.45*	*	.40*
AN9	1.15*	2.13*	3.2*	22*	88*	190*	.39*	.71*	.93*
GN9	1.15*	2.13*	3.2*	22*	88*	190*	.39*	.71*	.93*
CN3	.9	1.37	.90	25	72	147	.54	.75	.82
AN3	.6	1.22	1.9	16	145	168	.94	.72	.93
AN1	.8	1.14	1.1	16	117	168	.49	.85	.87
AN7	.6*	.71	.7*	25	*	*	.73	.71	.99*
GN7	.6*	.82	.7*	25	*	*	.57	.71	.99*

* = overlapping peaks

†E. coli DNA 200 +/- 50 base pairs (27 mg/ml); 40 mM sodium cacodylate, pH 7.2, 40 mM NaCl, 0.1 mM EGTA 20° ±/-0.5°C; T_1 and linewidth measurements made without NOE.

a) corrected for applied linebroadening

b) average of 3 determinations using peak heights

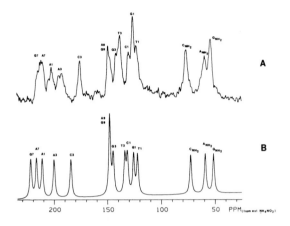

Figure 3. (A) ^{15}N NMR spectrum (24.4 MHz) of ^{15}N-labeled DNA (27 mg/ml) in 40 mM sodium cacodylate, pH 7.2 and 40 mM NaCl at 20°C. Spectral parameters: 10,000 transients, 70° pulses, 0.65 sec acquisition time, 3.0 sec repetition time, 10 Hz exponential line broadening, broadband proton decoupling on during acquisition. (B) ^{15}N spectrum of DNA simulated from chemical shifts of individual nucleotides.

COMPARISON OF CLOSED, DUPLEX DNA AND LINEAR DNA

A significant amount of naturally-occurring genetic material is in the form of covalently closed, or superhelical DNA, which exhibits some unique properties of chemical and biological reactivity (19). We recently obtained results of the first NMR experiments on intact plasmid DNA in closed duplex, supercoiled form (cdDNA) (6). The species studied was pIns36 (7200 bp, 5 megadaltons), which has the human insulin gene cloned into plasmid pBR322. ^{31}P NMR relaxation parameter measurements indicated that the internal motions in the phosphodiester moiety are about the same for linear, nicked circular, and supercoiled forms of pIns36. However, the slower motion which dominates T_2 relaxation was found to be faster in circular DNA and faster still in the supercoiled form, being about two orders of magnitude faster than for the linear form. All available evidence indicates that the motion governing T_2 processes in linear DNA is bending on the microsecond timescale (1,2,6). It was suggested that the increased that the increased effective frequency of these bending motions in cdDNA results mainly from coupling to higher frequency torsional motion and from an excess of conformational free energy. Fluorescence depolarization experiments on intercalated ethidium in DNA indicate a timescale of about 10 ns for torsional motions (20,21). As mentioned above, twisting motions _per se_ cannot account

Table 3. Plausible Models for ^{15}N Relaxation in DNA Bases

Internal Motion Model	^{15}N data fit?	1/jump rate(ns)
Isotropic Overall (Spherical Reorientation)	no	
(a) Free diffusion	no	
(b) Restricted diffusion	no	
(c) Jump	no	
Anisotropic Overall (Cylindrical Reorientation)		
(a) Free diffusion	no	
(b) Restricted diffusion	no	
(c) Jump ($> \pm 35°$)	yes	10-70
Lattice Model		
(a) Base twist	no	
(b) Base tilt	yes	10-70
(c) Base propeller twist	yes	10-70
Wobble	no	
Segmental (Wobble) + Internal Jumps	no	

for ^{31}P relaxation; but if the twisting motions drive bending motions at a faster rate, faster T_2 relaxation can be effected.

The large T_2 value observed for ^{31}P in cdDNA suggested that proton NMR spectra might be obtained with relatively high resolution in spite of the high molecular weight of pIns36. Figure 4 confirms that hypothesis for the imino proton resonances.

As seen in Fig. 4, the linewidths of the imino proton resonances of pIns36 (Fig. 4c-e) are narrower than those of linear DNA ~ 20 times shorter (Fig. 4a). The width of the resonances from linear DNA with molecular weight comparable to pIns36 are broadened nearly to oblivion (Fig. 4b). It can be seen in comparing (Figs. 4c-3) that changing the temperature or magnetic field strength has little influence on the linewidth of the cdDNA.

On the basis of previous studies (9,22), the band centered at 12.6 ppm in the supercoiled DNA spectrum can be assigned to N1 protons of the G-C base pairs and the signals at 13.2 ppm, 14.0 ppm and 14.3 ppm to N3 protons of the A-T base pairs. In the 200-700 bp linear DNA (Fig. 4a), these resonances are located at 12.7 ppm and 13.7 ppm,

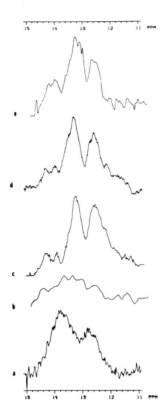

Figure 4. Proton NMR spectra of (a) 200-700 bp calf-thymus DNA, 240 MHz, 24°C, (b) 0.6-10 Kbp calf-thymus DNA, 240 MHz, 24°C, (c) super-coiled pIns36, 240 MHz, 24°C, (d) supercoiled pIns36, 240 MHz, 11°C, and (e) supercoiled pIns36, 360 MHz, 24°C. The spectra were obtained using modified Redfield pulse sequences.

respectively. The G-C imino resonance is little changed in its position implying negligible structural variation in these base pairs. However, the A-T resonances suggest some structural differences (or perhaps different dynamical averages) between linear and supercoiled DNA entailing the A-T base pairs. We note that the resonances at 14.0 ppm and 14.3 ppm are downfield too far to be attributed to any G-C base pairs of known structure (22,23). Gel electrophoresis, run after obtaining spectra, revealed no linear DNA and a small amount (< 10%) of nicked, circular DNA in the sample. It is suggested that the small peak at 14.0 ppm may be due to the relaxed circular form.

The imino proton spectra clearly show that there is more than one A-T environment in the cdDNA and, on average, the A-T environment differs from that in linear DNA.

Our previous molecular mechanics calculations on [d(C-G-C-G-A-A-T-T-C-G-C-G)]$_2$ and [d(A)]$_{12}$·[d(T)]$_{12}$ indicated that some A-T base pairs, but no G-C base pairs, were trapped in structures about 4 kcal/mol higher in energy than the starting B-DNA structure when base pairs were opened (14); these structures entailed a "base-opened" state with non-Watson-Crick hydrogen bonds formed. Tritium exchange experiments reveal that base opening occurs more readily in supercoiled DNA than in linear DNA (24). Since supercoiled DNA is underwound relative to linear DNA, we have calculated the consequences of opening the A5.T20 base pair in [d[C-G-C-G-A-A-T-T-C-G-C-G)]$_2$ using energy-refined starting structures forced to have 9, 10, 11, or 12 bp/turn. In each case, the duplex relaxed to a structure different from the starting Watson-Crick structure, one in which a hydrogen bond between the thymine carbonyl and adenine amino group was formed. The energy differences (kcal/mol) between the Watson-Crick base-closed structures and non-Watson-Crick "base-opened" structures are 0.0, 1.6, 6.2, and -3.0 for 9, 10, 11, and 12 bp/turn, respectively. Clearly, these results imply that the formation of such "base-opened" states is more favorable in underwound DNA such as cdDNA than in linear DNA with 10-10.5 bp/turn. For the titratable superhelical density of pIns36 (6), we can estimate that on average the helix is underwound by 1.3 bp/turn if all of the "supercoiling" is attributed to twist; the helix is at least underwound by 0.8 bp/turn if the "supercoiling is partitioned to twisting and writhing according to Benham (25).

Using the energy-refined [d(C-G-C-G-A-A-T-T-C-G-C-G)]$_2$ structure, the chemical shift contribution of the hydrogen bond protons from ring current and atomic magnetic anisotropy effects (23) have been calcula-ted. Compared to the normal B-DNA structure, the chemical shifts for the "base-open" state were about 1.0 ppm downfield for the hydrogen bond proton in the base-opened A-T pair and about 0.5 ppm downfield for the imino proton of the adjacent base pair. Although we do not have definitive experimental evidence for the existence of such "base-opened" A-T states, they would be formed more readily in supercoiled DNA and could give rise to the downfield peak at 14.3 ppm in the cdDNA spectra of Fig. 4.

ACKNOWLEDGMENT

Financial support for this research was provided by research grants GM25018 and CA27343 and Research Career Development Award AM00291 from the National Institutes of Health.

REFERENCES

1. Bolton, P.H. and James, T.L.: 1979, J. Phys. Chem. 83, pp. 3359-3366.

2. Bolton, P.H. and James, T.L.: 1980, J. Am. Chem. Soc. 102, pp. 25-31.
3. Bolton, P.H. and James, T.L.: 1980, Biochemistry 19, pp. 1388-1392.
4. James, T.L., James, J.L., and Lapidot, A.: 1981, J. Am. Chem. Soc. 103, pp. 6748-6750.
5. Keepers, J.W. and James, T.L.: 1982, J. Am. Chem. Soc. 104, pp. 929-939.
6. Bendel, P., Laub, O., and James, T.L.: 1982, J. Am. Chem. Soc. 104, pp. 6748-6754.
7. Klevan, L., Armitage, I.M., and Crothers, D.M.: 1979, Nucl. Acids Res. 6, pp. 1607-1616.
8. Hogan, M.E. and Jardetzky, O.: 1980, Biochemistry 19, pp. 3460-3468.
9. Early, T.A. and Kearns, D.R.: 1979, Proc. Natl. Acad. Sci. U.S.A. 76, pp. 4165-4169.
10. Shindo, H.: 1980, Biopolymers 19, pp. 509-522.
11. Rill, R.L., Hilliard, P.R., Jr., Bailey, J.T., and Levy, G.C.: 1980, J. Am. Chem. Soc. 102, pp. 418-420.
12. Opella, S.J., Wise, W.B., and DiVerdi, J.A.: 1981, Biochemistry 20, pp. 284-290.
13. Olson, W.K.: 1979, Biopolymers 18, pp. 1235-1260.
14. Keepers, J.W., Kollman, P.A., Weiner, P.K., and James, T.L.: 1982, Proc. Natl. Acad. Sci. U.S.A. 79, pp. 5537-5541.
15. Lipari, G. and Szabo, A.: 1982, J. Am. Chem. Soc. 104, pp. 4546-4559.
16. Berne, B.J. and Pecora, R.: 1976, "Dynamic Light Scattering", Wiley, New York.
17. Kollman, P.A., Weiner, P.K., and Dearing, A.: 1981, Biopolymers 20, pp. 2583-2681.
18. Woessner, D.E., Snowden Jr., B.S., and Meyer, G.H.: 1969, J. Chem. Phys. 51, pp. 2968-2976.
19. Bauer, W.R.: 1978, Annu. Rev. Biophys. Bioeng. 7, pp. 287-313.
20. Millar, D.P., Robbins, R.J., and Zewail, A.H.: 1980, Proc. Natl. Acad. Sci. U.S.A. 77, pp. 5593-5597.
21. Barkley, M.D. and Zimm, B.H.: 1979, J. Chem. Phys. 70, pp. 2991-3007.
22. Kearns, D.R.: 1977, Annu. Rev. Biophys. Bioeng. 6, pp. 477-523.
23. Arter, D.B. and Schmidt, P.G.: 1976, Nucl. Acids Res. 3, pp. 1437-1447.
24. Jacob, R.J., Leibovitz, J., and Printz, M.P.: 1974, Nucl. Acids Res. 1, pp. 549-558.
25. Benham, C.J.: 1979, Proc. Natl. Acad. Sci. U.S.A. 76, pp. 3870-3874.

^{31}P NMR STUDIES ON THE STRUCTURE AND DYNAMICS OF DNA IN HYDRATED FIBERS

H. Shindo, U. Matsumoto, H. Akutsu* and T. Fujiwara*

Tokyo College of Pharmacy, Hachioji, Tokyo and *Osaka University, Yamadaoka, Suita, Osaka, Japan

Solid state ^{31}P n.m.r. was applied to elucidate the structure and dynamics of DNA in the solid state. In the analyses of the line shapes of ^{31}P n.m.r. spectra and the relaxation parameters, it was demonstrated that A-DNA has a single backbone conformation but that B-DNA is considerably heterogeneous in its backbone probably depending on base sequences, and furthermore that the motion of DNA in hydrated fibers can be descrived in terms of rapid internal motion and rotational and wobbling motions.

1. INTRODUCTION

There is a considerable current interest in the nature of the structure and dynamics of DNA. Among many methods, especially the x-ray diffraction method has been employed to elucidate the molecular structure of highly oriented DNA fibers. However, the lack of crystallinity of natural DNA and molecular motions at high relative humidities almost always prevent this technique from resolving the structure at the atomic level, although it is the only technique to deduce the overall picture of DNA. In favorable cases other methods can be particularly useful for individual problems. We have previously demonstrated by ^{31}P n.m.r. that natural DNA is considerably heterogeneous in its backbone structure [1]; this feature was not detected by the x-ray fiber diffraction method. This finding was supported by recent x-ray diffraction studies of single crystals of synthetic oligonucleic acid duplexes.[2-4] We will describe here a brief review on this structural aspect deduced by ^{31}P n.m.r. spectroscopy.

The n.m.r. method is an extremely useful means to yield information on molecular motions in the time scale from a few tenths of second to 10^{-10} sec. In the past five years DNA in solution have been subjected and is still subjected to extensive studies of ^{1}H, ^{13}C and ^{31}P n.m.r. and most of the results have been essentially consistent with each other in the respect that a rapid internal motion occurs in DNA in solution.[5-9] Notably, the motion was found to be very fast in the order of a few

B. Pullman and J. Jortner (eds.), Nucleic Acids: The Vectors of Life, 169–182.
© *1983 by D. Reidel Publishing Company.*

nanosec and to have a considerably large amplitude.[5-8] By examining n.m.r.
results of DNA's with different base sequences, we have shown the direct
evidence that their structural nature is very similar between DNA in solution and highly hydrated solid DNA.[10] It is interesting, therefore, to
see if the same is true of the molecular dynamics of DNA's in the hydrated fibers. We will describe a preliminary analysis of the anisotropic
[31]P relaxation times, T_1 and T_2, of DNA in the solid state and will report the indication that the internal motion of DNA in hydrated fibers
is very similar to that in solution. We will also demonstrate that the
anisotropy effects on relaxation times can promise additional information
on molecular dynamics.

2.   EXPERIMENTAL PROCEDURES

Salmon sperm DNA was purchased from PL Biochemicals and was purified as
before.[1] The purified DNA was disolved into 10 mM NaCl aqueous solution
and followed by ultracentrifugation for 15 hrs at 45,000 r.p.m. A drop
of the viscous DNA pelett was put between the edges of tooth picks several milimeters apart and dried in the atomosphere. The quality of molecular orientation in the fibers was judged by polarizing microscopy and
the x-ray diffraction method. For n.m.r. measurements several fibers
were aligned parallel with an aid of a teflon holder with many drilled
holes. Relative humidity was controlled by using saturated aqueous solutions of several kinds of salts.
     [31]P n.m.r. measurements were performed under cross-polarization and
[31]P-^{1}H dipolar decoupling conditions at observing frequencies of 24.3
and 40.1 MHz for ^{31}P nuclei. The relaxation times, T_1 and T_2, were measured by the cross-polarization inversion recovery[11] and cross-polarized
spin echo methods, respectively.

3.   PHOSPHODIESTER CONFORMATION OF DNA

We have previously shown how to determine the orientation of the phosphodiester in DNA by solid state [31]P n.m.r. spectroscopy.[1] The method is
based on the knowledge of the nature of ^{31}P chemical shielding anisotropy
i.e., its principal values and the principal axis system relative to the
phosphodiester atomic group. Due to the tensor characteristics of the
shielding anisotropy the spectral patterns obtained from highly oriented
DNA fibers display a strong angular dependence as is shown in Fig. 1.
The spectra also demonstrate a conformational transition induced by humidity (or hydration of the fibers), which is no doubt attributed to the
transition from the A form to the B form of DNA.

     A-DNA. As for the spectra of the A form of DNA at humidities below
87%, the parallel spectrum exhibits a singlet and the perpendicular displays a trimominal pattern. Such a singlet pattern provides evidence
that the orientation of the chemical shielding tensor relative to the
fiber axis is identical for all the phosphodiesters within the DNA fibers,
namely, the A form of DNA has a single backbone conformation.

Using the three observed chemical shift values, one from the parallel and the other two from the perpendicular spectrum, the orientation

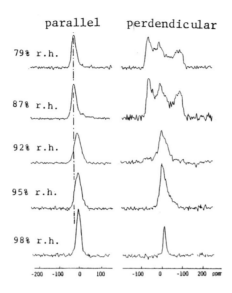

Fig. 1. Solid state ^{31}P n.m.r. spectra (40.1 MHz) of oriented
DNA fibers at various relative humidties.

of the tensor and hence phosphodiester was determined for various DNA's
[12-14] and summarized in Table 1. As can be seen in this Table, the direction cosine angles obtained for different samples are nearly identical
with each other within errors, suggesting that the A form of DNA has a
single backbone conformation independent of its base sequences.

Table 1. Orientation of the shielding tensor of the A-DNA[a]

	θ_1	θ_2	θ_3	ref.
Exptl.				
poly(dAdT)·poly(dAdT)	55	138	110	12
salmon sperm DNA	61	144	109	13
calf thymus DNA	64	149	105	14
Calcd.				
A-DNA model[b]	79	163	103	21

[a]Angles θ_1, θ_2 and θ_3 in degree describing the tensor orientation relative to the helical axis.
[b]Calculated by assuming the orientation of the tensor having
a symmetry relative to the atomic coordinates of the phosphodiester.

B-DNA. The ^{31}P spectra of the B form of DNA are distinguishable
from those of the A form of DNA in two respects, the peak position of

the parallel spectra and the patterns of the perpendicular spectrum (Fig.
1). Notably, the parallel spectrum for the B-DNA fibers at 92% relative
humidity is broader than that for the A-DNA fibers at 87% at which A-DNA
is considered to be virtually rigid. Such a broadening, in spite of increased motions, implies the presence of conformational dispersion of the
phosphodiester backbone, namely, the B form of DNA is considerably heterogeneous in its backbone conformation, whereas A-DNA has a single conformation. This is the first evidence that the conformation of B-DNA is
heterogeneous.

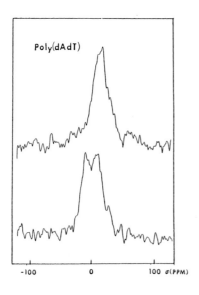

Fig. 2. Solid state [31]P n.m.r. spectra
(40.1 MHz) of the B form
poly(dAdT)·poly(dAdT) fibers at
98 % r.h.

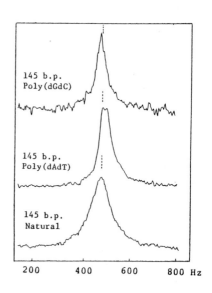

Fig. 3. Solution [31]P n.m.r. spectra
(109.3 MHz) of DNA's with
145 b.p.

Figure 2 shows the ^{31}P spectra of the B form of synthetic polynu-
cleic acid, poly(dAdT)·poly(dAdT) fibers.[10] The doublet pattern of the
parallel spectrum clearly indicateds that the B-form of this polymer has
at least two distinct conformations; this stems from an alternating base
sequence of this polymer. The above results manifest that the conformation of B-DNA varys depending upon the base sequences.
    Figure 3 shows solution ^{31}P spectra (109.3 MHz) obtained from poly(
dGdC)·poly(dGdC), poly(dAdT)·poly(dAdT) and natural DNA fragments.[15] All
the DNA fragments were prepared so as to have the same length of 145 base
pairs. Therefore, it is reasonable to consider that individual line
widths would be the same with each other if the B form of DNA were a single conformation. However, the spectra shown in Fig. 3 indicate that
this is not the case. The fragment, poly(dGdC), exhibits a sharp, single
line, and the fragment, poly(dAdT), yields a doublet just as found for
the B form of this polymer in the fiber and the two distinct conformations

may be attributed to the doublet. Remarkably, natural DNA exhibits a
very broad resonance. Such an extra broadening, therefore, indicates
heterogeneity in the backbone conformation of DNA.

4. ANISOTROPIC MOLECULAR MOTION IN HYDRATED DNA FIBERS

Rapid internal motions with a large amplitude in relatively short DNA
fragments in solution have been detected by solution n.m.r. spectroscopy,
[5-8] and they were often attributed to the puckering motions of the deoxyribose rings in DNA.[6,8] On the contrary, the analysis of the ^{31}P line
widths of high molecular weight DNA suggested negligible or almost no internal motion.[16] This discrepancy among the investigations may arise
because the rapid motions are significantly damped in the middle of a
long DNA chain compared to that of short DNA fragments. It is, therefore, interesting to investigate that molecular dynamics of DNA in highly
condenced and hydrated conditions.

Solid state n.m.r. provides a favorable opportunity for such studies
on the anisotropy effects which can not be directly detected in solution.
The overall motion of DNA in hydrated fibers may be neglected; only internal or local motions can be considered since DNA molecules are aligned
along the fiber axis. We present here the results from solid state ^{31}P
n.m.r. of DNA fibers at high relative humidity that clearly indicate the
rapid internal motions occurring in the DNA fibers.

4.1. Anisotropy effects observed for the B form of DNA fibers.

^{31}P n.m.r. line shape and intensity. In Fig. 4, the ^{31}P spectra of
B-DNA fibers at 98% relative humidity are recorded under the same measurement conditions, and they bear a strong dependence on the orientation of
the fibers relative to the magnetic field. Both the parallel and perpendicular spectra are singlets, but differ from each other in the line
width and intensity. Remarkably, the spectrum for 45° is very broad.

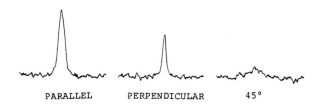

PARALLEL   PERPENDICULAR   45°

Fig. 4. Solid state ^{31}P n.m.r. spectra (40.1 MHz) of the B form
of DNA at different fiber orientations.

As mentioned in the previous section, upon increasing the relative humidity, a change in the perpendicular spectrum of the DNA fibers is most
likely due to the occurrence of the rotational motion about the helical

axis. But, theoretical consideration shows that a singlet pattern should
be observed at any direction of the fibers relative to the magnetic field
when a rapid rotational motion about the helical axis takes place; this
is inconsistent with the spectrum at $45°$ (Fig. 4). Thus, there must be
present another motion which is relatively slow compared to the n.m.r.
time scale of 10^{-4} sec. Detail line shape analysis of ^{31}P spectra of
the B-DNA will be reported elsewhere.

Figure 5 shows the dependence of the integral amplitude of the ^{31}P
signals for the parallel and perpendicular orientations as a function of
the cross-polarization contanct time under Hartmann-Hahn conditions in a
single contact experiment.[17] The initial rise on the left side of each
plot is due to the growth of ^{31}P magnetization induced by the cross-polarization by protons, and the latter decline is a manifestation of relaxation of the spin locked ^{1}H magnetization. The former rate is characterized mainly by the cross relaxation time, T_{IS}, and the latter by the
relaxation time of the protons in the rotationg frame, $T_{1}\rho$. It is clear

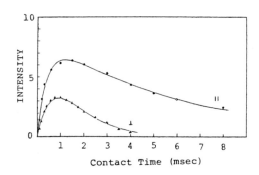

Fig. 5. The integral amplitude of the ^{31}P signal vs. cross-polarization contact time; ● for the parallel and ▲ for the
perpendicular.

from Fig. 5 that the efficiency of the cross polarization for the parallel spectrum is about twice higher than that for the perpendicular spectrum; this is due to the fact that the parallel fibers have shorter T_{IS}
and longer $T_{1}\rho$. Because the more motional averaging of dipolar interactions results in the less efficiency of the cross polarization, the
experimental results indicate the occurrence of more motional averaging
in the direction normal to the fiber axis.

Anisotropic relaxation times, T_1 and T_2. Relaxation time T_1 was
measured by the modified inversion recovery method, cross polarization-
$90°-\tau-90°$ pulse sequence,[11] and relaxation time T_2 was measured by the
spin echo method, corss polarized-$\tau-180°-\tau$ pulse sequence. The plots of
the signal intensity vs. pulse interval,τ, gave a straight line and the
results are summarized in Table 2. Both T_1 and T_2 values for the parallel spectrum are shorter by factors of 1.5 to 2 than those for the perpendicular spectrum; clearly, relaxation times also exhibit a strong anisotropy.

All the ^{31}P n.m.r. parameters observed for the B-DNA fibers revealed
the effects of anisotropy probably due to anisotrpic motions of DNA in
the fiber. We present here a preliminary analysis of relaxation times,
T_1 and T_2.

Table 2. Observed relaxation parameters of ^{31}P n.m.r.
of DNA in hydrated fibers (40.1 MHz)

	parallel	perpendicular
T_1	1.17 (sec)	2.10 (sec)
T_2	1.78 (msec)	2.61 (msec)
T_{IS}	0.39 (msec)	0.70 (msec)
$T_1\rho$	6.38 (msec)	1.25 (msec)

4.2. Motional modes of DNA in hydrated fibers

In order to interpret our ^{31}P n.m.r. data, molecular motions in the B-DNA
fibers may be inferred from the results obtained for DNA in solution.
Figure 6 compares plausible models for the motion of DNA in solution and
that in hydrated fibers. For DNA in solution, the overall motion has
often been treated as being rotation about the long axis and end-to-end
motion about the short axis, and a rapid internal motion has been approximated as being two-site jump. For DNA in hydrated fibers, however,
the space available for the endo-to-end motion may be significantly restricted and thus the motion is better described by wobbling as is shown in
Fig. 6B. Thus, the molecular motion of DNA in oriented fibers may be

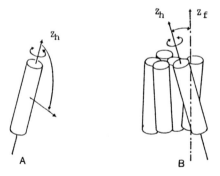

Fig. 6. Schematic drawing of motional modes in solution (A)
and in the oriented fibers (B).

described by two-site jump, rotational reorientation about the helical
axis and wobbling motion.

4.3. Calculations of ^{31}P relaxation parameters

In order to estimate ^{31}P relaxation parameters, T_1 and T_2, spectral density $J_m(\omega)$ must be found for the possible motional modes and it is given
by Fourier transformation of the autocorrelation function $C_m(t)$ as

follows,

$$J_m(\omega) = 2\int_0^\infty C_m(t)\cos\omega t \; dt \qquad (1)$$

Because of very complex and extremely laborious calculations of the auto-correlation functions for the wobbling motion together with the other two motions, we will derive an explicit autocorrelation function for the two-site jump and rotation, and then, if necessary, we will introduce the wobbling motion into it by approximations.

Autocorrelation functions for two-site jump and rotation. Figure 7 shows the schematic representation of the coordinate systems of $^{31}P-^1H$ dipolar and ^{31}P chemical shielding tensors. Z_{p1} and Z_{p2} represent the principal axes of a given tensor at the sites 1 and 2, Z_m refers to the molecular frame in accord with the bisector of the above two principal axes, Z_h is taken along the helical axis of DNA, and Z_L is the laboratory frame (i.e., the direction of the magnetic field). We do not distingui-

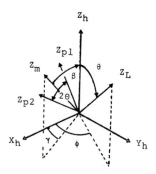

Fig. 7.   Schematic representation of the coordinate systems; Z_L, the laboratory frame; Z_h the helical axis; Z_m, the molecular frame; Z_{p1} and Z_{p2}, the principal axes.

sh   Euler angles for the dipolar tensor from those for the chemical shielding tensor. If necessary, for example, β may be redefined by β_{dd} and β_{cs} for the dipolar and chemical shielding tensors, respectively. According to these notations, the two-site jump can be expressed by interconversion between Z_{p1} and Z_{p2}, and the rotational motion is described by rotation about the axis Z_h. Accordingly, the autocorrelation function is given by the following expression by using Wigner rotation matrices,

$$C_m(t) = (\rho_{20}^{(2)})^{-2} \sum_{nn'aa'bb'=-2}^{2} \rho_{2n}^{(2)} \rho_{2n'}^{(2)}$$

$$x \langle D_{na}^{(2)*}(\Omega_{PM}(0)) \; D_{n'a'}^{(2)}(\Omega_{PM}(t)) \rangle$$

$$x \; \{D_{ab}^{(2)*}(\Omega_{MH}) \; D_{a'b'}^{(2)}(\Omega_{MH})\}$$

$$x \langle D_{bm}^{(2)*}(\Omega_{HL}(0)) \; D_{b'm}^{(2)}(\Omega_{HL}(t)) \rangle \qquad (2)$$

where $< >$ represents the ensamble average and $\rho_{2n}^{(2)}$ are the components of the $\ell=2$ irreducible spherical tensor operators in their principal axis system. Ω are the Euler angles for the coordinate transformations from the principal axis system through the laboratory frame, and they are defined as (see Fig. 7),

$$\Omega_{PM} = (\alpha_{PM}, \Theta, \pi-\Phi)$$

$$\Omega_{MH} = (\alpha, \beta, \pi-\gamma)$$

$$\Omega_{HL} = (\phi, \theta, 0) \tag{3}$$

For the axial symmetry of the spherical tensor operators, ρ_{2n} is not zero only if $n=n'=0$. With use of the same treatments as for the two-site jump model[9,18] the second and third terms in eq. (2) lead to ,

$$\langle D_{0a}^{(2)*}(\Omega_{PM}(0))\ D_{0a'}^{(2)}(\Omega_{PM}(t)))\rangle\{D_{ab}^{(2)*}(\Omega_{MH})\ D_{a'b'}^{(2)}(\Omega_{MH})\}$$

$$= \frac{1}{4}\left[d_{0a}^{(2)}(\Theta)d_{0a'}^{(2)}(\Theta)\right]\left[(1+(-1)^a)(1+(-1)^{a'})+e^{-\lambda t}(1-(-1)^a)(1-(-1)^{a'})\right]$$

$$\times[d_{ab}^{(2)}(\beta)d_{a'b'}^{(2)}(\beta)]e^{i(a-a')\alpha+i(b-b')(\pi-\gamma)} \tag{4}$$

where the equal probability is assumed at the two sites, and $d_{ij}^{(2)}(a)$ are the reduced Wigner rotation matrices and λ is the sum of the interconversion velocities between the sites 1 and 2. In the case of the free diffusional rotation about the axis Z_h, using the result derived by Wallach,[19] the fourth term in eq. (2) can be expressed as follows,

$$\langle D_{bm}^{(2)*}(\Omega_{HL}(0))\ D_{b'm}^{(2)}(\Omega_{HL}(t))\rangle$$

$$= \delta_{bb'}[d_{bm}^{(2)}(\theta)\ d_{b'm}^{(2)}(\theta)]e^{-b^2 D_r t} \tag{5}$$

Combining eqs. (4) and (5), we have an exact form of the autocorrelation function for the two-site jump and rotational motions,

$$C_m(t) = \sum_{aa'b=-2}^{2}[d_{bm}^{(2)}(\theta)]^2\ e^{-b^2 D_r t}$$

$$\times \frac{1}{4}[d_{0a}^{(2)}(\Theta)d_{0a'}^{(2)}(\Theta)]\{(1+(-1)^a)(1+(-1)^{a'})$$

$$+ e^{-\lambda t}(1-(-1)^a)(1-(-1)^{a'})\}$$

$$\times[d_{ab}^{(2)}(\beta)d_{a'b}^{(2)}(\beta)]\ e^{i(a-a')\alpha} \tag{6}$$

Here again, the Euler angle θ represents the orientation of the fiber axis with respect to the magnetic field (e.g., the parallel for $\theta=0°$ and the perpendicular for $\theta=90°$, see Fig. 7). For a given θ, the autocorrelation function can be expressed as functions of jumping frequency λ, diffusion constant D_r, fluctuation amplitude Θ for the two-site jump, and α which is the angle between the axis Z_h and the plane made by the axes Z_{p1} and Z_{p2}.

Generally, $C_m(t)$ consists of time dependent terms, but it contains time independent terms as well when b=0. Therefore, $C_m(t)$ can be expressed as follows,

$$C_m(t) = C_m(\infty) + C_m'(t) \tag{7}$$

Calculation of relaxation times, T_1 and T_2. There are two mechanisms contributing to ^{31}P relaxation processes, ^{31}P-^{1}H dipolar interactions and ^{31}P chemical shielding anisotropy. Thus, the observed relaxation times, T_1 and T_2, become the sum of these two mechanisms,

$$\frac{1}{T_1} = \frac{1}{T_1 dd} + \frac{1}{T_1 cs} \tag{8}$$

$$\frac{1}{T_2} = \frac{1}{T_2 dd} + \frac{1}{T_2 cs} \tag{9}$$

where the superscripts, dd and cs, stand for dipole-dipole interactions and chemical shielding terms, respectively. The general expressions for $T_1 dd$ and $T_1 cs$ are,[20]

$$\frac{1}{T_1 dd} = \frac{1}{4} C^{dd} \{J_0(\omega_I - \omega_S) + 3J_1(\omega_I) + 6J_2(\omega_I - \omega_S)\} \tag{10}$$

$$\frac{1}{T_1 cs} = \frac{1}{3} C^{cs} J_1(\omega_I) \tag{11}$$

and

$$C^{dd} = \hbar \gamma_I^2 \gamma_S^2 / r^6 \quad \text{and} \quad C^{cs} = (\gamma_I H_0 \Delta\sigma)^2$$

where all the parameters in eqs. (10) and (11) are the usual notations. For two cases of $\theta=0°$ and $90°$, T_1 values are plotted in Fig. 8 as a function of $1/\lambda$, assuming $\lambda \gg D_r$ and using the following parameters; $\Theta=40°$, $\alpha=0°$ both for the dipolar and chemical shielding tensors, and $\beta_{dd}=60°$ and $\beta_{cs}=77°$ which were calculated based on the atomic coordinates of the B form of DNA.[21] The observed T_1's with error limits are found to be nearly the minima of the T_1-$1/\lambda$ curves both for the parallel and perpendicular cases. We checked dependences of T_1 on fluctuation amplitudes, Θ and α. As decreased Θ from $45°$ to $0°$, T_1 minimum rose slowly in the range of $45°$ to $30°$, and then sharply increased to the finite. On the other hand, T_1 minimum was found to vary only by factors of about 3 over the entire range of α from $0°$ to $180°$, suggesting that the curve is far less sensitive to α compared with Θ. So we set $\alpha=0°$ for simplicity. The shape itself of the T_1 curves, however, remained unchanged regardless of Θ and α.

Since $\lambda \gg D_r$ was assumed, T_1 values are exclusively determined by the terms of λ but not those of D_r; this is likely the case for DNA.

With use of the usual definition of correlation time, $\tau_c \approx 1/\lambda$, about 5 nanosecond of τ_c can uniquely be obtained for the two-site jump motion.

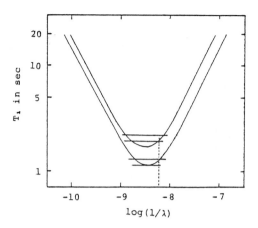

Fig. 8. The plots of T_1 vs. $\log(1/\lambda)$.

As for the transverse relaxation time, T_2, its formula are written as follows,[20]

$$\frac{1}{T_2 dd} = \frac{1}{8}\ C dd\{4J_0(0)+J_0(\omega_I-\omega_S)+3J_1(\omega_I)+6J_1(\omega_S)+6J_2(\omega_I+\omega_S)\} \qquad (11)$$

$$\frac{1}{T_2 cs} = \frac{1}{3}\ C cs\{4J_0(0)+3J_1(\omega_I)\} \qquad (12)$$

The plots of T_2 vs. $1/D_r$ are shown in Fig. 9, where we used $\lambda=2\times10^8$ sec^{-1} which was previously obtained from th T_1 data, and we also assumed the same Euler angles as before. Observed T_2's are shown in Fig. 9 by horizontal lines. The curve for the perpendicular orientation of the fibers seems reasonable but it can be clearly seen that the curve for the parallel yields very high T_2 values and runs out of the range of the observed T_2. This discrepancy between the theory and the experimental data requires the presence of another motion in addition to the two-site jump and rotational motions.

Consider wobbling motion as the third motional mode whose diffusion constant may be denoted as D_{eff}. This motion is described in such a way that the helical axis can incline or bend within a corn made by a solid angle χ_0 about the fiber axis. Lipari and Szabo[22] derived a closed form of the autocorrelation functions for such a wobbling motion,

$$G_m(t) = G_m(\infty) + (G_m(0) - G_m(\infty))e^{-D_{eff}\,t} \qquad (14)$$

For simplicity, consider the most dominant term, i.e., the case for m=0, and then we have,

$$G_0(0) = \frac{1}{20}\{\cos\chi_0(1+\cos\chi_0)(9\cos^2\chi_0-1) + 4\} \tag{15}$$

$$G_0(\infty) = \frac{1}{4}\{\cos^2\chi_0(1+\cos\chi_0)^2\} \tag{16}$$

Assuming $\lambda \gg D_r > D_{eff}$, it is readily shown that the frequency independent terms of the spectral density, $J_0(0)$, are the determinant terms in T_2 values, and the other terms can be neglected (see eqs. 12 an 13).

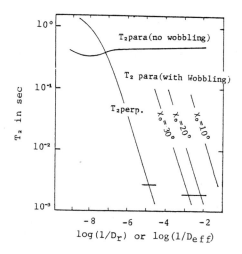

Fig. 9. The plots of T_2 vs. $\log(1/D_r)$ or $\log(1/D_{eff})$

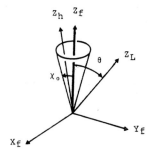

Fig. 10. Schematic drawing of wobbling motion.

Using the relation of eq. (7), $J_0(0)$ can be calculated from the following equations,

$$C_0(t)_{para} \simeq C_0(\infty)_{para}\{G_0(\infty) + (G_0(0)-G_0(\infty))\}e^{-D_{eff}t} \qquad (17)$$

$$C_0(t)_{perp} \simeq \{C_0(\infty)_{perp}+C_0'(t)\}\{G_0(\infty)+(G_0(0)-G_0(\infty))e^{-D_{eff}t}\} \qquad (18)$$

where subscrips, para and perp, represent the parallel and perpendicular orientations of DNA fibers, respectively, and D_{eff} is the averaged diffusion constant. The values of T_2 were calculated from eqs. (12) through (19) by fixing the parameters as before. The T_2 curves are shown as a function of $1/D_{eff}$ at a different corn angle χ_0 in Fig. 9. The parameters D_{eff} and χ_0 to be determined are of rather arbitrary choice since T_2 strongly depends on both D_{eff} and χ_0 as is seen in Fig. 9. Roughly speaking, χ_0 >30° might be too large because of the rather narrow line width in ^{31}P spectra observed for DNA fibers (Fig. 4), whereas the curve at χ_0<20° yields D_{eff}<$10^2 sec^{-1}$ which seems too slow for the wobbling motion. Thus, it seems reasonable that the value of χ_0 is in the range of 20~30°, and hence $1/D_{eff}=10^{-2}$~10^{-3} is obtained. All the results discussed in this section are summarized in Table 3. Agreement between the observed and calculated values for each of T_1 and T_2 is of very good, partly owing to a larger number of parameters compared with a less number of the observed.

Table 3. Comparison of the observed T_1 and T_2 with the Calculated

	parallel		perpendicular	
	T_1 (sec)	T_2 (msec)	T_1 (sec)	T_2 (msec)
Observed	1.2	1.8	2.1	2.6
Calculated	1.2	1.8	2.0	2.6

$\beta_{dd}=60°$, $\beta_{cs}=77°$, $\Theta=40°$ and $\chi_0=20°$~30°
Two-site jump rate $\lambda=2\times10^8$ sec^{-1}
Rotational diffusion $D_r=6.2\times10^4$ sec^{-1}
Wobbling motion $D_{eff}=10^2$~10^3 sec^{-1}

Nevertheless, we can conclude that a rapid internal motion occurs in DNA in hydrated fibers just as in solution, and that the wobbling or bending motion must be taken into account for the T_2 data of the DNA fibers perpendicularly oriented with respect to the magnetic field.

Acknowledgement. We gratitude Prof. M. Kyogoku of Osaka University for granting one of us (H.S.) to use his n.m.r. facilities and for his useful discussions. A part of this work was supported by Grant-in-aid for Scientific Research (56470132).

REFERENCES

1. H. Shindo et al.:1980, Biochemistry, 20, pp.518-526.
2. A.H.-J. Wang et al.:1979, Nature, 282, pp.680-686.
3. R.M. Wing et al.:1980, Nature, 287, pp.755-758.
4. Z. Shakked et al.:1981, Pro. R. Soc. London Ser. B213, pp.479-487.
5. T.R. Tritton and I.M. Armitage:1978, Nucl. Acids Res., 5, pp. 3855-3869.
6. L. Klevan et al.:1979, Nucl. Acids Res., 6, pp.1607-1616.
7. M.E. Hogan and O. Jardetzky:1980, Biochemistry, 19, pp.3460-3468.
8. P.H. Bolton and T.L. James:1979, J. Phys. Chem., 83, pp.3359-3366; J.W. Keepers and T.L. James:1982, J. Amer. Chem. Soc., 104, pp.929-939.
9. G. Lipari and A. Szabo:1980, Biochemistry, 20, pp.6250-6256.
10. H. Shindo and S.B. Zimmerman:1980, Nature, 283, pp.690-691.
11. D.A.Torchia:1978, J. Magn. Reso., 30, pp.613-616.
12. H. Shindo et al.:1981, Biochemistry, 20, pp.745-750.
13. H. Shindo et al., unpublished.
14. B.T. Nall et al.:1980, Biochemistry, 20, pp.1881-1887.
15. R.T. Simpson and H. Shindo:1980, Nucl. Acids Res., 8, pp. 2093-2103.
16. S.J. Opella et al.:1981, Biochemistry, 20, pp. 224-290.
17. A. Pine et al.:1973, J. Chem. Phys., 59, pp.569-590.
18. D.A. Torchia and A. Szabo:1982, J. Magn. Reso., 49, pp.107-121.
19. D. Wallach:1967, J. Chem. Phys., 47, pp.5258-5268.
20. R.J. Wittebort and A. Szabo:1978, J. Chem. Phys., 69, pp.1722-1736.
21. S. Arnott and D.W.L. Hukins:1972, BBRC, 47, pp.1504-1509.
22. G. Lipari and A. Szabo:1981, J. Chem. Phys., 75, pp.2971-2976.

NUCLEIC ACID JUNCTIONS: THE TENSORS OF LIFE?

Nadrian C. Seeman
Center for Biological Macromolecules
SUNY at Albany, Albany, New York   12222

Neville R. Kallenbach
Department of Biology
Univ. of Pennsylvania, Philadelphia, PA., 19104

ABSTRACT

    Nucleic acids which interact to generate structures in which three
or more double helices emanate from a single point are said to form a
junction.   Such structures arise naturally as intermediates in DNA
replication and recombination.   It has been proposed that stable
junctions can be created by synthesizing sets of oligonucleotides of
defined sequence that can associate by maximizing Watson-Crick
complementarity (Seeman, 1981, 1982).   In order to make it possible to
design molecules that will form junctions of specific architecture, we
present here an efficient algorithm for generating nucleic acid
sequences that optimize two fundamental properties: fidelity and
stability.   Fidelity refers to the relative probability of the junction
complex relative to all alternative paired structures.   Calculations
are described which permit approximate prediction of the melting curves
for junction complexes.

INTRODUCTION

    The existence of DNA as a stable extended double helix is by now a
concept that is familiar to all.   Base-paired duplexes involving
oligonucleotide model systems have provided a major source of detailed
conformational information (Seeman, 1980; Kallenbach and Berman, 1977)
on the state of the bases and backbones in various forms of double
helical structure.   While it is known that triply and even quadruply
branched structures of DNA have a transient existence as intermediates
in the replication or recombination of DNA molecules (Broker and Lehman,
1971; Kim, Sharp and Davidson, 1972), it has not been possible to
investigate these forms structurally at high resolution in terms of
short chain molecules, where the region of chain at the junction
provides a significant component of the signal.   Forked replicative
intermediates or four-stranded recombinational structures of the type
proposed by Holliday (1964) provide examples of what we define as
nucleic acid junctions, i.e., structures in which three or more double

B. Pullman and J. Jortner (eds.), Nucleic Acids: The Vectors of Life, 183–200.
© 1983 by D. Reidel Publishing Company.

helices emanate from a single point. Both replicative and
recombinational intermediates are normally unstable due to internal
sequence symmetries, which allow their resolution to double helices,
via the process of branch point migration (Thompson, Camien and Warner,
1976; Warner, Fishel and Wheeler, 1979; Nilsen and Baglioni, 1979;
Seeman and Robinson, 1981). Since this can be a very rapid process
(Thompson, Camien and Warner, 1976; Warner, Fishel and Wheeler, 1979),
these forms have not been tractable to physical characterization at the
oligonucleotide level.

It has recently been suggested that the range of migration
available to junctions can be severely restricted, to form semi-mobile
junctions, or eliminated altogether, to form immobile junctions from
oligonucleotides (Seeman, 1981, 1982). The idea is that oligonucleotides can be constructed which will preferentially associate to form
junctions via Watson-Crick base pairing, while the sequences of these
molecules do not possess the symmetry necessary to permit branch point
migration. Semi-mobile junctions have recently been constructed by Hsu
and Landy (Nash, 1981). An example of an immobile junction is
illustrated in Fig. 1. A set of rules has been formulated (Seeman,
1981, 1982) which must be obeyed by oligonucleotide sequences, if they
are to form stabilized junction structures. These conditions, however,
must be supplemented by thermodynamic criteria in order to optimize the
sequence, and assure the stability of a given designated junction.
Here, we present the details of these thermodynamic criteria. We
further present a rapid junction generation algorithm and a procedure
for sequence selection for nucleic acid junctions. Using this
procedure it is possible to optimize sequences for a given junction
architecture before embarking on the formidable chemical effort
necessary to actually synthesize the component oligonucleotide
molecules.

The rules indicated in the earlier publications (Seeman, 1981,
1982) can be outlined as follows: The construction of immobile and
semi-mobile junctions relies on unique base pairing patterns. These,
in turn, are a function of the free energy of association of the
individual strands involved. Each strand which is chosen to
participate in the formation of an immobile junction may be considered
to be composed of a series of overlapping segments of a given criterion
length, Nc. For example, each hexadecameric strand in the immobile
junction shown in Fig. 1 is a series of 13 overlapping segments of
length 4. Each of these segments is termed a 'criton'. A given value
of Nc implies a diversity of 4^{Nc} critons available with which to
construct a given junction. Watson-Crick pairing arrangements which
compete with the desired pairing must be considered from a thermodynamic point of view for lengths less than Nc. However, if the rules
indicated below are obeyed, there will be no competing Watson-Crick
pairing interactions for segments of length Nc or longer. Clearly, Nc
is a number to be minimized, since this in turn minimizes the strengths
of competing interactions, by shortening the lengths involved. The
generation of junctions containing more and more bases implies that

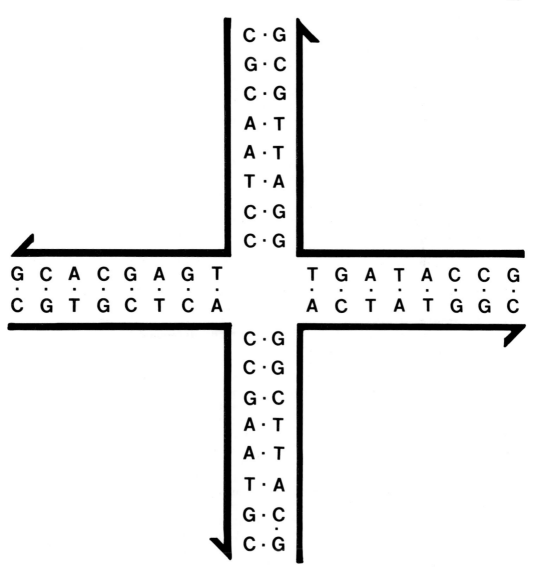

Figure 1. A Fourth Rank Junction Composed of Four Hexadecameric Fragments. This fragment has four arms, each designated as being composed of eight base pairs. Since Nc = 4 (see text), the fidelity of this junction is very high. It contains no repeating G sequence longer than two, and is predicted to have a uniphasic melting profile. The melting temperature is estimated to be 65°C, at 1 M salt, with concentrations of 0.1 mM for each strand. This is the best sequence which was derived from the optimization procedure (see text).

more and more critons are necessary to supply the requisite sequences.
However, while the lengths of strands necessary to generate longer and
more stable arms grows arithmetically, the diversity of sequences
available with each increase in criton length grows geometrically.

Two further terms need to be defined: A 'bend' is a
phosphodiester linkage which is flanked by bases paired to different
strands; for semi-mobile junctions, the bend includes the mobile
nucleotides. The 'rank', R, of a junction is the number of base pairs
which directly abut it. Thus, the junction shown in Fig. 1 has R = 4.
In order to generate uniquely paired structures with nonmigratory
junctions (for length Nc or greater), the following rules must be
obeyed within the designated pairing regions:

1. Each criton in the individual strands forming the
    junction must be unique throughout all strands.
2. The complement to any criton which spans a bend in a
    strand must not be present in any strand.
3. Self-complementary critons are not permitted. If Nc
    is an odd number, this holds for all critons of size
    (Nc + 1).
4. The same base pair can only abut the junction twice.
    If it is present twice, those two occurrences must
    be on adjacent arms.

The practical problem of choosing specific sequences for synthesis
as model junctions demands a procedure to optimize the sequence of
junctions with a defined architecture, subject to thermodynamic criteria
for both stability and fidelity, as well as any additional constraints
that may be imposed by the investigators. The previous suggestion
(Seeman, 1981, 1982) that immobile and perhaps semi-mobile junctions
might be used as building blocks for the construction of rigid and
semi-rigid geometric figures makes optimization of these attributes
particularly desirable. A rapid algorithm is necessary, because each
independent base in the junction otherwise increases the extent of
calculation by a factor of 4. It is desirable to be able to include
investigator-imposed constraints on the generation of junction
sequences. This permits the implementation of criteria which are more
complex than the straightforward thermodynamic and numerical criteria
which result in junction fidelity, uniqueness and stability. For
example, these constraints may be used to eliminate certain sequence
possibilities which engender non-Watson-Crick alternative pairing
structures (such as G-G pairs) which have not yet been adequately
characterized with thermodynamic data. Similarly, end-fraying may be
minimized by requiring the 3' and 5' terminal bases to be G's or C's.

OVERALL STRATEGY FOR OPTIMIZED JUNCTION SEQUENCE GENERATION

Figure 2 indicates the fundamental logic associated with the
junction optimization procedure which we have developed. The double

boxes represent the steps done by the investigator in this procedure, while the rest of the logic has been coded in a FORTRAN computer program series. Clearly, not all steps have been indicated; rather, we have only shown those steps which are most critical to understanding the logic of the procedure. It should be noted that it is possible to establish within the computer a numerical sequence, in base 4, which obeys the same qualitative rules as Watson-Crick base pairing complementarity. The sequence may be represented in this numerical fashion. For example, a complementarity relationship of the sort $c = k - i$ is possible, where the independent base is represented by the number i, the complementary base by the number c, and k is a constant. There exist eight different permutations of bases corresponding to this numerical encryption. Sequences may be screened for adherence to the rules at the numerical level before proceeding to thermodynamic calculations which involve specific base identities. This treatment of the problem, in turn, can result in a large saving in computer time.

The architecture of a junction requires the specification of both covalent connectivity and base pairing relationships. Because of the complementary nature of the Watson-Crick double helices which constitute the junction structure, only half of the bases must be treated as independent variables; those bases complementary to them are treated as dependent variables. In the case of semi-mobile junctions, only one out of four of the mobile bases is independent. With the computer, new sequences may be generated simply by the process of counting in base 4. If all of the arms of a junction have the same length, it is possible to fix one independent base at the numerical level, thereby decreasing the number of independent variables by one. (This is analogous to specifying the origin in crystallographic phasing procedures (Hauptman and Karle, 1956)). Even when these considerations have been taken into account, a large number of tasks must still be done by the computer, since N independent bases imply 4^N individual sequences to be tested. This number can be reduced by use of the following procedures.

The junction rules are so constituted (Seeman, 1981, 1982), that it is possible to develop a rapid algorithm centered on them. The division of each strand into critons facilitates this algorithm. The independent bases may be ordered by the rapidity of the rates of change of the digits representing their identities within the program. By 'order' we mean an inverse measure of the rate at which the digit representing the base is incremented. Thus, the lowest ordered base will be changed on every pass, the next lowest ordered base will change on every fourth pass, the next lowest ordered base on every sixteenth pass, and so on. The critons themselves may be ordered according to the lowest ordered base within the criton. The critons are then tested for adherence to the rules sequentially, from highest to lowest order. Thus, if a given criton violates one of the rules, the base corresponding to the order of that criton is advanced, rather than the base of lowest order. Until a base at the order of violation has been changed, no changes at lower orders would correct the existence of the violation. When a base of any order is incremented, those bases of

OPTIMIZED JUNCTION
SEQUENCE GENERATION

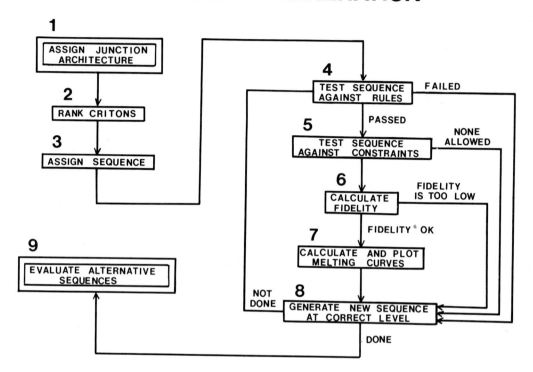

Figure 2. A Flowchart for Optimizing the Sequence of a Given Junction.
The nine logical steps in this procedure are indicated schematically.
The two steps in double boxes must clearly be done by the investigator,
while the other steps are done automatically by the programs. In the
first step, the covalent connectivity and desired base pairing are
selected by the investigator. Furthermore, specific constraints can be
imposed at this stage. In the second step, the critons are ranked by
the order of the most rapidly changing base which they contain (see
text). After that, an initial numerical sequence must be assigned
(step 3). This numerical sequence is tested against the junction
rules, and if it fails, a new sequence is generated by the fast
algorithm. If the sequence obeys the rules, its base permutations are
then tested (step 5) against investigator selected constraints. If any
of the eight sequences implied by the numerical procedure are
acceptable, their fidelities are calculated (step 6), and if these are
acceptably high, melting curves are calculated and plotted (step 7).
New sequences are then generated (step 8) and tested iteratively until
all possibilities have been exhausted. The investigator may then
evaluate the alternatives presented by the programs.

RAPID JUNCTION SEARCH

ALGORITHM	UNORDERED	ORDERED
START	0 0 0 0 0 0	0 0 0 0 0 0
FIRST TRY	0 0 0 0 0 1	0 1 0 0 0 0
FIRST SUCCESS	0 1 2 3 4 5	0 1 2 3 4 5
TOTAL STEPS	12,345	15

Figure 3. The Odometer Analogy to the Rapid Junction Algorithm. The object in this example is to generate configurations of numbers with an odometer, in which each digit is unique. This is similar to the criton rules for junction formation. Two alternative pathways are indicated. On the left, the odometer is incremented in the ordinary fashion, from right to left, until the first number which fulfills the criteria, viz., 012345, is discovered. On the right, we start at the same point, but the digits are ranked. The highest ranking digit which violates the uniqueness rule, that in the 10,000's place, is incremented in the first step. The next step will increment the digit in the 1,000's place from 0 to 1. It will still violate uniqueness relative to the digit in the 10,000's place, so a second incrementation of the 1,000's place will take place to yield 012000. In like fashion, the 100's place will be incremented 3 times, the 10's place 4 times and the 1's place 5 times. The fifteen steps are a great saving in time over the 12,345 steps needed by always incrementing the 1's place, as shown on the left.

order lower than that of the incremented base are, of course, set to
their lowest value.

This algorithm will be easier to understand if we note that the
procedure is analogous to the generation of configurations of numbers
with defined properties, using an odometer or crowd counter, as
indicated in Fig. 3. In that figure, the uniqueness of each digit is
the specific property required for the numerical configurations. This
property is similar to the properties involved in the criton rules for
junction formation. If we start at the top of the figure, with six
zeros as our initial configuration, and increment the most rapidly
changing digit sequentially, as shown on the left, it will take 12,345
steps to get the first successful numerical configuration. On the
other hand, if we correct the highest ordered digit which is violating
the uniqueness rule, that indicated in the 10,000's place, and then
proceed accordingly, as shown on the right, it will only take 15 steps
to reach the same point. The way in which this algorithm is applied to
the junction generation is indicated in Fig. 2.

Fidelity

Once a junction fulfills the uniqueness and nonmobility rules,
thermodynamic criteria must be applied to all sequences of length less
than Nc. The first question to consider is the pairing fidelity: Is
the desired base pairing configuration the most probable configuration
in which these particular sequences are to be found in solution? If
so, what is its probability relative to other pairing configurations?
We have treated this problem in a pairwise fashion; the program
routinely considers all alternative binary base pairing configurations
for lengths less than Nc.

The stability of an oligonucleotide duplex depends on its chain
length, sequence and concentration, as well as on environmental
variables, such as pH, ionic strength and temperature (Kallenbach and
Berman, 1977). Data on the relative stabilities of oligoribonucleic
acids in conditions equivalent to 1 M NaCl, pH 7, have been accumulated
by Tinoco and his coworkers (Borer, Dengler, Tinoco and Uhlenbeck,
1974). The effects of sequence can be evaluated in terms of units
representing adjacent sets of two base pairs; the equilibrium constants
corresponding to the association within each unit are available, as is
the nucleation constant for initial strand interaction. This is denoted
by β, with units M^{-1}. A given sequence will then be paired with its
complement by a weighting factor that depends on the product of a set
of numbers:

$$K_{AB} = \beta K_1 K_2 K_3 \ldots K_{N-1}, \qquad (1)$$

where N is the chain length of complementary sequences between chains A
and B, and β is the nucleation constant. The values of the K_i are
tabulated at 25°C by Borer et al. (1974) as:

$$K_i = \exp(-\Delta G_i / RT).$$

To illustrate the use of equation (1), consider the tetramer

(5') AGCU (3')

(3') UCGA (5')

to be decomposed into the three subunit "pairs",

(5') AG GC CU (3')

(3') UC CG GU (5'),

each of which has an approximate equilibrium constant assigned (Borer et al., 1974).

In this way, the maximum concentration of paired molecules of a duplex of arbitrary sequence can be predicted. The situation for oligodeoxynucleotides is unfortunately not so completely defined as for oligoribonucleotides. However, thermodynamic data are available from which primitive sets of K_i's can be created, together with rough values of the ΔH°_i's (Marky and Breslauer, 1980). Despite the uncertainties, these data make it possible to estimate the relative contributions of different sequences with reasonable accuracy, particularly if appropriately scaled values from oligoribonucleotides are used. For scaling, we alter the ΔG° values of Borer et al. (1974), so as to lower the stability of the corresponding oligodeoxynucleotide by 20°C. Comparison of the values of K_{AB} for each set of interactions below the criton length then permits us to estimate the relative contributions of the base pairing in each case. Junctions of maximum fidelity will be those that contain sub-criton pairing sequences of minimal stability, relative to the stability of the complete arms. All binary Watson-Crick alternatives are checked by the program, and their stabilities are compared with those calculated for the double helices chosen for the architecture of the junction. The highest probability junctions above a selected fidelity minimum are retained for further processing. It should be pointed out that fidelity is a function of temperature. Clearly, sequences must be compared for relative fidelity at a standard temperature, for which we use 25°C. We have found that when pairing lengths are 4 or 5 residues greater than Nc, fidelity approaches unity very closely (0.999).

Estimation of Junction Melting Curves

Junction sequences whose fidelities are sufficiently high must next be considered for stability in solution over a range of temperatures. High fidelity is necessary for a monodisperse junction complex, but this criterion is not sufficient. For example, Fig. 4 shows a pair of 4th rank junctions assembled from hexanucleotides. These junctions were generated with the constraint that the two base pairs furthest from the designed junction be G-C pairs, while GpG sequences were prohibited. The sequence in Fig. 4a has very interesting symmetry properties, but its fidelity is not particularly high. The fidelity of the sequence in Fig. 4b is certainly acceptable. However, it is necessary to estimate the melting curves for these materials in order to make sure that they are likely to exist in intact form under the conditions of interest. It is clear from the thermal transition profiles shown in Fig. 4 that neither of the junctions illustrated there is likely to be a stable structure in solution at convenient temperatures.

The information contained in the estimated equilibrium constants for pairing specific sequences can be used to predict approximate transition profiles for junctions. In order to do this, enthalpy values, ΔH_i° corresponding to the equilibrium constants K_i, used to assess fidelity, are required. These are considerably less certain for oligodeoxynucleotides than for oligoribonucleotides, but nonetheless reasonable estimates are available, and missing values can be filled in by scaling the corresponding RNA data, as described.

In the case of pairing between sequences on two non-identical strands, A and B, the value of K_{AB} and the starting concentrations of the two species uniquely characterizes the equilibrium; for starting concentrations, C_A and C_B, and paired complex concentration C_{AB} (moles per liter),

$$K_{AB} = \frac{C_{AB}}{(C_A - C_{AB})\,(C_B - C_{AB})}. \qquad (2)$$

We have discussed how to approximate K_{AB}; thus, C_{AB} can be calculated (Zimm, 1960). This can be done at any temperature if the ΔH_i° for each K_i is known.

Consider next the interaction of four oligomers, A, B, C and D, which contain uniquely complementing half sequences that can lead to formation of a 4th rank junction complex. Since at equilibrium the concentrations must be independent of reaction pathway, it is sufficient to calculate the junction concentration resulting from any one pathway. For example, one might select:

$$A + B = AB \qquad (i)$$

$$C + D = CD \qquad (ii)$$

$$AB + CD = ABCD \qquad (iii)$$

From the values of K_{AB}, K_{CD}, and introducing a new factor, σ_R to describe the statistical weight of the central junction "loop" structure of rank R, the concentrations of junction can be expressed in

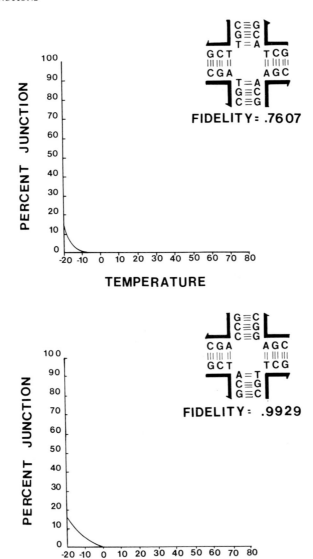

Figure 4. Hexadecamers Forming Junctions with Arms 3 Base Pairs Long.
Both junctions were generated with the constraint that the two base
pairs furthest from the junction be G-C, while GpG sequences were
prohibited form this part of the structure. (a) A junction with
interesting symmetry properties. The junction and estimated melting
curve for pH 7, 1 M NaCl is shown. The fidelity is unacceptable, as is
the melting temperature which is well below 0°C. (b) The hexadecameric
junction with the highest fidelity. The fidelity is acceptable, but the
melting curve is insufficiently improved to indicate the usefulness of
trying to form a junction with these molecules.

terms of known quantities. That is, C_{AB} and C_{CD} can be calculated by solving equation (2), and these values can be introduced into reaction (iii) above to give:

$$K_{ABCD} = \frac{C_{ABCD}}{(C_{AB} - C_{ABCD})(C_{CD} - C_{ABCD})}.$$

The value of K_{ABCD} is estimated as:

$$K_{ABCD} = \beta^{-1}((\sigma_R K_{BC} K_{DA}) + K_{BC} + K_{DA}).$$

This is very nearly equal to $\beta^{-1} \sigma_R K_{BC} K_{DA}$, since σ_R is not expected to be very different from unity, while the K's are large at low temperature. It is expected that $\beta^{-1} \sigma_R \gg 1$. If $\beta^{-1} \sigma_R < 1$, only negligible concentrations of the complete junction will be detectable, as discussed more fully below. If a junction entails no strain, we anticipate that only a simple Jacobson-Stockmayer (1950) term is involved:

$$\sigma_R \sim (R(U + 1))^{-3/2},$$

where U is the number of unpaired bases abutting the junction on each arm.

More generally, we can write:

$$\sigma_R = \sigma_o (R(U + 1))^{-3/2},$$

where the factor σ_o reflects the difficulty of forming the junction.

Concentrations of the ternary and higher (for $R > 4$) intermediates can also be calculated, using stepwise paths such as:

AB + C = ABC,

ABC + D = ABCD.

Thus, the equilibrium concentration of each intermediate, as well as the junction itself can be calculated; a series of relations exists among these intermediates of the form:

$$C_{AB} + C_{BC} = C_{ABC} + C_B,$$

which simplifies the problem considerably for this approximate treatment.

For values of $\beta^{-1} \sigma_R$ much greater than unity, intermediate forms are much less favored than the junction. Therefore, the following relationship holds approximately for any rank:

$$K_J = C_J (\prod_{i=1}^{R} (C_i - C_J))^{-1}.$$

In this equation, K_J is the equilibrium constant for the system, C_J is the concentration of junction, and C_i is the initial concentration of the i'th component strand.

RESULTS

Since the problem of fidelity of synthetic junctions has been dealt with before (Seeman, 1982), we focus here on the calculated melting curves for these structures. Short duplex oligonucleotides tend to denature in all-or-none fashion, and significant populations of intermediates arise only in longer chains (Kallenbach and Berman, 1977). The transition behavior of very short chains is such that (1) $1/Tm$ is found to be a linear function of log C, where C is the strand concentration of each species of interacting molecule:

$$C = C_A = C_B,$$

and (2), for homogeneous sequences, $1/Tm$ is expected to be an approximately linear function of $1/(N - 1)$, where N is the chain length.

Because junctions are inherently inhomogeneous in sequence, only the first of these relations applies. To investigate the behavior of the model, as well as the method of calculation, a computation was carried out. The concentration of all four strands of the junction of rank 4 shown in Figure 1 was varied over a 10,000-fold range; the resultant Tm values were graphed, as shown in Fig. 5, as $1/Tm$ vs. log C, where $C = C_A = C_B = C_C = C_D$. As can be seen, even for junctions with N = 16, with arms of length 8, the reciprocal plot is linear. Thus, the nucleation characteristic of short duplexes is preserved in the junctions, with a similar tendency to favor an all-or-nothing transition.

The conclusion from this calculation is that the quaternary nucleation process required for forming a stable junction is not innately different from that for a duplex in its concentration dependence other than in the apparent enthalpy. If one arbitrarily sets all the K_i value equal, it can be shown that the linearity of $1/Tm$ vs. $1/(N - 1)$ is also preserved in the quaternary complexes.

A final problem to consider is proper closure of the junction. Two alternatives exist: (1) pairing of the 5' end of the R'th strand

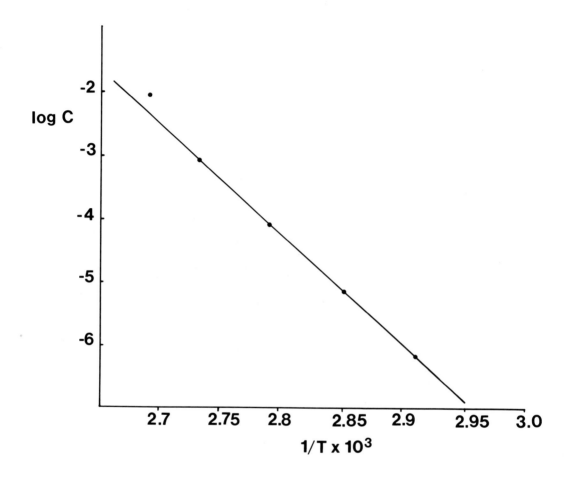

Figure 5.   The Concentration Dependence of the Melting Temperature.
The logarithm of the strand concentration has been plotted against the
reciprocal of the melting temperature. Note that the linear
relationship expected for duplex formation also holds for junction
formation under the theory propounded in the text.

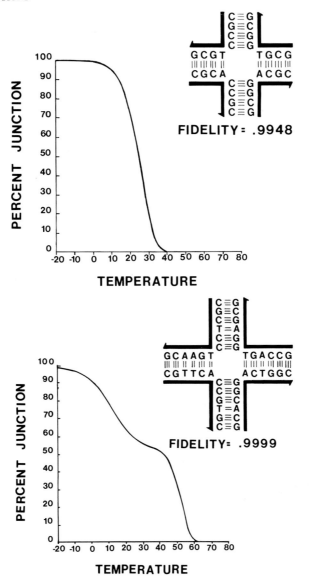

Figure 6.   Octameric and Dodecameric Intermediate Structures Generated
in the Course of Generating the 'Best' Hexadecameric Structure.   (a)
The Octameric structure with the highest fidelity.   This structure had
the same constraints as the structures indicated in Figure 4, except
that no GpGpG sequences were permitted.   Both the fidelity and melting
temperature are vastly improved.   (b) The Dodecameric Structure with
the highest fidelity.   This structure was generated by inserting two
bases into the middle of each arm of the structure shown in (a).   The
same constraints were applied.   The biphasic melting curve results from
the fact that there are more A–T base pairs in the horizontal arms than
in the vertical arms.

to the 3' end of the first strand (proper closure) and (2) pairing of
the 5' end of the R'th strand to the 3' end of another molecule of the
same species as strand 1 (concatenation). The second alternative leads
to large aggregates, since the system is not closed. If $\beta^{-1} \sigma_R \gg 1$,
the first alternative will be favored. A combination of spectroscopic
and ultracentrifugal or light scattering experiments should readily
distinguish these alternatives.

DISCUSSION

        The purpose of this paper has been to present and clarify the
procedures for choosing the sequences from which to construct immobile
and semi-mobile junctions. In view of this, it is useful to review the
steps by which an actual sequence for the immobile junction shown in
Fig. 1 was chosen. This case is particularly germane, because there
are 31 independent bases in this junction, and limitations on computing
time render blind application of the above procedure impossible; no
matter how efficient the algorithm, it is impossible to scan all 4^{31}
possible sequences which the design of this junction implies. As with
most procedures of this sort, this sequence was generated in a stepwise
fashion, optimizing the sequence at each step along the way.

        The first step in this procedure involved using the program in an
automatic fashion to generate a junction composed of four octameric
strands. These strands were generated in accord with the following
constraints: only GpC or CpG sequences could form the ends of the
double helices furthest from the junction, and no GpGpG sequences were
permitted. These constraints ensured the stability of the ends of the
double helices, while excluding the possibility of G-G non-Watson-Crick
pairing which could interfere with junction formation. Application of
the program generated the junction shown in Fig. 6a as the one with the
highest fidelity. The next pass involved the insertion of two bases in
the middle of each tetrameric double helical segment, to yield a
junction composed of four dodecameric strands. Again,
the constraints against G-G non-Watson-Crick pairing were applied. The
sequences with the highest fidelities were considered optimal at both
the octameric and dodecameric stages.

        The sequence shown in Fig. 6b had the highest fidelity index of
those sequences generated, although it clearly showed a biphasic
melting curve. This feature of the junction composed of dodecameric
strands was not a serious impediment in using it as a base for
generating a junction composed of hexadecameric fragments. The same
constraints were applied and again the two new bases in each double
helical fragment were inserted in the middle of the arm. However, the
criteria for selecting the final sequence were different at this stage.
This was because all sequences generated by the program had fidelities
greater than 0.999. Here, we selected the sequence shown in Fig. 1 as
'optimized', on the basis of both its sharp uniphasic melting curve
(not shown) and the fact that it had the highest melting temperature

(65°C) we encountered while scanning the 126 sequences generated by the last pass of the program. If we calculated the melting curve without the bases nearest the junction being paired, we still got a sharp uniphasic melting curve, this time with T_m = 48°C. For initial experimental studies, these features are of paramount importance. The fidelity of this junction is not the highest found by the program, but it is well above 99.9% at room temperature.

It should be emphasized that the initial step in this process, generating the octameric sequences, with 8 independent variables, took much more computer time than the subsequent steps. This is because of the nature of the application of the constraints in the two subsequent steps. The two extensions of the initial octameric sequence took only a few seconds on a Univac 1100/82. Thus, it can be seen that it is possible to generate a sequence for a junction with arms of moderate length, with only a nominal investment in computer time. This is due to the application of both the rapid algorithm and of constraints which are based on optimization of physical parameters at each stage in junction sequence generation. At the same time, however, it has to be recognized that in terms of either fidelity or stability, the resulting structure does not necessarily represent a global optimum. A similar strategy should be applicable to junctions of any architecture.

We are presently synthesizing the sequences in the structure shown in Fig. 1.

ACKOWLEDGEMENTS

We would like to thank Leonard Lerman for valuable discussions. This research has been supported by grant GM-26467, GM-29554 and ES-00117 from the N.I.H. N.C.S. is a Research Career Development Award Recipient. We thank Susan Fitzsimmons and Robert Speck for help with the figures and Linda P. Welch for preparation of the manuscript. Computational facilities were generously provided by the S.U.N.Y./Albany computer center.

REFERENCES

Borer, P.N., Dengler, B., Tinoco, I., and Uhlenbeck, O.C.: 1974, "Stability of Ribonucleic Acid Double Helices" J. Mol. Biol. 86, pp. 843-853.
Broker, T.R., and Lehman, I.R.: 1971, "Branched DNA Molecules: Intermediates in T4 Recombination" J. Mol. Biol. 60, pp. 131-149.
Hauptman, H., and Karle, J.: 1956, "Structure Invariants and Seminvariants for Non-Centrosymmetric Space Groups" Acta Cryst. 9, pp. 45-55.
Holliday, R.: 1964, "A Mechanism for Gene Conversion in Fungi" Genet. Res. 5, pp. 282-304.
Jacobson, H., and Stockmayer, W.: 1950, "Intramolecular Reaction in

Polycondensations. I: The Theory of Linear Systems" J. Chem.
Phys. 18, pp. 1600-1606.

Kallenbach, N.R., and Berman, H.M.: 1977, "RNA Structure" Quart. Rev.
Biophysics 10, pp. 138-236.

Kim, J., Sharp, P.A., and Davidson, N.: 1972, "Electron Microscopic
Studies of Heteroduplex DNA from a Deletion Mutant of
Bacteriophage ϕX 174" Proc. Nat. Acad. Sci. (USA) 69, pp.
1948-1952.

Marky, L.A., and Breslauer, K.J.: 1980, "Calorimetric and Spectroscopic
Investigation of the Helix-To-Coil Transition of the Self-Complementary d(G-G-A-A-T-T-C-C) duplex" Fed. Proc. 39, p. 1880.

Nash, H.: 1981, "Integration and Excision of Bacteriophage λ" Ann.
Rev. Genet. 15, pp. 143-167.

Nilsen, T., and Baglioni, C.: 1979, "Unusual Base Pairing of Newly
Synthesized DNA in HeLa Cells" J. Mol. Biol. 133, pp. 319-338.

Seeman, N.C.: 1980, "Crystallographic Investigation of Oligonucleotide
Structure" IN Nucleic Acid Geometry and Dynamics, R.H. Sarma,
editor, Pergamon Press, New York, pp. 109-148.

Seeman, N.C.: 1981, "Nucleic Acid Junctions: Building Blocks for
Genetic Engineering in Three Dimensions" IN Biomolecular
Stereodynamics, Vol. 1, R.H. Sarma, editor, Adenine Press, New
York, pp. 269-278.

Seeman, N.C.: 1982, " Nucleic Acid Junctions and Lattices" J. Theor.
Biol. 99, pp. 237-247.

Seeman, N.C., and Robinson, B.H.: 1981, "Simulation of Double Stranded
Branch Point Migration" IN Biomolecular Stereodynamics, Vol. 1,
R.H. Sarma, editor, Adenine Press, New York, pp. 279-300.

Thompson, B.J., Camien, M.N., and Warner, R.C.: 1976, "Kinetics of
Branch Migration in Double Stranded DNA" Proc. Nat. Acad. Sci.
(USA) 73, pp. 2299-2303.

Warner, R.C., Fishel, R., and Wheeler, F.: 1978, "Branch Migration in
Recombination" Cold Spring Harbor Symp. Quant. Biol. 43, pp.
957-968.

Zimm, B.: 1960, "Theory of Melting of the Helical Form in Double Chains
of the DNA Type" J. Chem. Phys. 33, pp. 1349-1356.

NMR STUDIES ON SHORT HELICES

P.O.P Ts'o, D.M. Cheng, D. Frechet, and L.S. Kan
Division of Biophysics
School of Hygiene & Public Health
The Johns Hopkins University
Baltimore, Maryland 21205

This NMR study on short DNA helices has five general objectives: (1) to develop effective approaches in the assignment of proton and phosphorus resonances of short helices; (2) to calibrate the NMR theory as applied to nucleic acid; (3) to obtain the helical DNA conformational parameters as accurately as possible; (4) to serve as a model system for defected DNA studies, and (5) to serve as a model for protein-nucleic acid interactions.

I. Complete Assignments of Exchangeable and Non-Exchangeable Proton Resonances as well as Phosphorus Resonances of Short Helices d-CGCG and d-CGCGCG (Figures 1 and 2).

I.1 Assignments of the Non-Exchangeable Proton Resonances.

We have employed six different methods in order to complete the assignments of the base and sugar proton resonances of d-CGCG and d-CGCGCG helices. These methods are listed below with examples:

1. Chemical shift and coupling constant values--H_6 resonances (doublet with $J = 7.8$ H_z) of C^1 and C^3 of d-CGCG were assigned by comparison to the d-CG dimer.

2. NOE technique--for helix of B conformation, the molecular model (1) shows that the H_8 of G nucleotidyl unit is closer to H_2' of its own sugar residue (~ 2.0 Å) rather than its H_2" (~ 3.0 Å). In addition, the H_2" of the residue on the 3' side of the phosphodiester linkage is close to the H_8 of G base on the 5' side of the phosphodiester linkage (~ 2.5 Å). Therefore, as individual G-H_8 resonance was selectively irradiated, a major and a minor NOE effect was observed by H_2' and H_2" resonances. If H_2' and H_2" resonances are assigned, then G-H_8 can be unambiguously assigned to the appropriate nucleotidyl unit.

3. Sequential homodecoupling technique--the complete set of H_1', H_2', H_2", H_3', H_4', H_5' and H_5" resonances of individual

B. Pullman and J. Jortner (eds.), Nucleic Acids: The Vectors of Life, 201–216.
© *1983 by D. Reidel Publishing Company.*

sugar in a nucleotidyl unit can be interconnected and identified by sequential homodecoupling technique (2,3). Thus, when the identity of any one resonance from each set is known, this knowledge would then permit the assignment of the complete set to the appropriate residue correctly.

4. Unique coupling pattern--H_5' and H_5" of the first residue (i.e., C^1 of d-CGCG and d-CGCGCG) at the 5'OH terminal and H_3' of the last residue (i.e., G^4 of d-CGCG and G^6 of d-CGCGCG) at the 3'OH terminal can be easily identified from H_5' and H_5" resonances of other internal residues by their unique coupling pattern due to the absence of ^{31}P splitting.

5. Incremental procedure--the H_2' resonances of G^2 and C^3 residues in d-CGCG were assigned by comparing the chemical shifts of H_2' resonances with those from the corresponding residue in the dimer. The H_1' resonances of G^2, C^3, G^4, C^5 residues of d-CGCGCG were assigned by the application of the "incremental procedure" to the assignment of the H_1' resonances (3,4). After the H_1' resonances in d-CGCGCG are identified, then the other sugar proton resonances can be assigned through the sequential homodecoupling technique (3) as described in section 3. These assignments will then allow the assignment of H_8 resonances by NOE procedure as described in Section 2.

6. Computer simulation--actual chemical shift and coupling constant values of all sugar proton resonances of d-CGCG and d-CGCGCG were analyzed by computer simulation and are summarized in Tables I and II.

TABLE I.   Chemical shifts of the sugar protons of d-CGCG and d-CGCGCG in D_2O, pD 7.4, 0.02M NaHPO$_4$ (ppm from DSS).

Compounds		Temp.°C	H_1'	H_2'	H_2"	H_3'	H_4'	H_5'	H_5"
d-CGCG	C^1	9	5.85	1.98	2.40	4.71	4.08	3.74	3.73
	G^2		5.94	2.73	2.75	5.00	4.38	4.14	4.00
	C^3		5.78	2.00	2.45	4.87	4.22	4.12	4.10
	G^4		6.21	2.67	2.40	4.73	4.21	4.10	4.09
d-CGCGCG	C^1	25	5.78	1.99	2.43	4.72	4.07	3.74	3.71
	G^2		5.90	2.65	2.73	4.99	4.36	4.21	4.00
	C^3		5.76	2.04	2.43	4.87	4.21	4.16	4.16
	G^4		5.92	2.69	2.76	4.99	4.37	4.11	4.04
	C^5		5.78	1.90	2.34	4.41	4.16	4.20	4.19
	G^6		6.17	2.61	2.37	4.68	4.18	4.12	4.11

TABLE II. Coupling constants of the sugar protons of d-CGCG and d-CGCGCG in D_2O, pD 7.4, 0.02M $NaHPO_4$ (J_{H-H} or J_{H-P} in H_z)

Compounds	Temp °C	1'2'	1'2"	2'2"	2'3'	2"3'	3'4'	4'5'	4'5"	5'5"	3'P	4'P	5'P	5"P
d-CGCG														
C^1	9	8.0	6.1	-14.0	6.7	2.6	3.0	3.1	4.0	-12.1	5.7	-	-	-
G^2		8.2	6.6	-14.0	6.4	2.7	2.0	2.4	2.0	-11.8	5.5	1.0	3.0	3.8
C^3		7.9	6.1	-14.0	6.8	2.5	2.7	3.0	3.0	-11.3	5.5	2.1	3.2	3.2
G^4		8.0	6.3	-14.0	6.0	2.6	2.5	3.1	3.1	-11.3	-	1.4	3.5	3.5
d-CGCGCG														
C^1	25	8.1	6.2	-14.0	6.2	2.9	3.2	3.6	5.0	-11.8	5.9	-	-	-
G^2		8.6	5.5	-14.3	6.3	1.7	2.8	~3.5	2.0	-11.6	5.8	2.0	3.8	4.0
C^3		8.5	5.6	-14.0	6.7	3.1	2.7	~2.8	~2.8	-11.3	5.8	2.0	~3.0	~3.0
G^4		8.5	5.5	-14.3	6.3	1.7	2.8	~3.5	2.0	-11.5	5.8	1.9	~3.8	~4.2
C^5		8.3	6.4	-14.0	6.9	2.5	2.6	~2.8	~2.8	-11.3	5.8	2.0	~3.2	~3.2
G^6		8.0	6.0	-14.0	6.3	3.1	2.7	~3.0	~3.0	-11.3	-	2.0	~3.3	~3.3

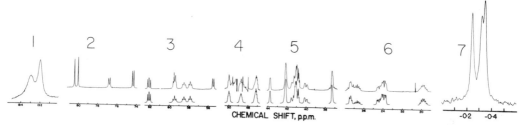

Fig. 1. Complete NMR spectrum of d-CGCG: NH-N resonances, proton resonances and phosphorous resonance. The line-shape simulation based on values listed in Tables I and II is shown at the bottom of the observed spectrum. 1 = NH hydrogen bonded resonances; 2 = non-exchangeable base proton resonances; 3 = H_1' resonances; 4 = H_3' resonances; 5 = H_4' H_5' H_5" resonances; 6 = H_2'H_2" resonances; 7 = phosphorus resonances.

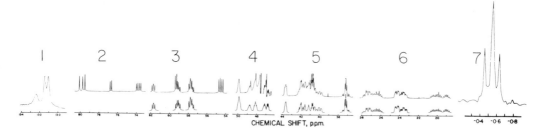

Fig. 2. Complete NMR spectrum of d-CGCGCG (same as Fig. 1).

I.2   Assignments of the NH-N Hydrogen Bonded Resonances.

As shown in Figures 1 and 2, only two and three NH-N hydrogen bonded resonances of d-CGCG and d-CGCGCG respectively (instead of four and six) were observed at 1°C due to their two-fold symmetry. The assignment of these exchangeable proton resonances is based on the expected thermal stability of the base pairs in the helix. This expectation, in turn, is based on the assumption that the melting of the helices initiates from the two symmetrical ends toward the center; this assumption is most likely to be true for a short and sequence-homogenous helix. This assumption has been verified by NOE technique in the study of d-CCAAGCTTGG helix (5).

I.3   Assignments of the ^{31}P Resonances.

As shown in Figures 1 and 2, two ^{31}P signals (1:2 ratio) of d-CGCG and three ^{31}P signals (1:3:1 ratio) of d-CGCGCG are observed. Through specific heterodecoupling techniques (6) between ^{31}P and ^{1}H, and the assistance from the assigned $H_{3'}$, $H_{4'}$, $H_{5'}$ and $H_{5''}$ resonances, the ^{31}P resonances of these two helices were unambiguously assigned.

I.4   Concluding Remark

The achievement in the complete assignment of all exchangeable and non-exchangeable proton resonances and ^{31}P resonances of d-CGCG and d-CGCGCG helices in B form establishes these two short CG helices as a valuable model system for the study on NMR and polymorphism of nucleic acids (7).

II.   A New Strategy for the Assignment of Non-Exchangeable Proton Resonances in Short DNA Helices.

The currently available assignment method described above relies in part on the "incremental procedure" (3,4), and therefore can be applied only if a series of sequence related shorter oligonucleotides is available (3,4) or if there is no repetition in the sequence (8,9). Thus, a new strategy for the assignment of the non-exchangeable proton resonances in a short helix is proposed (10). This method is based on: (1) the use of sequential homodecoupling (1D) or COSY (2D) (11) for the interconnection of the sugar resonances (12-15) pertaining to the same sugar residue; (2) the measurement of NOE (1D or 2D) between critically located protons in order to establish the connectivity between the base protons and sugar protons of the same nucleotidyl subunit and of the adjacent subunit.

The reliability of this assignment procedure was tested using the self-complementary helix d-CGCGCG. The assignment of the non-exchangeable base and sugar protons obtained by this new procedure was found to be in perfect agreement with the assignment made previously by the incremental procedure (7).

SUGAR   BASE

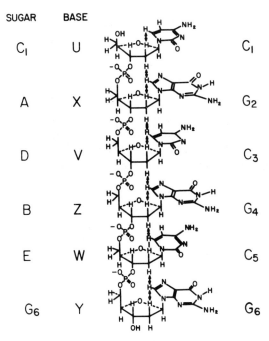

C_1	U		C_1
A	X		G_2
D	V		C_3
B	Z		G_4
E	W		C_5
G_6	Y		G_6

Fig. 3.   Schematic representation of a single strand of the d-CGCGCG
          molecule. The arrows connect the protons which are located at a
          distance of less than 3 Å from each other.

        As shown in Figure 3, and described in I, the $H_{3'}$, $H_{5'}$ and $H_{5''}$
resonances of the first and last residues of d-CGCGCG helix can be
identified because of the unique coupling pattern of these protons in
the terminal residues in the absence of ^{31}P splitting. The H_8 of
purines or the H_6 of pyrimidines are close to the $H_{2'}$ of their own
sugar residue (1) (~ 2.3 Å and 2.0 Å respectively). The $H_{2''}$ of the
residue on the 3' side of the phosphodiester linkage is close to the H_8
of the purine (or the H_6 of the pyrimidine) on the 5' side of the
phosphodiester linkage (~ 2.5 Å). Therefore, the interconnectivity
between base protons and sugar protons of the same nucleotidyl subunit
and of the adjacent subunit can be established by NOE measurement
between H_8 of purine or H_6 of pyrimidine and the $H_{2'}$ and $H_{2''}$ protons of
the deoxyribose. NOE can be measured either by 1D (16) or by 2D
technique (12-15). However, if the resonances are well resolved, then
the 1D NOE procedure can be applied in a straightforward sequential
manner (for example, see Ref. 7) starting from the 3' end ($H_{3'}$
resonance) or the 5' end ($H_{5'}$, $H_{5''}$ resonances) of the strand, walking
through the strand in the assignment of interconnectivity. The 2D
method will be preferred if the resonances of interest are closely
spaced or overlapping. Under this condition, it is very difficult
to irradiate selectively one resonance at a time. However, quantitative
measurement of peak intensity is difficult in 2D experiments (14).   In
this case, the assignment has to be obtained from the data matrix.

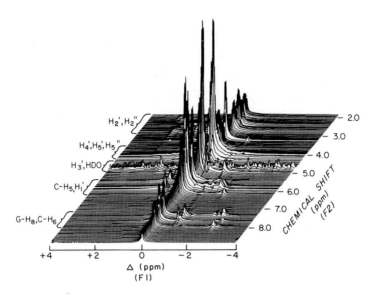

Fig. 4.   300 MHz ^{1}H 2D NOESY stacked plot of d-CGCGCG helix.

Figure 4 shows the stacked plot of the 300 MHz 2D NOESY spectrum of d-CGCGCG at 25°C. The intense peaks along the middle axis (0 ppm, F_1) correspond to the normal one dimensional ^{1}H NMR spectrum of d-CGCGCG and arise from protons that did not cross relax from other protons during the mixing time. As expected, this spectrum shows less resolution than the one shown in Figure 2. For example, the three H_6 of C are clustered together in a single envelope, but they can be located by their chemical shifts.

Figure 5 shows the cross sections of the three H_6 proton resonances of C: U, V and W. From the inter-proton distances described in the preceding section, we would expect that each H_6 of C will have a major NOE effect to the H_2' of its own sugar residue (~ 2.0 Å) and smaller NOE effect to the H_2" of the adjacent residue on the 3' side of the phosphodiester linkage (~ 2.5 Å) and to the H_2" of the sugar from its own sugar residue (~ 3-3.5 Å). Resonance U (Fig. 5a) is associated with a major cross peak at 1.99 ppm and a minor one at 2.44 ppm (1.635 ppm peak due to ammonium salt impurity). Resonance V (Fig. 5b) is associated with a major peak at 2.07 ppm and two smaller ones at 2.81 and 2.48 ppm. Resonance W (Fig. 5b) has a major cross peak at 1.87 ppm and two minor ones at 2.73 and 2.37 ppm. From the homodecoupling experiments, we know that in the region between 1.9 and 2.1 ppm, the H_2' resonances of sugar D, C_1 and E appear at low, intermediate, and high field position respectively. Thus, base U is connected to H_2' in sugar C_1, base V is connected to H_2' in sugar D and base W is connected to H_2' in sugar E. As far as the minor signals are concerned, they should belong to the H_2" of the same residue and the H_2" of the adjacent residue in the 5' direction. Base U (sugar C_1) does not have any adjacent residue in the 5' direction, and it therefore has only one minor cross peak at 2.44 ppm which arises from the H_2" of its own sugar, as confirmed by homodecoupling experiments. Base V (sugar D) has two minor cross peaks: one at 2.81 and one at 2.43. The 2.43 ppm peak belongs to the H_2" of its own sugar residue (D) as seen by homodecoupling experiments. Therefore, the peak at 2.81 ppm is associated with the residue adjacent to it in the 5' direction. Base W (sugar E) has minor cross peaks at 2.73 and 2.37 ppm. By similar arguments, we can show that the H_2" resonance at 2.37 ppm belongs to the same sugar residue (sugar E) and the one at 2.73 ppm belongs to the adjacent residue in the 5' direction.

From the cross section of three $G-H_8$ resonances, similar interconnectivity between proton bases (X, Y, Z) and sugar protons (A, G_6, B) can be established. Together with the cross sections in the H_2' and H_2" regions, we were able to conclude that the sequence is (10): Sugar: C_1 A D B E G_6
          Base: U X V Z W Y
                  ↑ ↑ ↑ ↑ ↑ ↑
                  C_1 G_2 C_3 G_4 C_5 G_6

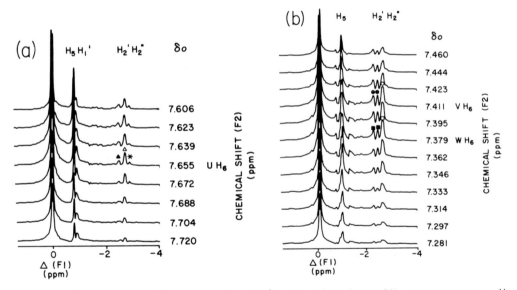

Fig. 5. F_1 cross sections corresponding to the three CH_6 resonances: U (Fig. 5a) V and W (Fig. 5b). $\delta_{cp} = \delta_0 + 2\Delta$.   ▲ = 2.44 ppm, Δ = 1.99 ppm, * = 1.635 ppm, ◐ = 2.81 ppm, ● = 2.43 ppm, 0 = 2.07 ppm, ▨ = 2.73 ppm, ■ = 2.37 ppm and □ = 1.87 ppm.

III.   A New Procedure for the Assignment of the [31]P Resonances of an Oligonucleotide.

Recently, our laboratory has demonstrated an effective method for unambiguous assignment of the ^{31}P resonances of any oligonucleotide whose H_3', H_4', H_5' and H_5" resonances from the furanose-backbone region can be completely resolved and identified (6). However, this method is quite tedious and demanding since the proton spectra of the backbone regions of the oligonucleotides is not easy to be completely characterized (3). Therefore, a new procedure has been developed for the assignment of ^{31}P resonances of an oligonucleotide based on the chemical shift values of ^{31}P resonances of its constitutive dimeric units (17).

III.1   Temperature Dependence of ^{31}P Chemical Shifts of Deoxy Oligonucleotides.

Through the specific hetero-decoupling technique, the ^{31}P resonances of eight short oligonucleotides have been unambiguously assigned. Table III shows the chemical shifts of assigned ^{31}P resonances of d-CCA, d-CCAA, d-CCAAG, d-CCAAGA, d-CCAAT, d-TGG, d-TTGG and d-CTTGG as well as their constitutive dimers. The data indicates that (i) there is a linear temperature dependence of the chemical shifts of ^{31}P resonances of these deoxy oligomers, except d-CCAAT and d-CTTGG (Figures 6 and 7); (ii) the spectral position of ^{31}P resonances of these deoxy oligomers are shifted upfield relatively by 0.2-0.3 ppm as compared to their constitutive dimeric units; (iii) the 3'-OH terminal phosphorus resonance in these deoxy oligomers tend to resonate at a position close to their constitutive dimeric units.

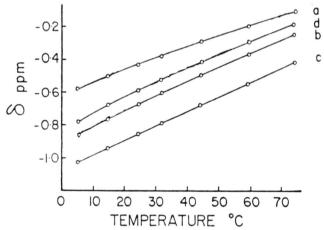

Fig. 6. ^{31}P chemical shift versus temperature profile of $d-C_aC_bA_cA_dG$. The symbols a,b,c represent the designation of different phosphate group.

As shown in Figure 7, four linear curves are observed for the ^{31}P resonances of d-CTTGG at temperatures above 30°C. At temperatures below 30°C, one upward, concave curve for ^{31}P resonance f and another downward, convex curve for ^{31}P resonance h are observed, while the remaining two ^{31}P resonances i and g are nearly linear. These non-linear lines observed for d-CTTGG at low temperature may be due to the intermolecular interactions between C and G residues. Therefore, the

concentration dependence of the ^{31}P chemical shifts of d-CTTGG was
studied. At 20-fold dilution, no pronounced convex or concave curves
are observed. This observation suggests that the non-linear temperature
dependence of chemical shifts observed for d-CTTGG in Figure 7 is due to
the intermolecular interactions.

TABLE III.   Chemical shifts (δ) of ^{31}P resonances of deoxyoligomers in
              D$_2$O, 0.01 M phosphate buffer, pD 7.4 at different
              temperature (ppm from 0.2 M H$_3$PO$_4$ in 15% aqueous solution
              as reference)^a

		δ 30°C	δ 65°C
d-CCA	C$_p$C	-0.35 (-0.16)	-0.09 (+0.06)
	C$_p$A	-0.39 (-0.35)	-0.12 (-0.07)
d-CCAA	C$_p$C	-0.38 (-0.16)	-0.13 (+0.06)
	C$_p$A	-0.62 (-0.35)	-0.29 (-0.07)
	A$_p$A	-0.62 (-0.61)	-0.29 (-0.28)
d-CCAAG	C$_p$C	-0.39 (-0.16)	-0.16 (+0.06)
	C$_p$A	-0.62 (-0.35)	-0.32 (-0.07)
	A$_p$A	-0.81 (-0.61)	-0.50 (-0.28)
	A$_p$G	-0.55 (-0.56)	-0.24 (-0.24)
d-CCAAGA	C$_p$C	-0.42 (-0.16)	
	C$_p$A	-0.66 (-0.35)	
	A$_p$A	-0.85 (-0.61)	
	A$_p$G	-0.76 (-0.56)	
	G$_p$A	-0.42 (-0.46)	
d-CCAAT	C$_p$C	-0.39 (-0.16)	-0.13 (+0.06)
	C$_p$A	-0.62 (-0.35)	-0.30 (-0.07)
	A$_p$A	-0.77 (-0.61)	-0.44 (-0.28)
	A$_p$T	-0.62 (-0.46)	-0.24 (-0.18)
d-TGG	T$_p$G	-0.43 (-0.50)	-0.12 (-0.17)
	G$_p$G	-0.62 (-0.45)	-0.27 (-0.17)
d-TTGG	T$_p$T	-0.41 (-0.36)	-0.11 (-0.08)
	T$_p$G	-0.64 (-0.50)	-0.36 (-0.17)
	G$_p$G	-0.67 (-0.45)	-0.34 (-0.17)
d-CTTGG	C$_p$T	-0.36 (-0.31)	-0.12 (-0.05)
	T$_p$T	-0.64 (-0.36)	-0.31 (-0.08)
	T$_p$G	-0.64 (-0.50)	-0.28 (-0.17)
	G$_p$G	-0.37 (-0.45)	-0.17 (-0.17)

^{a} Data in parenthesis correspond to the chemical shifts of constitutive
    dimers

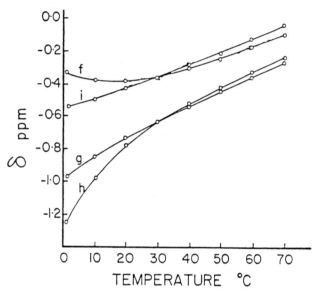

Fig. 7. ^{31}P chemical shift versus temperature profile of d-C$_f$T$_i$T$_g$G$_h$G.
(Same as Fig. 6).

III.2 Comparison of Chemical Shifts of ^{31}P Resonances of Deoxy
       Oligomers with their Constitutive Dimeric Units.

        Figure 8 shows the chemical shift values of the assigned ^{31}P
resonances of d-CCA, d-CCAA, d-CCAAG and their constitutive dimers at
30°C. A careful analysis of the data shown in Figure 8 reveals a
general pattern of change in chemical shifts of the ^{31}P resonances of
these deoxy oligomers as compared to the chemical shifts of the
constitutive dimeric units. The ^{31}P resonance d-CpC is shifted upfield
by 0.18 ppm in d-CpCA, by 0.21 ppm in d-CpCAA, by 0.22 ppm in d-CpCAAG.
The ^{31}P resonance d-CpA is shifted upfield slightly by 0.03 ppm in
d-CCpA, while shifted upfield significantly by 0.26 ppm in d-CCpAA and
in d-CCpAAG. Similarly, the ^{31}P resonance d-ApA is shifted upfield
slightly by 0.03 ppm in d-CCApA, while shifted upfield significantly
by 0.22 ppm in d-CCApAG. The ^{31}P resonance d-ApG is shifted downfield
slightly by 0.01 ppm in d-CCAApG.

        From these observations, we can draw three general rules: (1)
the 3'-end terminal phosphorus resonance in an oligomer tends to locate
at a position relatively close to its constitutive dimer; (2) upon chain
elongation (from 5' - toward 3'-end), the phosphorus resonance will be
shifted upfield by 0.2-0.3 ppm as compared to its constitutive dimer.
Addition of base residue onto 5' direction has little effect on the
chemical shift value of phosphorus resonance; and (3) the relative
positions of phosphorus resonances in an oligomer tend to remain in the
same order as their constitutive dimeric units, which is particularly
true at higher temperature.

Therefore, these rules may suggest an effective procedure to assign the ^{31}P resonances in single-stranded oligomers, based on the chemical shift values of the constitutive dimers (17).

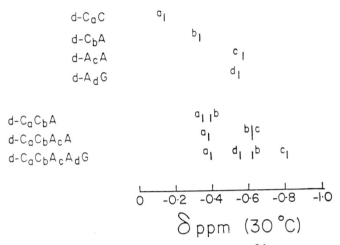

Fig. 8.   Comparison of the chemical shift of ^{31}P resonance of the dimer to the same dimeric unit within the oligomers d-CCA, d-CCAA and d-CCAAG.

IV.   Conformational Analyses.

Table IV shows the sugar conformation of d-CGCG and d-CGCGCG helices (7).   At 9°C in ~100% helical form, d-$(CGCG)_2$ shows a clear preference to populate in ^{2}E-conformation, especially the internal G^2 and C^3 residues.   When the temperature was elevated to 70°, i.e. d-CGCG is now in single-stranded state, ^{2}E population is reduced and approaches to its monomeric value, such as 66% ^{2}E for d-Cp and 70% ^{2}E for pdG (2). Similarly, there is a clear preference for d-$(CGCGCG)_2$ helix to populate in ^{2}E-type conformer.   Internal C^3 and C^5 residues have a slightly higher %^{2}E than the terminal C^1 residue (~5-7%).

The population distribution of conformers about the $C_{4'}-C_{5'}$ bond (%gg) and $C_{5'}-O_{5'}$ bond (%g'g') for d-CGCG and d-CGCGCG are also compiled in Table IV.   Generally, the data show a clear preference for gg and g'g' conformers for both d-CGCG and d-CGCGCG helices.   Elevation of temperature causes a reduction in gg populations.   This is clearly indicated in C^1 residue of d-CGCG, whose $J_{4'5'}$ and $J_{4'5''}$ coupling constants can be accurately determined since $H_{5'}$ and $H_{5''}$ resonances of C^1 residue are clearly resolved at the most upfield region (Figures 1 and 2, Ref. 7).

The rotamer distribution about the $C_{3'}-O_{3'}$ bond (ϕ') for d-CGCG and d-CGCGCG is listed in Table IV (7).   Data shows that both deoxyoligomers have similar ϕ' values, i.e., ~195°/285°.   Since no $J_{H2'}$-P four-bond

couplings were observed in both short oligomers, it appears that the
torsion about the $C_{3'}-O_{3'}$ bond is restricted to a domain around $\phi' \cong$
195°. Temperature or helix-coil transition has little effect on the
average torsion angle of this $C_{3'}-O_{3'}$ bond.

For the decamer helix, d-CCAAGCTTGG, only $H_{1'}$ resonance
region is well-resolved at 600 MHz (5). Therefore, only sugar conformation of that decamer helix can be computed and is listed in Table IV.
The sugar puckering of that helix is predominately in the $C_{2'}$-endo form
ranging from 66% to 87% with an indication of a gradation which is
related to conformational stability from both ends toward the center
of the helix (5).

TABLE IV.   Conformation and population distribution of conformers of
the sugar backbone of d-CGCG, d-CGCGCG and d-CCAAGCTTGG at
different temperatures[a]

Compounds		Temp. °C	% Helix	% ^{2}E	% gg	% g'g'	ϕ'
d-CGCG	C^1	9	100	72	68	-	195°/285°
	G^2			81	96	88	194°/286°
	C^3			75	79	89	194°/286°
	G^4			76	77	87	-
	C^1	70	0	68	51	-	196°/284°
	G^2			74	93	85	195°/285°
	C^3			68	73	87	196°/284°
	G^4			69	71	86	-
d-CGCGCG	C^1	25	100	71	53	-	196°/284°
	G^2			75	85	83	195°/285°
	C^3			76	83	91	195°/285°
	G^4			75	85	82	195°/285°
	C^5			78	83	89	195°/285°
	G^6			76	79	88	-
d-CCAAGCTTGG	C^1	20	100	66			
	C^2			72			
	A^3			82			
	A^4			84			
	C^6			87			
	T^7			77			
	G^{10}			71			

[a] $\%^2E = 100 - [J_{3'4'}/(J_{1'2'} + J_{3'4'})] \times 100$; accurate to $\pm$ 1-2%.
$\%gg = [(13.7 - \Sigma)/9.7] \times 100$; $\Sigma = J_{4'5'} + J_{4'5''}$; accurate to $\pm$ 6-7%.
$\%g'g' = [(25-\Sigma')/20.8] \times 100$; $\Sigma' = J_{5'p} + J_{5''p}$; accurate to $\pm$ 6-7%.
$^3J_{HP} = 18.1 \cos^2 \theta_{HP} - 4.8 \cos \theta_{HP}$; $\phi' = 240° \pm \theta$; accurate to $\pm$ 3°.

These results suggest that it is very unlikely that the furanose conformation of these helices will ever assume a 100% ^2E form. Therefore, while these data clearly do not indicate that these helices assume an A form with a ^3E sugar conformation, the data suggest that the sugar conformation of these helices in solution may differ from those of a "typical" B form derived from fiber diffraction pattern (1).

V. Theoretical Computation.

The through-space magnetic field effect on the base protons in the helices due to the local environment can be calculated in terms of ring-current effect, the local atomic magnetic susceptibility effect, and the polarization or electric field effect, when a set of the coordinates for the helix is provided, such as the coordinates from A, A', B or Z conformation, etc. (18). Among these three contributions of the through-space magnetic field effect, careful experimental calibration and validation have been made for the ring-current effect (RC) and the local atomic magnetic susceptibility effect (LA) using purine stacking in solution as a model system with a mean deviation of about 0.2 ppm for 8 atoms in purine between the calculated and the observed values (19). The polarization effect (P), on the other hand, has not been calibrated and preliminary data (20) indicated that the computed contribution may be much too large, due to the neglect of the positive counter ion effect in the calculation. The experimental evaluation and the theoretical calculation of the through-space magnetic effects are based on the differences in chemical shifts between the monomeric units and the corresponding unit in the tetramer, hexamer or decamer helices (δ_{H-M}).

Table V shows the mean deviation between the observed $\Delta\delta_{H-M}$ and the computed $\Delta\delta_{H-M}$, as well as the standard deviation of the mean deviation in selected cases for both base proton and NH-N resonances of d-CGCG, d-CGCGCG and d-CCAAGCTTGG helices (5,7). For the calculation based on the B form and for all these three short helices, the mean deviation is smallest for the RC + LA treatment (0.15 ppm for tetramer, 0.17 ppm for hexamer and 0.20 ppm for decamer), though in general, these three different treatments (i.e., RC, RC + LA, RC + LA + P) did not yield significantly different results. The mean deviations derived from calculation based on B conformation are within the expected range of 0.2 ppm error and therefore are in support of the conformation of the assignment of these three short helices based on other experimental data. As for the mean deviations derived from the calculation on A and A' conformations, the values become larger when LA treatments are added to the RC treatment, and become largest when both LA + P treatment are added to the RC treatment. Since these helices are not in A or A' form, a larger mean deviation would indicate a better agreement between observed values and computed values. This comparison so far indicates that the agreement between observed values and the compared values would be increased if the LA treatment is added to the RC treatment, while the significance of P treatment is not yet certain. In all cases, however,

not only the differences between the observed and computer mean
deviation values are small for the B form, but also the differences
between the computed values among the three conformations (A, A' and B)
are barely significant.

TABLE V.   Comparison of the observed and computed $\Delta\delta_{H-M}$ (ppm) in three
helical conformations: average error (X) and standard
deviation of averaged errors (S) of d-$(CGCG)_2$, d-$(CGCGCG)_2$
and d-$(CCAAGCTTGG)_2$[a]

		A			A'			B		
		RC	RC+LA	RC+LA+P	RC	RC+LA	RC+LA+P	RC	RC+LA	RC+LA+P
CGCG Base Protons	X	<u>0.17</u>	0.27	0.31	0.21	0.34	0.46	0.19	<u>0.15</u>	0.17
	S								0.18	
NH-N Protons	X	0.31	0.71	0.52	0.25	0.62	0.40	<u>0.16</u>	0.55	0.55
	S							0.11		
CGCGCG Base Protons	X	<u>0.16</u>	0.27	0.37	0.22	0.36	0.45	0.21	<u>0.17</u>	0.21
	S	0.08							0.17	
NH-N Protons	X	0.29	0.66	0.42	0.23	0.75	0.53	<u>0.19</u>	0.66	0.66
	S							0.12		
CCAAGCTTGG Base Protons	X	0.26	0.30	0.36	0.26	0.41	0.59	0.25	<u>0.20</u>	0.27
	S								0.20	
NH-N Protons	X	0.65	0.68	0.70	0.28	0.61	0.66	<u>0.18</u>	0.57	0.54
	S							0.16		

[a] $\bar{X}$ = Average error = $\dfrac{1}{n} \sum\limits_{i=1}^{n} \left| X_{i_{calc.}} - X_{i_{obs}} \right|$

S = Standard deviation of average error = $\left[\left(\sum\limits_{i=1}^{n} \left(\left| X_{i_{calc.}} - X_{i_{obs}} \right| -\bar{X} \right)^2 \right) / (n-1) \right]^{1/2}$ where $\bar{X}$ is the average error.

As for the NH-N resonances, Table V shows that the mean
deviation for all these three helices is smallest for the B form when
only RC treatment is used (0.16 ppm for tetramer, 0.19 ppm for hexamer
and 0.18 ppm for decamer).   The mean deviations for the A and A' forms
are detectably larger.   However, the mean deviations become
significantly larger for all A, A' and B conformations when LA treatment

is added to the RC treatment (LA + RC), while the P treatment is of less significance. This result does not support the application of LA treatment or P treatment to the calculation of the through-space magnetic effect on NH-N protons in the helix at the current stage of understanding.

In summary, sufficient data exist to establish that these short helices in dilute salt solution assume a conformation close to B form and not A, A' or Z form. This information can now be used to calibrate the existing NMR theory and procedure currently available for the calculation of the through-space magnetic effects on the base protons and NH-N protons. The comparison between observed values and the computed values yields two general conclusions: (1) for the base protons, application of both ring current effects and local anisotropic effects is preferred; but for the NH-N protons, the application of ring current effect alone is preferred over the application of both ring current effects and local anisotropic effects; and (2) while the computed values in the best theoretical treatment cited above did support the conclusion that these helices are in the B conformation, the differences among the computed values for these three conformations, A, A' and B, are not very large, at least for these helices (5,7).

References

1. Arnott, S., Hukins, D.W.L. and Dover, S.D (1972) Biochem. Biophys. Res. Commun. 48, 1392.
2. Cheng, D.M. and Sarma, R.H. (1977) J. Am. Chem. Soc. 99, 7333-7348.
3. Cheng, D.M., Kan, L.S., Leutzinger, E.E., Jayaraman, K., Miller, P.S. and Ts'o, P.O.P (1982) Biochemistry 21, 621-630.
4. Borer, P.N., Kan, L.S. and Ts'o, P.O.P (1975) Biochemistry 14, 4847-4863.
5. Kan, L.S., Cheng, D.M., Jayaraman, K., Leutzinger, E.E., Miller, P.S. and Ts'o, P.O.P. (1982) Biochemistry 21, 6723-6732.
6. Cheng, D.M., Kan, L.S., Miller, P.S., Leutzinger, E.E. and Ts'o, P.O.P. (1982) Biopolymers 21, 697-701.
7. Tran-Dinh, S., Neumann, J.M., Huynh-Tinh, T., Allard, P., Lallemand, J.Y. and Igolen, J. (1982) Nucleic Acids Res. 10, 5319-5332.
8. Tran-Dinh, S., Neumann, J.M., Huynh-Dinh, T., Igolen, J. and Kan, S.K. (1982) Org. Mag. Res. 18, 148-152.
9. Cheng, D.M., Kan, L.S., Frechet, D., Ts'o, P.O.P., Uesugi, S., Shida, T. and Ikehara, M. (1983) submitted to Biopolymers.
10. Fretch, D., Cheng, D.M., Kan, L.S. and Ts'o, P.O.P. (1983) Biochemistry (in press).
11. Aue, W.P., Bartholdi, E., Ernst, R.R. (1976) J. Chem. Phys. 64, 2229-2246.
12. Kumar, A., Ernst, R.R. and Wuthrich, K. (1980) Biochem. Biophys. Res. Commun. 95, 1-6.

13. Kumar, A., Wagner, G., Ernst, R.R. and Wuthrich, K. (1980) Biochem. Biophys. Res. Commun. 96, 1156-1163.

14. Kumar, A. Wagner, G., Ernst, R.R. and Wuthrich, K. (1981) JACS 103, 3654-3658.

15. Bosch, C., Kumar, A., Baumann, R., Ernst, R.R. and Wuthrich, K. (1981) J. Mag. Res. 42, 159-163.

16. Noggle, J.H. and Schirmer, R.E. (1971) The Nuclear Overhouser Effect: Chemical Applications, Academic Press, New York.

17. Cheng, D.M., Kan, L.S., Iuorno, V. and Ts'o, P.O.P. (1983) submitted to Biopolymers.

18. Giessner-Prettre, C., Ribas-Prado, F., Pullman, B., Kan, L.S., Kast, J.R. and Ts'o, P.O.P. (1981) Comp. Prog. in Biomed. 13, 167-184.

19. Cheng, D.M., Kan, L.S., Ts'o, P.O.P., Giessner-Prettre, C. and Pullman, B. (1980) J. Am. Chem. Soc. 102, 525-534.

20. Cheng, D.M. and Ts'o, P.O.P., Giessner-Prettre, C. and Pullman, B. unpublished data.

THEORETICAL PROBES OF NUCLEIC ACID CONFORMATION

Wilma K. Olson
Rutgers, The State University of New Jersey
Department of Chemistry
New Brunswick, New Jersey 08903, U.S.A.

ABSTRACT

The computations described briefly here illustrate the interrelationship between the local structure and macroscopic behavior of the
DNA helix. Statistical mechanical studies help to identify the most
likely morphological arrangements of the polynucleotide backbone and
to understand the macroscopic flexibility of the DNA as a whole. Model
building and potential energy calculations uncover the detailed local
geometries of the chain and clarify the likely pathways between the
multitude of allowed spatial forms.

INTRODUCTION

The application of polymer chain statistics, or configurational
statistics, to the treatment of DNA provides an essential link between
the subtle features of chemical architecture and the unique physical and
biological properties of this chain molecule. Information regarding
structure and function arises from the ability of detailed chain models
to reproduce measured values of various experimentally determined chain
properties. The models, as outlined below, depend jointly upon the
structural geometry (i.e., chemical bond lengths and valence bond angles)
of the 2'-deoxynucleotide repeating unit and the potentials impeding free
rotation about its single bonds. The spatial arrangements of lowest
energy dominate the computed chain averages and ostensibly account for
the observed experimental behavior of the chain.

Comprehension of the spatial properties of DNA, however, is
complicated by the multidimensionality of its conformation space. Even
when "structural" variables are held fixed, the nucleic acid chain is
capable of assuming an enormous number of three-dimensional conformational arrangements. Individual bases, for example, can be found in both
anti and syn arrangements with the bulky portions of their structures
(i.e., the 2'-keto group of a pyrimidine or the six-membered ring of a
purine) directed respectively away from or toward the sugar ring (1).
The sugars themselves are known to interconvert between two principal
217

B. Pullman and J. Jortner (eds.), Nucleic Acids: The Vectors of Life, 217–227.
© *1983 by D. Reidel Publishing Company.*

Figure 1. Computer generated
segment of a DNA chain showing
chemical bonds and internal rotation
angles. Torsions are named in
terms of the chemical bonds with
which they are associated (χ for C-
N, ϕ' and ϕ for C-O, ψ' and ψ for
C-C, ω' and ω for P-O). The
$\alpha\beta\gamma\delta\epsilon\zeta$-angle nomenclature used in
other chapters of this volume is
noted in parentheses.

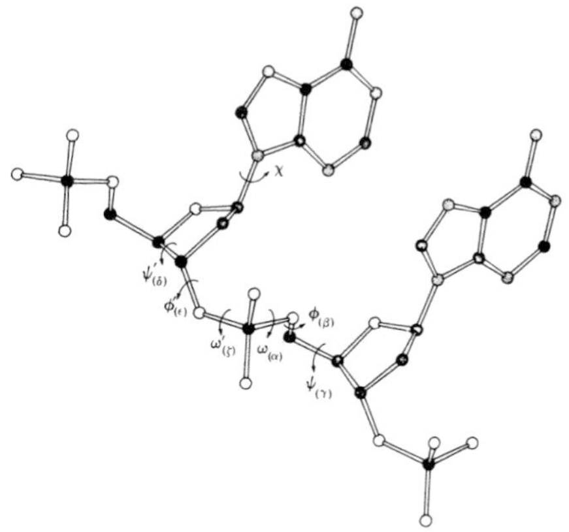

puckered forms, commonly termed C2'-endo and C3'-endo, where the specified atoms are displaced out of mean-planes described by the remaining
ring atoms (2). The five exocyclic backbone torsions are generally
confined to staggered angular arrangements in the so-called <u>trans</u> (<u>t</u>)
and <u>gauche</u> (g) ranges (where the torsion angles, defined relative to a
value of 0° for the planar <u>cis</u> state, are roughly 180°±30° and ±60°±30°,
respectively) (3). The P-O bonds, the exocyclic C5'-C4' rotation, and
the O5'-C5' torsion are able to adopt all three possible conformers
(<u>t</u>, g$^+$, g$^-$). The C3'-O3' angles, however, are restricted by severe
steric contacts to only two (<u>t</u> and g$^-$) angular ranges. (See Figures 1
and 2 for a more detailed description.)

On the basis of such flexibility there are accordingly at least
2x2x2x3x3x3x3=648 distinct conformational arrangements per mononucleotide repeating unit and approximately $648^X = 10^{2.8X}$ potential geometric
combinations in a single-stranded DNA chain of x repeating units. In
addition, the potential barriers separating the various torsion angle
ranges are low enough that a single repeating unit, and hence the chain
as a whole, is capable of assuming a much larger number of conformational
arrangements than cited here. Fortunately, the rotational states
available to a single nucleotide unit are roughly independent of those
associated with neighboring units (3) so that polynucleotide chain
flexibility can be treated in terms of an assembly of independent repeating units.

CONFIGURATIONAL STATISTICS

Among the various properties that depend upon the spatial conformation of the polynucleotide are the chain extension and cyclization
probability. These properties are determined by vector or tensor
contributions from individual skeletal bonds and groups that must be
first summed over all contributing residues and then averaged over all
arrangements of the chain. The average unperturbed end-to-end separation

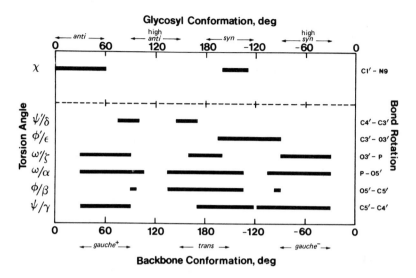

Figure 2. Diagram illustrating the allowed (shaded) and disallowed rotational ranges of the nucleic acid chain.

of the chain ends $\langle r^2\rangle_0^{\frac{1}{2}}$, for example, is obtained from the average scalar product of the vector connecting the ends of the chain

$$\langle r^2\rangle_0^{\frac{1}{2}} = \langle \underset{\sim}{r}\cdot\underset{\sim}{r}\rangle^{\frac{1}{2}} \tag{1}$$

where the end-to-end vector $\underset{\sim}{r}$ is constructed from the n constituent chemical bond vectors $\underset{\sim}{\ell}$.

$$\underset{\sim}{r} = \sum_{i=1}^{n} \underset{\sim}{\ell}_i = \underset{\sim}{\ell}_1 + \underset{\sim}{T}_1\underset{\sim}{\ell}_2 + \underset{\sim}{T}_1\underset{\sim}{T}_2\underset{\sim}{\ell}_3 + \cdots + \underset{\sim}{T}_1\cdots\underset{\sim}{T}_{n-1}\underset{\sim}{\ell}_n \tag{2}$$

The $\underset{\sim}{T}_i$ appearing in this expression are transformation matrices that relate coordinate systems of adjacent chemical bonds (4). The mean-square radius of gyration $\langle s^2\rangle_0$ is related to the average separation distances of all atom pairs (i,j) in the chains.

$$\langle s^2\rangle_0 = (n + 1)^{-2} \sum_{i<j} \langle r_{ij}^2\rangle_0 \tag{3}$$

The $\langle r_{ij}^2\rangle_0$ are determined using expressions analogous to those outlined above for $\langle r^2\rangle_0$ where $i=0$ and $j=n$.

Distribution functions provide more comprehensive descriptions of the flexibility of the polynucleotide chain than the average values associated with the complete array of its internal conformations. The averages simply monitor the extension of a system rather than its macromolecular stiffness. A distribution function, on the other hand, describes the probability per unit volume of space that a polynucleotide molecule adopts a particular end-to-end arrangement $\underset{\sim}{r}$, the variations of probability with chain displacement determining both the relative extension and the stiffness of the chain.

The distributions may alternatively be regarded as three-dimensional clouds of chain termini that cluster in some specific pattern (5). The positions, shapes, sizes, and densities of the clouds are functions of both the chain length and the detailed conformational behavior of the chain backbone. The centers of the distributions are located at the so-called persistence vectors, a (6). This parameter is the mean vector location of the remote end of a polynucleotide chain with respect to a reference frame embedded within its backbone. Because the nucleic acid molecule is subject to structural constraints of fixed bond lengths and valence angles as well as to limitations in internal rotations, a is a non-null vector for all real chains, including those of infinite length.

The distribution of end-to-end vectors connecting the terminal atoms of an infinitely long polymer chain free of excluded volume effects is well known to mimic an ideal random walk (4,7) and can be described by a spherically symmetric Gaussian function with maximum probability density at its center, a,

$$W_a(\rho)d\rho = (3/2\pi<\rho^2>)^{3/2} \exp(-3\rho^2/2<\rho^2>)d\rho \tag{4}$$

In this expression $\rho^2 = (\rho \cdot \rho)/3$ is the mean-square displacement of vector $\rho = r-a$ along any one of its orthogonal coordinate axes. At shorter chain lengths, the distributions of end-to-end vectors are distorted by the restrictions of chemical architecture and internal bond rotations in DNA to nonspherical forms that are difficult to describe in concise mathematical terms. The probability densitities must then be estimated on the basis of a sample of computer-generated chains or approximated by the first few terms of an infinite series expansion (8). The former Monte Carlo method, however, is feasible only in relatively small molecular systems. With DNA chains greater than 50-100 nucleotide repeating units, it is generally impractical to obtain a reliable sample of Monte Carlo chains (9). At such chain lengths, where the Gaussian limit does not generally hold, it is necessary to estimate the distribution by a series expansion. Because it is desirable to choose a mathematical function that will converge to Gaussian behavior at long chain lengths, the expansion is frequently expressed in terms of Hermite polynomials (4). The spatial density distribution of a polynucleotide chain of x = n/6 nucleotide repeating units is then given by

$$W_a(\rho) = W_a^0(\rho)\left[1 + \sum_{\nu=3}^{\infty} (\nu!)^{-1} <H_\nu> H_\nu(\tilde{\rho})\right] \tag{5}$$

where $W_a^0(\rho)$ is the Gaussian expression detailed in Equation 4, $<H_\nu>$ is the configurational average of the ν-th Hermite polynomial tensor $H_\nu(\tilde{\rho})$, and $\tilde{\rho}$ is a reduced displacement vector given by

$$\tilde{\rho} = <\rho \cdot \rho>^{-\frac{1}{2}} \rho \qquad\qquad\qquad (6)$$

For polynucleotide chains not much below the Gaussian limit only a few Hermite terms involving moments through $\nu = 4$ are required to reproduce the Monte Carlo estimates of probability density (5,9,10). For much shorter chains, however, a large number of Hermite correction terms is required to describe the skewed distribution functions.

The facility with which the polynucleotide chain forms condensed end-to-end structures is related to the distribution functions describing the locations of the remote ends of the molecule (10). In order that the terminal residues of the chain undergo cyclization reactions, the ends of the molecule must come within some appropriate distance of one another. The likelihood of closed circular structures, for example, is measured by $W_a(\rho)$ near $\rho = -a$ (i.e., where $r = 0$). The occurence of such structures is also contingent upon angular correlations between terminal residues of the chain (10). The fulfillment of such requirements can be monitored by orientational factors $\Gamma(\gamma)$ that describe the probability of finding terminal bond vectors in some ideal angular arrangement $\xi = \cos^{-\frac{1}{2}}\gamma$. Both Monte Carlo and series expansion approximations analogous to those described above for chain displacement are used to evaluate the angular terms (10,11).

CONFORMATIONAL INFERENCES

The observed experimental extension of double helical DNA is indicative of a locally stiff structure (9,10,12,13). Values of the mean-square radius of gyration $<s^2>_0$ reported in light scattering studies (14-16) of calf thymus DNA can be matched only by models that severly limit the motions of adjacent residues. According to Figure 3, the experimental dimensions of DNA duplexes containing 200-2000 base pairs (cross symbols) are consistent with theoretical models that restrict the tilting of adjacent base planes Λ to angles of 10-12° and the average vertical base-base separation distance Z to values of 3-4 Å. The bases of the polynucleotide chain are well known to align in energetically favored parallel arrays that resemble a stack of coins. An overlapping, stacked geometry is also known to place a large number of nonbonded atom pairs at or near their preferred van der Waals separation distance (17), thereby providing a sizable attractive contribution to the total potential energy of the system. No such components are found in unstacked arrangements where the interatomic distances are large and the van der Waals energies die out. According to thermodynamic and direct calorimetric measurements of base association (18), each single-stranded pairwise stacking interaction is responsible for 2-5 kcal/mole stabilization of the duplex structure. As evident from Figure 3, theoretical models that disrupt the normal parallel alignment of stacked bases are less extended structures than those observed experimentally. Chains with greater local flexibility are also found to approach their characteristic asymptotic limit $C_\infty = <s^2>_0/n\overline{\ell^2}$, where $\overline{\ell^2}$ is the mean-square chemical bond length, more

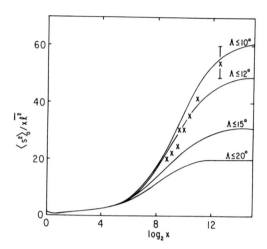

Figure 3. Characteristic ratio of the mean-square unperturbed radius of gyration of DNA as a function of the number of repeating units (x). Curves are distinguished by the local flexibility of internal units. Experimental data on calf thymus DNA (14-16) are noted by crosses.

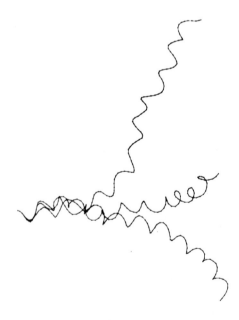

Figure 4. Computer generated perspective representations of a flexible DNA helix. Three 128-segment Monte Carlo structures are drawn in a common coordinate system embedded within the first unit of each strand. For purposes of clarity, the chains are represented by the sequences of virtual bonds connecting successive phosphorus atoms.

rapidly than stiffer chains. Moreover, the dimensions are dramatically depressed by the occasional introduction of sharp kinks or bends into the DNA chain (9). Bends of 90° must be limited to less than one per thousand to reproduce the observed extension of the DNA duplex.

As illustrated in the perspective drawings of Figure 4, there is little deviation of the DNA backbone from ideal helical geometry at short chain lengths. The nucleic acid as a whole can be described as a rigid rod at chain lengths as great as 50 residues. The individual turns of these short pseudohelical chains, however, are seen to differ appreciably in size as a consequence of local variations in the nucleotide repeating units. As chain length increases and the number of such structural distortions build up, the helical chains are found to deviate significantly from regular helical geometry. The ends of these longer chains (~100 residues) are described by the mushroom-shaped density distribution functions typical of wormlike flexibility (9). With further increase in chain length the molecules are able to bend gradually back upon themselves. Indeed, at chain lengths greater than 200 residues, the stiff helical backbones modeled in Figure 4 are found to condense without the occurrence of severe bends or kinks into circular structures (10). Chain of 200 residues, however, are much more likely to form hairpin loops where terminal chain units are anti-parallel than ring structures where the same bonds are parallel to one another. Orientational correlations between terminal nucleotide units are not found to vanish until the DNA chain attains lengths of 1,000 residues or more. In fact the DNA duplex is not found to achieve ideal Gaussian behavior until it contains more than 8,000 repeating residues (9,10).

The very slow attainment of Gaussian behavior in the DNA molecule stems from the severe restrictions upon local bending and twisting of adjacent chain units. These limitations upon local morphology, however, do not necessarily restrict the DNA backbone to a single conformational arrangement. The relatively mobile nature of the sugar-phosphate backbone together with the relatively long stretches of chemical bonds between adjacent residues introduce an unusual local mobility in the macroscopically stiff chain (3).

As evident from the compilation of torsion angles in Table I, there are more than 25 distinct conformational arrangements of the deoxyribose-phosphate backbone that can describe a regular highly stacked, right-handed DNA helix. In these examples the chain is restricted to ordered arrangements with exactly eleven residues per helical turn and vertical displacements of 3.0 Å between adjacent units. The phosphodiester torsions associated with each helix in the table are further illustrated in Figure 5 where the C-C and C-O rotational changes primarily responsible for the observed variations between examples are designated by variously styled connecting lines. The continuous sequences of torsional changes along these lines are likely models of the smooth rotational transitions between major helical forms of DNA. The route between points a and o, for example, is a probable pathway linking A- and B-DNA while that between a and b is a possible transformation from A- to Watson-Crick DNA. All of the torsion angle combinations in the table as well as the intervening transition states described in Figure 5 are consistent with the macroscopic properties of the DNA chain. Any combination of these conformational states is found to reproduce the observed chain extension (14-16) and the loop closure rates (19) of the DNA duplex. All of the arrangements are also roughly equivalent in terms of non-bonded van der Waals' interaction energies (3). The inability to discriminate among these structures notwithstanding, the straightforward computations of chain extension and helical parameters are able to eliminate the majority (more than 90%) of the potential monomeric conformations theoretically available to a DNA chain.

Table I   Backbone conformations of right-handed 11_3 DNA helices

Phosphodiester state (Figure 5)	Backbone bond rotations (deg.)					
	C4'-C3'	C3'-O3'	O3'-P	P-O5'	O5'-C5'	C5'-C4'
a	78	-155	-66	-72	-179	59
b	78	-155	-56	159	-179	180
c	78	-155	-48	50	-179	-60
d	78	-155	-141	148	-90	180
e	78	-155	22	177	90	180
f	78	150	-156	-135	-90	180
g	78	-90	-143	99	-90	180
h	78	150	71	-82	-90	-60
i	78	-90	-28	-130	90	180
j	78	150	63	123	90	180
k	78	150	-168	74	90	59
l	78	-155	-146	10	90	59
m	78	-90	-145	-38	90	59
n	78	-90	-37	58	-90	-60
o	137	-133	-123	-80	180	42
p	137	-133	-109	139	180	180
q	137	-133	-118	38	180	-60
r	137	-90	-150	-109	180	42
s	137	150	-80	-86	-90	42
t	137	-90	-137	124	180	180
u	137	-133	-52	180	90	180
v	137	150	12	130	90	180
w	137	150	168	-162	-90	180
x	137	150	-96	89	136	-60
y	137	150	-162	41	136	42
z	137	150	179	68	90	42
zz	137	-82	-153	-80	136	42

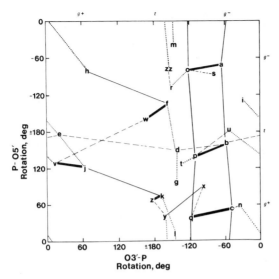

Figure 5.   Diagram that illustrates the various conformations (a-zz) of
the phosphodiester angles that generate right-handed 11_3 DNA helices
with internal base stacking. The style of the lines connecting pairs
of states is used to illustrate the accompanying rotational changes,
if any, in the four sugar torsions: C4'-C3'(———), C3'-O3'(···), O5'-C5'
(---), and C5'-C4'(▬▬).

   The precise blend of local conformational states in the DNA helix
must be determined through potential energy studies. The computed
preferences are generally found to reflect the conformational preferences of the simple low molecular weight nucleic acid analogs around
which the potentials are parameterized (3). Accordingly, the 2'-deoxy-
ribose ring is more apt to adopt a C2'-endo conformation than a C3'-endo
form and the exocyclic C5'-C4' bond is more likely to adopt a g$^+$ than
either a t or a g$^-$ conformational arrangement. <u>Gauche</u> combinations of
the P-O torsions are also slightly favored over <u>trans</u> arrangements in
potential energy calculations (20). The latter studies are approximate
and in need of further investigation. The extent to which the interactions of DNA with surrounding counterions and aqueous solvent alter
these intrinsic torsional preferences is also not understood and an
important area of future research.

   At present there is also no simple mathematical scheme to account
for the complicated long-range angular correlations describing the
various stacked arrangements of the DNA helix. Such a model will, of
course, be required in the treatment of DNA once the potential energies
of the sugar-phosphate backbone are better understood. Until now,
almost all analyses of polynucleotide behavior have been based upon the
interactions of adjacent (i.e., second-neighbor) rotations, although
computations that include the third-neighbor correlations of C4'-C3'
and O3'-P rotations and of P-O5' and C5'-C4' rotations have been offered

(21). The computations of macromolecular stiffness described in Figure 3 and 4, for example, are based upon a virtual bond scheme (22) that permits interdependent conformational mobility in the phosphodiester rotations alone. All other bonds in the chain are held fixed so that it is impossible to introduce appreciable local flexibility in the double helical backbone (9,10). Such mobility can be achieved only by varying pairs of distant bonds, such as third-, fourth-, or even higher neighbors. The correlated rotations associated with DNA base stacking can, however, be treated with a series of virtual bonds (21-23) of different torsion angle composition. Properties of the nucleic acid as a whole are determined in terms of averaged virtual bond lengths and the averaged transformation matrices that relate coordinate systems affixed to adjacent virtual bonds. The averages are obtained by weighting the virtual bonds and transformation matrices associated with each rotational state combination of backbone angles by an appropriate statistical (i.e., potential energy) factor.

ACKNOWLEDGMENTS

This research was sponsored by grants from the U.S. Public Health Service (CA 25981 and GM 20861) and the Charles and Johanna Busch Memorial Fund of Rutgers University. Computer time was supplied by the Rutgers University Center for Computer and Information Services.

REFERENCES

1. Haschemeyer, A.E.V. and Rich, A.: 1967, J. Mol. Biol. 27, pp. 369-384.
2. Altona, C. and Sundaralingam, M.: 1972, J. Am. Chem. Soc. 94, pp. 8205-8212.
3. Olson, W.K.: 1982, in "Topics in Nucleic Acid Structures: Part 2" (S. Neidle, Ed.), Macmillan Press, London, pp. 1-79.
4. Flory, P.J.: 1969, "Statistical Mechanics of Chain Molecules," Interscience Publishers, New York, Chapters 2 and 4.
5. Yevich, R. and Olson, W.K.: 1979, Biopolymers 18, pp. 113-145.
6. Flory, P.J.: 1973, Proc. Natl. Acad. Sci. USA 70, pp. 1819-1823.
7. Flory, P.J.: 1953, "Principles of Polymer Chemistry," Cornell University Press, Ithaca, New York, Chapter 12.
8. Yoon, D.Y. and Flory, P.J.: 1974, J. Chem. Phys. 61, pp. 5358-5380.
9. Olson, W.K.: 1979, Biopolymers 18, pp. 1213-1233.
10. Olson, W.K.: 1979, in "Stereodynamics of Molecular Systems" (R.H. Sarma, Ed.), Pergamon Press, New York, pp. 297-314.
11. Flory, P.J., Suter, U.W., and Mutter, M.: 1976, J. Am. Chem. Soc. 98, pp. 5733-5739.
12. Schellman, J.A.: 1974, Biopolymers 13, pp. 217-226.
13. Schellman, J.A.: 1980, Biophys. Chem. 11, pp. 321-328 and 329-337.
14. Godfrey, J.E. and Eisenberg, H.: 1976, Biophys. Chem. 5, pp. 301-318.
15. Jolly, D. and Eisenberg, H.: 1976, Biopolymers 15, pp. 61-95.
16. Kam, Z., Borochov, N., and Eisenberg, H.: 1981, Biopolymers 20, pp. 2671-2690.
17. Olson, W.K.: 1978, Biopolymers 17, pp. 1015-1040.

18. Ts'o, P.O.P.: 1974, in "Basic Principles in Nucleic Acid Chemistry,"
     vol. 1 (P.O.P. Ts'o, Ed.), Academic Press, New York, pp. 453-584.
19. Shore, D., Langowski, J., and Baldwin, R.L.: 1981, Proc. Natl.
     Acad. Sci. USA 78, pp. 4833-4837.
20. Sheridan, R., Krogh-Jespersen, K., Levy, R.M., and Olson, W.K.:
     unpublished data.
21. Olson, W.K.: 1980, Macromolecules 13, pp. 721-728.
22. Olson, W.K.: 1975, Macromolecules 8, pp. 272-275.
23. Olson, W.K. and Flory, P.J.: 1972, Biopolymers 11, pp. 1-23, 25-56,
     and 57-66.

A NOVEL CONFORMATIONAL EQUIVALENCE BETWEEN THE HEMINUCLEOTIDE BLOCKS
OF THE NUCLEOTIDE REPEAT. ITS IMPLICATIONS IN CONFORMATION ANALYSIS
OF NUCLEIC ACIDS [x]

N.Yathindra and R.Malathi
Department of Crystallography and Biophysics
University of Madras, Guindy Campus, Madras 600025, India

ABSTRACT

   A conformational equivalence between the heminucleotides in the
repeating nucleotide backbone exists which permits to regard the sugar-
phosphate-sugar backbone chain of nucleic acids to be comprised of these
minimum energy compact blocks themselves as the repeating moieties, at
least in the first approximation, instead of the conventional variable
nucleotide unit spanning the successive phosphorous atoms. The heminucleotide scheme brings forth naturally the known variability in the
sugar residue and phosphodiester conformations and the near neighbour
longrange correlations that exist between them. These correlations which
refer to the alignment of adjacent bases for optimal stacking and
hydrogen bonding interactions in nucleic acid helices have been examined
in terms of the torsions around the virtual bonds which span the
successive heminucleotides. The relevance of these in describing the
dynamical conformational movements and even local helical parameter
variations in DNA helices is indicated. The constant and independent
nature of the heminucleotides to the rotations which most often govern
nucleic acid chain folding affords a common principle for description
and hence comparison of different nucleic acid conformations. This has
enabled us to obtain distance diagonal plots for different DNA helices
as well as yeast tRNA[phe]. Analyses of these bring out visualisation of
several features characteristic to each of them. Foremost among them
concerns possible interpretation in terms of chain folding patterns
into distinct domains which may provide an understanding of the apparent
conservation of L shape structure for tRNAs lacking D stem. The diagonal
plot also seems to suggest a correlation between the tRNA structure
and the exon pattern of its gene.

B. Pullman and J. Jortner (eds.), Nucleic Acids: The Vectors of Life, 229–252.
© 1983 by D. Reidel Publishing Company.

INTRODUCTION

In a polymeric chain, especially biopolymers, the recognition of the existence of a common feature with respect to the conformation of the repeating chemical moiety has proved to be crucial in the simplification of not only the understanding of their three dimensional architecture but have also been instrumental in providing stereochemical basis for the development of methods to probe, analyse and predict secondary and tertiary structures. Conformational equivalence of the successive repeating peptides in the _trans_ form in proteins and of the pyranoses in the C_1 chair form in polysaccharides are well known. These forms indeed also correspond to their potential energy minimum. Nucleic acids, in comparison to these, present a challenging task for the application of theoretical approaches to study their molecular conformations, dynamics, energetics and interactions. The presence of six rotatable single bonds in the repeating nucleotide moiety and the associated conformational variability do not exhibit any such conspicuously strict and well characterised conformational equivalence thus elusive of any simple minded approach or treatment that bring down the level of complexity at least to those prevailing in protein and polysaccharide backbones. Such a conformational equivalence for every repeating nucleotide moieties cannot be invoked considering even the energetically most preferred states of the repeating units since there exist two energetically nearly equivalent but distinguishable conformations resulting from the most predominant C3'endo and C2'endo sugar puckers (of course it is recognised such conformational equivalence of the repeating moiety is mandatory for ideally ordered helical poly mono- or poly di-nucleotide helical structures). Nevertheless on careful observation of the stereochemical features of the backbone bonds it becomes recognisable that there indeed exists within the nucleotide unit novel conformational equivalence between the heminucleotide blocks. These conformationally equivalent blocks which also correspond to the energetically most preferred states can themselves be implicated as the repeating structural blocks at least as a first approximation. We describe below details of this scheme and its implications in facilitating the understanding and description of nucleic acid conformations.

Treatment of polynucleotide backbone chain in terms of these conformationally equivalent blocks or through the virtual bonds spanning them readily brings forth most naturally the intrinsic nature of the near-neighbour bond longrange correlated movements that exist between the bonds in the heminucleotides but most importantly provides a convenient handle for their analysis through torsions around them. The overall effect of subtle (such as found in helical conformations) and seemingly large (such as found in tRNA) variations in the chemical bonds on the polynucleotide folding can be meaningfully judged or even better expressed from the analysis of virtual bond torsions. They enable to follow effectively the base sequence induced helical parameter variations in the polynucleotide backbone since they reflect the cumulative effect of chemical bond rotations. The local helicity for example may be caused due to the cumulative effect of sugar pucker as well as P-O3' bond variations rather than sugar pucker alone. Other descriptions like diagonal plots and radial projections which become feasible with the heminucleotide concept because of its uniform applicability are also described for DNA helices and yeast tRNA^{phe}. Application to probe random coil conformations of polynucleotide chains is briefed.

Apparent similarity of heminucleotide blocks with peptides raises intriguing possibility of existence of complementarity relating to their interactions and also conformational similarities between the backbones of polypeptide and polynucleotide chains especially in ordered conformations.

FORMULATION OF THE BLOCKED OR HEMINUCLEOTIDE SCHEME

Chemical bond rotations of the sugar-phosphate-sugar chain of nucleic acid backbone are generally confined[1] to the three staggered states of gauche$^+$(g$^+$), trans(t) and gauche$^-$(g$^-$). Among the three pairs of bonds, namely, P-O, O-C and C-C, the polynucleotide chain most often derives its flexibility required for variant three dimensional structures from the P-O and C-C bond pairs which form next nearest neighbours in the nucleotide repeat. Distinct variations in the double helical secondary structures[2] including the left handed poly(dinucleotide) Z types[3] as well as the tertiary structures found in yeast tRNA^{phe} involve changes (sometimes correlated) in either one of these pairs or both.[4-6] Base-phosphate and sugar-phosphate interactions constrain the other pair of C-O bonds almost always to the trans or near trans conformational state. This readily suggests that the six bond backbone of the nucleotide unit, P-O5'-C5'-C4'-C3'-O3'-P, may be represented[7-10] as a first approximation by two blocks comprising three bonds each, P-O5'-C5'-C4' on the 5'-side, and C4'-C3'-O3'-P on the 3'-side, as shown in Fig.1. The two heminucleotide blocks which are chemically and conformationally symmetric may in turn be represented by a realistic virtual bond[7,8,11,12] joining the atoms P-C4' and C4'-P (Fig.2). The mutual orientation of the virtual bonds are governed by the same two pairs of torsions (ψ, ψ') and (ω',ω) which control the folding mechanics of nucleic acids while their magnitudes are independent of them. The nucleotide and internucleotide conformations can therefore be effectively and conveniently described by virtual bond parameters. Analysis of crystal structure data of nucleotides, oligonucleotides and tRNA show that the magnitude of these blocks is about 3.9 Å (Fig.3). A variation of 0.5 Å is found rarely when the two C-O bonds ($\emptyset$ and $\emptyset'$) exhibit marked deviations from the energetically preferred trans character. The C3'-O3' torsion ($\emptyset'$) associated with the 3'-nucleotide is found[1,13] more often to assume values around 270°. The variation is much smaller compared to the order of distances relevant to the present study and their effect would average out in random coils. Thus the polynucleotide backbone for practical purposes may actually be regarded as comprised of the smaller and conformationally nearly invariant heminucleotides as the effective repeating structural moieties (Fig.1) instead of the conformationally variable nucleotide unit. There are several advantages in this compart-

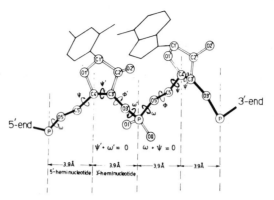

Fig.1   Section of a nucleic acid chain showing the conformationally
        equivalent heminucleotide blocks and their magnitudes.

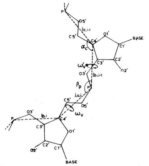

Fig.2   Virtual bond representation of heminucleotides. Torsions around
        them are indicated.

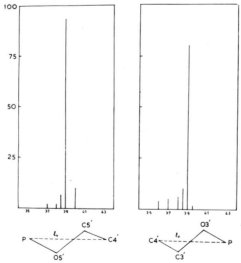

Fig.3   Histograms of virtual bond lengths P-C4' and C4'-P obtained from
        X-ray data of a number of nucleotides, oligonucleotides and yeast
        tRNA^{phe}.

mentalisation: 1) feasibility to incorporate directly sugar pucker dependent conformational variations in conformational calculations, 2) visualisation of the near-neighbour and long range (ω, ψ) and (ω', ψ') correlations and 3) effectiveness in providing a common premisis for stereochemical comparison of all types of nucleic acid conformations.

NEAR-NEIGHBOUR LONG RANGE CORRELATIONS AND CONCERTED MOVEMENTS IN HELICAL NUCLEIC ACID BACKBONE

Conformational energy calculations[14-18] showed that variations in sugar residue conformation (ψ, ψ') by way of either sugar pucker (ψ') or C4'-C5' bond (ψ) torsions affect the energetically favored phospho-diester P-O3' (ω') and P-O5' (ω) bond rotations in a correlated fashion. Helical parameter analysis[19,20] with "mononucleotide" as the helix repeat indicated a strong interrelationship between the helix forming phosphodiester domain (ω', ω) and the sugar residue (ψ, ψ') conformation. The results showed most importantly of a specific and pronounced correlation between the pair of bonds P-O5' (ω) and C4'-C5' (ψ) and P-O3' (ω') and C4'-C3' (ψ'). For instance a change in C4'-C5' bond rotation (ψ) by 120° from the gauche$^+$ (60°) conforma-tion to the trans conformation (180°) in a 5'-heminucleotide brings forth changes only in the P-O5' bond (ω) of the phosphodiester by exactly the same amount but in the opposite direction. Increment by another 120° again induces an identical change. Similar scheme was also found between P-O3' (ω') and C4'-C3' (ψ') bonds of the 3'-heminucleotide. The physical significance of these two independent correlations or movements represents the optimal changes that restore alignment of the adjacent bases for optimum intra-strand (stacking) and inter-strand (base pairing) interactions in the least energetic pathway among the several other conformational possibilities that exist involving rotations around a number of chemical bonds. In other-words this is suggestive of the backbone propensity for helical conformations which always lead to adjacent base overlap and facilitate base pairing interactions. Because of this it was concluded[19] that stacking represents an inherent feature of nucleic acid helices.

It is recognised that the near neighbour bond longrange correlations in polynucleotide backbone arise as a consequence of parallel disposition of the correlated bonds caused by the propensity of C5'-O5' ($\emptyset$) and C3'-O3' ($\emptyset'$) bonds to occur in the trans or near trans conformations. Such pronounced correlations exist probably in all ordered polymer chain conformations and they may govern the chain folding process. Polynucleo-tide backbone folding and consequently the dynamics may depend on neighbourhood correlations described above since there are many more conformational possibilities of bringing together the bases which are distant along the chain by at least eight chemical bonds compared to the highly limited number of possibilities of bringing together, say, bonds which are proximal along the chain. Since the relative disposition of successive bases in a single stranded chain and base pairs in double stranded is left nearly unaltered by the correlated changes, the

movements of the backbone could be associated with the dynamical
properties of DNA and in general of a polynucleotide backbone. They may
also be relevant and even contribute to the local variation of helicity
especially since it is observed that any conformational perturbations
or wripples introduced at any bond site due to sequence or external
effects could be absorbed by first neighbour bonds and they do not
propagate along the polynucleotide backbone beyond second neighbour bond
at the most.[21]

The (ω', ψ') correlation is more subtle and the (ω, ψ) correlation is best seen when large changes involving switch from one staggered
conformation to another occurs. The former brings forth "fine tuning" of
conformational movements and therefore assumes greater relevance. To
probe the effect of sugar ring conformations within the broad family
of C3'endo and C2'endo as well as those in between them helical parameters
analysis with a flexible sugar ring has been made[22] (Fig.4). The effect
of even subtle variations in the sugar pucker on the P-O3' bond torsions
and also on the helical twist is apparent. It is clear from the plot
that sugar puckers like O1'endo and others can also form helices of
required (n, h) values with appropriate correlated variations in the
P-O3' bond but without sacrifice to either base stacking or base pairing
interactions. The correlation is obviously related to the movement of
successive phosphorous atom and hence measures helical twist inherent
to the sugar-phosphate backbone of polynucleotide chain. Experimentally
observed (ω', ψ') angles from the native BDNA dodecamer[23] are plotted
in Fig.4 to indicate the presence of (ω', ψ') correlation in the
structure. All the experimental points (with the exception of two for
which no correlation is expected due to non trans character around the
C-O bonds) lie between n = 6 ∿ 13 in the theoretical plot corresponding
to twist angles in the range 60° ∿ 27° but occur at slightly different
P-O3' bond torsions correlated by sugar ring conformations. The twist
angles are found to vary between 23 ∿ 45° in the crystal structure[23]
indicating a fair agreement. This prompted us to suggest that the
correlated movements of the P-O3' bond torsion wrought by variations
in sugar pucker may also contribute to the observed sequence dependent
helical parameter variations in the BDNA dodecamer.[23] Further evidence
for the linearly coupled movements is provided by a fairly high value
of -0.8 for the calculated circular correlation coefficient (Fig.5).
Further analysis of BDNA data[24] shows that helical twist of the backbone
in fact followsvariations as determined by base parameters. While we
realise that accurate and more data are required to firmly establish
the above results, the data are however sufficient to provide some
confirmatory evidence for the near-neighbour bond long range correlations suggested from theoretical calculations.[19,21,22,25,26] These
correlations are rechristened lately as crankshaft linkage[24,25] and are
also recognised in the molecular mechanical studies.[27,28]

The nature of these concerted movements obviously[29] is always such
that the sum of the torsions ω' + ψ' = $\xi' \simeq$ 0 and $\omega + \psi = \xi \simeq$ 0. These
probably reflect better the backbone structural correlations and such
values obtained from native BDNA dodecamer structure[23] are shown in

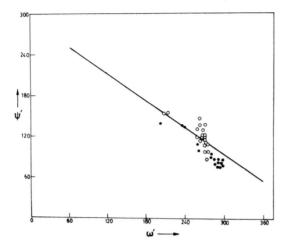

Fig.4 Curves of iso n (6, 8, 10, 12, 20), number of nucleotides per turn
obtained for a flexible sugar ring. h = 0 line is shown by a
continuous line. Approximate ranges of endocyclic (τ_3) and exocyclic (ψ') bond torsions for different sugar puckers are marked
using the relation[49] $\psi' = 121° + 1.11 \ \tau_3$. Experimental points[23]
are indicated by dots.

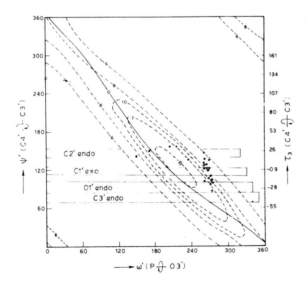

Fig.5 Near neighbour longrange correlation between the C4'-C3' bond of
sugar pucker and the phosphodiester P-03' bond. X-ray data of a
few oligonucleotides including those of B DNA dodecamer[23] (0)
are used to obtain the circular correlation coefficient.[49]

Fig.6. The continuous and discontinuous curves represent the ξ and ξ' variation and both of them are close to zero indicating the suggested coupled movements between the torsions of the phosphodiester and the sugar residue bonds. These subtle variations in the nucleotide backbone geometry should indeed also be effectively reflected in the rotations around the virtual bonds that span across the correlated chemical bonds in the 5' and 3'-heminucleotides. These are also shown in Fig.6. Notice a large similarity in the variations of virtual bond torsions ω'_V and ω_V and ξ' and ξ respectively demonstrating that cumulative effect in the variations of chemical bond torsions are indeed expressed in virtual bond torsions. In fact it can be shown that $\omega'_V = \omega' + \psi'$ and $\omega_V = \omega + \psi$, the values would be approximately zero in nearly stacked helices. Further it is found that $\xi' = \omega'_V + \phi'$ and $\xi = \omega_V + \phi$. Dark dots in Fig.6 correspond to ω'_V values obtained from this relation. It is clear from both the ξ curve as well as virtual bond torsion ω_V that P-O5' and C4'-C5' bonds do not exhibit much variations indicating that 5'-heminucleotide is practically rigid and the backbone variability is embedded in the 3'-heminucleotides. This is in accord with the observation of Dickerson et al.[24] that ω' and ψ' are responsible for the variability in the backbone of BDNA dodecamer. Virtual bond torsions ω'_V show pronounced variations. Calculations[30] of helical twist and residue height as a function of the virtual bond torsions themselves suggest that higher values of ω'_V result in increased helical twist quite sharply, and surprisingly the nature of variation of ω'_V closely represents the helical parameter variation[24] found from consideration of either backbone or side chain bases of BDNA dodecamer. Further it is noticed that large values of ω'_V correspond to the region where a bend is initiated in the structure [23,24] suggesting that increased helical twist is synonymous of a bend in the polynucleotide backbone.[31] Thus the torsional flexibility in DNA helices is guided by fine tuning of sugar pucker dependent correlated conformational fluctuations. The virtual bond properties may provide a measure of these and hence serve as a probe for understanding internal motions of DNA helices.

Similar near neighbour long range conformational interdependence between the bonds separated by a trans chemical bond or even a virtual bond are also found from theoretical investigations [32,33] of poly(di-nucleotide) helices especially of Z-type. These correlations, as in poly(mononucleotide) helices, are related to the restoration of the alignment of the adjacent bases for optimal intra and interstrand interactions involving conformationally heterogeneous "dinucleotide" repeat. Additional correlations are found[26,33] between C3'-O3' torsions (ϕ') and the P-O3' bond torsions (occurring between the repeating di-nucleotides) due to the large values around 270° for ϕ' found in certain Z helical forms. Thus P-O3' bond favors g^+ for ϕ' near trans (t) and g^- for ϕ' near g^- bringing forth significantly large varia-tions.

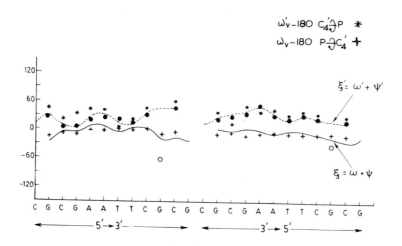

Fig.6   Plot showing the calculated values of $\omega' + \psi'$ (---) and $\omega + \psi$
(——) in B DNA dodecamer.[23] Virtual bond torsions are shown by
+ and ✗. Dark dots correspond to $\omega'_V - \phi'$.

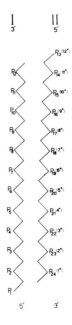

Fig.7   Schematic sketch of double helix indicating the number of heminucleotides considered for computing diagonal plot. The 3'end of
strand I and 5'end of strand II are treated as contiguous.

Application of Heminucleotide Scheme for Backbone Diagonal Plots

An important feature of the heminucleotides is their constancy to
rotations (ψ, ψ') and (ω',ω) which govern the mechanics of polynucleotide backbone folding. This paves the way for comparison and
description of all types of nucleic acid structures on a common footing.
We have exploited this unique aspect to obtain distance diagonal plots
for all the known types of secondary helical structures and yeast tRNA[phe].
An additional feature here is that they also reveal interactions with
the sugar residue through P...C4' and C4'...C4' separations.

Distance Diagonal Plots for DNA Helices. The distinguishing features
in A, B, C, D and Z type structures are brought forth by variations in
sugar residue and phosphodiester rotations. Minor variations in these
strongly affect the helical parameters which cause distinct variations
in the magnitude of helical grooves. These are extremely important in
understanding the interactions with drugs, solvents, metal ions and
hydration effects of helical transitions. A double helix of 12 base
paired residues (48 heminucleotides) which represent greater than one
full turn is considered. The 3'end of strand I and the 5'end of strand
II (Fig.7) are regarded as contiguous and distances of P and C4' atoms
of every heminucleotide in the chain are computed and plotted as shown
in Fig.8 which depicts diagonal plot for A and B DNA helices. The
coordinates used for the A, B, C, D, Z and poly A helices for obtaining
diagonal plots and   radial projections are taken from references 34-
38 respectively. The coordinates for the native BDNA dodecamer[23] were
obtained from Brookhaven protein data bank. Only phosphorous atoms are
numbered and the constant value contours of 10 Å, 15 Å and 20 Å are
drawn. The central region marked H represents the base paired separations of 17-19 A. Patterns below and above are due to non base paired
interactions and represent major (M) and minor (m) grooves respectively.
Major groove in A DNA$_o$consists of distances less than 10 Å with the
shortest P...P $\doteq$ 8.5 A. The P...P separations in the minor groove are
found to be 17 A and appear as hooks extending into 15 A contour. The
shaded region (m) in B DNA plot also corresponds to separations less
than 10 Å but are due to C4$_o$...C4' and C4'...P interactions. The closest
P...P separations are 11.5 A and appear as hooks extending into 10 A
contours. The major groove (M) is not distinctly seen since distances
are of the same order as base paired residues. The reversal in the
width of the major and minor grooves, caused due to variation in the
orientation of the 3'-heminucleotide (i.e. sugar pucker and correlated
changes), is conspicuous.

The reduced size of both the major and the minor grooves in D DNA
by about 3 Å compared to B DNA as a direct consequence[39] of increased
twist (45°) is clearly noticeable in the distance plot shown in Fig.9.
Other features are similar to B DNA. The enlarged shaded region is
indicative of many closer interactions between C4'...C4', P...C4' and
P...P in the minor groove of$_o$D helix. The shortest P...P separation is
about 7.3 A instead of 11.5 A seen for B DNA while the closest C4'...C4'

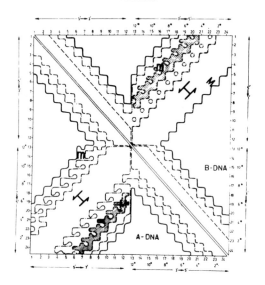

Fig.8   Comparison of diagonal plots of A and B DNA. Constant value
        contours of 10, 15, 20 Å in this and other are indicated by
        (···), (−·−·−) and (——) respectively. The phosphorous atoms are
        numbered. Central region marked H represents distance separations
        of base paired residues. M and m denote major and minor grooves
        respectively. Note the reversal of the widths of major and minor
        grooves.

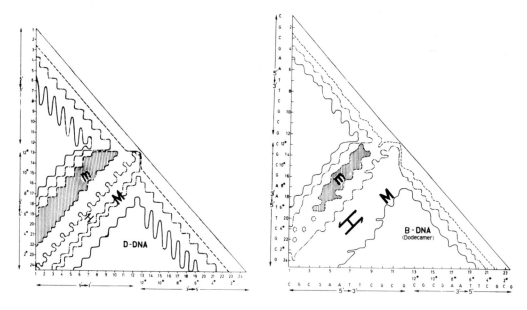

Fig.9   Diagonal plot for D DNA.   Fig.10   Diagonal plot for B DNA
                                                 dodecamer.

and C4'...P separations are of the order of 5 Å
    Distance plot for BDNA dodecamer. It is of interest to obtain the
distance plot of the BDNA dodecamer[23] and compare with the ideal BDNA
helix. The sugar-phosphate backbone variability found to be associated
primarily with the correlated ω' and ψ' rotations in the 3'heminucleotide and the other modifications observed in the structure may
be reflected by the features in the distance plot (Fig.10). Comparison
of Fig.10 with Fig.8 (BDNA) shows that the contours are not regular
revealing the nonuniformity of backbone conformation especially at the
termini.The irregularity is most conspicuous in the minor grooves where
separations lesser and greater than found in BDNA are found. The shortest
separation is about 9 Å which is significantly smaller than 11.5 Å found
in BDNA. The narrowing of the minor groove separations in the AATT
regions as revealed by the X-ray structure[23] can be seen distinctly.
Thus the shaded region (< 10 Å) encompasses a larger and nonuniform
area (which includes P...P separations in addition to P...C4' and
C4'...C4') compared to uniform and narrow pattern found in Fig.8. Also
the shaded region does not run through the entire length (Fig.10)
indicating the larger narrow separations in BDNA dodecamer especially
at the terminal regions compared to BDNA (Fig.8). The irregularity in
the major groove separations is also noticeable by the kinks in the
contours. The subtle variations in the chemical bond torsions in the
3'-heminucleotide and the consequent effect on the helical surface
properties observed are well reflected in the diagonal plot.
    Distance plot for poly A. Fig.11 ilustrates the diagonal plot for
polyadenylic acid which is shown[38] to assume a parallel double helical
structure at acidic pH. The most distinguishing feature in Fig.11 is
the appearance of a broad rectangular domain parallel and away from
diagonal reflecting the parallel nature of the two strands. Region
marked H characterises distances of 11-12 Å reflecting the smaller
diameter of the helix. Shaded disk shaped regions disposed on either
side of H are related by a mirror symmetry and represent closest P..P
separations of 10 Å throughout the double helical groove. No distinction
such as minor or major groove can therefore be made.
    Distance plot for poly(dinucleotide) Z type helices. The constancy
of the heminucleotide irrespective of whether it is a poly(mononucleotide)
or a poly(dinucleotide) helix has permitted the description of the left
handed Z type helical structures also through distance plots. It may be
noted that suggestions[40] were made purely from theoretical considerations
that the flexible framework of conformationally heterogeneous "dinucleotide" repeat may be important for probing helical conformations of
polysequential nucleotides. As in other diagonal plots H in Fig.12
represents base pair residue separations of 13-15 Å reflecting the
smaller diameter of Z helices.[3] Notice the repetitive feature after
every other residue in all the contour levels reflecting the characteristic "dinucleotide" helical repeat. The shaded L shaped pattern
represents the closest separations between the 5'-heminucleotides of
guanine residues involving j, j-1, etc of one strand and i+3, i+4 etc
of the complementary strand. This in fact represents the minor groove
(m) and interestingly it occurs on the side where major groove is found
in BDNA (Fig.8). Note that these are flanked on either side by patterns

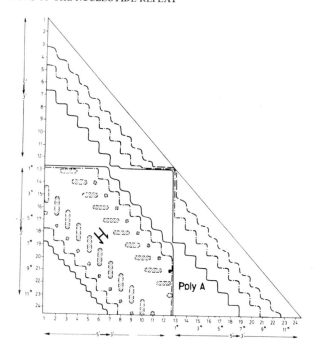

Fig.11 Diagonal plot for polyadenylic acid.

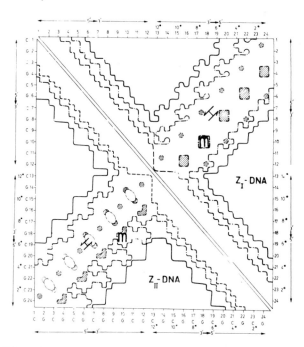

Fig.12 Diagonal plot for Z_I and Z_{II} DNA.

of 15 Å which characterise separations between the 5'-heminucleotides
of cytosines of the two strands separated by two base pairs. The
repetitive occurrence of shortest intervened by large P...P separations,
is indicative of the nonuniformity in the width of the minor groove.
Consideration of alternating phosphate separations here of course leads
to two uniform grooves of different widths and depths in the same
region. The dumbbell pattern and striated circles flanking them and
also the clover leaf patterns seen for Z_{II}DNA refer to certain similarities in the nature of interactions between heminucleotides. It is
seen that the distance plots of Z_I and Z_{II} forms are very similar. The
more compact nature of Z_I is reflected in the slightly larger area of
shaded region representing the minor groove (m) and the conspicuous
absence of dumbbell shaped pattern of 15 Å. It may be of interest that
the shortest interstrand phosphate separations in the minor grooves
occur between phosphates having the g^+g^+ conformation which incidentally
also gives rise to the shortest intrastrand P...P separations.

Distance diagonal plot for yeast tRNAPhe. Probably the distance
matrix together with the heminucleotide scheme find their most useful
application in probing the structure of molecules like transfer RNAs
since not only do they possess a fairly large number of nucleotides and
hence heminucleotides but also they occur in secondary helical as well
as tertiary bend and kink conformations. Fig.13 shows the diagonal plot
obtained for monoclinic yeast tRNAPhe. The coordinates used here are
those supplied along with the Labquip molecular model kit in 1979. The
patterns representing various structural domains, the interactions
between them and the approximate two fold symmetry relating them are
indicated. Four distinct patterns marked 1, 2, 3, 4 with recurring
features corresponding to the four major domains of cloverleaf
secondary structure are seen. Each of these domains may be viewed as
composed of three subdomains with the central narrow strip (20 Å
contour) representing the base paired stem. The patterns (15 Å contour)
just above and below this represent interaction of 5' strand with
residues preceding and succeeding the base paired 3' strand. These
patterns reflect medium and long range interactions and are regular in
certain regions and distorted in some others depending on the degree of
orderliness of the substructures involved in interactions and are
characterised by either helix-helix or helix-loop type interactions.
For instance patterns 5 and 6 along with 1a and 1b are due to long
range tertiary interactions while similar patterns 3a and 4a represent
medium range tertiary interactions. The similarity in the shape of these
patterns reflect similar type of interactions between polynucleotide
stretches having similar conformations. Careful analysis shows that
these tertiary structural patterns arise due to interactions between a
loop and a substructure comprising a loop and a few helical residues,
although the predominant interactions are between loop residues. The
size of these characteristic patterns reflect the number of residues
involved in such interactions. Manifestation of the P10 loop, the
approximate two fold symmetry even between the tertiary structures
probably not hither to obvious, and the noninvolvment of anticodonloop
in any tertiary interactions are the other features explicit in the
plot.

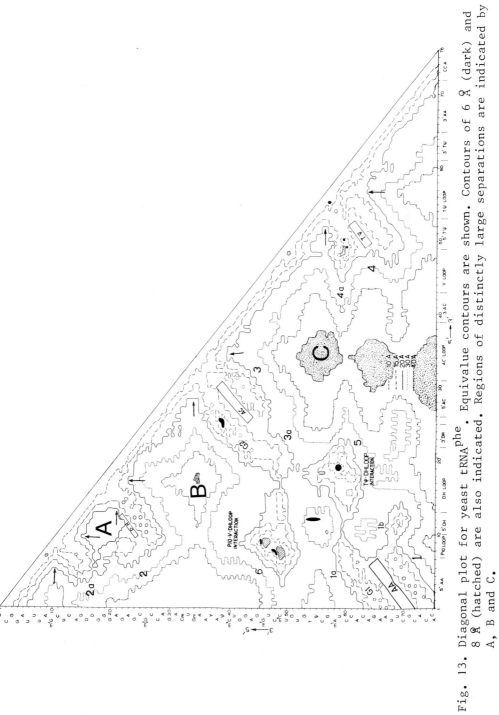

Fig. 13. Diagonal plot for yeast tRNA^Phe. Equivalue contours are shown. Contours of 6 Å (dark) and 8 Å (hatched) are also indicated. Regions of distinctly large separations are indicated by A, B and C.

Interestingly separations $>$ 20 Å occur close to diagonal (marked A) between the patterns representing P_{10} loop and D stem. This indicates that the domain comprising residues 9-26 is sufficiently segregated from the aminoacid and anticodon stems to be regarded as an independent domain. This then clearly explains that those tRNAs lacking the D stem[41] can conserve the L shaped structure. Residue 10 and 26 for example are separated by about 15 A and can easily be linked by a stretch of 3-5 nucleotides. Indeed Sundaralingam has shown[42] the stereochemical feasibility of L framework for such tRNAs. Hence the basic structural functional unit of tRNAs may indeed be this compact form. Probably the ancestral tRNAs did not have the D stem. Separations in region A would get even more amplified in the diagonal plot of such tRNAs. Similarly large separations $>$ 40 Å (marked C) not far from diagonal may be the amplified signal of the segregated nature of the variable stem domain from anticodon and pseudouridine stems. A corollary of this is that the basic framework of tRNAs will not be disturbed by the long variable arm.[43] A diagonal plot of such tRNAs would manifest a domain replacing region C with large separations similar to A flanking it.

Most interestingly a small region B where distances $>$ 40 Å appear and this surprisingly coincides with the anticodon region where an intron is found[44] in the gene of yeast $tRNA^{phe}$. Apparently this separates the two structural domains for which most definite evidence is available with regard to their role in tRNA function. Does this therefore represent a correlation between structure and the exon pattern similar to that ascribed[45] to haemoglobin?

Distance Plots with Side Chain Bases

Distance plots similar to those described above can also be obtained using side chain glycosyl nitrogen atoms since their dispositions are also governed by (ψ, ψ') and (ω', ω) rotation pairs. Hence the results obtained from backbone and sidechain are correlatable. Fig.14 shows such a plot computed for yeast $tRNA^{phe}$. Correlation of the different patterns occurring due to helices, loops and bends are readily seen. Besides predominant stacking (5 Å contour close to diagonal) tertiary base pairing (2b in Fig.14) and tertiary stacking or inter- calative (6a and 6b in Fig.14) interactions are vividly revealed. Consideration of contours of 15, 20 and 30 Å results in virtually identical patterns observed from backbone distance plot (Fig.13) including the two fold symmetry relating the tertiary domains and distinct large separations close to diagonal. The latter may be interpreted as the segregated nature of the chain folding pattern into distinct domains and provide rationale for the conservation of the basic L shape framework for all tRNAs notwithstanding considerable variability in the primary structure such as deletion of complete D stem or presence of a long variable stem.

Similar base diagonal plots obtained for the native B DNA dodeca-

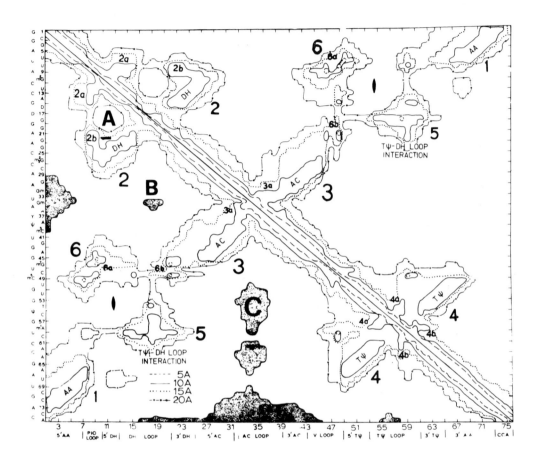

Fig.14 Diagonal plot for yeast tRNA^{phe} obtained using glycosyl N...N separations. Equivalue contours are shown. Manifestation of P10 loop, approximate two fold symmetry and longrange interactions are highlighted. Note the complete resemblance with Fig.13 at higher contour levels. Regions A, B and C indicate distinctly large separations.

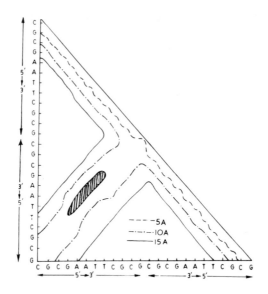

Fig.15   Diagonal plot for B DNA dodecamer obtained with glycosyl N...N
separations. Note the displacement of 10 A contour at the AATT
centre.

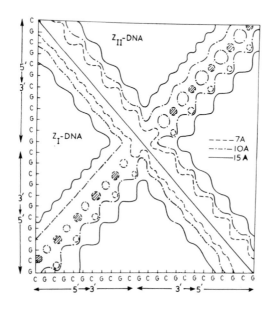

Fig.16   Diagonal plot for Z_I and Z_{II} DNA obtained with glycosyl N...N
separations.

mer[23] (Fig.15) and Z DNA[3] (Fig.16) reveal interesting features. The glycosyl nitrogen separations between paired bases are 8 Å in B DNA[23] and 9 Å in Z DNA.[3] The narrowing (shaded region showing < 8 Å separations) of the minor groove at the AATT centre with concomitant broadening at the termini is clear. Displacement of 10 Å contour inside for residues 4-9 probably reflect the nonlocalised nature of bending[24] in the B DNA dodecamer. Characteristic differences in the glycosyl nitrogen separations involving unpaired bases and their alternating nature found in Z DNA are apparent in Fig.16. The shaded regions represent distances < 7 Å (4 to 5 Å) between cytosines within the base paired dinucleotide repeat. In general it is found that separations due to similar type are larger in Z_{II} DNA compared to Z_I indicating the compact nature of the former.

Application to Obtain Radial Projections

Yet another description to follow the sugar-phosphate backbone especially of helical structures is the radial projections which provide better visualisation of variations in groove widths, orientation of anions, etc. due to differences in helical winding or type of helices. The radial projections are drawn by cutting open the helix along a line parallel to Z axis passing through Z = 0 and projecting on a cylindrical surface. One full turn of a double helix is considered so as to delineate all the relevant features. Atoms P and C4' of the heminucleotides, anionic oxygens and the C1' of sugar are plotted. P and C4' atoms are indicated by closed and open circles and joined by thick lines. The anionic oxygens from P are shown by thin lines. Appropriate base orientations are indicated by dashed lines between C4' and C1'.

Radial projections shown in Fig.17 and Fig.18 for different helices are characterised by a central line representing one of the strands with sections of the other strand appearing above and below. These represent nonbase paired interactions and hence result in major (M) and minor (m) grooves. The separations between the lines provide a measure of groove widths. The reversal in the magnitude of groove widths in A and B DNA and their reduction in C and D DNA are apparent. Note that anionic oxygens face each other in major groove in all these while in Z DNA they do so along minor groove which occurs below the central line. The similarity in Z_I and Z_{II} DNA helices, alternating features typifying characteristic dinucleotide repeat, closest separations between guanine phosphates, proximity of cytosines, sugar-guanine stacking are indicated in the projections.

Application to Probe Random Coil Conformations

Heminucleotide scheme has recently been utilised[7,8,11,12] to compute unperturbed end-to-end dimensions of polynucleotide chains because of its ability to incorporate simultaneously in the calculations

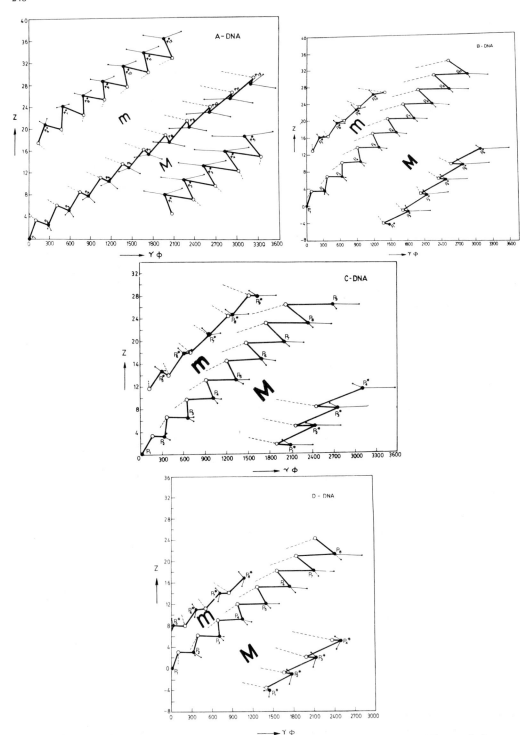

Fig.17 Radial projections obtained for A, B, C and D DNA. Minor (m)
        and major (M) grooves are indicated.

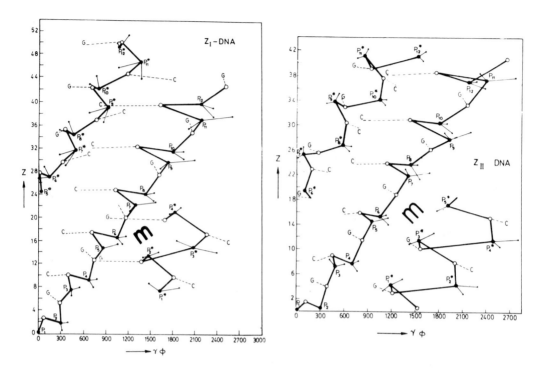

Fig.18   Radial projections obtained for Z_I and Z_{II} DNA.

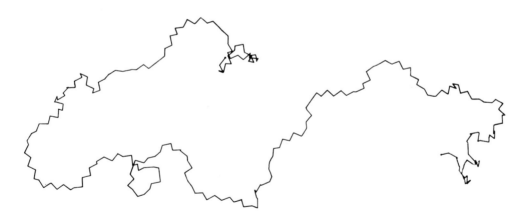

Fig.19   Perspective drawing of a Monte Carlo polynucleotide chain which
         reproduce the experimentally observed characteristic ratio.
         Successive lines are virtual bonds across heminucleotides.

the experimentally observed variability in sugar residue conformations.
With a view to provide visualisation of the chain trace or the local
conformation of a random coil, calculations using Monte Carlo approach
have been performed.[46] Perspective drawing of a representative Monte
Carlo polynucleotide chain that reproduce the experimental value of C_∞
is shown in Fig.19 through virtual bonds joining the successive heminucleotides. We are currently using the heminucleotide scheme to
determine the hydrodynamic parameters such as end-to-end dimensions and
persistence lengths of Z DNA helices.

Heminucleotide-peptide Similarity

The most interesting feature of the heminucleotide is its qualitative
similarity with peptides in dimensions as well as conformation. It is
tempting to infer that nucleotide backbone nearly mimics a dipeptide
and the polynucleotide backbone is governed by two sets of "$(\emptyset, \psi)$"
namely (ψ, ψ') and (ω', ω). The 3'-heminucleotide has conformational
properties akin to proline. This prompted us to make an intentional
search for near neighbour bond correlations discussed above in proteins.
By extrapolation the only correlation should be $\psi_i + \emptyset_{i+1} \simeq 0$. Indeed
protein data analysis[46] reveal this for antiparallel and parallel beta
sheets. The physical significance here of course refers to a horizontal
alignment of carbonyl and amino functions to optimise interstrand
hydrogen bonds. The correlation has been noted earlier[48] from other
considerations. It is of interest to probe whether dipeptide equivalence
to nucleotide has any bearing on interactions between them and their
polymer themselves.

ACKNOWLEDGEMENT

We thank Mrs.Fabris for her excellent cooperation in typing the
manuscript. R.M.thanks C.S.I.R. India for a fellowship.

⌘ Contribution NO:626 from this Department.

REFERENCES

1. Sundaralingam, M.: 1969, Biopolymers 7, pp 821-860.
2. Arnott, S.: 1970, Prog.Biophys.Mol.Biol. 21, pp 265-319.
3. Wang, A.H.J., Quigley, G.J., Kolpak, F.J., Crawford, J.L., van der
   Marel, G. and Rich, A.: 198L, Nature, 282, pp 680-686.
4. Quigley, G.J., Seeman, N.C., Wang, A.H.J., Suddath, F.L. and Rich,
   A.: 1975, Nucl.Acids.Res., 2, pp 2329-2341.
5. Jack, A., Ladner, J.E. and Klug, A.: 1976, J.Mol.Biol., 108, pp
   619-650.
6. Sundaralingam, M., Mizuno, H., Stout, C.D., Rao, S.T., Liebman, M.
   and Yathindra, N.: 1976, Nucl.Acids Res., 3, pp 2471-2484.

7. Malathi, R. and Yathindra, N.: 1980, Curr.Sci., 49, pp 803-807.
8. Malathi, R. and Yathindra, N.: 1981, Int.J.Biol.Macromol., 4, pp 18-24.
9. Malathi, R. and Yathindra, N.: 1982, Biochem.J., 205, pp 457-460.
10. Malathi, R. and Yathindra, N.: 1982, Biopolymers, 21, pp 2033-2047.
11. Olson, W.K.: 1980, Macromolecules, 13, pp 721-726.
12. Malathi, R. and Yathindra, N.: 1981, Int.J.Quan.Chem., 20 pp 241-257.
13. Olson, W.K.: 1982. In Topics in Nucleic Acid Structures, Ed.Neidle, N., pp 1-79.
14. Yathindra, N. and Sundaralingam, M.; 1975, Biopolymers, 14 pp 2387-2399.
15. Perahia, D., Pullman, B., Vasilescu, D., Cormillon, R. and Broch, H.: 1977, Biochim.Biophys.Acta, 478, pp 244-259.
16. Broch, H. and Vasilescu, D.: 1979, Biopolymers, 18, pp 909-930.
17. Broyde, S. and Hingerty, B.: 1979, Nucl.Acids Res., 6, pp 2165-2178.
18. Kumar, N.V. and Govil, G.: 1979, Ind.J.Biochem.Biophys., 16, pp 414-427.
19. Yathindra, N. and Sundaralingam, M.: 1976, Nucl.Acids Res., 3, pp 729-747.
20. Olson, W.K.: 1976, Biopolymers, 15, pp 859-878.
21. Yathindra, N.: 1981. In Biomolecular Structure, Conformation, Function and Evolution, Eds. Srinivasan, R., Subramanian, E. and Yathindra, N., pp 379-401.
22. Jayaraman, S. and Yathindra, N.: 1980, Biophys.Biochem.Res.Comm., 97, pp 1407-1419.
23. Dickerson, R.E. and Drew, H.R.: 1981, J.Mol.Biol., 149 pp 761-786.
24. Dickerson, R.E., Kopaka, M.L. and Drew, H.R.: 1982. In Conformation in Biology, Ed.Srinivasan, R. and Sarma, R.H., pp 227-257.
25. Olson, W.K.: 1982, Nucl.Acids Res. 3, pp 777-787.
26. Olson, W.K.: 1981. In Biomolecular Stereodynamics, Ed.Sarma, R.H., pp 327-343.
27. Keepers, J.W., Kollman, P.A., Weiner, P.K. and James, T.L.: 1982, Proc.Natl.Acad.Sci., 79, pp 5537-5541.
28. Kollman, P., Keepers, J. and Weiner, P.: 1982, Biopolymers, 21, pp 2345-2376.
29. Sundaralingam, M. and Westhof, E.: 1981. In Biomolecular Stereo-dunamics, Ed.Sarma, R.H., pp 301-326.
30. Malathi, R. and Yathindra, N.: unpublished results.
31. Zhurkin, V.B., Lysov, Y.P., Florentiev, V.L. and Ivanov, V.I.: 1982, Nucl.Acids.Res., 10, pp 1811-1830.
32. Jayaraman, S. and Yathindra, N.: 1981, Int.J.Quan.Chem., 20, pp 211-230.
33. Jayaraman, S.: 1982, Ph.D.Thesis. Madras University, India.
34. Arnott, S. and Hukins, D.W.L.: 1972, Biophys.Biochem.Res.Comm., 47, pp 1504-1509.
35. Marvin, D.A., Spencer, M., Wilkins, M.H.F. and Hamilton, L.D.: 1961, J.Mol.Biol., 3, pp 547-565.
36. Arnott, S. and Selsing, E.: 1974, J.Mol.Biol., 88, pp 551-555.
37. Wang, A.H.J., Quigley, G.J., Kolpak, P.J., Marel, G.V., van Boom, J.H. and Rich, A.: 1981, Science, 211, pp 171-176.

38. Rich, A., Davies, D.R., Crick, F.H.C. and Watson, J.D.: 1961, J.
     Mol.Biol., 3, pp 71-80.
39. Ivanov, V.I., Minchenkova, L.E., Schyolkina, A.K. and Poletayev,
     A.I.: 1973, Biopolymers, 12, pp 89-110.
40. Yathindra, N.: 1978. Paper presented at the Int.Sympo.on Bio-
     molecular Structure, Conformation, Function and Evolution. Jan.4-8,
     Madras, India.
41.   Gauss, D.H. and Sprinzl, M.: 1983, Nucl.Acids Res., 11, pp r1-r53.
42. Sundaralingam, M.: 1982. In Conformation in Biology, Ed.Srinivasan,
     R. and Sarma, R.H., pp 191-225.
43. Brennan, T. and Sundaralingam, M.: 1976, Nucl.Acids Res. 3, 3235-
     3246.
44.   Gauss, D.H. and Sprinzl, M.: 1983, Nucl.Acids Res., 11, pp r55-
     r103.
45. Go, M.: 1981, Nature, 291, pp 90-92.
46. Malathi, R. and Yathindra, N.: unpublished results.
47. Eisenberg, H. and Felsenfeld, G.: 1967, J.Mol.Biol. 30, pp 17-29.
48. Peticolas, W.L. and Kurtz, B.: 1980, Biopolymers, 19, pp 1153-1166.
49. Kitamura, K., Wakahara, A., Mizuno, H., Baba, Y. and Tomita, K.:
     1981, J.Amer.Chem.Soc., 103, pp 3899-3906.

PRESENCE OF NONLINEAR EXCITATIONS IN DNA STRUCTURE AND THEIR RELATIONSHIP
TO DNA PREMELTING AND TO DRUG INTERCALATION

Asok Banerjee and Henry M. Sobell
Department of Radiation Biology and Biophysics
The University of Rochester School of Medicine and Dentistry
Rochester, New York 14642

Department of Chemistry, The University of Rochester
River Campus Station, Rochester, New York 14627

ABSTRACT

We propose that collectively localized nonlinear excitations (solitons)
exist in DNA structure. These arise as a consequence of an intrinsic nonlinear ribose inversion instability that results in a modulated βalter-
nation in sugar puckering along the polymer backbone. In their bound
state, soliton-antisoliton pairs contain β premelted core regions capable of undergoing breathing motions that facilitate drug intercalation.
We call such bound state structures -- β premeltons. The stability of a
β premelton is expected to reflect the collective properties of extended
DNA regions and to be sensitive to temperature, pH, ionic strength and
other thermodynamic factors. Its tendency to localize at specific
nucleotide base sequences may serve to initiate site-specific DNA premelting and melting. We suggest that β premeltons provide nucleation
centers important for RNA polymerase-promoter recognition. Such nucleation
centers could also correspond to nuclease hypersensitive sites.

The possibility that nonlinear excitations (solitons) exist in biopolymers and play a central role in energy transfer was first advanced by
Davydov in his classic series of papers (1-3). In addition, a different
class of solitons that give rise to localized conformational changes in
DNA structure has been proposed by Englander et al. to explain DNA
breathing phenomena (4).

Solitons are intrinsic locally coherent excitations that move along a
polymer chain with a velocity significantly less than the speed of sound
(they may even be stationary). They are combinations of intramolecular
and deformational excitations that appear as a consequence of an intrinsic nonlinear instability in the polymer structure.

253

B. Pullman and J. Jortner (eds.), Nucleic Acids: The Vectors of Life, 253–263.
© *1983 by D. Reidel Publishing Company.*

Extensive research on solitons in many physical systems has shown that this nonlinearity gives the spacially localized conformational excitation a robust character (5,6). Solitons do not significantly interact with conventional normal mode excitations (i.e., phonons). They have their own identity and can be treated by Newtonian dynamics as heavy Brownian-like particles, each having an "effective mass". Solitary structures -- as sites for biochemical activity -- behave like independent species and can be treated by statistical mechanics and chemical thermodynamics. They can arise from equilibrium or nonequilibrium processes.

Here, we propose that localized nonlinear excitations -- solitons -- exist in DNA. These arise as a consequence of an intrinsic nonlinear instability in DNA structure which is primarily associated with interconversions between the two predominent sugar-pucker conformations, C2' endo and C3' endo. In their bound state, soliton-antisoliton pairs surround β premelted core regions -- these regions can undergo breathing motions that facilitate the intercalation of drugs and dyes into DNA. Similar bound state structures could act as phase boundaries that connect different DNA forms.

Several communications describing our ideas have already appeared (7-9). Here, we develop these ideas in greater detail.

DRUG INTERCALATION INTO DNA

Structural features of our theory have been suggested by information obtained through X-ray crystallographic studies of model drug-DNA complexes. These studies have indicated that DNA undergoes a highly specific conformational change at the immediate intercalation site when binding a class of intercalating agents known as simple intercalators. This (localized) structural state is characterized in part by the presence of a mixed sugar-puckering pattern connecting adjacent base-paired nucleotides -- C3' endo (3'-5') C2' endo -- and, in addition, involves less pronounced changes in the torsional angles that describe the glycosidic and sugar-phosphate bond linkages. This allows DNA to stretch and to unwind when accomodating an intercalator.

More generally, we have proposed this (base-paired dinucleotide) structure to belong to a family of structures, each of which corresponds to an element of organized β structure (see discussion below)(10). All β elements process the same mixed-sugar puckering pattern and have similar backbone conformational angles -- but vary in the degree of base-unstacking. Lower energy forms contain base-pairs partially unstacked, while higher energy forms contain base-pairs completely unstacked. Steroidal diamines, such as irehdiamine A, bind to the lower energy β element through partial intercalation, while planar drugs and dyes bind to higher energy β elements by complete intercalation (11).

A DNA structure of particular interest is stabilized at saturating concentrations of drugs such as irehdiamine and ethidium. This is an organized helical structure that contains the β element as the asymmetric
unit. We call this neighbor-exclusion structure βDNA, or, for reasons
to be described shortly, β premelted DNA (12).

Irehdiamine stabilizes a low energy form of β premelted DNA, while
ethidium stabilizes a higher energy form. We have described the molecular nature of these structures elsewhere.

KINK-ANTIKINK BOUND STATES IN DNA STRUCTURE -- THE β PREMELTON

How do β structural elements arise in DNA -- and what are the surrounding structural features on either side of these elements?

We propose that β elements arise as part of more organized β structure,
whose appearance reflects the presence of a soliton-antisoliton (i.e.,
kink-antikink) bound state in DNA structure (see Figures 1a and b).
Such a structure contains a modulated β alternation in sugar puckering
about the central β premelted core region which gradually merges into
B DNA on either side -- typically, a structure of this kind would give
rise to an energy density profile similar to that shown in Figure 1c
(13). We call this type of composite structure a β premelton (14).

We have already described the methods used to construct this β premelton
(9). The B DNA to β DNA transition was first computed as a homogeneous
transition involving the entire polymer length. This was accomplished in
a series of steps in which the sugar puckering of alternate deoxyribose
residues was altered and the structure then energy minimized subject to
a series of constraints and restraints. To simulate the bound state
structure, base-paired dinucleotide elements from each structure in this
sequence were then pieced together using a least squares procedure. The
decision to compute twelve intermediate structures and to then use these
to construct the soliton-antisoliton bound state (without a detailed
knowledge of the total energies involved) is somewhat arbitrary, but
does not alter the basic conclusions presented here. More complete computations are in progress to obtain a family of minimum energy bound
state structural forms.

We cannot say at the present time whether the center of the β premelton
is open enough to accomodate an intercalator such as ethidium. However,
structures such as these have an intrinsic ability to undergo low frequency breather motions and it is possible that such motions facilitate
the intercalation process.

MECHANISM OF DRUG INTERCALATION AND ITS RELATIONSHIP TO DNA PREMELTING

We envision intercalation to begin with the loose (external) binding by
an intercalator to the B helix. A β premelton then appears in the region
-- this either immediately permits the drug molecule to intercalate, or
a breathing motion of sufficient amplitude is first required. Simple
intercalators may pin (immobilize) the β premelton. They could also prevent soliton-antisoliton anihilation (a spontaneous event that returns
DNA in the immediate region to its ground state structure, i.e., B DNA).
This is because simple intercalators bind the β premelted core region
tightly. Complex intercalators may vary in this regard. Actinomycin
could stabilize the β premelted core within the β premelton, pinning the
structure and also preventing its self-anihilation (15). Daunomycin,
however, may be unable to do this and ends up, therefore, binding to a
conformation very similar to B DNA (16).

The term "soliton" is synonymous with "kink", and refers to a gradation
of structural change either within a single structural type (i.e., a
nontopological soliton) or connecting two different structures (i.e., a
topological soliton). The structure shown in Figure la consists of two
topological solitons - a (B→β) soliton and a (β→B) antisoliton.
These two structures surround the β premelted core region and are
related by 2-fold symmetry.

In Figure la, we show a β premelton containing a small β premelted core.
β premeltons having higher energies can occur. Bifunctional intercalators, such as echinomycin, require β premeltons with larger cores.
Intercalators that necessitate the transient rupture of hydrogen bonds
connecting base-pairs to gain entrance into (and out of) DNA (i.e.,
meso-tetra-(4-N methylpyridyl)porphine) (17) require β premeltons with
still larger cores. We term distortions that combine base unstacking
with the transient rupture of hydrogen bonds connecting base-pairs --
DNA breathing distortions. We expect such (dynamic) distortions to take
place in the center of extended regions of β premelted DNA.

The β premelted core region within soliton-antisoliton bound state
structures can be thought of as a transition state that nucleates the
DNA melting process (hence, the name β premelted DNA). At lower temperatures, soliton-antisoliton pairs surround small β premelted cores. As
the temperature rises, these bounding soliton-antisoliton pairs move
apart leaving central regions of the growing β premelted cores more and
more disordered. Finally, with increasing temperature, denaturation
bubbles appear that contain regions of denatured DNA connected to native
DNA. These kink and antikink structures, therefore, act as structural
boundaries connecting denatured DNA with native DNA in the helix to coil
phase transition. They also act as (partial) phase boundaries in a
variety of DNA structural phase transitions -- we will describe the
nature of one such transition below.

B-A JUNCTION

Soliton concepts allow us to propose the molecular nature of the junction that connects B with A DNA during the B to A structural phase transition. Such a junction (shown in Figures 2a and b) has been constructed in a manner similar to that described above.

A homogeneous β DNA to A DNA change was computed as a sequence of transitions involving the entire polymer length. Dinucleotide elements from this sequence and the B to β sequence were then connected together using the least squares joining procedure. The final structure consists of two _different_ topological solitons -- a $(B \rightarrow \beta)$ soliton and a $(\beta \rightarrow A)$ antisoliton and vice versa. Such a junction constitutes a phase boundary transforming B DNA to A DNA. In the global sense, this kink-antikink bound state is topological since it alters the structure of the polymer as it moves along its length.

How does the B to A transition begin?

We envision the B to A transition to begin at the centers of β premeltons within B DNA structure. Nucleation could be site specific, and involve growing regions of A DNA joined to B DNA on either side as $[(B \rightarrow \beta)]$: $(\beta \rightarrow A)]$ and $[(A \rightarrow \beta): (\beta \rightarrow B)$ β premelton pairs (see Figures 3a-d). Each β premelton contains a small β premelted core, and moves along the polymer with minimal activation energy. Its high free energy and ease of movement could account for the cooperative nature of the phase transition.

DISCUSSION

Soliton concepts provide a strong rationale for expecting coherent nonlinear excitations to extend over multiple base pairs to provide either transient or permanent conformational changes in specific DNA regions. These will always be present to a certain extent at normal thermal energies, but their concentrations will depend on temperature, pH, ionic strength, hydration, extent of superhelicity and other thermodynamic factors.

The stability of soliton-antisoliton pairs in naturally occuring DNA is further expected to reflect the detailed nucleotide base sequence. This is partly because the ease with which β premelted core regions form primarily reflects the magnitude of localized base stacking energies. Nucleotide sequences with minimal base overlap (i.e, alternating purine-pyrimidine sequences) are likely to be favored to form β premelted core regions, along with sequences that contain high A-T/G-C base ratios (18). Equally important are the energetics in the kink and antikink regions. As seen in Figure 1c, the stability of a kink-antikink bound state reflects the depth in the energy minimum within the β premelted core coupled with the height and separation of

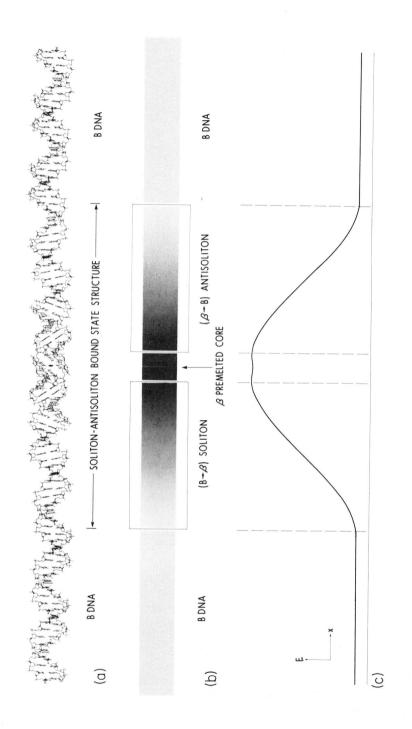

Figure 1. (a) Molecular representation of the soliton-antisoliton bound state structure in B DNA. (b) Schematic illustration of the gradation of structural change that connects the β premelted core with B DNA on either side through soliton-antisoliton pairs. (c) A typical energy density profile of the kink-antikink bound state structure -- the dip in the center of the energy density profile signifies the presence of a metastable structural state. We call this type of composite structure a β premelton.

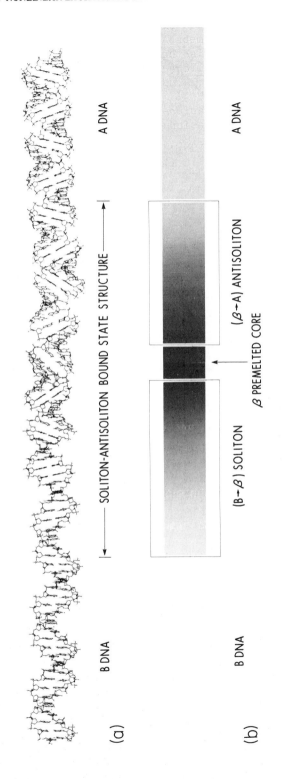

Figure 2. (a) Phase boundary connecting B with A DNA. (b) Schematic illustration of the gradation of structural change that connects the β premelted core region with B DNA on one side and A DNA on the other side through kink–antikink pairs. See text for discussion.

A. BANERJEE AND H. M. SOBELL

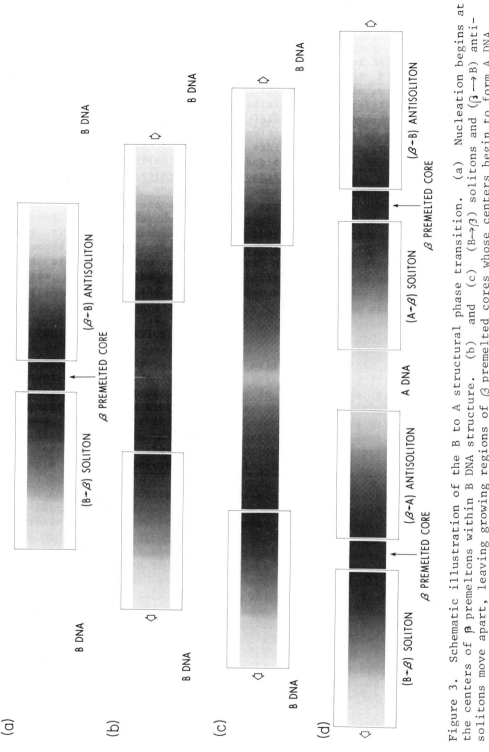

Figure 3. Schematic illustration of the B to A structural phase transition. (a) Nucleation begins at the centers of β premeltons within B DNA structure. (b) and (c) (B→β) solitons and (β→B) anti- solitons move apart, leaving growing regions of β premelted cores whose centers begin to form A DNA, (d) Completion of the [(B→β):(β→B)] and [(A→β):(β→B)] β premelton pairs with intervening A DNA structure.

the domain walls on either side.

It is evident that sequentially homogeneous DNA polymers are not good
models to understand the properties of naturally occurring DNA since
they lack sufficient information in their nucleotide base sequence to
give rise to this site specificity. In the context of soliton models,
this poses the problem of describing soliton behavior in the presence
of locally altered potentials -- a general theory for this has been
developed (19). This theory shows that (nontopological) solitons either
move nonuniformly or are trapped by locally favorable potentials. It
remains to extend this theory to DNA structure to predict the localiza-
tion of kink-antikink bound states at specific nucleotide sequences.
Since the kink-antikink pair is multiple base-pairs in extent, it may
be that the effective trapping potential involves the recognition of an
extended sequence rather than being determined by any single base-pair
energetics or its immediate neighbors.

We note experimental evidence that indicates the presence of nuclease
hypersensitive sites in eukaryotic DNA, many of these located at 5' ends
of genes (20,21). These same sites are sensitive to cleavage by a
1,10-phenanthroline-copper(I) complex, a known intercalating agent. It
is possible that these sites correspond to the centers of β premeltons
localized in these regions.

The presence of β premelted core regions within kink-antikink pairs
could provide a key component in the recognition of the promoter by the
RNA polymerase enzyme. RNA polymerase-promoter recognition could begin
with the loose binding to β premelted nucleation sites within promoter
regions, followed by a sequence of conformational transitions that progressively lead to a more open tight binding complex. Such an auto-
catalytic process can be viewed as one that creates "an avalanche of
kinks", whose energies come from the stepwise interactions between the
RNA polymerase and the promoter. This results in the formation and propagation of a new DNA phase -- such a phase could resemble the β
structure, but be more completely unwound and single-stranded.

The process of RNA transcription could begin within this more open β
premelton structure. The RNA polymerase enzyme then acts to catalyze
the stepwise polymerization of ribonucleoside triphosphates into RNA,
using one or two β denatured DNA chains as a template. Normal chain
termination either occurs spontaneously, or requires other factors.
Actinomycin and other related antibiotics are known to cause premature
chain termination (22). We suggest that these antibiotics intercalate
tightly into β premelted DNA regions within the transcriptional complex.
This tends of immobilize the complex, increasing the probability of premature chain termination.

In recent years, increasing attention has focused on understanding the
full range of conformational flexibility in DNA structure. This has
culminated in the discovery of the Z DNA structure, a conformational
state radically different from either B or A structures (23). The

detailed nature of soliton excitations in Z DNA will be different from these described here. However, concepts of kink-antikink bound states as phase boundaries interconverting these forms will still be appropriate. We are attempting to model the molecular nature of the phase boundary conneting B with Z DNA with this in mind.

ACKNOWLEDGEMENTS

This work has been supported in part by the National Institutes of Health and the Department of Energy. The paper has been assigned report no. UR-000-0000 at the DOE, the University of Rochester. Denise Alexander and James A. Krumhansl are thanked for their stimulating discussions.

REFERENCES

1. Davydov, A.S., Physica Scripta 20, 387-394 (1979).

2. Davydov, A.S. and Kislukha, N.I., Phys. Stat. Sol. B 75, 735-742.

3. Davydov, A.S., Physica 3D 1, North Holland Publishing Company, pp. 1-22 (1981)

4. Englander, S.W., Kallenbach, N.R., Heeger, A.J., Krumhansl, J.A., and Litwin, S., Proc. Nat. Acad. Sci. USA 77, No. 12, 7222-7226 (1980).

5. Scott, A.C., Chu, F.Y.F. and McLaughlin, D.W., Proc. IEEE 61, 1443-1483 (1973).

6. Barone, A., Esposito, F., Magee, C.J. and Scott, A.C., Riv. Nuovo Cimento 1, 227-267 (1971).

7. Alexander, D. and Krumhansl, J.A., in Structure and Dynamics: Nucleic Acids and Proteins, Adenine Press, Inc., pp. 61-80 (1983).

8. Sobell, H.M., Lozansky, E.D. and Lessen, M., Cold Spring Harb. Symp. Quant. Biol. 43, 11-19 (1978).

9. Sobell, H.M., Sakore, T.D., Jain, S.C., Banerjee, A., Bhandary, K.K., Reddy, B.S., and Lozansky, E.D., Cold Spring Harb. Symp. Quant. Biol. 47, 293-314 (1983).

10. In previous communications, we have called this structure the "kink". However, the word "kink" has a broader meaning in the soliton physics area and, to avoid confusion, we have decided to rename this structure the β element. Its precise definition and full meaning is described in the text.

11. Sobell, H.M, Reddy, B.S., Bhandary, K.K., Jain, S.C., Sakore, T.D., and Seshadri, T.P., Cold Spring Harb. Symp. Quant. Biol. 42, 87-102 (1977).

12. The terms βDNA and βpremelted DNA replace βkinked DNA used earlier.

13. The dip in the center of the energy density profile signifies the presence of a metastable structure state within the soliton-anti-soliton bound state structure. This reflects the relaxation of strain energy in the sugar-puckering within the β premelted core region.

14. We use the term βpremelton in the most general sense to describe kink-antikink bound states in double-helical DNA and RNA structure. Thus, β premeltons can arise in B DNA and A DNA (or A RNA) and can act as phase boundaries transforming B DNA to A DNA during the B to A structural phase transition.

15. Sobell, H.M. and Jain, S.C., J. Mol. Biol. 68, 21-34 (1972).

16. Quigley, G.J., Wang, A.H.J., Ughetto, G., van der Marel, G., van Boom, J.H. and Rich, A., Proc. Nat. Acad. Sci. USA 77, 7204-7207 (1980).

17. Fiel, R.J. and Munson, B.R., Nucleic Acids Res. 8, 2835-2842 (1980).

18. Bloomfield, V.A., Crothers, D.M. and Tinoco, Jr., I., Physical Chemistry of Nucleic Acids, Harper and Row, Publishers (1974).

19. Fogel, M.B., Trullinger, S.E., Bishop, A.R. and Krumhansl, J.A., Phys. Rev. Lett. 36, 1411-1414 (1976).

20. Jessee, B., Gargiulo, G., Razvi, F. and Worcel, A., Nucleic Acids Res 10, No. 19, 5823-5834 (1982).

21. Cartwright, I.L. and Elgin, S.C.R., Nucleic Acids Res. 10, No. 19, 5835-5852 (1982).

22. Reich, E. and Goldberg, I.H., in Progress in Nucleic Acid Research and Molecular Biology, 3, pp. 183-234 (1964).

23. Wang, A.H.J., Quigley, G.J., Kolpak, F.J., van der Marel, G., van Boom, J.H., and Rich, A., Science 211, No. 9, 171-176 (1981).

RIBONUCLEASE T_1: MECHANISM OF SPECIFIC GUANINE RECOGNITION AND RNA HYDROLYSIS.

Udo Heinemann and Wolfram Saenger
Max-Planck Institut für experimentelle Medizin, Hermann-Rein-Str. 3, 3400 Göttingen, FRG
and Institut für Kristallographie, Freie Universität Berlin, Takustr. 6, 1000 Berlin 33, FRG.

Dedicated to Prof. Friedrich Cramer on the occasion of his 60. birthday.

The crystal structure analysis of the complex RNase $T_1 \cdot$ 2'-guanylic acid has provided insight into specific protein-nucleic acid interaction and into the mechanism of RNA hydrolysis catalyzed by RNase T_1. Recognition of guanine is via hydrogen bonding of main chain dipeptide Asn43-Asn44 to O_6 and N_1-H of the base and additional stacking of Tyr45 with the guanine heterocycle and of Tyr42 with guanine O_6. RNA hydrolysis is initiated by removal of a proton from O_2,H by Glu58 followed by formation of a cyclic 2',3'-guanylic acid, assisted by Arg77 and by His92. In a second step the cyclic intermediate is hydrolized by water to yield an RNA fragment with terminal 3'-guanylic acid.

SOME GENERAL REMARKS ON PROTEIN-NUCLEIC ACID INTERACTIONS.

If considering complexes between nucleic acids and proteins, one has to distinguish between two major structural principles. In one, the nucleic acid occurs as complementary, Watson-Crick type duplex with the two polynucleotide chains running antiparallel and intertwined in the form of a righthanded double helix. The overall geometry of this helix is rather well defined. The base-pairs are located in its center whereas sugar-phosphate chains are at the periphery. Recognition by a protein can be either unspecific with sugar-phosphate residues as main target or specific, with direct interactions between protein and nucleic acid bases. The latter can take place only in minor or major groove sites and have been proposed to occur with repressors *cro* and *lambda* (1,2) and with catabolite activator protein (CAP) (3), the crystal structures of which have been determined recently. Model building studies suggested that there is a common motif of binding to DNA in the double helical B-form, with dyad related α-helices of protein dimers inserted into two adjacent (and dyad related) major groove sites of DNA. In addition *cro* exhibits a pleated sheet struc-

265

B. Pullman and J. Jortner (eds.), Nucleic Acids: The Vectors of Life, 265–276.
© *1983 by D. Reidel Publishing Company.*

ture located between the two α-helices which could anchor in the minor groove flanked by the major grooves accommodating the α-helices (1). Since in all these cases only crystals of the proteins are available thus far, information on mutual recognition between the partners is based solely on model building studies. Therefore, close details of the specific interactions actually occurring in these complexes are still lacking and we do not know how an amino acid side chain (or main chain) recognizes and binds to the DNA in major or minor groove site.

In the other structural principle governing protein-nucleic acid complexes, the nucleic acid is single stranded. This means that it does not have to obey geometrical restrictions imposed by Watson-Crick type double helix geometry and is free to adjust its configuration according to spatial requirements. In single stranded polynucleotides, main torsional flexibility in the sugar-phosphate chain is found in P-O diester bonds, and in sugar conformation with preferential $C_2,$-endo or $C_3,$-endo puckering. In addition, the base can rotate about the glycosyl link and is mainly oriented anti in the pyrimidine series whereas for purines, both syn and anti are possible. These degrees of freedom provide the polynucleotide with an overall and local mobility so that the complementary binding site of a protein interacting with it can have a variety of shapes, in contrast to the thus far observed stereotype α-helix of proteins binding to the DNA double helix.

The crystal structure of a protein binding to single stranded DNA has been reported (4). If oligonucleotides are cocrystallized with it in order to determine the pattern of mutual recognition, aggregation takes place and very complex structural motifs are created which could not yet be analyzed in detail. Therefore, direct information on protein-nucleic acid interaction is not available and model building is less informative than with double helical DNA because single stranded DNA is so flexible.

In another case, ribonuclease T_1 (RNase T_1) complexed with guanylic acid, however, details of protein-nucleic acid recognition were obtained. RNase T_1 is an enzyme of molecular weight 11.400 daltons (5,6). It interacts with single stranded RNA and hydrolyzes it specifically at the 3'-end of guanosine. Cocrystallization of RNase T_1 with the inhibitor 2'-guanylic acid (2'-GMP) produces a complex which was subjected to X-ray crystal structure analysis. It provided for a first example of specific protein-nucleic acid interaction (7,8) and the active site geometry suggested a reaction mechanism which will be discussed in relation to earlier chemical studies.

THREE DIMENSIONAL STRUCTURE OF RNASE T_1 COMPLEXED WITH 2'-GUANYLIC ACID

The amino acid sequence of RNase T_1 was determined by Takahashi (9) and is displayed in Figure 1. It contains four cysteins at positions

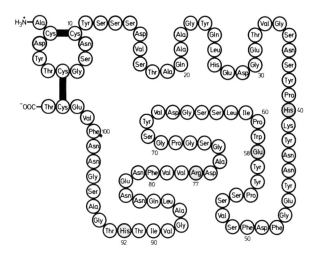

Figure 1. Covalent structure of RNase T$_1$ (ref. 9).
Residues shown shaded are critical for enzymatic
activity, i.e. His40, Glu58, Arg77, His92.

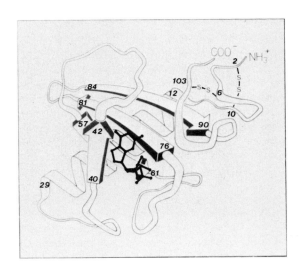

Figure 2. Schematic drawing of the RNase T$_1$·2'–guanylic
acid complex. Sequential numbers of some of the amino
acids are given.

2, 6, 10 and 103. A disulfide bridge between cysteins 2 and 10 closes
a small peptide loop and a bridge between cysteins 6 and 103 serves
to hold the two ends of the polypeptide chain close together. Chemical modification studies (10-12), NMR spectroscopy (13,14) and results
from kinetic experiments (15) suggested that amino acid side chains
of His40, Glu58, Arg77 and His92 are involved in the catalytic action
of this enzyme, i.e. cleavage of RNA at the 3'-side of guanylic acid.

The chain folding of RNase T_1 is illustrated in the schematic Figure
2. Starting with the small peptide loop closed by disulfide bridge
Cys2-Cys10, the polypeptide chain enters an α-helix of 4.5 turns followed by two wide loops separated by a stretch of extended ß-struc-
ture. This stretch provides the recognition and binding site for the
guanine base. Then three strands of antiparallel ß-pleated sheet are
observed, with another larger loop inserted after the first strand.
The tertiary structure folding is completed by a wide loop, bringing
the two termini of the chain together so that disulfide Cys6-Cys103
can be formed.

The secondary structure of the α-helix is very regular, whereas the
four-stranded ß-pleated sheet is severely bowed and twisted in a left-
handed sense. The intra-strand hydrogen bonding pattern between the
last two strands (Asp76-Asn81 and Asn84-Ile90) is disturbed by insertion of extra residues in the last strand, Val79 forming a classical
"ß-bulge" with the dipeptide Ala87-Gly88.

THE TWO HYPOTHETICAL STAGES OF PROTEIN-NUCLEIC ACID INTERACTION

If considering general principles that could govern the interaction
of a protein with a single stranded nucleid acid, two stages have to
be envisaged. The first is approach and binding of the protein to the
sugar-phosphate backbone. This process should not be sequence (or base)
specific. It involves formation of salt bridges between positively
charged amino acid side chains (Lys, Arg, His) and negatively charged
phosphate groups as well as (weaker) hydrogen bonds between functional
amino acid side chains or peptide groups and phosphate or sugar oxygens. These latter interactions have to be invoked in order to distinguish between DNA and RNA on the basis of the unique 2'-hydroxyl in
RNA.

This initial, primary and rather unspecific binding is augmented by
specific interactions between the protein RNase T_1 and nucleic acid
bases, with favourable binding to guanine. What are the mechanisms
utilized in that recognition? Let us look at Figure 3 where conceivable hydrogen bonding patterns between guanine on one side and protein constituents on the other side are illustrated. The latter com-
prise amino acid side chains of Arg, Asp, Glu, Asn, Gln as well as a
peptide link in *cis* configuration and a dipeptide in usual *trans* form.
Only those configurations are displayed where at least two (cyclic)
hydrogen bonds are formed simultaneously. Comparable schemes are also

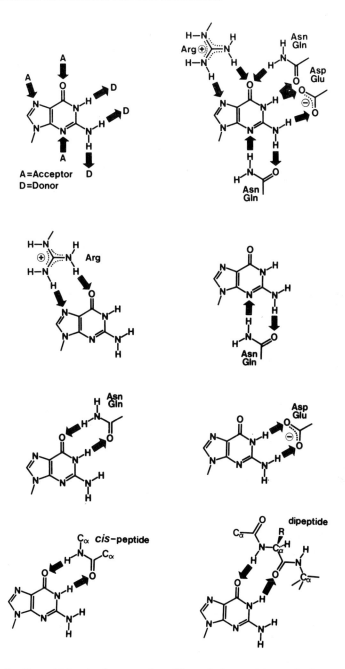

Figure 3. Possible hydrogen bonding interactions between protein main chain and side chain groups with guanine. Top left, guanine's donor and acceptor groups, top right amino acid side chains clustering around guanine, with individual interactions given in illustrations in second and third rows. Pictures in bottom line indicate recognition of guanine by the unusual *cis*-peptide and by and all-*trans* dipeptide.

observed in base-base interactions and suggest that cyclic bonds are
more specific with respect to single, individual hydrogen bonds (16).
Figure 3 shows clearly that amino acid side chains have the potential
to specifically recognize a nucleic acid base (in this case guanine),
and if several of these interactions are combined, error rates in recognition could be reduced dramatically. Similar schemes can also be
derived for adenine and, to a lesser extent, for uracil (thymine) and
cytidine, because these have only three functional groups. It must be
emphasized that specificity in base-amino acid side chain recognition
is not only based on purely geometric fitting as considered here but
also involves fine tuning due to different pK-values of carboxylate
and amide groups.

In addition, the "inert" polypeptide backbone can also bind to guanine.
The *cis*-peptide group is rather improbable because it represents a
higher energy form of the usually observed *trans*-peptide and is therefore only rarely found in protein structures. On the other hand, it
could provide for an excellent recognition site just because it is so
unusual. Proper folding of the main chain could induce and stabilize
the *cis*-form, yielding a unique and specific site in an otherwise all-
trans polypeptide. Concerning the *trans*-dipeptide, Figure 3, or any
other longer peptide sequence, they all can be folded into a geometry
suitable for base recognition.

Let us now look at the topological features of the actual protein-
nucleic acid complex provided by RNase T_1·2'guanylic acid.

IN RNASE T_1, SURFACE IS NEGATIVELY CHARGED AND POSITIVE CHARGES
CLUSTER AT ACTIVE SITE

RNase T_1 is an acidic protein with a pI of 3.8 (17). Therefore, charge
distribution on its surface is predominantly negative at physiological
conditions or at pH ∿5 where it displays maximum binding affinity for
small substrate analogs (18). As a consequence, it should repel the
negatively charged polyelectrolyte RNA. In fact, the three dimensional
structure of RNase T_1 indicates that negative charges located on Asp
and Glu residues are distributed over the surface of the molecule. In
contrast, positive charges carried by His27, His40, His92, Arg77 and
Lys41 are clustered in a narrow zone around the active site.

This characteristic charge distribution of RNase T_1 suggests that the
enzyme repels a negatively charged RNA when approaching it at random.
If the mutual orientation of the two partners is such that the RNA can
enter into the active site, it is bound and recognized by the cluster
of positively charged amino acid side chains. The mechanism of initial
binding of the enzyme to RNA still remains obscure. The question remains whether RNase T_1 binds first to RNA and then slides along the
polymer or whether it binds, gets off and binds in another place in a
random walk? Because RNase T_1 attaches exclusively to single stranded
RNA, it can also be considered as an RNA double helix destabilizing

protein (or RNA single strand binding protein), in analogy to this class of proteins known for DNA.

RECOGNITION OF GUANINE BY RNASE T_1 IS VIA MAIN CHAIN HYDROGEN BONDING SUPPORTED BY SIDE CHAIN STACKING

If we look at the protein-guanine recognition observed in the RNase $T_1 \cdot 2'$-guanylic acid complex (Figure 4), it becomes obvious that none of the interactions between amino acid side chains and guanine displayed in Figure 3 are actually employed. There is, however, contact between guanine and a main chain dipeptide with hydrogen bonding between peptide NH of Asn43 to O_6 of guanine and from N_1-H of guanine to peptide oxygen of Asn44. This interaction appears rather unspecific because it is to the "inert" protein backbone. Attached to the dipeptide, however, is Tyr45, the phenolic moiety of which is stacked on guanine in 3.5Å distance to confer at least some specificity due to an amino acid side chain. In addition, the phenolic ring of Tyr42 is in van der Waals (stacking) contact with O_6 of guanine. Otherwise, there are no contacts between nucleotide and protein except a salt bridge formed between His40 and the nucleotide phosphate.

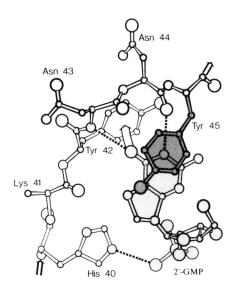

Figure 4. Recognition of 2'-guanylic acid by RNase T_1. Shown is 2'-GMP together with a section of the polypeptide chain from His40 to Tyr45. Hydrogen bonds indicated as broken lines between base and main chain atoms are N(1)-H...O(peptide Asn44) and O(6)...H-N(peptide Asn43). The side chain of Tyr45 stacks over guanine and guanine O(6) interacts with the phenolic ring of Tyr 42 (curved arrow). A hydrogen bond links His40 with phosphate, shown by dashed line.

These results are in overall agreement with chemical and spectroscopic data suggesting that guanine N_1 and O_6 are essential for recognition by RNase T_1 (5). However, protonation of guanine N_7 upon binding
as proposed from an NMR study (14) must clearly be rejected because
(i) the pK_a of this nitrogen, 2.8, is too low to induce protonation by
a protein side chain and (ii), there is no suitable residue in the vicinity which could transfer a proton. In another NMR study (19), a
strong interaction of guanine N_2-amino group with RNase T_1 was proposed. Looking at our model, such kind of interaction can also be
ruled out because no functional group of the protein is close enough
for hydrogen bonding or stacking to guanine N_2. Also, chemical studies
indicate that substrates with inosine substituting for guanine are
converted at almost the same rate so that guanine N_2 cannot be crucial for binding (5).

MAIN CHAIN VERSUS SIDE CHAINS SPECIFICITY

Comparing Figures 3 and 4, the difference between expected and observed interactions involving guanine and RNase T_1 becomes clear. Why
are our predictions so wrong and why did nature choose an unpredictable scheme for specific recognition of guanine by RNase T_1?

In all schemes outlined in Figure 3 except for *cis*- and dipeptide, the
functional groups of amino acid side chains are involved in hydrogen
bonding to guanine. These should be quite strong if charged groups are
involved, viz. Arg or Asp/Glu. However, between the rather rigid, well
defined main chain and the functional groups, there are several C-C
single bonds which are rather flexible. This might be favourable for
certain cases of interaction between different partners. However, if
a residue with well defined geometry such as guanine is to be recognized, it will be advantageous to have a preformed template (a matrix)
where all complementary hydrogen bonding acceptors and donors are held
in the right place for rapid complex formation.

This is observed in RNase T_1. The geometry of the polypeptide stretch
between Tyr42 and Tyr45 responsible for guanine binding is conserved
in this type ribonucleases because it is also found in RNase isolated
from Bacillus amyloliquefaciens (20) and from Streptomyces erythreus
(21) which exhibit specificity of cleavage of RNA at 3'-ends of
guanine or of purine. Given this template of hydrogen bonding acceptors and donors, rapid recognition of guanine and of uracil with the
same distribution of keto oxygen (O_6 guanine $\equiv O_4$ uracil) and amide
N-H (N_1-H guanine $\equiv N_3$-H uracil) is conceivable. How, then, is it possible to distinguish between these two possibilities? This could be
achieved either on the basis of stereochemistry because ribose and
base are attached differently in case of pyrimidine and purine nucleotide (N_1 and N_9). Or it could be by stacking interactions because
both guanine and tyrosine are good stackers whereas uracil is very
poor in this respect.

Another feature of the RNase $T_1 \cdot$ 2'-guanylic acid crystal structure de-
serves mention. The side chain of Tyr45 is not only in the position
displayed in Figure 4. The electron density distribution suggests that
it is also (to a lesser extent) present in another location and not
stacked with guanine, Figure 5. This indicates that the side chain of
Tyr45 is flexible, rotated "away" from the active site if none or a
false base other than guanine is bound to the main chain stretch
Asn43-Asn44. If a guanine is recognized, Tyr45 can swing over into
stacking position and the enzyme acts by cleaving the P-$O_{5'}$ bond via a
mechanism to be discussed in the following.

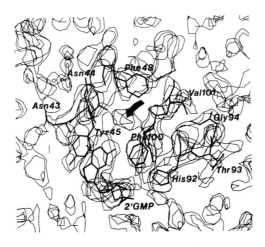

Figure 5. Section of electron density map with superimposed molecular model of RNase $T_1 \cdot$ 2'-GMP complex.
Tyr45 as shown is stacked over guanine, but a second
position with lower occupancy is found nearby as
indicated by arrow.

MECHANISM OF RNA HYDROLYSIS BY RNASE T_1 AS DEDUCED FROM THE CRYSTAL
STRUCTURE

As mentioned at the outset, chemical, spectroscopic and kinetic studies suggested that His40, Glu58, Arg77 and His92 are involved in the
catalytic hydrolysis of RNA by RNase T_1. The process follows a two-
step mechanism, as indicated in Figure 6, with a terminal 2',3'-cyclic
guanylic acid as reaction intermediate.

The geometry of the active site is clearly revealed by the crystal
structure of the RNase $T_1 \cdot$ 2'-guanylic acid complex. Since 2'-guanylic
acid is not a true substrate, the 3'-analog was fitted to the active
site, with only slight rearrangements of the sugar necessary to allow
for a good fit. The result is displayed in Figure 7. We see that
Glu58 interacts with $O_{2'}H$ and activates it by removing the proton. In

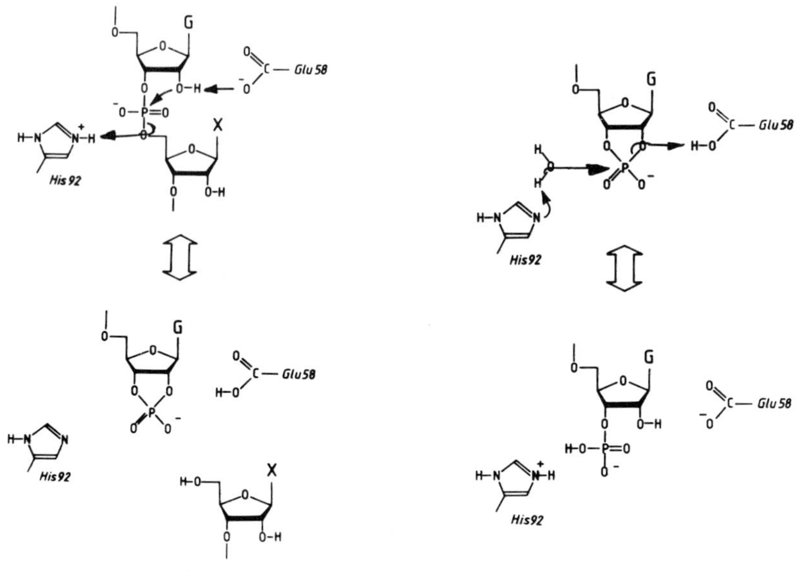

Step I   Step II

Figure 6. Two-step reaction mechanism of RNase T$_1$ as
elucidated first by chemical and spectroscopic methods
and corrected on the basis of crystalligraphic results.

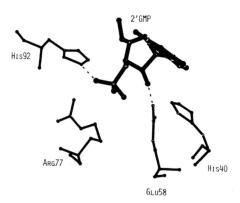

Figure 7. Active site geometry of RNase T$_1$, with 3'-GMP
fitted into the location of 2'-GMP of the *actual*
RNase T$_1$·2'-GMP complex.

the reaction mechanism proposed in Figure 6, O_2^- attacks the phosphate and displaces O_5' of the adjacent nucleotide to yield 2',3'-cyclic guanosine phosphate as reaction intermediate. This step can be facilitated by His92 which comes in handy to give a proton to the leaving O_5^-. In the second step, His92 removes a proton from an incoming water molecule, HO^- attacks the cyclic phosphate, opens it to yield 3'-guanylic acid. Arg77 would be in the right place to stabilize the pentacovalent phosphate occurring as intermediate. His40 cannot take part directly in the reaction because it is too far away to interfere with the bound RNA. It is possible, however, that it activates Glu58. Comparing amino acid sequences of RNase T_1-type ribonucleases, we find that Glu58, Arg77, His92 are conserved, and His40 occurs only in RNase T_1. Therefore, His40 cannot be essential for the reaction mechanism.

REFERENCES

1) Matthews, B.W., Ohlendorf, D.H., Anderson, W.F., Fisher, R.G., and Takeda, Y.: 1983, Trends in Biochem. Sci. 8, pp. 25-29.
2) Pabo, C.O., and Lewis, M.: 1982, Nature 298, pp. 443-447.
3) McKay, D.B., and Steitz, T.A.: 1981, Nature 290, pp.744-749.
4) McPherson, A., Jurnak, F., Wang, A., Kolpak, F., Milineux, I., and Rich, A.: 1979, Cold Spring Harb. Symp. Quant. Biol. XLIII, pp. 21-29.
5) Egami, F., Oshima, T., and Uchida, T.: 1980, Mol. Biol. Biochem. Biophys. 32, pp. 250-277.
6) Takahashi, K., and Moore, S.: 1982, The Enzymes 15, pp.435-467.
7) Heinemann, U., Wernitz, M., Pähler, A., Saenger, W., Menke, G., and Rüterjans, H.: 1980, Eur. J. Biochem. 109, pp. 109-114.
8) Heinemann, U., and Saenger, W.: 1982, Nature 299, pp. 27-31.
9) Takahashi, K.J.: 1971, J. Biochem. Tokyo 70, pp. 945-960.
10) Takahashi, K.J., Stein, W.H., and Moore, S.: 1967, J. Biol. Chem. 242, pp. 4682-4690.
11) Takahashi, K.J.: 1970, J. Biochem. Tokyo 67, pp. 838-839.
12) Takahashi, K.J.: 1976, J. Biochem. Tokyo 80, pp. 1267-1275.
13) Fülling, R. and Rüterjans, H.: 1978, FEBS Lett. 88, pp.279-282.
14) Arata, Y., Kimura, S., Matsuo, H., and Narita, K.: 1979, Biochem. 18, pp. 18-24.
15) Ostermann, H.L., and Walz, F.G. jr.: 1979, Biochem. 18, pp. 1984-1988.
16) Kyogoku, Y., Lord, R.C., and Rich A.: 1967, Proc. Natl. Acad. Sci. USA 57, pp. 250-257.
17) Kanaya, S., and Uchida, T.J.: 1981, J. Biochem. Tokyo 89, pp. 591-597.
18) Takahashi, K.: 1972, J. Biochem. Tokyo 72, pp. 1469-1482.
19) Kyogoku, Y., Watanabe, M., Kainosho, M., and Oshima, T.: 1982, J. Biochem. Tokyo 91, pp. 675-679.
20) Mauguen, Y., Hartley, R.W., Dodson, E.J., Dodson, G.G., Bricogne, G., Chothia, C., and Jack, A.: 1982, Nature 297, pp. 162-164.
21) Nakamura, R.T., Iwahashi, K., Yamamoto, Y., Iitaka, Y., Yoshida, N., and Mitsui, Y.: 1982, Nature 299, pp. 564-566.

Acknowledgement. The authors are grateful for financial support by Sonderforschungsbereich 9 of the Deutsche Forschungsgemeinschaft.

DNA BENDING AND PROTEIN-DNA INTERACTIONS

Donald M. Crothers, Stephen D. Levene and Hen-Ming Wu
Department of Chemistry, Yale University, New Haven, CT.

ABSTRACT

We show how the site of bending in a DNA fragment can be identified by gel electrophoresis of circularly permuted analogs of the
fragment. A molecular model which incorporates bending at ApA residues
yields a shape for a bent kinetoplast DNA fragment which is in excellent
agreement with the electrophoretic analysis. Hydrodynamic theory is
used to estimate the extent of bending. A similar analysis of the
complex of CAP protein with the lac operator DNA fragment shows that
the protein bends the DNA molecule.

INTRODUCTION

Bending of DNA in response to protein binding is one of the
possible conformational alterations of the double helix as a result of
interaction with a specific protein. Other known examples include
torsional winding or unwinding of the duplex, possibly accompanied by
base pair melting as in the case of RNA polymerase-promoter interactions (1). Other proteins, such as the gene 32 product of T4 bacteriophage, bind with strong preference to single stranded nucleic
acids (2), and therefore can act to disrupt the double helix entirely.

The most common form of DNA bending by proteins is in assembly of
the nucleosome from core histones and an appropriate length of DNA.
Another example is provided by the "gyrasome", formed from DNA gyrase
and 140 bp of DNA (3). So far, however, DNA bending has not been
experimentally implicated in the control of gene expression by
repressor or gene activating proteins. Our objective in this paper is
to explore the involvement of DNA bending in activation of the E. coli
lac operon by the cAMP binding protein CAP (also called CRP). The
properties of bent DNA molecules needed for the analysis are deduced
from study of certain naturally bent DNA sequences.

CAP provides a classic example of a protein which can regulate
gene activity by positive control. Emmer et al (4) and Zubay et al

B. Pullman and J. Jortner (eds.), Nucleic Acids: The Vectors of Life, 277–282.
© 1983 by D. Reidel Publishing Company.

(5) first showed that the effect of cAMP on gene expression is regulated by a cAMP binding protein (CRP or CAP). (For a review, see de Crombrugghe & Pastan (6).) The crystal structure of CAP, a dimer of 22,500 subunit molecular weight (7), is known to 2.9 Å resolution (8). The dimer binds two cAMP molecules, with a binding constant of about 1 x 10^5 M^{-1} (9). Direct evidence of CAP action at the level of transcription has come from the studies of de Crombrugghe et al (10), Majors (11), and Nissely et al (12), who showed that CAP + cAMP stimulated in vitro transcription of the lac and gal operons by 20-50 fold. CAP is, however, also capable of acting as a repressor (13).

de Crombrugghe et al (10) and Beckwith et al (14) proposed that CAP might stimulate transcription by binding to a specific lac promoter site, a suggestion that was supported by binding studies reported by Riggs et al (15), Nissely et al (16), Majors (17) and Mitra et al (18). Binding to specific fragments was found to be cAMP dependent. Protection (19) and footprinting (20) studies have revealed the location of the specific binding sites, along with an additional CAP-specific but apparently non-functional site between -4 and +25 in the lac operator region. Fried and Crothers (21) and Garner and Revzin (22) showed that each site in the lac operon is occupied by a single CAP dimer.

Some general facts and speculation are available on possible structural changes which may occur when the specific CAP-promoter complex is formed. Dickson et al (23) originally suggested a DNA structural change which propagated into the polymerase binding sites. McKay and Steitz (8) advanced the structurally specific hypothesis that the DNA switches locally to left handed form. CD measurements which we performed on an oligonucleotide-CAP complex indicate that there is some alteration of DNA structure, but the results do not support a switch to a left handed helix (24). Kolb and Buc (25) examined the topological effect of CAP binding to closed circular DNAs, and concluded that the unwinding predicted by the McKay and Steitz model does not occur. Some unknown structural influence of CAP on DNA is suggested by the footprinting studies of Taniguchi et al (26), who found CAP-enhanced DNAse I cleavage at sites well outside the region occupied by CAP. A conformational change of the protein which differs for specific and non-specific binding is supported by the fluorescence experiments of Wu et al (27).

A major barrier to study of possible bending of DNA fragments by proteins has been the lack of experimental methods for detecting the phenomenon. Fortunately, DNA fragments isolated from the kinetoplast body of trypanosome parasites have recently provided a model system for exploring the properties of bent DNA (28). Especially striking is the anomalously slow gel electrophoretic mobility exhibited by the fragment, along with a much-accelerated rotational correlation time, the latter as expected for a fragment rendered more compact by bending. The structural basis for bending appears to reside in periodicities in the base sequence (28). Specifically, the sequence ApA is repeated with a period of 10.5 bp, in phase with the double helix screw repeat.

Presumably, although the underlying structural reason remains unclear,
the sequence ApA produces a small helix bend (29), which, because of
the phase relationship, yields bending in a plane. Our objective in
this paper is to explore further the properties of bent DNA fragments,
and to exploit the evident similarities between kinetoplast DNA properties and the electrophoretic characteristics of the CAP-DNA complex.

RESULTS

<u>Location of the bend in the kinetoplast fragment</u>. The objective
of these experiments was to identify the position of primary bending
in the 424 bp kinetoplast (K) DNA fragment. The logic of the experiment relies on the expectation that the anomalous properties conferred
by a DNA bend will be much more pronounced when the bend is located in
the middle (Figure 1a), as compared to the end (Figure 1b) of the
molecule

(a)

(b)

Figure 1

This hypothesis was initially verified by placing the K-DNA fragment
at the end or at the center of a longer fragment. Significantly
slower mobility was observed in the latter case.

DNA fragments analogous to those shown in Figure 1 were prepared
by cloning a tandem repeat of the K-DNA fragment, which was subsequently cut with a series of restriction enzymes each of which cleaves
at a single site in the repeat. The result is a series of DNA molecules containing the K-DNA sequence in circularly permuted form. The
gel mobility of these fragments was estimated to reach a maximum when
the fragment end is located 150 bp from the natural end of the fragment
(Figure 2).

natural fragment   circularly permuted fragment

Figure 2. Diagram showing the natural K-DNA fragment and its circularly
permuted form which displays maximum gel mobility. The distance from
the left end (E_1,) to the bend region B,B' is estimate to be 150 ± 5 bp.

Molecular models for bending. We have explored the relationship between the K-DNA base sequence and its bent geometric structure. The only simple satisfactory model we have found assumes bending of the helix direction by an angle θ at every ApA residue. We used this model (and other models tested) to generate helix axis trajectories for specific DNA sequences. The results, as displayed on a molecular graphics system, showed excellent agreement with the conclusion of the gel electrophoresis experiment, namely that the main bending anomaly is centered about 150 bp from the left end of the fragment.

Use of hydrodynamic theory to investigate the extent of bending. In spite of some recent theoretical advances (30, 31), it is still impossible to provide a quantitative interpretation of gel electrophoretic mobility changes which result from DNA bending. Fortunately, hydrodynamic theory is considerably more advanced, and we have been able to compute the expected rotational relaxation times of the K-DNA fragment and its circularly permuted isomers. The method used is similar to that described by Hagarmann (32), except that the chain contains systematic bends at ApA residues. Thermal flexibility is allowed to an extent which is consistent with the persistence length of normal linear DNA. A bend of about 9° at ApA yields a decrease by a factor of 2 in the K-DNA rotational correlation time, compared to linear controls. This is approximately the experientally observed ratio (28); the experiments to determine this quantity more precisely are still in progress.

CAP protein-induced DNA bending. We and others have earlier described the use of gel electrophoresis to examine the properties of protein-DNA complexes (33,34). Of particular relevance for the present discussion is the observation that mutation in the CAP-DNA binding site yields a protein-DNA complex of altered gel electrophoretic mobility (21), implying that the conformation of the complex is sensitive to the details of the protein-DNA interaction.

In exploring the possible nature of DNA conformational alteration by CAP binding, we tried an experiment analogous to that carried out on the K-DNA fragment. Circularly permuted 203 bp lac operator fragments were prepared, and their gel electrophoretic mobility was examined upon addition of CAP protein. Lac repressor protein binding to the same fragments served as control for the experiment. We observed, as with K-DNA, a dramatic variation of electrophoretic mobility in the circularly permuted CAP-DNA complexes, and virtually no variation was seen in the lac repressor controls. This result argues strongly for bending of the lac operator fragment by CAP protein. We estimate that the center of the bending site is near the center of quasi two-fold symmetry of the CAP binding site, but the experiments to identify the position more precisely are still in progress.

We discovered an additional empirical similarity between bent kinetoplast DNA and the CAP-DNA complex. Very low levels(about 1 drug per 50-100 bp) of the oligopeptide antibiotic distamyci were found to normalize the gel mobility of both species. In particular, the drug removed the dependence of gel mobility on circular permutation of the sequence, and caused K-DNA to migrate at nearly the position expected for a 424 bp DNA fragment. These observations reinforced the sense of similarity between K-DNA and the CAP-DNA complex.

DISCUSSION

Our results open several areas for speculation, including the structural origin of bending of the CAP-DNA complex, and the relationship of the bend to CAP function. Recent models for the structure of the CAP-DNA complex have independently included as part of their focus the possibility that DNA may bend at the protein contact site (T.A. Steitz, private communication). We cannot yet specify from our work the extent of the bend, but we estimate by analogy with the gel electrophoretic and hydrodynamic properties of the K-DNA fragment that the total bending angle may exceed 45°. Presumably this results from optimization of specific protein-DNA contacts in the complex.

One plausible functional consequence of DNA bending in the complex is to facilitate CAP-RNA polymerase contacts in the bent complex which would not be possible in the linear form, as illustrated in Figure 3:

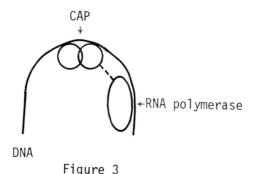

Figure 3

Perhaps different extents of DNA bending could help explain the variation in linear distance between CAP and the polymerase initiation site which is seen in different CAP-activated operons.

REFERENCES

1. Siebenlist, U.: 1977, Nature 279, 651-652.
2. Kowalczykowski, S.C., Lonberg, N., Newport, J.W. & von Hippel, P.H.: 1981, J. Mol. Biol. 145, 75-104.
3. Klevan, L. & Wang, J.C.: 1980, Biochemistry 19, 5229.
4. Emmer, M., deCrombrugghe, B., Pastan, I. & Perlman, R.: 1970, Proc. Natl. Acad. Sci. USA 66, 480-487.
5. Zubay, G., Schwartz, D. & Beckwith, J.: 1970, Proc. Natl. Acad. USA 66, 104-110.

6. deCrombrugghe, B. & Pastan, I.: 1978, in: The Operon, Miller, J. & Reznikoff, W.S., eds. pp 303-324.
7. Anderson, W.B., Schneider, A., Emmer, M., Perlman, R. & Pastan, I.: 1971, J. Biol. Chem. 246, 5929-5937.
8. McKay, D.B. & Steitz, T.A.: 1981, Nature 290, 744-749.
9. Takahashi, M., Blazy, B. & Baudras, A.: 1980, Biochemistry 19, 5124-5130.
10. deCrombrugghe, B., Chen, B., Anderson, W., Nissley, P., Gottesman, M., Perlman, R. & Pastan, I.: 1971, Nature New Biology 231, 139-144.
11. Majors, J.: 1975, Proc. Natl. Acad. Sci. USA 72, 4394-4398.
12. Nissely, P., Anderson, W., Gottesman, M., Perlman, R. & Pastan, I.: 1971, J. Biol. Chem. 246, 4671-4678.
13. Mouva, R., Green, P., Nakamura, K. & Inouye, M.: 1981, FEBS Letters 128, 186-190.
14. Beckwith, J., Grodzicker, T. & Arditti, R.: 1972, J. Mol. Biol. 69, 155-160.
15. Riggs, A., Reiness, G. & Zubay, G.: 1971, Proc. Natl. Acad. Sci. USA 68, 1222-1225.
16. Nissely, S., Anderson, W., Gallo, M., Pastan, I. & Perlman, R.: 1972, J. Biol. Chem. 247, 4265-4269.
17. Majors, J.: 1975, Nature 256, 672-674.
18. Mitra, S., Zubay, G. & Landy, A.: 1975, Biochem. Biophys. Res. Commun. 67, 857-863.
19. Majors, J.: 1977, Dissertation, Harvard University.
20. Schmitz, A.: 1981, Nucl. Acids Res. 9, 277-292.
21. Fried, M.G. & Crothers, D.M.: 1983, Nucl. Acids Res. 11, 141-158.
22. Garner, M.M. & Revzin, A.: 1982, Biochemistry 24, 6032-6036.
23. Dickson, R., Abelson, J., Johnson, P., Reznikoff, W. & Barnes, W.: 1977, J. Mol. Biol. 111, 65-79.
24. Fried, M., Wu, H.M. & Crothers, D.M.: 1983, Nucl. Acids Res. in press.
25. Kolb, A. & Buc, H.: 1982, Nucl. Acids Res. 10, 473-485.
26. Taniguchi, T., O'Neill, M. & deCrombrugghe, B.: 1979, Proc. Natl. Acad. Sci. USA 76, 5090-5094.
27. Wu, F., Nath, K. & Wu, C.: 1974, Biochemistry 13, 2567-2572.
28. Marini, J.C., Levene, S.D., Crothers, D.M. & Englund, P.T.: 1982, Proc. Nat. Acad. Sci. USA 79, 7664-7668.
29. Trifonov, E.N. & Sussman, J.L.: 1980, Proc. Nat. Acad. Sci. USA 77, 3816-3820.
30. Lerman, L.S. & Frisch, H.L.: 1982, Biopolymers 21, 995-997.
31. Lumpkin, O.J & Zimm, B.H.: 1982, Biopolymers 21, 2315-2316.
32. Hagarman, P. 1981: Biopolymers 20, 1503-1535.
33. Fried, M. & Crothers, D.M.: 1981, Nucl. Acids Res. 9 6505-6525.
34. Garner, M.M. & Revzin, A.: 1981, Nucl. Acids Res. 9, 3047-3059.

DYNAMICS AND SPECIFICITY OF PROTEIN-NUCLEIC ACID INTERACTIONS IN MODEL SYSTEMS.

Dietmar Porschke

ABSTRACT

The dynamics of protein–nucleic acid interactions is analysed by the field jump technique using simple model systems. Oligopeptide-oligonucleotide-complexes are formed in a diffusion controlled reaction with rate constants around 10^{10} $M^{-1}s^{-1}$. The rate increases with chain length N of coiled polymers approximately by a factor $\sqrt{N}$ following the increase in the mean square radius of the polymer sphere. Variations in stability are mainly reflected in the rate constants of dissociation. A separate slow step is found for the binding of aromatic peptides, probably representing an insertion reaction of aromatic residues.

The affinity of various amino acid residues to polynucleotides is analysed by measurements of melting temperatures t_m in the presence of amino acid amides. It is shown that the amides of hydrophobic amino acids induce a decrease of melting temperatures for poly(A)*poly(U) and poly(I)*poly(C) in a defined concentration range, whereas the amides of hydrophilic amino acids continuously increase t_m-values. The physical basis of these effects is discussed together with some biological implications for melting proteins and the evolution of the genetic code.

INTRODUCTION

The interactions between proteins and nucleic acids have been studied in a great number of different systems, both with natural and synthetic components [1-4]. These investigations provided a wealth of information. Nevertheless it is often difficult to separate general features of protein–nucleic acid interactions from the ones observed only for special systems. The results described in the present contribution are from model experiments designed to obtain general information on the dynamics and specificity of protein nucleic acid interactions.

B. Pullman and J. Jortner (eds.), Nucleic Acids: The Vectors of Life, 283–293.
© 1983 by D. Reidel Publishing Company.

DYNAMICS

It is known that the elementary steps of base pairing and of various conformational transitions in nucleic acids are particularly fast [5]. This is apparently the result of an evolutionary pressure for fast processing of the genetic information. Since proteins are directly involved in the information processing, it may be expected that their interactions with nucleic acids are also optimised for high response rates.

We have analysed some 'elementary steps' of protein-nucleic acid interactions using the field jump technique. This technique is particularly useful for the investigation of complexes with electrostatic interactions [6,7]. Due to the high charge density of nucleic acids, electrostatic contacts are involved in virtually all protein nucleic acid complexes. High electric field pulses weaken the electrostatic contacts and lead to dissociation of the complexes. The amplitudes and the time constants of the relaxation process provide information about thermodynamic and kinetic parameters of the complexes ([8], cf. Fig. 1).

The results obtained for simple oligonucleotide-oligopeptide complexes [9,10] may be summarised as follows: The rate constants for complex formation are in the range of 1 to $2*10^{10}$ $M^{-1}s^{-1}$ and indicate a diffusion controlled process with a reaction between nucleotide and peptide, whenever they approach each other. Under these conditions differences in the binding affinity are reflected in the rate constants of dissociation. As expected the affinity increases with the number of electrostatic contacts or with increasing charge density on positively charged peptides and negatively charged nucleotides. A particularly high affinity is found for a complex between Arg_3 and $I(pI)_5$. The binding constant for this complex is higher by a factor of 4 than that for the complex between Lys_3 and $I(pI)_5$. Comparison with various other complexes suggests, that the high affinity is due to specific hydrogen bonding between the guanidino group of arginine and the inosine base. Corresponding hydrogen bonds may be formed with guanine bases.

The contribution to the binding affinity from amino acids with non-ionic side chains remains rather small. For example, the affinity of LysGlyLys and LysTyrLys to oligonucleotides like $A(pA)_5$ is only slightly different. Furthermore, the binding kinetics for peptides having aromatic amino acids with oligonucleotides closely corresponds to that for 'non-aromatic' peptides; a separate relaxation effect indicating some insertion reaction of aromatic acids was not observed for the binding of 'aromatic' peptides to short oligonucleotides.

The rate constant for a diffusion controlled reaction is dependent upon the sum of the diffusion coefficients (D_p+D_n) of the reaction partners and the reaction radius r, which represents the closest approach at which spontaneous reaction occurs [11]. Thus the rate of

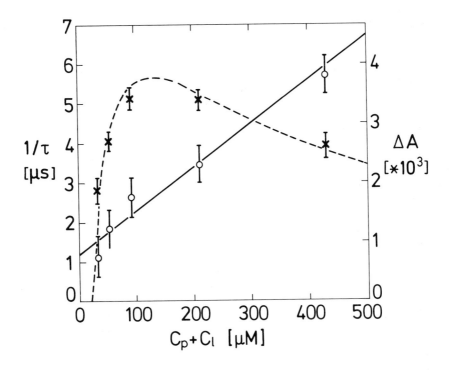

Fig. 1 Reciprocal relaxation times $1/\tau$ (o) and amplitudes Δ(x) from field jump experiments for $U(pU)_7$ + Lys-Thr-Arg-Glu-Lys-Val as a function of the sum of free oligomer concentrations (measured at constant $U(pU)_7 = 20.9\ \mu M$). The peptide is a fragment of Lac-repressor (residues 33 to 38), it does not show specificity in binding to single strands. Association constants 9.3 mM^{-1} for $U(pU)_7$ (10.0 mM^{-1} for $I(pI)_7$); association rate constant for $U(pU)_7$ $1.1*10^{10}\ M^{-1}s^{-1}$ (from ref. [10]).

peptide binding to polynucleotides of increasing chain length is influenced by two different factors. The decrease in the diffusion coefficient D_n with increasing chain length has only a minor influence, since D_n enters as a sum with the diffusion coefficient of the peptide D_p. The increase of the reaction radius r with chain length has a more important influence. It is not simple, however, to calculate r as a function of the chain length, since the effective value of r depends, among other factors, upon the conformation of the polymer chain. For a coiled polymer statistics predicts that the mean square radius increases with the square root of the chain length [12]. This prediction is verified by experimental data e.g. on the binding of simple peptides to polynucleotides. The rate constant for the binding

Table I: Stability and rate constants from field jump experiments in 1 mM Na-cacodylate pH 5.9, 50 µM EDTA at 20°C.

Nucleotide	Peptide	Stability constant $[mM^{-1}]$	Rate constant $[10^{10}M^{-1}s^{-1}]$
$(I)_6$	$(Arg)_3$	49	2.5
$(I)_6$	$(Lys)_3$	12	1.5
$(U)_6$	$(Arg)_3$	15	2.0
$(U)_6$	$(Lys)_3$	8.9	2.0
$(A)_6$	$(Arg)_3$	38	1.3
$(A)_6$	$(Lys)_3$	17	1.1
$(A)_5$	$(Arg)_3$	12	1.5
$(A)_5$	$(Lys)_3$	7.9	0.94
$(A)_4$	$(Lys)_3$	1.3	–
$(A)_6$	$(Arg)_2$	3.7	0.5
$(A)_6$	$(Lys)_2$	2.7	0.35
$(A)_6$	LysPheLys	2.4	0.7
$(A)_6$	LysTyrLys	2.9	0.56
$(A)_6$	LysGlyLys	1.8	0.46
$(U)_6$	LysPheLys	1.8	0.62
$(U)_6$	LysTyrLys	2.6	0.78
$(U)_6$	LysGlyLys	1.1	–

of Lys_3 to polyribouridylates increases by a factor of 15, when the chain length increases from 5 to 1000 [13]. Thus the experiments show that the reaction radius is not e.g. a linear function of the chain length, but approximately increases with the square root of the chain length ($\sqrt{1000/5} \sim 14$).

The binding of peptides with tyrosine or tryptophane residues to polynucleotides may be studied conveniently by fluorescence measurements [14]. The kinetics of the binding process has been investigated again by the field jump technique. The concentration dependence of relaxation time constants observed for the binding of LysTrpLys to poly(A) [15] is consistent with a two step mechanism

$$\text{peptide + polynucleotide} \underset{k_d^I}{\overset{k_a^I}{\rightleftharpoons}} \text{complex I} \underset{k_d^{II}}{\overset{k_a^{II}}{\rightleftharpoons}} \text{complex II} \quad (1)$$

The formation of complex I is controlled by diffusion and is fast relative to the second step, which may represent an insertion reaction of tryptophane residues between adenine bases. The observed rate constant $k_a^{II} = 1.5*10^5 s^{-1}$ is in the order of magnitude expected for such an insertion reaction, whereas the equilibrium constant $k_a^{II}/k_d^{II} \sim 55$ is unexpectedly high.

The kinetics of LysTrpLys binding to DNA has also been studied by

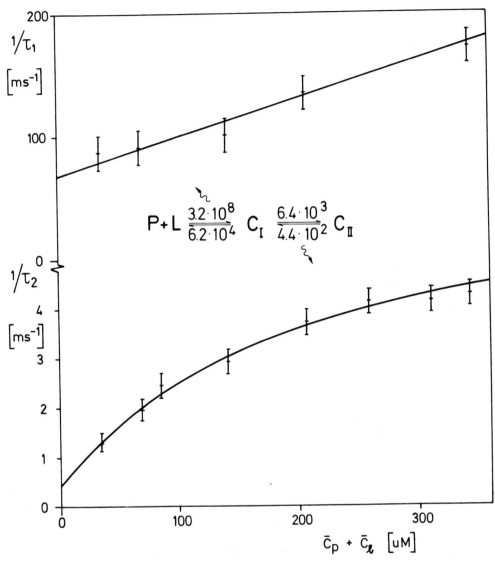

$$P+L \underset{6.2 \cdot 10^4}{\overset{3.2 \cdot 10^8}{\rightleftharpoons}} C_I \underset{4.4 \cdot 10^2}{\overset{6.4 \cdot 10^3}{\rightleftharpoons}} C_{II}$$

Fig. 2 Reciprocal relaxation times $1/\tau_1$ and $1/\tau_2$ as a function of the free reactant concentration for LysTrpLys + DNA (calf Thymus; 30000 base pairs; 1 mM NaCl, 1 mM Na-cacodylate pH 7.0, 0.2 mM EDTA; 20°C) $k_a^I = 3.2*10^8$ $M^{-1}s^{-1}$, $k_a^{II} = 6.4*10^3 s^{-1}$, $k_d^{II} = 440$ s^{-1} (from ref. [19]).

field jump measurements with fluorescence detection [16]. Two separate relaxation processes with opposite amplitude were observed for the binding of LysTrpLys to high molecular DNA. The concentration dependence of the time constants is consistent again with a two step reaction (cf. eq. 1 and Fig. 2). The rate constant for the formation of complex II $k_a^{II} = 6.4*10^3 s^{-1}$ is clearly lower than that observed in the case of single stranded poly(A). The relatively low rate for DNA apparently reflects the rather high stability of base stacks in the cooperative double helical structure.

SELECTIVE INTERACTIONS OF AMINO ACID RESIDUES WITH POLYNUCLEOTIDES

The interactions of amino acids having non-ionic side chains with nucleic acids are notoriously weak. Any contribution of such amino acids to the binding affinity with polynucleotides may only be detected by particularly sensitive methods. It is known that the cooperative melting transition of double helices is quite sensitive to the binding of ligands [17, 18]. A rather low extent of ligand binding may lead to a considerable change in the melting temperature t_m. Owing to the cooperativity of the helix-coil transition t_m-values can be measured with high accuracy. In the past the advantage of this approach has not been really exploited yet. This may be partly due to the fact that the number of parameters affecting the helix-coil transition is relatively large. Usually it is not possible to evaluate all the parameters involved in the transition. However, this problem may be overcome, at least partly, by t_m-measurements at many different ligand concentrations. A crucial step is, of course, the selection of an appropriate model system. In the following a set of data is described for the binding of amino acid amides to polynucleotides. These model compounds were selected for various reasons. First of all the amide derivatives are commercially available for most of the amino acids. The positive charge of these compounds in aqueous solution at pH $\leq$ 7 provides a general electrostatic driving force for the association with polynucleotides, which may be modulated by other types of interactions. The structure of the amide function is close to that of the amide linkage in proteins and does not introduce any major unnatural perturbation into the amino acids.

Melting data were collected both for deoxyribopolymers [19] and for ribopolymers [20, 21]. The data obtained for the ribopolymers are particularly instructive. Addition of amino acid amides to solutions of poly(A)*poly(U) or poly(I)*poly(C) does not always lead to an increase of t_m-values, but in some cases induces a clear decrease of the melting temperature (Fig. 3). When usual monovalent electrolytes are added to solutions of double helices, an increase of melting temperatures is observed in virtually all cases. This increase is in agreement with polyelectrolyte theory [22, 23]. An exception has been documented for special mixtures of mono- and bivalent ions and is in line with counterion condensation theory on the basis of merely electrostatic interactions [22]. The present observations cannot be explained by electrostatic interactions alone. The melting data clearly indicate the contribution of other interactions and suggest the existence of binding sites. A simple formalism to describe the experimental data is provided by the following site model [17, 18]

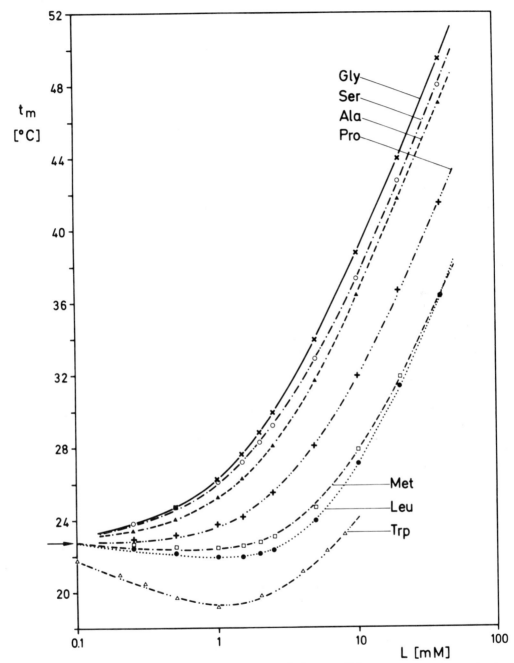

Fig. 3 Melting temperatures of poly(A)*poly(U) as a function of the concentration of various amino acid amides (1 mM NaCl, 1 mM Na-cacodylate pH 7, 0.2 mM EDTA). The melting temperature in the absence of ligand is indicated by the arrow. The continuous lines represent least squares fits according to the site model (from ref. [21]).

$$HL \rightleftharpoons H + L \rightleftharpoons C + L \rightleftharpoons CL \qquad (2)$$

The ligand L may bind both to the helix H and the coil C. The melting temperature T_m (in degrees Kelvin) is then given by

$$\frac{1}{T_m^0} - \frac{1}{T} = \frac{R}{\Delta H} \frac{(1 + K_h*L)^{2/n_h}}{(1 + K_c*L)^{2/n_c}} \qquad (3)$$

where t_m^0 is the melting temperature in the absence of ligand, ΔH the enthalpy change associated with melting of a base pair, $K_h (K_c)$ binding of the ligand to the helix (coil), $n_h (n_c)$ number of bases in the helix (coil) per binding site and R gas constant. Using this equation the experimental data may be represented to a high degree of accuracy. A special example is given in Fig. 4 showing a joint fit of data obtained for 3 different ion concentrations. The dependence of the equilibrium constants K_h and K_c upon the ion concentration has been described according to the expectation from polyelectrolyte theory for a unit charge compensation upon complex formation (for details cf. ref. 20). The evaluation of all the binding parameters is possible for cases with a minimum of t_m-values as a function of the amide concentration. In the other cases the absolute values are subject to a rather large uncertainty, whereas the ratio K_h/K_c remains a rather well defined quantity.

Table II: Parameters for the binding of amino acid amides to polynucleotides according to a site model (definition see text; from ref. [20] and [21]; 1 mM NaCl, 1 mM Na-cacodylate pH 7, 0.2 mM EDTA)

	polyA*polyU			polyI*polyC		
	K_h/K_c	$K_h [M^{-1}]$	n_c	K_h/K_c	$K_h [M^{-1}]$	n_c
Ala	1.30	150	1.14	1.20	210	1.15
Gly	1.15	350	1.17	1.15	480	1.16
Ser	1.07	380	1.17	1.16	410	1.16
Asn	1.00	260	1.17	1.00	360	1.17
Pro	0.84	510	1.19	0.95	480	1.18
Val	0.80	420	1.20	0.85	490	1.19
Met	0.80	460	1.18	0.76	380	1.21
Ile	0.76	440	1.20	0.78	470	1.20
Leu	0.75	430	1.20	0.75	440	1.19
Phe	0.70	480	1.15	0.59	290	1.19
Tyr	0.71	750	1.18	0.71	730	1.13
Trp	0.75	1480	1.11	0.71	1060	0.95

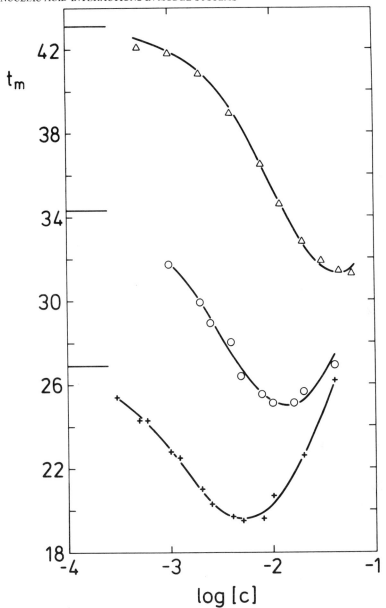

Fig. 4 t_m-values of poly(I)*poly(C) as a function of the logarithm of the Phe-amide concentration in 1 mM NaCl, 1 mM Na-cacodylate, 0.2 mM EDTA (+), 2 mM NaCl, 2 mM Na-cacodylate, 0.4 mM EDTA (o) and 10 mM NaCl, 1 mM Na-cacodylate, 0.2 mM EDTA (Δ). The solid lines show a simultaneous fit of all the data by equation (2) with an ionic strengtn dependence expected for unit charge compensation (for details cf. ref. [20]; in the legend to the corresponding figure of ref. [20] Phe-amide was changed by mistake to Tyr-amide.). The bars give the t_m-values in the absence of ligand in the 3 different buffers. Binding parameters are compiled in Table II.

The data obtained from the melting curves demonstrate a clear gradation in the binding properties of amino acids. Hydrophilic amino acids induce an increase of t_m-values at all concentrations, whereas hydrophobic amino acids induce a decrease of helix stability in a certain concentration range. The gradation in the binding properties is reflected in high ratios K_h/K_c for hydrophilic amino acids whereas, hydrophobic amino acids are characterised by relatively low ratios K_h/K_c. This effect may be explained by the hydrophilic exterior of double helices with their high density of phosphate groups and sugar residues. The parts of the bases exhibited to the grooves are probably also more hydrophilic than hydrophobic. This environment is expected to favour association of hydrophilic ligands. Although the single strands have, of course, the same molecular composition, the hydrophobic bases are clearly more exposed to the outside in the single stranded state. Moreover, the larger flexibility of the single strands may allow the formation of some hydrophobic pockets, which favour the association of hydrophobic ligands. This effect is most obvious for the aromatic amino acids, which may be inserted between adjacent bases. In fact, the decrease in helix stability is particularly large for the amides of the aromatic amino acids.

BIOLOGICAL IMPLICATIONS

The analysis of various melting proteins demonstrated that aromatic amino acids are directly involved in the binding to single stranded polynucleotides [14]. In some cases evidence has been presented that the binding involves insertion between the nucleic acid bases. Our observation of a decrease in the helix stability upon binding of aromatic amino acids suggest a direct link. It is possible, or even likely, that the evolution of melting proteins started from simple peptides containing aromatic residues.

Another connection may be postulated with respect to the evolution of the genetic code. It has been observed previously that there is a striking similarity in the codons for hydrophobic amino acids [25]. The codons for valin, isoleucine, leucine, methionine and phenylalanine all have a U at their second position. Two other hydrophobic amino acids, tyrosine and tryptophane, have a U at the first position of their codons. It is easily possible to construct adaptor molecules with a similar overall structure, but a different conformation at the 3'end, which is loaded with amino acids. Adaptors with a GC rich anticodon may have their 3'end preferentially in a double helical conformation, which favours the association of hydrophilic amino acids. In contrast the 3'end of adaptors with a AU rich anticodon may be preferentially in a single stranded conformation providing a pocket for hydrophobic amino acids. Although the difference in the binding affinity is not large, it may have been useful at an early stage of evolution.

REFERENCES

1. Vogel, H.R. ed.: 1977, Nucleic Acid–Protein Recognition, Academic Press, New York
2. Schimmel, P.R.: 1980, CRC Crit. Rev. Biochem. $\underline{9}$, pp. 207–251
3. Gabbay, E.J.: 1977, Bioorganic Chemistry III Macro- and Multimolecular Systems, Ed. van Tamelen, Academic Press, New York, pp. 33–70
4. Helene, C., Maurizot, J.C.: 1981, CRC Crit. Rev. Biochem. $\underline{10}$, pp. 213–258
5. Porschke, D.: 1977, Molecular Biology, Biochemistry and Biophysics $\underline{24}$, pp. 191–218
6. Eigen, M., DeMaeyer, L.: 1963, Techniques of organic chemistry, Eds. Friess, S.L., Lewis, E.S., Weissberger, A. New York Vol. 8 part 2, pp. 895–1054
7. Onsager, L.: 1934, J. Chem. Phys. $\underline{2}$, pp. 599–615
8. Bernasconi, C.F.: 1976, Relaxation Kinetics, Academic Press, New York
9. Porschke, D.: 1978, Eur. J. Biochem. $\underline{86}$, 291–299
10. Porschke, D., Gutte, B.: 1981, FEBS Letters $\underline{127}$, 63–66
11. von Smoluchowsky, M.: 1916, Physik, Z. $\underline{17}$, pp. 585–599
12. Bloomfield, V.A., Crothers, D.M., Tinoco, I.: 1974, 'Physical Chemistry of Nucleic Acids', Harper + Row, New York
13. Porschke, D.: 1979, Biophys. Chem. $\underline{10}$, 1–16
14. Brun, F., Toulme, J.J., Helene, C.: 1975, Biochemistry $\underline{14}$, pp. 558–563
15. Porschke, D.: 1980, Nucleic Acids Res. $\underline{8}$, 1591–1612
16. Porschke, D., Ronnenberg, J.: 1981, Biophys. Chem. $\underline{13}$, 283–290
17. Lazurkin, Y.S., Frank-Kamenetskii, M.D., Trifonov, E.N.: 1970, Biopolymers $\underline{9}$, pp. 1253–1306
18. McGhee, J.D.: 1976, Biopolymers $\underline{15}$, pp. 1345–1375
19. Porschke, D., Ronnenberg, J.: 1983, Biopolymers, in press
20. Porschke, D., Jung, M.: 1982, Nucleic Acids Res. $\underline{10}$, 6163–6176
21. Porschke, D., J. Mol. Evol., submitted
22. Manning, G.S.: 1978, Quart. Rev. Biophys. $\underline{11}$, pp. 179–246
23. Record, M.T., Anderson, C.F., Lohman, T.M.: 1978, Quart. Rev. Biophys. $\underline{11}$, pp. 103–178
24. Kowalczykowski, S.C., Bear, D.G., Hippel, P.H.: 1981, in 'The Enzymes', Vol XIV, pp. 373–444
25. Volkenstein, M.V.: 1966, Biochem. Biophys. Acta $\underline{119}$, pp. 421–424

THE recA GENE PRODUCT FROM *E. coli*. BINDING TO SINGLE-STRANDED AND DOUBLE-STRANDED DNA.

Christian Cazenave, Marie Chabbert, Jean-Jacques Toulmé
and Claude Hélène.
Laboratoire de Biophysique, Unité INSERM 201,
ERA CNRS 951,
Muséum National d'Histoire Naturelle, 61 Rue Buffon
75005 PARIS (France).

INTRODUCTION

The recA gene product from *E. coli* plays a central role in two major functions which are essential to the bacterium life : i) it catalyzes general genetic recombination by promoting DNA strand exchange reactions. *In vitro* it has been shown that RecA protein catalyzes the formation of duplex DNA from complementary single-stranded molecules (annealing), the assimilation of linear single-strands into duplex circular DNA to form D-loops, the pairing of gapped circular DNA with either superhelical or nicked circular DNA, the complete exchange of strands between full length linear duplex and homologous circular single-stranded DNA. D-loop formation requires ATP but not its hydrolysis whereas extension of an heteroduplex via branch migration requires ATP hydrolysis and proceeds with a unique polarity (for review see references 1,2) ; ii) the recA gene product plays a crucial role in DNA repair. When DNA replication is perturbed either as a consequence of damages induced in DNA by physical (radiations) or chemical agents, or by an alteration of the replication fork, a series of genes called the "SOS genes" are derepressed. These genes are under the control of a common repressor, the lexA gene product, which is cleaved by the RecA protein. This proteolytic activity of the RecA protein is induced when it binds to single-stranded DNA in the presence of ATP, even though ATP hydrolysis is not required for the cleavage reaction (for review see reference 3).

Binding of RecA protein to both double-stranded and single-stranded DNA is required as a first step in all reactions involved in genetic recombination. Furthermore, two enzymatic activities (ATPase and protease) are induced when the RecA protein binds to single-stranded DNA. To understand how the different functions of the *E. coli* recA gene product are activated a better knowledge of its DNA binding properties is clearly requested. We have therefore started an investigation of the interactions between the RecA protein and both double-

295

B. Pullman and J. Jortner (eds.), Nucleic Acids: The Vectors of Life, 295–304.
© 1983 by D. Reidel Publishing Company.

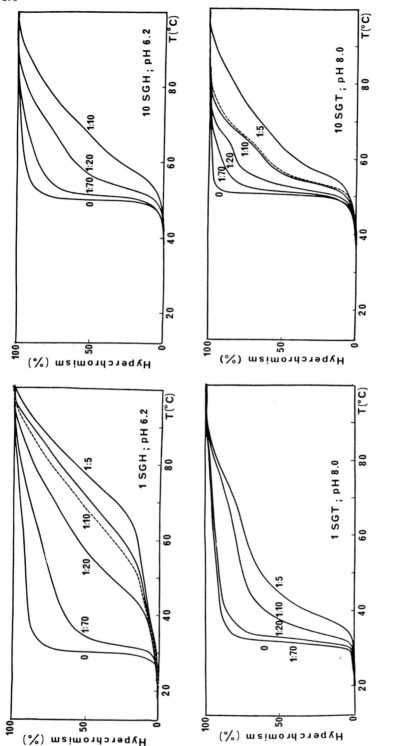

Figure 1 : Melting curves of polydAT in the presence of different concentrations of RecA protein (numbers on the curves indicate the ratio of RecA monomer to nucleotides). The dotted lines represent experiments carried out in the presence of polydT (at the same concentration in nucleotides as polydAT) and for a ratio of RecA monomer to polydAT phosphates of 1 : 10. Buffers used in these experiments : 1 SGH is 1 mM Na cacodylate, 1 mM NaCl, 0.2 mM EDTA at pH 6.2 ; 1 SGT is 1 mM Tris HCl, 1 mM NaCl, 0.2 mM EDTA at pH 8.0 ; 10 SGH contains 10 mM Na cacodylate and 10 mM NaCl ; 10 SGT contains 10 mM Tris HCl and 10 mM NaCl.

stranded and single-stranded DNAs using physicochemical techniques
(absorption, fluorescence, circular dichroism). A summary of the
results presently available is presented below.

RESULTS

The RecA protein binds preferentially to double-stranded DNA at low
ionic strength

It has been shown previously that a complex between RecA
protein and duplex DNA can be detected by nitrocellulose filter only
if ATP or its non-hydrolyzable analog ATPγS is present in the mixture (4).
Upon binding to closed circular double-stranded DNA in the presence
of ATP or ATPγS the RecA protein unwinds the double helix and produces
positive superhelical turns that can be relaxed by eukaryotic topoiso-
merase I (5,6).

If RecA protein binds both duplex and single-stranded DNA
the melting temperature of duplex DNA is expected to increase or
decrease depending on whether binding is stronger for the double helix
or for the single-strands, respectively. At low ionic strength RecA
protein strongly stabilizes the double helix against melting in the
absence of ATP or ATPγS (Figure 1). RecA protein is known to promote
the formation of double strands from complementary single-stranded molecules. Therefore the apparent stabilization of the double helix could
be due to the fact that the two single strands are maintained in
register during the heating process, especially when alternating
polyd(AT) is used as a substrate. To determine whether the stabilization
of duplex DNA was due to preferential binding of RecA protein to the
double helix the melting of polyd(AT) and duplex DNA was measured in
the presence of single-stranded polydT. This polynucleotide was chosen
because it has the highest affinity for the RecA protein among all
single-stranded polynucleotides tested. The presence of polydT (one or
two thymines per A-T base pair) did not change markedly the increase
of the melting temperature observed for polyd(AT) alone. Therefore it
can be concluded that RecA protein does bind preferentially to double-
stranded DNA at low ionic strength in the absence of ATP or ATPγS.

The stabilization of duplex DNA afforded by RecA protein
binding depends on the pH of the solution. A low pH (6.2) leads to a
higher stabilization than a higher pH (8.0) (Figure 1). The low ionic
strength buffers used in these experiments did not contain Mg^{2+} ions.
Neither ATP nor ATPγS were required to observe these effects. Therefore
the pH dependence of RecA protein binding to duplex DNA previously observed in the presence of Mg^{2+} ions and ATP or ATPγS (4) is a property of
the protein and not of the RecA-ATP (ATPγS) complex.

It has been previously reported that RecA protein binding
to duplex DNA in the presence of ATPγS unwinds DNA (5,6). Electron
microscopy investigations have shown that duplex DNA is stretched

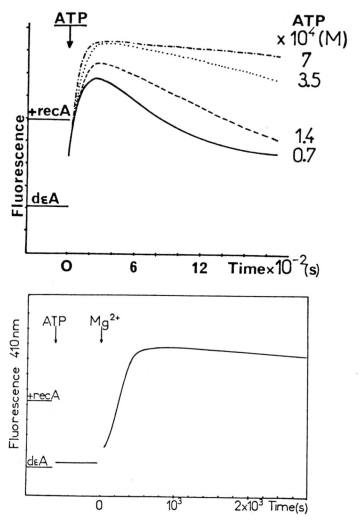

Figure 2 : Fluorescence studies of the effects of ATP and Mg^{2+} ions
    at 20°C on RecA binding to polydɛA (λ_{exc} = 305 nm, λ_{em} = 410 nm)
Upper : RecA (0.5 µM) was added to polydɛA (4 µM) in a pH 7.5 buffer
containing 20 mM Tris HCl, 4 mM $MgCl_2$, 1 mM β-mercaptoethanol. Then
ATP was added and the fluorescence of polydɛA followed as a function
of time. ATP concentrations are indicated on the Figure
Lower : RecA (0.5 µM) was added to polydɛA (4 µM) in a pH 7.5 buffer
containing 20 mM Tris-HCl, 1 mM β-mercaptoethanol. Then 1 mM ATP
was added which led to a dissociation a the recA-polydɛA complex
(the small fluorescence difference as compared with polydɛA alone is
due to the contribution of RecA which is still excited at 305 nm).
Then 4 mM Mg^{2+} was added and the fluorescence of polydɛA monitored as
a function of time (the fluorescence of polydɛA is slightly increased
in the presence of Mg^{2+} ions).

by a factor 1.5 in these complexes which appear as helical filaments
with a pitch of ≃ 100 Å and 6.2 RecA monomers per turn covering
18.6 base pairs (6). The absorbance measurements reported above demonstrate that at low ionic strength, in the absence of Mg^{2+} and ATPγS,
the stacking of DNA base pairs remains intact in RecA-duplex DNA
complexes since no hyperchromism is observed before melting starts
(Figure 1).

Increasing the ionic strength of the buffer leads to a
decrease of the stabilization afforded by RecA protein binding. The
effect is more important at pH 6.2 than at pH 8.0 (Figure 1). In the
presence of Mg^{2+} ions (1 mM) a different behavior is observed. As
the temperature is raised formation of large aggregates leads to an
important light scattering. The temperature at which aggregates start
to form corresponds to the beginning of the melting of free polydAT
or DNA. This phenomenon probably reflects the formation of ternary
complexes. As single-stranded regions appear in duplex DNA RecA
protein binds both single strands and duplex regions thus inducing
the formation of a network of branched structures.

Binding of RecA protein to single-stranded fluorescent polynucleotides

The binding of RecA protein to denatured DNA or polydT leads
to a quenching of the RecA intrinsic tryptophan fluorescence which
is too weak to be conveniently used to determine binding parameters
(7). Therefore we have relied upon chemical modifications of
adenine bases by chloroacetaldehyde to produce fluorescent polynucleotides.
In single-stranded DNA both adenine and cytosine bases are modified but
only 1,N^6-ethenoadenine (εA) is fluorescent. PolyrA and polydA also
give fluorescent derivatives, polyrεA and polydεA. Under our
experimental conditions nearly 100 % of adenine bases were modified.

Addition of RecA protein to εDNA or polydεA in a buffer
containing or not containing Mg^{2+} ions induced an increase of εA
fluorescence which leveled off when one RecA monomer was added per
about 7 bases as already reported in the case of εDNA (8). If the
buffer contained both Mg^{2+} ions and ATP or ATPγS, the increase in
fluorescence of εDNA or polydεA was higher but the same stoichiometry
was obtained. On the contrary the fluorescence of polyrεA was not affected upon addition of RecA protein, independently of the presence or
absence of Mg^{2+} ions and ATP or ATPγS suggesting that no binding
occurs.

Addition of ATP or ATPγS in the absence of Mg^{2+} ions induced
a dissociation of the complexes formed by RecA protein with εDNA or
polydεA. Further addition of Mg^{2+} ions led again to RecA binding in
a time-dependent process (Figure 2). When ATPγS was added to the RecA-
polydεA complex formed in the presence of Mg^{2+} first a decrease and
then an increase of εA fluorescence were observed. This last result
indicates that ATPγS first induces a dissociation of the RecA-polydεA
complex both in the presence and in the absence of Mg^{2+} ions and then

a slow process allows binding of a RecA–ATPγS–Mg^{2+} complex to polydεA. The slow binding process however depends on the order of addition of the two ligands ATPγS and Mg^{2+}. When Mg^{2+} is added after ATPγS a lag period is observed which is not detected when ATPγS is added after Mg^{2+} ions (within the limits of our mixing experiments < 5s). This lag phase might reflect a conformational change induced by Mg^{2+} binding to the RecA–ATPγS complex as already observed when Mg^{2+} binds, e.g., to G–actin (for a review, see reference 9) or to an exchange between bound ATPγS and an ATPγS–Mg^{2+} complex.

When ATP replaces ATPγS the first two steps are identical to those obtained with ATPγS : dissociation of the RecA–polydεA complex then binding of the RecA–ATP–Mg^{2+} complex. An additional slow process is then observed which is characterized by a decrease of εA fluorescence intensity. This process results from ATP hydrolysis which is known to be quite slow as measured from the rate of ADP production using ^{3}H– labelled ATP. The product of ATP hydrolysis, ADP, also binds to the RecA protein. When ADP binds to the RecA–polydεA complex in the absence of Mg^{2+} ions a dissociation of the complex is induced as described above with ATP or ATPγS (it should be noted that AMP has no effect). In the presence of Mg^{2+} ions, ADP allows some binding of RecA protein to polydεA. The increase in fluorescence intensity is much smaller than that induced by ATP or ATPγS which probably reflects a much weaker binding.

On the basis of these observations the different steps involved in the binding of RecA protein to single–stranded polynucleotide chains can be summarized by the scheme presented below :

$$\text{Rec A} + \text{Poly N} \rightleftharpoons \text{RecA–Poly N}$$

ATP · ATP

$$\text{RecA–ATP} + \text{Poly N}$$

Mg^{2+} · · · · · · · · · · · · · · · $\downarrow\uparrow$

$$\text{RecA–ATP·Mg}^{2+} + \text{Poly N}$$

$\updownarrow$

Nucleation

$\updownarrow$

Elongation

$\downarrow$

$$\left[\text{RecA–ATP–Mg}^{2+}\right]_n\text{–Poly N}$$

$\downarrow$

ATP hydrolysis

$\downarrow$

Dissociation

In this scheme, the rate-limiting processes would be the nucleation and polymerization reactions with a contribution of conformational changes in RecA protein induced by simultaneous binding of Mg^{2+} and ATP. It should be noted that, in the absence of Mg^{2+} ions, neither ATP nor ATPγS are able to induce the slow binding process even though they bind to the RecA protein.

Several experiments reveal that RecA protein self-associates in solution : i) sedimentation studies of RecA protein have already demonstrated the formation of RecA oligomers. RecA protein can form long filaments in the absence of any nucleic acid at low pH (6.2) (10). It can be precipitated in a low salt buffer containing 20 mM $MgCl_2$ and this precipitation step has been included in a RecA purification procedure (11) ; ii) binding of the fluorescent probe ANS to RecA protein is a slow process which is characterized by an increase in ANS fluorescence and a decrease of the light scattered by the protein solution (unpublished results). It seems likely that ANS binds to hydrophobic regions of the protein which are involved in self-association and therefore induces a dissociation of RecA aggregates ; iii) binding of ATP or ATPγS to the RecA-ANS complex in the presence of Mg^{2+} ions leads to a release of ANS (unpublished results). This is likely due to an induced reassociation of RecA monomers ; iv) binding of ATP or ATPγS to the RecA-polyN complex measured from the increased fluorescence of εDNA is a cooperative process whose Hill coefficient varies between 3 and 5 depending on the initial concentrations of reactants (8). The Hill coefficient increases when the ratio of εDNA to RecA protein increases which indicates that the ATP binding process is dependent on DNA concentration.

On the basis of these observations it may be concluded that the formation of RecA-ATP-Mg^{2+} polymers is catalyzed by DNA binding and that a minimum size of RecA-ATP-Mg^{2+} oligomers is required for DNA binding (nucleation process). This hypothesis is supported by the results of experiments carried out with oligonucleotides of different lengths. Different oligodeoxyadenylates were modified by chloroacetaldehyde to obtain fluorescent $(dεA)_n$ where n indicates the number of monomers per oligonucleotide. No binding of RecA or RecA-ATP-Mg^{2+} was observed with $(dεA)_8$. With $(dεA)_{18}$ a very slow binding of the RecA-ATP-Mg^{2+} ternary complex was observed, much slower than that described above for polydεA. The rate of complex formation increased in the order $(dεA)_{18} \ll (dεA)_{19-24} < (dεA)_{40-60} <$ poly dεA.

To determine the relative strengths of binding to different single-stranded structures, competition experiments were carried out. In the absence of ATP or ATPγS, RecA protein was added to a mixture of εDNA and the competing nucleic acid. The increase of εA fluorescence intensity reflects the binding of RecA protein to εDNA. Therefore the difference between the fluorescence increase observed with εDNA alone and that observed when it is mixed with the competing species reflects the distribution of RecA protein between the two nucleic acids. The same experiments were carried out with polydεA instead of εDNA. No

competition was observed with double-stranded DNA with either εDNA or
polydεA. The following order of RecA binding affinities for single-
strands was deduced :

polydT > εDNA ≃ polydU > polydεA > ssDNA > polydA,polydC

In the presence of ATPγS competition experiments were carried
out in a different way because slow binding processes are observed as
described above. RecA protein was incubated with the competing nucleic
acid at different concentrations in the presence of 100 μM ATPγS for
half-an-hour. Then εDNA or polydεA was added at a concentration such
that all RecA protein would be bound in the absence of a competing
nucleic acid. The increase of εA fluorescence intensity therefore reflect
the amount of RecA protein which can be transferred from the competing
nucleic acid to εDNA or polydεA. PolydT was again the most effective
in preventing RecA protein binding to εDNA. At a ratio of about
5 thymines per RecA monomer binding to εDNA was completely inhibited.
With denatured ssDNA five times more nucleic acid was required to
observe a behavior similar to that of polydT. When native DNA was
used in such competition experiments the results were quite pH dependent.
At pH 8.0 native dsDNA did not prevent RecA protein binding to εDNA
when both DNA concentrations were equal. At higher concentrations of
native DNA part of RecA molecules were prevented from binding to εDNA.
At pH 6.2 the behavior of dsDNA was quite similar to that of ssDNA until
about 50 % transfer was inhibited. Then a plateau was reached indicating
that 50 % of RecA protein could bind to εDNA even if it was pre-incubated
with a large excess of duplex DNA and that the other 50 % was trapped
in double-stranded DNA complexes.

The behavior of RecA protein with respect to its nucleic acid
binding properties is therefore quite dependent on the experimental
conditions. As described previously (paragraph 1 above), binding is
stronger to native DNA at low ionic strength in the absence of both
Mg^{2+} ions and ATPγS (or ATP). In the presence of Mg^{2+}, binding to
single-strands is favored especially at pH above 7 independently of
whether ATPγS is present or not. The binary RecA-ATP(ATPγS) complex
has only a very weak affinity for nucleic acids whereas the ternary
complex RecA-ATP(ATPγS)-Mg^{2+} binds very strongly to nucleic acids
in a slow process.

DISCUSSION

A model for RecA protein binding to nucleic acids can be deduced from the results presently available. In the absence of Mg^{2+} ions
and ATP, RecA protein binds to nucleic acids with a preference for
duplex structures at low ionic strength (1-10 mM NaCl). Higher concentrations of monovalent or Mg^{2+} cations reverse the binding preference.
ATP or ATPγS bind to RecA protein in the absence of Mg^{2+} ions and
dissociate RecA from single-stranded nucleic acids. The ternary complex
RecA-ATP-Mg^{2+} is the DNA binding species when both ATP and Mg^{2+} are

present. Mg^{2+} binding to RecA or RecA-ATP complex induces a conformational change in the protein. The ternary complex then undergoes a polymerization reaction whose rate depends on the presence of DNA. A minimum size of the ternary complex oligomers is required for binding to DNA (nucleation). Then elongation takes place on the DNA matrix.

This mode of RecA binding to DNA in the presence of ATP and Mg^{2+} ions might explain several results. For example if duplex DNA contains a single-stranded gap of sufficient size, the single-stranded region could act as a nucleation site to induce the polymerization of RecA and the propagation of RecA polymers along the duplex structure even though duplex DNA would not induce RecA polymerization in the absence of a nucleation process. This could explain why single-stranded gaps of about 30 nucleotides are sufficient to increase the ATPase activity to a level almost as great as that which would be oberved with a completely single-stranded DNA (12).

It has also been observed that RecA protein in excess over single-stranded DNA fragments leads to a dissociation of D-loops formed with superhelical closed circular DNA (form I DNA) and to the "inactivation" of form I DNA (5). Formation of a D-loop generates a single-stranded region in the circular DNA which could act as a nucleation site for Rec-ATPγS-Mg^{2+} polymerization. This reaction would result in duplex DNA being coated by RecA proteins through an "invasive" process with subsequent unwinding of duplex DNA (5,13)

The model proposed above rests upon the assumption of conformational changes induced in the RecA protein by effector binding (ATP, Mg^{2+}) and of a polymerization reaction whose nucleation is catalyzed by DNA binding. Further experiments should allow us to describe these different steps in a more quantitative way.

REFERENCES

1. Dressler, D. and Potter, H. : (1982) Ann.Rev. Biochem. 51 , 727-761.
2. Radding, C.M. : (1982) Ann. Rev. Genet. 16, 405-437.
3. Little, J.W. and Mount D.W. : (1982) Cell, 29 11-22.
4. Weinstock, G.M. : (1982) Biochimie, 64, 611-616.
5. Ohtani, T., Shibata, T., Iwabuchi, M., Watanabe, H., Iino, T. and Ando, T. (1982) Nature, 299, 86-89.
6. Stasiak, A. and Di Capua, E. : (1982) Nature, 299, 185-186.
7. Hélène, C., Toulmé, J.J., Behmoaras, T. and Cazenave, C. : (1982) Biochimie, 64, 697-705.
8. Silver, M.S. and Fersht, A.R. : (1982) Biochemistry, 21, 6066-6072.
9. Korn, E.D. : (1982) Physiol. Rev., 62, 672-737.
10. McEntee, K., Weinstock, G.M. and Lehman, I.R. : (1981) J. Biol. Chem., 256, 8835-8844.
11. Cotterill, S.M., Satterthwait, A.C. and Fersht, A.R. : (1982) Biochemistry, 21, 4332-4337.

12. West, S.C., Cassuto, E., Mursalim, J. and Howard-Flanders, P. (1980) Proc. Natl. Acad. Sci. U.S.A., 77, 2569-2573.
13. Wu, A.M., Bianchi, M., Das Gupta, C. and Radding, C.M. : (1983) Proc. Natl. Acad. Sci. U.S.A., 80, 1256-1260.

FLUORESCENCE DECAY STUDIES OF PEPTIDE-NUCLEIC ACID COMPLEXES

Thérèse Montenay-Garestier, Masashi Takasugi and
Trung Le Doan

Laboratoire de Biophysique, INSERM U201, ERA CNRS 951,
Muséum National d'Histoire Naturelle, 61, Rue Buffon,
75005 PARIS (FRANCE)

ABSTRACT

Fluorescence decay studies of two tryptophan-containing peptides
and one tyrosine-containing peptide reveal the existence of at least
two conformers for each peptide. When these peptides bind to poly(U),
denatured or native DNA changes are observed in both the lifetimes and
the respective contributions of the two components. A model is proposed
in which each peptide conformer forms two types of complexes, one involving only electrostatic interactions, the second one involving both electrostatic and stacking interactions. Moreover energy transfer from tyrosine to nucleic acid bases is shown to play an important role in the
quenching of the fluorescence of tyrosine-containing peptides bound to
nucleic acids.

INTRODUCTION

The specificity of protein binding to nucleic acids depends
on the interactions between functional groups of both molecules (1).
A study of oligopeptide-polynucleotide complexes should help characterize these interactions. Stacking involving aromatic amino acids has
been suggested to play an important role in the recognition of single
strands (2).

On the basis of fluorescence, nuclear magnetic resonance and
circular dichroism studies, a model has been proposed for the binding
of oligopeptides containing basic and aromatic residues to single-
stranded and double-stranded DNA (2). Binding has been assumed to involve
two successive steps : a purely electrostatic (outside) complex is in
equilibrium with a stacked complex. In the case of tryptophan-containing
peptides the analysis of steady-state fluorescence measurements has
been carried out by assuming that the fluorescence quantum yield of
the outside complex is identical to that of the free peptide, and that
the fluorescence of the stacked complex is completely quenched (3). In
the case of tyrosine-containing peptides, the analysis of fluorescence

305

B. Pullman and J. Jortner (eds.), Nucleic Acids: The Vectors of Life, 305–315.
© *1983 by D. Reidel Publishing Company.*

data is complicated by the fact that tyrosine emission in the outside
complex can be quenched by energy transfer to nucleic acid bases even
in the absence of any direct tyrosine-base interaction (4).

In order to provide more information on the fluorescence
properties of bound oligopeptides time-resolved fluorescence experiments
have been carried out. These studies have shown that a model for the
binding process must take into account the equilibrium between different
conformers of the peptides.

RESULTS

1. Fluorescence decay studies of tryptophan - and tyrosine - containing
   peptides

We have investigated in some detail the fluorescence decays
of three peptides, two containing tryptophan : Lys-Gly-Trp-Lys-OtBu
(KGWK), Lys-Trp-Gly-Lys-OtBu (KWGK) and one containing tyrosine :
Lys-Tyr-Lys (KYK). In all three cases, the decay curve can be quite
satisfactorily fit by the sum of two exponential
functions as already reported for the tripeptide Lys-Trp-Lys (5-6).

$$(1) \quad I(t) = a_1 \exp -t/\tau_1 + a_2 \exp -t/\tau_2$$

An average fluorescence lifetime $<\tau>$ can be calculated as :

$$(2) \quad <\tau> = (a_1 \ \tau_1 + a_2 \ \tau_2) / (a_1 + a_2)$$

In order to determine the origin of these two components, fluorescence decays were analyzed as a function of temperature and pH
for the two tryptophan-containing peptides. As can be seen from
figure 1, these two peptides behave differently. Both lifetimes
(τ_1 and τ_2) decrease as the temperature increases. Their respective
contributions (a_1 and a_2) remain constant in the case of KGWK
(Figure 1a) whereas the contribution of the shorter lifetime (τ_1)
decreases with temperature in the case of KWGK (Figure 1b). This
behavior does not seem to be affected by a change in ionic strength.

For KGWK τ_1 is observed to increase slightly as the pH is
raised from 4 to 7 and then to be constant from pH 7 to 9 ; neither
τ_2 nor a_1 and a_2 are affected between pH 4 and 9. It seems likely
that the two components in the fluorescence decay of these peptides
can be ascribed to two different conformers providing different environments for the aromatic ring. Since both decays can be measured ,
the exchange between the two conformers must be slow with respect to
the fluorescence lifetimes. In the case of KGWK, the respective weights
of the two conformers are independent of temperature indicating that
their energies are quite similar. The long lifetime (τ_2) is not affected
by pH variation when the α-amino group is titrated indicating that in
the corresponding conformer the indole ring does not interact with the

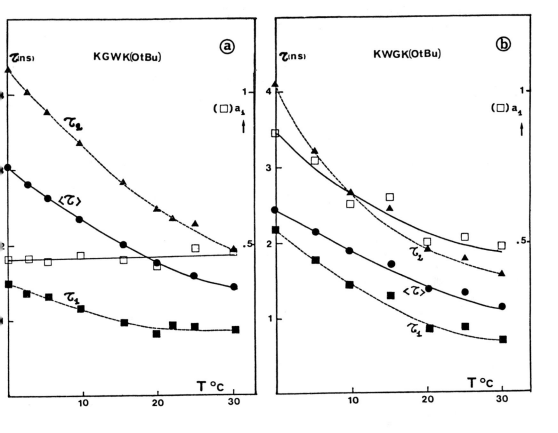

Figure 1 : *Dependence on temperature of the average fluores-*
cence lifetime <τ>, *and of the two component life-*
times τ₁ *and* τ₂ *and of the preexponential coefficient* a₁
(population of the short-lived species) of a) 10⁻⁵ M
KGWK(OtBu) and b) 10⁻⁵ M KWGK(OtBu) in a pH 6 buffer
(1 mM sodium cacodylate, 1 mM sodium chloride, 0.2 mM EDTA)
(λ_exc = 280 nm, λ_em = 350 nm).

α amino group. The slight variation of τ₁ suggests some kind of interaction for this conformer.

In the case of KWGK the respective contribution of the two
conformers changes with temperature suggesting a larger energy difference
than in the case of KGWK. Both components are affected by a change of
pH in the region 5-8 indicating that both conformers involve an
interaction between the α-amino group and the indole ring.

Even though no detailed investigation was carried out for
the tyrosine-containing peptide (KYK) the two components of the
fluorescence decay may also be ascribed to two conformers.

It should be noted that although fluorescence decay studies reveal only two components, this does not exclude more than two conformers. Some conformers might not be detected by this technique for several reasons : several conformers might have completely quenched fluorescence or the same fluorescence lifetime ; the interconversion between some conformers might be fast compared with the excited-state lifetime.

2. Oligopeptides bound to nucleic acids

When the tryptophan - and tyrosine - containing peptides bind to a nucleic acid, it is expected that each conformer will exhibit a different behavior. We have investigated the steady-state and time-resolved fluorescence parameters of the peptides bound to native or denatured DNA and to polyU in a low ionic strength pH 6 buffer under conditions such that complete binding occurs (7).

- Tryptophan-containing peptides

Table 1 summarizes the fluorescence decay results for KGWK and KWGK bound to polyU, denatured and native DNA at 3°C in a pH 6 buffer containing 1 mM sodium cacodylate, 1 mM sodium chloride and

	τ_1 (NS)	A_1	τ_2 (NS)	A_2	$<\tau>$	ϕ/ϕ_F
KGWK	1.37	0.42	3.92	0.58	2.85	1.00
+ POLY U	1.51	0.8	3.39	0.2	1.89	0.18
+ D.DNA	1.64	0.62	3.97	0.38	2.53	0.17
+ N.DNA	1.95	0.56	4.52	0.44	3.08	0.9
KWGK	1.93	0.72	3.47	0.28	2.36	1.00
+ POLY U	1.35	0.80	2.86	0.20	1.65	0.15
+ D.DNA	1.59	0.71	3.57	0.29	2.16	0.11
+ N.DNA	1.61	0.49	3.57	0.51	2.61	0.56

Table 1 : *Fluorescence decay analysis (average of two different experiments) for two tryptophan-containing peptides KGWK(OtBu) and KWGK(OtBu) at 2.5 x 10^{-5} M concentration in absence and in presence of 5 x 10^{-4} M polyU, denatured or native DNA. The measurements (λ_{exc} = 280 nm, λ_{em} = 350 nm) have been carried out in a pH 6 buffer at 3°C according to a two exponential decay function (see equation 1). The last column gives the relative fluorescence quantum yields.*

0.2 mM EDTA. With respect to the average lifetime $<\tau>$ both peptides behave quite similarly. An important decrease is observed with polyU (34 % and 30 % for KGWK and KWGK, respectively). However, this decrease is much smaller than that of the fluorescence quantum yield (82 % and 85 %, respectively). For native DNA a slight increase of the average, lifetime is observed (8 % and 10 % for KGWK and KWGK, respectively) whereas the respective fluorescence quantum yields are decreased by 10 and 44 % respectively. In the case of denatured DNA a slight decrease of the fluorescence lifetime is observed (11 % and 9 % for KGWK and KWGK, respectively) ; this decrease is very much smaller than that measured for the fluorescence quantum yield (83 % and 89 %, respectively). Since the fluorescence decay can be adequately represented in all cases by a superposition of two exponentials, the above results imply that part of the bound peptide is non fluorescent, especially in the case of denatured DNA and polyU. Previous studies (3) have already demonstrated that stacking of tryptophan with bases is responsible for this total quenching.

Experiments carried out with polyU and KGWK in the presence of 10 mM sodium chloride in the pH 6 buffer gave identical results to those observed in 1 mM sodium chloride (Figure 2).

The variations of the lifetimes and contributions of the two components are presented in Table 1. In most cases, changes are observed in the lifetimes but the most noticeable changes are in the respective weights of the two components when KGWK binds to polyU. If the two fluorescence decay components detected for the peptides bound to nucleic acids correspond to the two conformers described above for the free peptides, then the results presented in Table 1 show that the equilibrium between the two conformers can be shifted in the complexes. This is especially true in the case of KGWK bound to polyU. For KWGK the contributions of the two components are not markedly affected except in the case of native DNA.

The simplest scheme which can account for all the results so far obtained is the following :

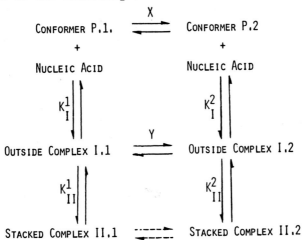

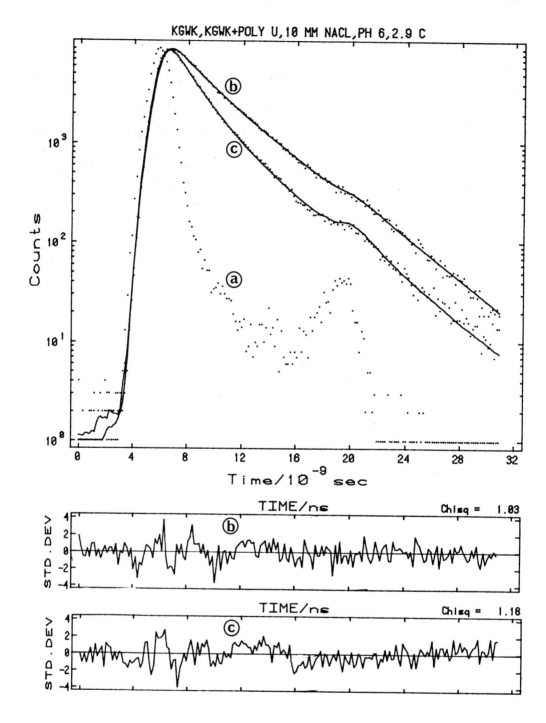

Figure 2 : *Fluorescence decays (λ_{exc} = 280 nm, λ_{em} = 350 nm) of 2.5 x 10^{-5} M KGWK(OtBu) in absence (b) and in presence (c) of 5 x 10^{-4} M polyU in a pH 6 buffer containing 10 mM NaCl at 2.9°C. The full curves are the fitted functions with two exponential components. Curve (a) is the flash profile. Standard deviations corresponding to curves (b) and (c) are given below (opposite page).*

In this scheme each conformer is assumed to bind in a two-step process as previously proposed (2). The outside complexes (I.1 and I.2) involve only electrostatic interactions whereas stacked complexes (II.1 and II.2) involve both electrostatic and stacking interactions. An interconversion between the two conformers can take place in the free peptide and in the outside complex whereas it is likely that the interconversion between the stacked complexes requires the intermediate involvement of the outside complexes (dotted arrows in the above scheme). The equilibrium constants X and Y are measured by the ratio of the respective populations of the two conformers determined from the fluorescence decay analysis (a_1 and a_2) in the free peptide and in the outside complex, respectively. The fluorescence quantum yield of the two conformers are proportional to the corresponding lifetimes as reported in Table 1. Steady-state fluorescence can then be analyzed according to the above scheme assuming that the two stacked complexes have their fluorescence completely quenched. A more complete treatment can be found in reference 7.

- Tyrosine-containing peptides

Three principal mechanisms may be responsible for the quenching of tyrosine fluorescence in peptide (protein) - nucleic acid complexes (8) : stacking with bases ; hydrogen bonding interactions with the hydroxyl group acting as hydrogen bond donor, and energy transfer to nucleic acid bases. The situation is therefore more complex than in the case of tryptophan where only one of these mechanisms (stacking) is efficient in quenching fluorescence (8). Previous studies (review in reference 2) have shown that binding of tyrosine-containing peptides to native DNA is accompanied by fluorescence quenching even though nuclear magnetic resonance data indicate that no stacking occurs (9). Therefore fluorescence quenching must arise from either one of the two other mechanisms, namely , hydrogen bonding or energy transfer. In the case of denatured DNA binding of tyrosine-containing peptides is accompanied by a stronger quenching of tyrosine fluorescence and NMR studies reveal that stacking is quite efficient in contrast to the observation with native DNA.

In order to provide some information on tyrosine interactions responsible for fluorescence quenching we have investigated the time-resolved fluorescence of KYK bound to native and denaturated DNA in a low ionic strength pH 6 buffer (1 mM Na cacodylate, 1 mM NaCl, 0.2 mM EDTA). The results are summarized in Table 2. A superposition of two exponentials satisfactorily accounts for the fluorescence decay in the

	τ_1 (NS)	A_1	τ_2 (NS)	A_2	$\langle \tau \rangle$	χ^2
KYK	1.14	0.87	2.66	0.13	1.34	1.04
KYK + D-DNA	0.55	0.92	2.40	0.08	0.70	1.03
KYK + N-DNA	0.62	0.94	2.68	0.06	0.74	1.08

Table 2 : *Fluorescence decay analysis of (2.5 x 10^{-5} M) Lysyl-Tyrosyl-Lysine in absence and in presence of 5 x 10^{-4} M denatured or native DNA. The measurements have been carried out at 11°C in a pH 6 buffer at a 270 nm excitation and a 315 nm emission wavelengths according to a two exponential decay function (see equation 1). The last column gives the χ^2 values (mean standard deviations).*

absence as well as in the presence of nucleic acids. The major change observed in the complexes concerns the short-lived component whose lifetime is reduced by about 50 % while its contribution is slightly increased (Figure 3). However the reduction of the average fluorescence lifetime $\langle \tau \rangle$ is smaller than that of the average fluorescence quantum yield (45 % compared with 66 % for native DNA, 48 % compared with 78 % for denatured DNA). Two conclusions can be drawn i) in both native and denatured DNA complexes part of the peptide has its fluorescence completely quenched. The relative contribution of the conformer with completely quenched fluorescence is more important with denatured than with native DNA ; ii) one of the two peptide conformers exhibits a strong perturbation of its fluorescent excited state (reduction of its fluorescence lifetime).

Either hydrogen bonding or total energy transfer could explain the larger reduction of the fluorescence quantum yield compared with that of the average lifetime when KYK is bound to native DNA. Stacking interactions contribute to a further reduction of the quantum yield in the case of denaturated DNA but are without any effect on the lifetime (since stacked complexes do not emit fluorescence at all). Experiments were previously carried out with the peptide KY(OMe)K where the hydroxyl group of tyrosine is substituted by a methoxy group therefore preventing any hydrogen bonding (9). The fluorescence of this substituted peptide is strongly quenched in the complex formed with native DNA even though NMR data reveal only limited stacking. This experiment demonstrates that hydrogen bonding is not required to explain

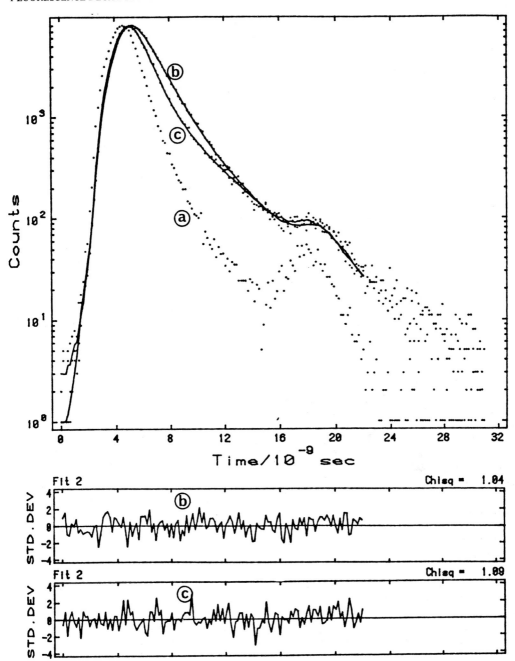

Figure 3 : *Fluorescence decay (λ_{exc} = 270 nm, λ_{em} = 315 nm) of 2.5 x 10^{-5} M KYK in absence (b) and in presence (c) of* 5 x 10^{-4} M native DNA in a pH 6 buffer (see legend of Figure 1) at 11°C. *The full curves are the fitted functions with two exponential components. Curve (a) is the flash profile. Standard deviations corresponding to curves (b) and (c) are given below.*

fluorescence quenching although this interaction cannot be excluded in the non-substituted tyrosine-containing peptide KYK.

The significant reduction of the shorter fluorescence lifetime could be accounted for by two mechanisms : i) the conformation of the peptide is changed upon binding and this brings a quenching group in the vicinity of the tyrosyl ring ; ii) energy transfer occurs from the tyrosyl ring to nucleic acid bases.

The excited-state parameters of tyrosine are much less sensitive to environmental perturbations than those of tryptophan. Moreover, phosphate groups in nucleic acid chains are not expected to affect the fluorescence of tyrosine since phosphodiesters have been shown to have no effect on tyrosine fluorescence in contrast to phosphate monoesters (10). Energy transfer from tyrosine to nucleic acid bases is a very efficient process. Förster critical distances ranging from 10.3 to 15.7 Å have been calculated depending on the nature of the nucleic acid base (4). It is therefore expected that some of the bound peptides will have their tyrosyl ring close enough to nucleic acid bases to allow energy transfer and therefore to reduce their fluorescence lifetime. However, energy transfer will reduce both quantum yield and lifetime and therefore cannot account for the difference in the decrease of these two parameters except if two kinds of complexes exist with one of them being completely quenched because tyrosine is very close to nucleic acid bases.

The long-lived fluorescence of KYK is not markedly affected by nucleic acid binding indicating that the corresponding conformer binds in such a way as to have either the distance or orientation of its tyrosyl ring such that energy transfer to the nucleic acid bases does not occur. The excited-state parameters of KYK are quite similar independently of whether it is bound to native or to denatured DNA. Therefore the structures of the fluorescent complexes should be quite similar in both cases. A much larger quenching of tyrosine fluorescence is observed with denatured DNA due to the existence of stacking interactions. In all cases we have checked that increasing the ionic strength to 0.2 M restores the fluorescence decay of the free peptide indicating that the complexes are dissociated under these high salt conditions.

CONCLUSION

Measurement of fluorescence decay curves of peptides bound to nucleic acids provides new information on the complexes that are formed. A general scheme involving - at least - two conformers has been proposed in the case of tryptophan-containing peptides. Each conformer binds in a two-step process to give two complexes which both involve electrostatic interactions. The second complex is charac-terized by additional stacking interactions. The detailed measurements of both fluorescence decays and overall quantum yields allow us to give a quite complete picture of the binding process (7).

In the case of tyrosine-containing peptides several types of complexes are detected from fluorescence decay analysis. Even though tyrosine does not stack with bases in duplex DNA, some of the complexes are not fluorescent. They might involve either hydrogen bonding of the tyrosyl ring or complete energy transfer to nucleic acid bases. This last phenomenon also seems to be responsible for the shortening of the fluorescence lifetime of part of the fluorescent complexes.

REFERENCES

1. Hélène, C. and Lancelot, G. : 1982, Prog. Biophys. Molec. Biol., 39, pp. 1-68.
2. Hélène, C. and Maurizot, J.C. : 1981, CRC Crit. Rev. Biochem., 10, pp. 213-258.
3. Montenay-Garestier, T. and Hélène, C. : 1971, Biochemistry, 10, pp. 300-306.
4. Montenay-Garestier, T. : 1975, Photochem. and Photobiol., 22, pp 3-6.
5. Szabo, A.G. and Rayner, D.M. : 1980, J. Am. Chem. Soc., 102, pp. 554-563.
6. Montenay-Garestier, T., Brochon, J.C. and Hélène, C. : 1981, Intern. J. Quant. Chem., 20, pp. 41-48.
7. Montenay-Garestier, T., Toulmé, F., Fidy, J., Toulmé, J.J., Le Doan, T. and Hélène, C. : 1983, in "Structure, Dynamics, Interactions and Evolution of Biological Macromolecules", C. Hélène (Ed), Reidel, pp. 113-128.
8. Hélène, C., Toulmé, J.J. and Montenay-Garestier, T. : 1982, in "Topics in Nucleic Acid Structures", S. Neidle (Ed.), pp. 229-258.
9. Mayer, R., Toulmé, F., Montenay-Garestier, T. and Hélène, C. : 1979, J. Biol. Chem., 254, pp. 75-82.
10. Alev-Behmoaras, T., Toulmé, J.J. and Hélène, C. : 1979, Biochimie 61, pp. 957-960.

ALTERATION OF THE B FORM OF DNA BY BASIC PEPTIDES

J.Portugal, J.Aymamí, M.Fornells & J.A.Subirana
Unidad de Química Macromolecular del C.S.I.C.
Escuela T.S.de Ingenieros Industriales
Diagonal, 647 - BARCELONA(28), Spain
M.Pons & E.Giralt
Department of Organic Chemistry, Faculty of
Chemistry, University of Barcelona, Spain

ABSTRACT

We show by fiber X-ray diffraction that the conformation of
DNA in some peptide-DNA complexes changes upon dehydration,
so that the number of base pairs per helical turn decreases
from about 10 at high humidity to a value between 8.4 and
9.6 upon complete dehydration, depending on the peptide
used. We have found this alteration of the B form of DNA
in complexes with nine different basic peptides, which do
not appear to have any obvious chemical relationship among
themselves. Two of the peptides used induce a smooth transition of DNA to the C form, we have not observed an intermediate A form. In the presence of the other seven peptides,
the DNA fiber pattern is of the B type and does not give
any clear evidence for the presence of the C form. The results obtained with all these peptides are in contrast
with other peptide-DNA complexes which stabilize the B
form of DNA with 10 base pairs per helical turn at all relative humidities (Fornells and Subirana, 1982; Fita et
al, 1983). In this context we also show in this paper that
Ala-Lys stabilizes the B form of DNA, whereas Lys-Ala and
acetyl-Ala-Lys-NHC$_2$H$_5$ alter it. A possible explanation for
this different behaviour is suggested by NMR studies of
the interaction, also reported in this paper. The stabilization of DNA with 10 base pairs per helical turn might
be due to a restriction of the conformational freedom of
DNA upon interaction.

INTRODUCTION

Considerable attention has been recently given to the
changes in conformation which may occur in DNA while preserving its B form. Keepers and collaborators (1982) have
reviewed the recent literature on conformational studies
and have also presented new calculations. Such investiga-

317

B. Pullman and J. Jortner (eds.), Nucleic Acids: The Vectors of Life, 317–330.
© *1983 by D. Reidel Publishing Company.*

tions clearly show that double-helical DNA in its B form
may retain a relatively rigid base stacking and hydrogen
bonding, while significant motions are occuring in the
conformation of the phosphodiester chain, including sugar
repuckering. In fact each of the five backbone torsion angles may be found in a conformational region (trans or
gauche) different from the standard situation without major
changes in the calculated energy.

From an experimental point of view, Fratini et al (1982)
have recently presented a detailed analysis of their work
with B-DNA dodecamers. They also find considerable changes
in the conformational angles for each nucleotide in the
structure. Most interestingly they find a correlation among
the variations of these angles, so that when one of them is
modified, the other vary in a predictable manner. In fact
they find the whole range of sugar conformations from
C2'-endo to C3'-endo in different nucleotides. Such changes
in sugar pucker are correlated with corresponding changes
in the phosphodiester backbone angles. However there are
a few bases which deviate from this behaviour. In such cases Fratini et al (1982) have observed that the main torsion
angles associated with the C(3')-O(3')-P elbow are gauche(-)
-trans instead of trans-gauche(-). The latter conformation
is found in most bases of the DNA fragments studied by them.
Fratini et al (1982) have called the standard conformation
B_I and the less common one B_{II}. They find that B_I has a considerable amount of conformational freedom, whereas B_{II} is
at one extreme of the conformational angles available for
the main chain and for sugar puckering, which is then restricted to the C-2'-endo, C-3'-exo region.

The experimental results that we have just reviewed should
be interpreted with caution, since they have been obtained
from crystals of a dodecamer with a single sequence, so
that packing forces in the crystal as well as sequence
and end effects may be important. On the other hand, the
recent studies of Arnott and collaborators (1983 and this
book) also show that different nucleotides may have different conformational angles and sugar puckering, thus confirming the inherent structural flexibility of the B form
of DNA. Most of these studies have been carried out under
partial dehydration, so that they may not apply to DNA in
solution.

In all the studies we have just discussed, little attention
has been paid to the influence of counterions on the conformation of DNA. In our laboratory we have used basic amino
acids and peptides as counterions for DNA, which may also
be considered as models for the study of protein-DNA interactions. Early studies showed that there are three main
types of DNA-peptide complexes (Campos et al, 1980):

- <u>stabilization</u> of the B form of DNA with 10 base pairs per helical turn under all humidity conditions, a behaviour which has been found in many cases, including all three basic amino acids (Fornells and Subirana, 1981) and many peptides and oligopeptides which contain arginine (Fita et al, 1983). Some peptides which contain lysine (Campos, 1983; Portugal, 1983) or histidine (Fornells, 1983) also show this behaviour.

- <u>alteration</u> of the B form of DNA, where the number of base pairs per helical turn diminishes as the sample is smoothly dehydrated.

- <u>destabilization</u> of the B form of DNA, where the structure of DNA becomes disordered upon a moderate dehydration of the sample. This behaviour is found in some basic peptides which have moderately hydrophobic groups attached.

In this paper we will describe in detail the behaviour of several peptide-DNA complexes in which the B form of DNA is altered, leading in some cases to the C form of DNA. We will compare this behaviour with that found in some peptides with a closely related structure in which the B form of DNA is stabilized. We will also discuss the eventual relationship which may exist between the stabilization-alteration effects of peptides and the inherent flexibility of the DNA backbone which we have reviewed in this introduction.

METHODS AND MATERIALS

The peptides used in this study were purchased from Bachem AG (Switzerland), Vega Biochemicals (USA) and Serva Feinbiochemica (Germany). Dr. R.Mayer (Orleans) synthesized acetyl-Ala-Lys-ethylamide and we are most thankful for a gift of this peptide. Calf thymus DNA was obtained by the detergent/chloroform method of Zamenhoff (1957). The peptide-DNA complexes were prepared by dialysis. The method used to prepare them and to obtain their X-ray diffraction patterns is described in detail elsewhere (Fita et al, 1983). The quantitative composition of the complexes was determined by standard biochemical methods (Portugal, 1983; Fornells, 1982).

For NMR we used sonified DNA at a concentration of 6.4 mg/ml in 1 mM phosphate buffer in D_2O. Peptide was added at a DNA/peptide ratio equal to 10. All reagents had been previously liofilized in D_2O in order to eliminate the exchangable hydrogens. The proton NMR spectra were obtained at 200 MHz in a Varian XL-200 spectrometer. Water resonance supresion was used in order to diminish the water signal. The assignment of the peaks was confirmed by spin decoupling.

RESULTS

Peptides which alter the B conformation of DNA

All the peptides described in this study, except Ala-Lys, alter the B-form of DNA, as we will show below. Most of the complexes were studied at a ratio of peptide net positive charges to DNA phosphates close to one, as shown on Table 1. When such a ratio has a value close to one or higher, we have never detected the A form of DNA. However, at lower ratios it is common to find transitions to the A form of DNA (Subirana et al, 1980; Fita et al, 1983). At low relative humidities (in general below 76 %) we found four different types of X-ray patterns which are shown on Fig. 1 and summarized on Table 1. In some cases mixtures of these four types appeared. The main features of each type can be summarized as follows:

a) It has strong reflections on the second layer line with streaks on the first layer line. The third layer line is very weak. A pattern related to this one has been atributed to the hexagonal C form of lithium DNA at 0 % relative humidity (Marvin et al, 1961).

b) Intensity distribution similar to that found in B form DNA (second layer line stronger than first and third; intensity maxima progressively displaced from the meridian as the layer line order increases).

c) Intensity distribution typical of the C form of DNA, with intensity maxima on the first layer line at about 1 nm-1 from the meridian and a broad meridional reflection.

d) Intensity maxima on the second and third layer lines with similar intensities and at the same distance from the meridian.

Inspection of table 1 shows that there is no apparent relationship between the chemical structure of the peptide used and the nature of the pattern obtained. All of them show a rather limited degree of order, only in some of the C form patterns some crystallinity can be detected, but only in the center of the pattern (Fig. 1c), which is an indication of screw disorder. The unit cell parameters of the latter pattern are given on table 2.

At high relative humidities all the complexes showed X-ray diffraction patterns of the B form with 10 ± 0.2 base pairs per helical turn, usually poorly ordered. Only in some cases

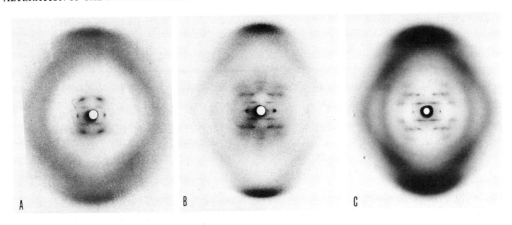

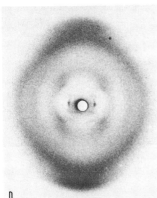

Fig. 1. Different types of X-ray diffraction patterns observed for the following peptide-DNA complexes at low relative humidities: a) Leu-amide at 0% r.h. (base pairs/repeat=9.3); b) Acetyl-Ala-Lys-ethylamide at 54% r.h. (base pairs/repeat=9.36); c) His-amide (pH=7.5) at 0% r.h. C form pattern (base pairs/repeat=8.42); d) $K_6A_3YA_3K_6$-ethylamide at 0% r.h. (base pairs/repeat=9.19).

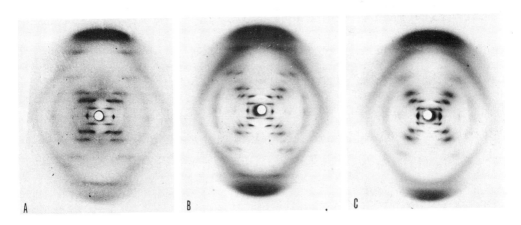

Fig. 2. X-ray diffraction patterns of the B type with about 10 base pairs/repeat: a) His-amide (pH=7.5) at 92% r.h.; b) Ala-Lys at 33% r.h.; c) Ala-Lys at 98 % r.h. The number of base pairs per repeat respectively are 9.80, 10.00 and 9.91.

Table 1. Characteristic parameters of the peptide-DNA
          complexes used in this study

Peptide	+/-	e_O (nm)	Pattern	Curve
Ala-Lys	0.96	1.79-1.87	B	5
Lys-Ala	1.03	1.75-1.78	b,d	4'
Acetyl-Ala-Lys-NHC$_2$H$_5$	0.93	1.76	b	4
Lys$_6$Ala$_3$TyrAla$_3$Lys$_6$NHC$_2$H$_5$*	ca1.2	1.87	d	3
Tyr-amide	0.85	1.84	b	4
Tyr-Tyr-amide	0.51	1.85	b	4
Leu-amide	1.35	1.78	a	3
Ala-amide	1.16	1.70	b,c	1
His-amide (pH=5.5)	0.99	1.73	c	2
His-amide (pH=7.5)	0.87	1.65	c	2
Acetyl-His-COOCH$_3$	0.95	1.74	a	4

The +/- column gives the ratio of peptide positive charges
to DNA phosphates in each complex. In the case of His-amide,
the peptide was assumed to contain two charges at pH=5.5
and one charge at pH=7.5. In the case of Lys-Ala, three complexes were studied, with +/- ratios in the range 0.97-1.10

The e_O column gives the equatorial spacing in the fiber
diffraction patterns of each complex at 0% relative humidity.

The "pattern" column gives the type of X-ray pattern observed at low relative humidities among those shown in Fig. 1.
In the case of the Ala-Lys complex, the pattern always
shows the B form of DNA with close to 10 base pairs/helical
repeat, as shown in Figs. 2 and 3.

The "curve" column indicates what type of curve among those
shown in Fig. 3 is found for each complex. This question
is further discussed in the text.

* The results obtained with this peptide had already been
  published elsewhere (Subirana et al, 1980)

Table 2. Unit cell parameters of some fiber patterns

Peptide	Relative humidity (%)	Form of DNA	Lattice	Unit cell dimensions (nm)		
				a	b	c
Ala-Lys	33	B	Orthorhombic	3.30	2.40	3.4
Ala-Lys	92	B	Orthorhombic	3.56	2.60	3.4
His-amide(pH 7.5)	0	C	Hexagonal	3.30	3.30	2.83

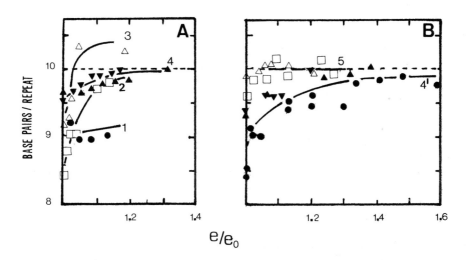

Fig. 3. Influence of the intermolecular distance on the pitch of the DNA helix. The data are given as a function of the equatorial spacing e divided by its value e_o at 0% relative humidity. The following complexes are represented on the graph: Curve 1, Ala-amide (●). For this complex at 98% relative humidity ($e/e_o=1.68$) a single point with a value of 10.0 was also found; Curve 2, His-amide at pH=7.5 (□); Curve 3, Leu-amide (△); Curve 4, Tyr-amide (▼) and Tyr-Tyr-amide (▲); Curve 4', Lys-Ala, full symbols; Curve 5, Ala-Lys, empty symbols. In the latter two cases, each type of symbol corresponds to a different complex.

well oriented patterns were observed. An example is shown on Fig. 2a.

In order to give a quantitative description of the alteration of the B form of DNA in these complexes, we have re-

presented in Fig. 3 the changes in the number of base pairs
per repeat unit as a function of the relative equatorial
spacing (e/e_o). The latter ratio gives a measure of the separation between neighbouring molecules and how this distance changes in different experiments. The influence of relative humidity on this ratio (not shown) is different for
every complex, depending on its hygroscopic characteristics.

All the complexes studied show a similar behaviour (curves
1 to 4 in Fig. 3), namely, the number of base pairs per
repeat unit increases as a function of relative humidity,
with a limiting value close to 10 base pairs per repeat unit
at the higher humidities studied. On the other hand, depending on the peptide, a value in the range 8.4 to 9.6 is
found at 0 % relative humidity. The changes in this ratio
are mainly due to differences in the value of the pitch of
the helix. The meridional spacing usually remains constant.
The largest variations in the latter spacing were found in
the peptides Leu-amide, Lys-Ala and Tyr-Tyr-amide. However
even in all these three cases, the increase in this spacing
from 0% to 98% relative humidity was smaller than 3%.

The increase in the number of base pairs per repeat unit as
the DNA molecules are moved further apart (e/e_o increases)
is usually gradual, but differences are found in every case,
as shown by the various curves represented on Fig. 3. Four
different types of behaviour are found (curves 1 to 4). Curve 4' is considered to belong to the same family as curve 4.
Again, there is no clear relationship between either the
chemical structure of the peptide or its diffraction pattern
(types a-d), and the shape of the curves obtained, as shown
on table 1. The alteration effect is small in some peptides
and very strong in others. Furthermore it appears that the
exact peptide to DNA ratio may have some influence on the
shape of the curve obtained. For example, in the case of
the Lys-Ala complex (curve 4' in Fig. 3), three independently prepared complexes were studied. One of them, represented
by circles in the figure, had a slightly higher peptide to
DNA ratio and gave consistently lower values.

In summary, all the peptides described in this section
alter the B form of DNA, in the sense that the number of
base pairs per helical turn gradually decreases upon a moderate dehydration of the sample.

The stabilization of the B form of DNA by Ala-Lys

In the course of our studies we have found many peptides
which maintain a stable B conformation of DNA with exactly
10 base pairs per helical turn under all humidity conditions.
This behaviour is found for example in the free amino acids

(Subirana et al, 1980; Fornells and Subirana, 1981), in many arginine peptides (Fita et al, 1983) and in quite a few different peptides (Fornells, 1982; Portugal, 1983; Campos, 1983). In this section we want to present the results obtained with the complex of DNA with Ala-Lys. We consider of interest to report these results here, since in the previous section we have shown that the closely related peptides acetyl-Ala-Lys-ethylamide and Lys-Ala induce an alteration of the B-form of DNA. On the other hand the Ala-Lys peptide stabilizes the B form of DNA, as shown on Figs. 2 and 3. Under all humidity conditions, the number of base pairs per helical turn remains in the range 10 ± 0.15, except for one point in one of the two complexes studied at 0% relative humidity.

NMR studies on the conformation of the peptides Ala-Lys and Lys-Ala in the presence of DNA

As shown above, these two peptides, in spite of their chemical simmilarity, interact in a different way with DNA. In order to understand the origin of this difference in behaviour we have studied the interaction by proton NMR. In both cases it is found that the bands are significantly broadened upon interaction with DNA as expected, but no precise measurements can be made of this effect due to the overlap of neighbouring bands. On the other hand the chemical shifts can be precisely measured. The results obtained

Table 3. NMR chemical shifts

Group	Ala-Lys		Lys-Ala	
	δ	Δ	δ	Δ
CH_3 (Ala)	1.52	.05	1.34	.01
$\beta,\gamma,\delta-CH_2$ (Lys)	1.73-1.78	.01	1.71-1.90	.00
$\varepsilon-CH_2$ (Lys)	3.03	.03	3.01	.01
CH (Ala)	4.06	.07	4.19	.01
CH (Lys)	4.15	.00	3.89	.12

δ gives the observed chemical shifts measured in ppm in a peptide solution and Δ the difference between these values and those observed in the presence of DNA. Dimethyl silapentane sulfonic acid was used as an internal reference in order to compute δ.

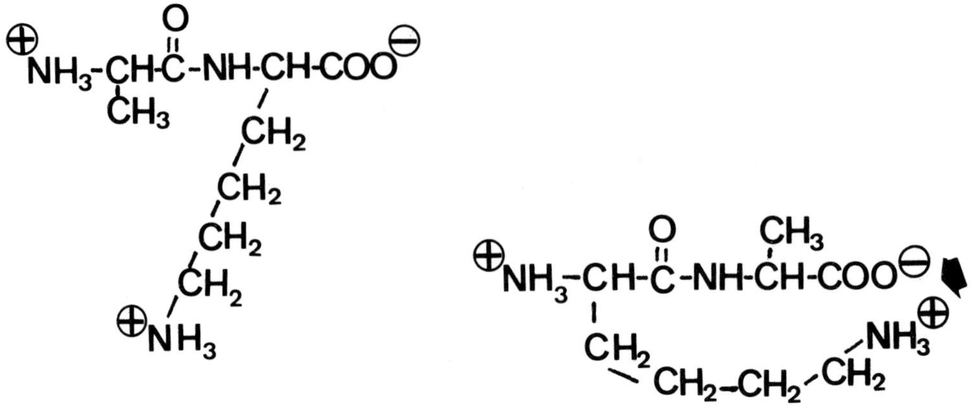

Fig. 4. Chemical formula of Ala-Lys and Lys-Ala, showing the internal salt bridge (arrow) suggested by the NMR experiments for Lys-Ala. According to this model, Ala-Lys has two amino groups available for interaction with DNA, whereas Lys-Ala has only one.

are summarized in table 3. It can be seen that upon interaction with DNA the Lys-Ala peptide shows a significant change in the CH residue of lysine, which is attached to the charged α-amino terminal group. In fact this is the only change detected upon interaction with DNA, the chemical shifts of the other hydrogen atoms in the molecule remain constant. It appears therefore that this peptide interacts with DNA only through its amino terminal charge. On the other hand the Ala-Lys peptide shows changes in the chemical shifts of the hydrogens attached to the two carbons of the molecule which support the two charged amino groups and also in the neighbouring methyl group of alanine. Therefore this peptide apparently interacts with DNA through both of its positively charged amino groups. In conclusion, the NMR experiments show that there is a clear difference in the way both peptides interact with DNA. In Fig. 4 we indicate a possible interpretation of such a difference, which might be due to the presence of an internal salt bridge in Lys-Ala. As a result this peptide would only have its amino terminal group available for interaction with DNA. Model building shows that such a bridge can not be easily formed in Ala-Lys, so that this peptide has both of its charged amino groups available for interaction with DNA, as found by NMR. This interpretation leads to an intriguing conclusion, namely that Ala-Lys, which has several posibilities of interaction with DNA, stabilizes the B form of DNA, whereas

Lys-Ala, with a single mode of interaction alters this form of DNA.

DISCUSSION

All the peptides studied in this paper, except Ala-Lys, produce an alteration of the B form of DNA. But the degree of change achieved varies significantly in each case, as shown in Fig. 3. In Lys-Ala and in His-amide we have found values as low as 8.4 for the number of base pairs per repeat unit at 0% relative humidity. At the other end, for the tyrosine peptides and for acetyl-histidine methyl esther, this value only decreases moderately to 9.6 at 0 % relative humidity. Intermediate values are found for the other peptides studied. Additional differences are found in the distribution of diffracted intensity as shown in Fig. 1 and Table 1. These differences may be due either to the contribution of the peptide to the diffracted intensity or to changes in the conformation of DNA. The limited amount of information available in the X-ray diffraction patterns prevents us to determine the relative contribution of both effects to the intensity distribution observed. In fact the alteration phenomenon may include quite different changes in the detailed conformation of DNA, given the versatility of its backbone (Keepers et al, 1982). The various peptides studied might give rise to different conformational changes in each case. In particular, the higher degree of order detected in the C type patterns (Fig. 1c) may indicate a more precisely ordered conformation in this case than in the complexes with the peptides which do not give this form of DNA. It is interesting to note that the two peptides which show the C form of DNA at low relative humidities have the general formula $NH_3^+CHRCONH_2$, a structure which might be instrumental in stabilizing this form of DNA.

We may ask if there is any general structural feature common to all the peptides studied which might be related to the alteration of DNA. No peptide containing arginine has been found to belong to this group. One feature that is shared by all of them is that they only have one positive charge, with the exceptions of Lys-Ala and the high molecular weight peptide $K_6A_3YA_3K_6NHEt$. In the case of Lys-Ala, which has two positive charges, our NMR studies show that one of them appears to be internally blocked, so that this peptide will also behave as having a single positive charge. In fact in many of the cases studied the single charge available corresponds to the amino terminal group.

The high molecular weight peptides may have a different type of interaction with DNA and we will not discuss them here.

In fact we have found that protamines also alter the B form
of DNA (Fita et al, 1983), as well as all the lysine-rich
proteins, including histone H1, that we are presently studying. In fact, the complexes formed by lysine-rich proteins
and DNA give a fiber diffraction pattern very similar to
that found in the complex with $K_6A_3YA_3K_6NHEt$, shown in Fig.
1d.

In order to gain some insight on the general features of
peptide-DNA interactions, it is useful to compare at this
point the results we have just discussed with those observed with amino acids and peptides which stabilize the B
form of DNA with 10 base pairs per helical turn, as described here for Ala-Lys. Many peptides which contain arginine
show this behaviour (Fita et al, 1983), as well as the three
standard basic amino acids (Fornells and Subirana, 1981).
All these peptides have in common that they either contain
more than one positive charge or an arginine residue. In
either case they may interact simultaneously with two
neighbouring phosphate groups, as the NMR data suggest for
example in the case of Ala-Lys. The guanidinium group of arginine may also interact with two neighbouring phosphate
groups (Fita et al, 1983). These peptides could therefore
act as a clamp binding together neighbouring phosphate groups
thus preventing changes of conformation in the DNA backbone.

Further support for the hypothesis we have just presented
can be obtained by comparing the behaviour of some related
peptides we have studied. The Lys-Ala peptide appears to interact with DNA only through its amino terminal group, whereas acetyl-Ala-Lys-ethylamide can only interact through the
amino group in the side chain of lysine. In both cases the
interaction with DNA occurs through a single amino group
and the conformation of DNA is altered. On the other hand
the Ala-Lys peptide, which has both groups available for interaction stabilizes the B form of DNA, thus reinforcing our
hypothesis that two charges may be required for such stabilization.

In the introduction we discussed the different kinds of
evidence which demonstrate the conformational versatility of
DNA in the B form. With this background in mind it is clear
how the conformation of DNA can be easily altered. On the
other hand it is not obvious why there are so many peptides
which stabilize the B form of DNA with precisely 10 base
pairs per turn of the helix. Furthermore, at high relative
humidities, DNA also has this pitch in the presence of any
of the peptides which alter the conformation of DNA as described in this paper. It appears therefore that the pitch
of 10 base pairs for the B form of DNA represents an intrinsic characteristic of this molecule. A possible explanation

for this fact would be that this pitch corresponded to one extreme conformation of those available for DNA in the B form. Tentatively this extreme conformation might be identified with the B_{II} conformation discussed by Fratini et al (1982). DNA in the relaxed state could be in the versatile B_I backbone conformation, but in the presence of certain peptides it could be stabilized in the more extreme B_{II} conformation, with 10 base pairs per turn.

A special case to consider are those samples in which the C form of DNA is clearly detected. In the cases we have studied, this form of DNA appears gradually upon dehydration, we have never observed an intermediate A form as found by Rhodes et al (1982). Our experiments rather support the view (Arnott and Selsing, 1975) that this form derives from the B form of DNA by a smooth deformation. On the other hand the much better definition of its diffraction pattern when compared with the other types of relaxed patterns shown in Fig. 1, suggests that the conformation of DNA is quite well determined under these conditions. In fact it may represent another limiting conformation among those available for B form DNA.

Another feature of interest of our results is that they show that the response of DNA towards dehydration is quite different when it is neutralized with peptides rather than with Na^+ or Li^+. When DNA is neutralized with basic peptides the B form is usually maintained upon dehydration. In some cases the pitch is stabilized at 10 base pairs per helical turn, whereas in the cases described in this paper this value decreases. On the other hand, in the alkaline salts of DNA dehydration usually promotes the formation of either the A or C forms. In the presence of peptides the C form is seldom found, whereas the A form appears only when DNA is neutralized with a mixture of peptides and alkaline ions (Subirana et al, 1980; Fita et al, 1983).

In summary, our results confirm the variability in conformations available for DNA in the B form, as shown from different points of view by Keepers et al (1982), Fratini et al (1982) and Arnott et al (1983). They also show that the conformation of DNA is not only susceptible to base sequence (Arnott et al, Dickerson, this book), but may be locally altered under the influence of its counterions. The stabilized B form with 10 base pairs per turn and the C form with 8.3 base pairs per turn may represent two extreme states within the conformational space available for DNA in the B form.

REFERENCES

Arnott, S. and Selsing, E.: 1975, J.Mol.Biol. 98, pp 265-269

Arnott, S., Chandrasekaran, R., Puigjaner, L.C., Walker, J.K., Hall, I.H. and Birdsall, D.L.: 1983, Nucl.Ac.Res. 11 pp 1457-1474

Campos, J.L.: 1983, Ph.D.Thesis. Faculty of Pharmacy, University of Barcelona

Campos, J.L., Subirana, J.A., Aymamí, J., Mayer, R., Giralt E. and Pedroso, E.: 1980, Studia Biophys. 81, pp 3-14

Fita, I., Campos, J.L., Puigjaner, L.C. and Subirana, J.A: 1983, J.Mol.Biol., in press

Fornells, M.: 1982, Ph.D.Thesis, Faculty of Biology, University of Barcelona

Fornells, M. and Subirana, J.A.: 1981, Studia Biophys. 84, pp 13-14

Fornells, M., Campos, J.L. and Subirana, J.A.: 1983, J.Mol. Biol, in press

Fratini, A.V., Kopka, M.L., Drew, H.R. and Dickerson, R.E.: 1982, J.Biol.Chem. 257, pp 14686-14707

Keepers, J.W., Kollman, P.A., Weiner, P.K. and James, T.L.: 1982, Proc.Natl.Acad.Sci.USA 79, pp 5537-5541

Marvin, D.A., Spencer, M., Wilkins, M.H.F. and Hamilton, L.D.: 1961, J.Mol.Biol. 3, pp 547-565

Portugal, J.: 1983, Ph.D.Thesis, Faculty of Biology, University of Barcelona

Rhodes, N.J., Mahendrasingam, A., Pigram, W.J., Fuller, W., Brahms, J., Vergne, J. and Warren, R.A.J.: 1982, Nature 296 pp 267-269

Subirana, J.A., Chiva, M. and Mayer, R.: 1980, in Biomolecular Structure, Conformation, Function and Evolution, V.1 (R.Srinivasan ed.), Pergamon Press, Oxford & N.York, pp 431-440

Zamenhof, S.: 1957, in Methods in Enzymology V.3 (S.P.Colowick and N.O.Kaplan), Academic Press, N.Y. pp 696-704

THE INTERACTION OF CERTAIN NUCLEIC ACIDS WITH PROTEINS AND METAL IONS IN SOLUTION

Ragnar Österberg
Department of Chemistry and Molecular Biology, Swedish University of Agricultural Sciences, S-750 07 Uppsala, Sweden
Dan Persson and Per Elias
Department of Medical Biochemistry, University of Göteborg, Box 33031, S-400 33 Göteborg, Sweden

ABSTRACT

When the ternary complex of the *E. coli* elongation factor Tu (EF-Tu), GTP, and aminoacyl-tRNA is formed, the conformation of the free molecules seems to be retained. This is indicated from neutron scattering data recorded from solutions of the complex, EF-Tu·GTP - valyl-tRNA$_{1A}^{Val}$, where the protein and tRNA are successively matched by the scattering density of the solvents, 40 % D_2O and 70 % D_2O, respectively; the gyration radii were found to be 2.5 and 2.2 nm, which are in agreement with those obtained for the corresponding free molecules. As further indicated by small-angle X-ray scattering, the gyration radius (R) of the EF-Tu·GTP - valyl-tRNA$_{1A}^{Val}$ complex is much larger than that of EF-Tu·GTP, ($\Delta R = 1.1$ nm). These data as well as the $p(r)$-curve are consistent with a multiellipsoid model for the complex, which indicates that the acceptor stem of tRNA is attached to EF-Tu·GTP and that the anticodon and loop protrude into the solution.

Chromium(III), one of the most potent inorganic carcinogens, induces condensation of DNA into a toroidal product at 37° and 30 mM ionic strength. This condensation occurs at a considerably lower molar ratio of Cr(III)/DNA than that of other trivalent ions, such as spermidine for instance. The reason for this may be the formation of stable complexes between Cr(III) and DNA as indicated from spectrophotometric data using ethidium as a competing ligand.

1. INTRODUCTION

This communication describes some recent results obtained from studies on the structure of nucleic acids in solution and their interactions with proteins and metal ions. One aspect involves the ternary complex formed by the *E. coli* elongation factor Tu (EF-Tu), GTP, and aminoacyl-tRNA, which is an intermediate in the binding of aminoacyl-tRNA to the ribosome (1). This step in the elongation cycle is functionally understood but there is only limited information regarding the structure of

331

B. Pullman and J. Jortner (eds.), Nucleic Acids: The Vectors of Life, 331–341.
© *1983 by D. Reidel Publishing Company.*

the ternary complex and how it correlates with function. It is of great advantage for the present X-ray and neutron scattering study that high-resolution structures for the components, EF-Tu and tRNA, are essentially known from X-ray crystallography (2-7). The present data could therefore be interpreted using the correct shapes of the molecules. In this investigation we present information on the orientation and conformation of EF-Tu·GTP and valyl-tRNA within the ternary complex (8). On the basis of the results a multiellipsoid model (9) for the complex is discussed.

Another aspect of our work involves the structure of DNA in solution; here, we have explored the capacity of combining flow orientation with small-angle X-ray scattering as a measuring method (10). These studies are now being extended to the analysis of the structure of DNA under various conditions; and, this study deals with the Cr(III) interactions of DNA. This particular Cr(III) study is also supposed to provide the necessary background information for cell biological studies on the molecular mechanisms of Cr(III)-induced cancer. As shown by electron microscopy (11) and absorption spectra, using ethidium bromide as a competing ligand, Cr(III) first forms a specific polynuclear complex and then induces condensation of DNA into a compact molecule.

2. EXPERIMENTAL

Materials

EF-Tu·GDP from *E. coli* was purified according to Lebermann *et al.* (12) and it was converted to EF-Tu·GTP as previously described (9). Valyl-tRNA$_{1A}^{Val}$ was prepared as reported in ref. (9). The formation of a ternary complex was ascertained in separate experiments using chromatography on AcA44 as described by Kruse *et al.* (13). The samples used for the small-angle scattering studies were prepared as described previously (9).

In the X-ray scattering experiments the complex formation was measured for a series of solutions having a constant concentration of valyl-tRNA$_{1A}^{Val}$ (39 and 81 µM) and varying concentration of EF-Tu·GTP (14 - 262 µM). In the neutron scattering experiments, the concentration of valyl-tRNA$_{1A}^{Val}$ was 5 - 10 % in excess of EF-Tu·GTP (80 and 620 µM) at 70 % D_2O, *cf.* (14), and 10 % less than EF-Tu·GTP (470 µM) at 40 % D_2O. We could therefore assume that the free concentrations of EF-Tu·GTP and valyl-tRNA$_{1A}^{Val}$ in the pertinent experiments were negligible.

Calf thymus DNA (Type I) was obtained from Sigma Chemicals Company and was purified from RNA and protein first by digestion with RNAse and proteinase then by phenol extraction.

Plasmid pBR 322 was isolated from an ampicillin resistant strain of *E. coli* HB 101 as described elsewhere (15). The supercoiled plasmid DNA was converted to linear molecules by the restriction enzyme Eco RI, which cuts the plasmid once.

Methods

Small-angle X-ray scattering. The data were recorded with the camera developed by Kratky and Skala (16). Monochromatization of the copper radiation was achieved with a nickel β-filter and pulse-height discriminator in conjunction with a proportional counter. The absolute scattered intensities were obtained using the Lupolen method (17).

Neutron scattering. The data were recorded with the camera recently built at the Risø Research Station, Denmark (18). Neutron scattering was recorded on the EF-Tu·GTP - valyl-RNA system using the reaction mixture mentioned above at a contrast of 70 % D_2O and 40 % D_2O. The wavelengths were 0.45 and 0.50 nm. That the contrast was close to the matching point of the protein was checked by measuring solutions of protein (0, 21, 64, 78, and 90 % D_2O) and then plotting $\sqrt{I(0)}$ against the D_2O concentration.

Spectrophotometric measurements. The readings were done with a Zeiss PMQII spectrophotometer. The Cr(III)-DNA-ethidium bromide samples were measured at 30 mM ionic strength ($NaNO_3$) in a 5 mM NaAc-HAc buffer of pH 4.8. The solutions were incubated for two days at $37^{\circ}C$. The spectra were measured with 1 cm or 5 cm quartz cuvettes. The data involved three sets of $E(B)_{C,A}$ data for A = 25, 50 and 100 µM and for four different values of C for each A-value. Here, E is the measured extinction, and A, B and C the total concentrations of nucleotide pairs, Cr(III), and ethidium. The molar extinction coefficient for ethidium bromide at 480 nm was found to be $\epsilon_{1\ cm}$ = 5,600. The measured sets of data $((E,B)_C)_A$ were used to calculate the data sets $((c,B)_C)_A$ using a procedure similar to Waring (19). (c is the free ethidium bromide concentration). These latter data are then being used for further analysis in the form of predominating species and their stability constants.

Electron microscopy. A JEOL JEM-100 CX microscope was used at 60 kV accelerating voltage for all studies. Clean copper grids (300 mesh) were layered with parlodian film. Plasmid DNA was mounted to the grid by a modified form of the aqueous technique by Davis $et\ al.$ (20) using cytochrome C in order to make the DNA visible (11).

3. RESULTS

The Interaction of Valyl-tRNA$^{Val}_{1A}$ with Elongation Factor Tu·GTP

Neutron scattering data. Fig. 1 shows the Guinier plots obtained for EF-Tu·GTP and its valyl-tRNA$^{Val}_1$ complex in 70 % D_2O. These data yield the gyration radii (R) of 2.2 and 2.1 nm for EF-Tu·GTP and its complex, respectively. Since the estimated experimental error is of the order ±0.1 nm, these values do not differ significantly. Thus, it is indicated that the gyration radius for EF-Tu·GTP is essentially the same

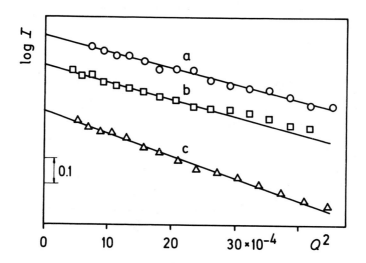

Figure 1. Guinier plots, log I *versus* Q^2, of neutron scattering data
for the complex EF-Tu·GTP – valyl-tRNA$_{1A}^{Val}$ in 70 % D_2O (a) and 40 % D_2O
(c) and for EF-Tu·GTP in 70 % D_2O (b); $Q = (2\pi \sin 2\theta)/\lambda$, where 2θ is the
scattering angle and λ the wavelength; from (8).

whether or not tRNA is bound. The R-value for EF-Tu·GTP in 70 % D_2O is
0.3 nm lower than that obtained from X-ray scattering in water (9).
Although this difference is close to the limits of experimental errors,
it might be due to scattering density fluctuations within EF-Tu·GTP (21).

Fig. 1 also shows the Guinier plot for the EF-Tu·GTP – valyl·tRNA$_{1A}^{Val}$
complex in 40 % D_2O yielding an R-value of 2.5 ± 0.1 nm; this value is
in excellent agreement with data reported for tRNA and aminoacyl-tRNA in
aqueous solution; Luzzatti *et al.* (22) reported 2.49 nm and Pilz *et
al.* (23) reported 2.44 nm.

X-Ray data. The X-ray scattering data were analysed by first cal-
culating the difference intensity, $\widetilde{\Delta I}(0)$ at zero angle after extrapolating the primary data *via* Guinier plots (24). The normalized form of
$\widetilde{\Delta I}(0)$, *i.e.* $\widetilde{\Delta I}(0)' = (\widetilde{\Delta I}(0)/[A]_{tot})/(\widetilde{\Delta I}(0)/[A]_{tot})_{max}$, was then plotted
against the EF-Tu·GTP concentration. Here, $[A]_{tot}$ is the total concen-
tration of valyl-tRNA. The data could be described by curves generated
for a 1:1 complex with log $K = 6.5$. A similar analysis of the data recorded for unesterified tRNA$_{1A}^{Val}$ and EF-Tu·GTP indicated much less stability for the complex, log $K \sim 4$, (9).

The X-ray scattering data for the 1:1 complex of valyl-tRNA$_{1A}^{Val}$ and
EF-Tu·GTP were further analysed from the weighted average of three scattering curves from which the contribution of free molecules had been
eliminated. These reduced experimental data were used to calculate the
distance distribution function $p(r)$ *via* a computer program (25); the

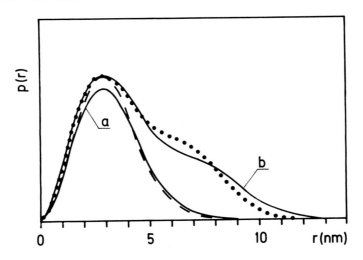

Figure 2. Distance distribution, $p(r)$ of EF-Tu·GTP (a) and of the EF-Tu· GTP - valyl-tRNA complex (b). (———) Experimental data; (- - - -) and (• • • •) calculated from the ellipsoid models described in the text.

result is shown in Fig. 2. In agreement with the increase of the gy-ration radius from 2.5 to 3.6 nm, when the tRNA is bound, there is an increase in D_{max} from 8.5 to 12.5 nm (9).

The $p(r)$-curve of the 1:1 complex shows one major peak with a pro-nounced shoulder indicating at least two major domains (Fig. 2). Further information on the shape of this complex was obtained when $p(r)$-curves were systematically calculated via the calculation of theoretical inten-sities (26) for various model complexes of the 1:1 type and then compared with the experimental data. The models used in the comparison were ellipsoids derived from the crystallographic analysis of EF-Tu·GTP (2-4) and tRNA^{Phe} (5,6). An ellipsoid of the dimensions $8.5 \times 5.0 \times 4.5$ nm rep-resented EF-Tu·GTP. The model of valyl-tRNA$^{Val}_{1A}$ was composed of two ellipsoids with their main axes at right angles to each other with the dimensions $7.7 \times 2.6 \times 2.6$ nm and $4.0 \times 2.6 \times 2.6$ nm. The best agreement with the experimental data was obtained for a curve generated for a model where the acceptor stem is attached to EF-Tu·GTP and the anticodon stem and loop protrude into the solution forming an angle of about 45° with the main axis of the EF-Tu·GTP model (Fig. 2), cf. (9).

The Cr(III) Interaction of DNA

Analysis of spectrophotometric data. As a first approximation we assume that the binding sites on the double helical DNA are equal for the Cr(III) binding and that they consist of a pair of nucleotides (= A). Then we assume that Cr(III) (= B) reacts with the binding sites, A,

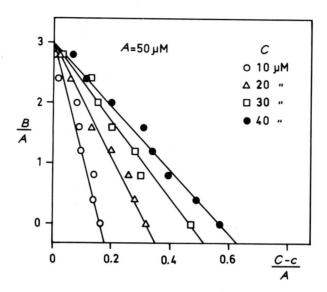

Figure 3.   Chromium(III) complexes of DNA.   Plot B/A *versus* $(C-c)/A$ of equation 5.

according to the general reaction:

$$pB \ + \ qA \ \rightleftharpoons \ B_pA_q \tag{1}$$

with the equilibrium constant, β_{pq}. Likewise, ethidium ($=C$) may react and form a series of complexes of the type A_qC_r with the stability constant, β^*_{qr}. If B, A, and C are the total concentrations of Cr(III), nucleotide pair, and ethidium and b, a, and c the corresponding free concentrations; we obtain:

$$B - b \ = \ \Sigma \ \Sigma \ p[B_pA_q] \tag{2}$$

$$A - a \ = \ \Sigma \ \Sigma \ q[A_qC_r] + \Sigma \ \Sigma \ q[B_pA_q] \tag{3}$$

$$C - c \ = \ \Sigma \ \Sigma \ r[A_qC_r] \tag{4}$$

Equations 2-4 can be considerably reduced if the measurements are done in the range where associations and aggregation of DNA do not have to be considered, *i.e.* for $A \leqslant 200$ µM. Then, if only one kind of each species B_pA_q and A_qC_r predominate in solution, assumed to be B_PA and A_QC species, we obtain from eqns. 2-4 after rearranging:

$$(B - b)/A \ = \ -PQ(C - c)/A + P(A - a)/A \tag{5}$$

For increasing values of B, $(A - a) \to A$ and $c \to C$. Thus, $\lim_{c \to C}(B-b)/A = P$, and the plot $(B-b)/A$ *versus* $(C-c)/A$ of eqn. 5 yields P as an intercept. In our study an estimate of the maximum possible value of P was obtained by plotting B/A against $(C-c)/A$; the result is shown in Fig. 3. As shown by Fig. 3 the intercept on the vertical axis is three and the main species in the solution may then have the composition BA, B_2A, or B_3A. So far, only a few Cr(III) – DNA solutions have been analysed for Cr(III) after ultracentrifugation. These data indicate that we cannot exclude the possibility that trinuclear complexes of the type B_3A are formed in solution.

At the present relatively low ionic strength, the ethidium binding of DNA is supposed to involve both strong and weak binding sites (19,28). At saturation the ratios of bound ethidium molecules per DNA phosphate are supposed to be 0.2 and 1.0, respectively (19,28). However, as long as the values of A and C are relatively far from the saturation value, 1.0, the binding may be considered to be of the 1:1 type in regard to a pair of nucleotides as binding sites, that is equal to AC species in our notations. Therefore, as a first approach AC species were assumed and it was found that a conditional constant K^* for the formation of the "AC" species fairly described the data within the present concentration ranges. A graphical analysis similar to the one described in ref. (27) gave log $K^* = 4.9$.

The constants K^* and β_3 of the possible existing complexes AC and B_3A are now being further analysed both graphically and by using the general least squares computer program Letagrop (29). In this analysis some other possible complexes, such as BA and B_2A, are also being tested in combination with the AC and B_3A complexes.

Electron microscopy. In Fig. 4 electron micrographs are shown, which were obtained from a plasmid DNA solution of Cr(III). As a control, DNA in the absence of Cr(III) is also shown (Fig. 4a). It follows from Fig. 4 that for a B/A ratio of 1:6 the DNA molecule becomes very compact, forming a so-called toroid (Fig. 4b).

4. DISCUSSION

The results described in the previous section have furnished information regarding nucleic acid interactions with a small ion, Cr(III), and with a protein, EF-Tu·GTP. In the first study a large nucleic acid, DNA, $(M_r \sim 1.4 \times 10^6)$ reacts with Cr(III) forming a condensed structure (Fig. 4). This condensation is what might have been expected from Manning's theory (30). Surprisingly, however, the present condensation proceeds at a much lower Cr(III) concentration than expected from studies on similar ions, such as $[Co(NH_3)_6]^{3+}$, (31). For instance, for 30 mM ionic strength (NaCl) condensation appears to occur first at a B/A ratio of 30 to 60, if $[Co(NH_3)_6]^{3+}$ ions are used (31). The reason for this difference may be due to the formation of Cr(III) – DNA complexes with a higher positive net charge than the trivalent Co(III) complex ion.

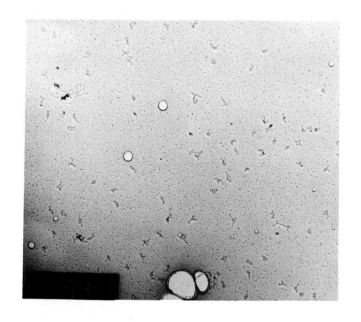

(b)

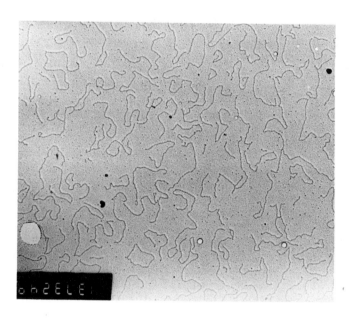

(a)

Figure 4. Electron micrographs of plasmid DNA (a) and plasmid DNA reacted with Cr(III) at a ratio B/A of 6:1 (b); from (11).

It is worth noting that chromium is a well recognized carcinogen as based on human studies, animal experiments, and short-term bioassays (32,33). The true environmental pollutant is believed to be Cr(VI), but Cr(VI) is readily reduced *in vivo* by various agents such as, for instance, glutathione, cysteine and ascorbic acid, forming Cr(III). Thus, Cr(III) appears to be the thermodynamically stable *in vivo* form of chromium and it must be considered as the ultimate carcinogen on the cellular level. Based on the present results, one may imagine how multicharged Cr(III) complexes are formed on naked parts of DNA. If they hit regulatory DNA sites it seems quite possible that they may interfere with regulatory events within the cell.

The Cr(III)-DNA system, where a large nucleic acid reacts with a small metal ion, is quite different from the other system of this work, the EF-Tu·GTP – valyl·tRNA$_{1A}^{Val}$ system, where a relatively small and well-defined nucleic acid reacts with a large molecule EF-Tu·GTP (M_r=46,000). In the DNA study, Cr(III) was able to dramatically change the nucleic acid structure, since at a certain Cr(III) concentration DNA condenses into a toroid. In the other part of the present investigation it was indicated from neutron scattering in 40 % and 70 % D$_2$O that the radii of gyration for EF-Tu·GTP and valyl-tRNA$_{1A}^{Val}$ did not change when the ternary complex was formed. This indicates that only minor conformational changes may occur as a result of the interaction between EF-Tu·GTP and aminoacyl-tRNA; a model of the ternary complex could therefore be composed of ellipsoids derived from the crystal structures of EF-Tu·GDP (2) and tRNA^{Phe} (5,6). A comparison of the $p(r)$-curve obtained from a small-angle X-ray scattering study of the ternary complex with the corresponding theoretical curves generated for various models shows that the best agreement with the experimental data is obtained when the acceptor stem is attached to EF-Tu·GTP and the anticodon stem and loop protrude into solution. In this model the anticodon stem can be directed away from or, alternatively, towards the EF-Tu ellipsoid (9). Within the limits of resolution there is a good agreement between our model and the results obtained from the nuclease digestion experiments reported by Wikman *et al*. (34) and the fluorescence studies of Adkins *et al*. (35).

One may finally ask how large the conformational change may be which is unrecognized by the present methods due to the limits of the experimental errors. For instance, it may be of interest to know how much the angle between the main axes of the two ellipsoids forming our tRNA model will change if we vary the gyration radius, R, of tRNA from 2.4 to 2.6 nm. Our calculations indicate that an angle of 90° yields R = 2.4 nm and an angle of 120° yields R = 2.6 nm. Thus, we may conclude by saying that although our data agree with the idea that the conformations of tRNA and EF-Tu·CTP are essentially retained within their complex, we cannot exclude the possibility of conformational changes which are comparable to a variation of the tRNA angle from 90° to 120°.

ACKNOWLEDGEMENTS

We are greatly indebted to the Swedish Work Environment Fund for financial support.

REFERENCES

1. Miller, D.L., and Weissbach, H.: 1977, "Nucleic Acid-Protein Recognition" (Vogel, H., ed.), Academic Press, New York, pp 409-440.
2. Morikawa, K., La Cour, T.F.M., Nyborg, J., Rasmussen, K.M., Miller, D.L., and Clark, B.F.C.: 1978, J. Mol. Biol. 125, pp 325-338.
3. Kabsch, W., Gast, W.H., Schultz, G.E., and Lebermann, R.: 1977, J. Mol. Biol. 117, pp 999-1002.
4. Jurnak, F., McPherson, A., Wang, A.H.J., and Rich, A.: 1980, J. Biol. Chem. 255, pp 6751-6757.
5. Kim, S.H., Suddath, F.L., Quigley, G.J., McPherson, A., Sussman, J.L., Wang, H.J., Seeman, N.C., and Rich, A.: 1974, Science 185, pp 435-440.
6. Robertus, J.D., Ladner, J.E., Finch, J.T., Rhodes, D., Brown, R.S., Clark, B.F.C., and Klug, A.: 1974, Nature 250, pp 546-551.
7. Woo, N.H., Roe, B.A., and Rich, A.: 1980, Nature 286, pp 346-351.
8. Österberg, R., Elias, P., Sjöberg, B., Kjems, J., and Bauer, R. "manuscript in preparation".
9. Österberg, R., Sjöberg, B., Ligaarden, R., and Elias, P.: 1981, Eur. J. Biochem. 117, pp 155-159.
10. Sjöberg, B., and Österberg, R., J. Appl. Crystallogr., "in press".
11. Österberg, R., Persson, D., and Bjursell, G., "manuscript in preparation".
12. Lebermann, R., Antonsson, B., Giovanelli, R., Gaurignata, R., Schuman, R., and Wittinghofer, A.: 1980, Anal. Biochem. 104, pp 29-36.
13. Kruse, T.A., Clark, B.F.C., and Sprinzl, M.: 1978, Nucleic Acids Res. 4, pp 1999-2008.
14. Giège, R., Jacrot, B., Moras, D., Thierry, J.C., and Zaccai, G.: 1977, J. Mol. Biol. 4, pp 2421-2427.
15. Birnbom, H.C., and Doly, J.: 1979, Nucleic Acids Res. 7, pp 1513-1523.
16. Kratky, O., and Skala, Z.: 1958, Z. Elektrochem. 62, pp 73-77.
17. Kratky, O., Pilz, I., and Schmitz, P.J.: 1966, J. Colloid Sci. 21, p 24.
18. Kjems, J., and Bauer, R., "manuscript in preparation".
19. Waring, M.J.: 1965, J. Mol. Biol. 13, pp 269-282.
20. Davis, R., Simon, M., and Davidsson, N.: 1971, "Methods in Enzymology" (Grossman, L., and Moldave, K., eds.), vol. 21, Academic Press, New York, pp 419-428.
21. Ibel, K., and Stuhrmann, H.B.: 1975, J. Mol. Biol. 93, pp 255-265.
22. Ninio, J., Luzzatti, V., and Yaniv, M.: 1972, J. Mol. Biol. 71, pp 217-229.
23. Pilz, I., Kratky, O., Cramer, F., von der Haar, F., and Schlimme, E.: 1970, Eur. J. Biochem. 15, pp 401-409.

24. Österberg, R.: 1975, J. Mol. Biol. 75, pp 394-400.
25. Glatter, O.: 1977, J. Appl. Crystallogr. 10, pp 415-421.
26. Pilz, I., Puchwein, G., Kratky, O., Herbst, M., Haager, O., Gall, W.E., and Edelman, G.M.: 1970, Biochemistry 9, pp 211-219.
27. Österberg, R.: 1965, Acta Chem. Scand. 19, pp 1445-1468.
28. Le Pecq, J.B., and Paoletti, C.: 1967, J. Mol. Biol. 27, pp 87-106.
29. Ingri, N., and Sillén, L.G.: 1965, Arkiv Kemi 23, p 97.
30. Manning, G.S.: 1978, Quart. Rev. Biophys. 11, pp 179-246.
31. Widom, J., and Baldwin, R.L.: 1980, J. Mol. Biol. 144, pp 431-453.
32. Flessel, C.P., Furst, A., and Radding, S.B.: 1980, "Metal Ions in Biological Systems" (Siegel, H., ed.), Marcel Dekker, New York, vol. 10, pp 23-54.
33. Sunderman, F.W.: 1979, Biol. Trace Element Res. 1, p 63.
34. Wikman, F.P., Siboska, G.E., Petersen, H.U., and Clark, B.F.C.: 1982, The EMBO Journal 1, pp 1095-1100.
35. Adkins, H.J., Miller, D.L., and Johnson, A.E.: 1983, Biochemistry 22, pp 1208-1217.

CHROMATIN HIGHER ORDER STRUCTURE AND GENE EXPRESSION

Gary Felsenfeld, James D. McGhee, Joanne Nickol, & Donald Rau
Laboratories of Molecular Biology and Physical Chemistry,
National Institute of Arthritis, Diabetes, and Digestive
and Kidney Diseases, National Institutes of Health,
Bethesda, Maryland 20205, U.S.A.

Recent studies from a number of laboratories have shown that the 30 nm chromatin fiber contains a solenoidal array of nucleosomes, with about six nucleosomes (chromatosome + spacer DNA) per turn of solenoid. We have used electric dichroism to determine the orientation of the chromatosomes within the solenoid. With reasonable assumptions about the path of the spacer DNA, we find that chromatosomes in a variety of chromatins with different spacer lengths have quite similar packing schemes. We show that hyperacetylation of histones (thought to be associated with transcriptional activity of genes) has little effect on the stability of the 30 nm solenoid. Furthermore, a typical active gene, the adult β globin gene of chicken erythrocytes, appears to be packaged in a structure similar to the 30 nm solenoid.

INTRODUCTION

The two lowest levels of organization of chromatin structure are now reasonably well understood. The first of these, the nucleosome, is a complex of DNA with an octamer of the basic proteins called histones. About 200 base pairs of DNA are associated with each nucleosome. Of this complement, 165 base pairs are wrapped tightly about the histone octamer in two superhelical turns (1). This structure has been named the chromatosome (2). The remaining DNA of the nucleosome repeat is called the spacer or linker DNA, and varies in length from zero base pairs to about 80 base pairs, depending upon the source of the chromatin. This serves to link one chromatosome to the next, forming the 10 nm filament or beaded string that can be observed if chromatin fibers are exposed to low ionic strength solvents. At higher ionic strength, the 10 nm filament is folded into a higher order structure, the 30 nm fiber. Evidence from a wide variety of experiments (3-7) indicates that the 30 nm fiber is constructed by winding the nucleosomes into a solenoidal array, with about six nucleosomes per turn of the solenoid, and with a pitch of 11 nm. The folding is a reversible process, which can be controlled by adjusting the ionic strength.

B. Pullman and J. Jortner (eds.), Nucleic Acids: The Vectors of Life, 343–352.
© 1983 by D. Reidel Publishing Company.

Although it is easy to show that most of the DNA within the eukaryotic nucleus must be packaged in this way, it is not at all certain that this structure is preserved near genes that are either being expressed, or have the potential for expression in a particular cell type. When such genes are being transcribed it seems likely that histones are displaced by the passage of RNA polymerase. It is less clear what the structure of such a region might be between rounds of transcription, particularly when the gene is one that is transcribed at a low frequency.

In this paper we describe some recent studies of the structure of the 30 nm fiber. We then show how such physical studies can be extended to examine the effects of histone modification, as well as to probe the structure and chemical properties of a particular gene, the adult beta globin gene of chicken erythrocytes.

The 30 nm Chromatin Fiber

During the past three years we have addressed the problem of nucleo-some packing within the 30 nm chromatin fiber. The technique of electric dichroism provides both an estimate of overall particle dimensions (by measurement of relaxation times) and of the geometry of nucleosome pack-ing (by measurement of the dichroism of the oriented fiber). In our earlier experiments (8), chromatin fibers were prepared from chicken erythrocyte nuclei by mild digestion with micrococcal nuclease, followed by size fractionation on sucrose gradients. The resulting particles, containing between 20 and 100 nucleosomes, were oriented in an electric field and the dichroism at 260 nm measured as a function of field strength. (The reduced dicroism, ρ, is defined as $\rho = (A_{\parallel} - A_{\perp})/A$, where $A_{\parallel}$ and $A_{\perp}$ are the absorbance of light polarized respectively parallel and perpendicular to the orienting field, and A is the absor-bance for unpolarized light (9). The limiting reduced dichroism, ρ_{∞}, corresponding to complete orientation, is obtained by extrapolating the field dependent dichroism to infinite field. In this way we find that the limiting dichroism of 30 nm fibers from chick erythrocyte chromatin is about -0.18.

The interpretation of dichroism in terms of structure depends upon the well understood properties of DNA as a chromophore, and knowledge of the path of DNA on the surface of the chromatosome. With this informa-tion it is possible to draw some important conclusions about possible arrangements of the chromatosomes in the 30 nm fiber. We began by neglecting contributions from the spacer DNA (which comprises only 21% of the DNA of the nucleosome repeat), and asking what classes of chromatosome packing were allowed by our data. Three possible orienta-tions of the chromatosomes within the 30 nm fiber are shown in Figure 1. Of these, packing arrangements related to the one shown in Figure 1C are entirely excluded by our data, since the large positive dichroism contributed by the chromatosomes could never be compensated by any possible negative contribution from the linker DNA. (The contributions of chromatosome and linker are additive, and must sum to give the

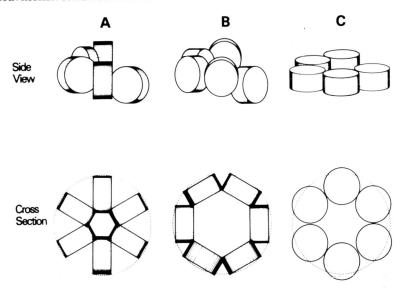

Figure 1.   Three possible simplified models for the packing of chromato-somes within the 30 nm chromatin solenoid. (A) is the starting point for the detailed models discussed in this paper. (From Ref. 8).

measured, negative, dichroism). The other two classes of packing, shown in Figures 1A and B, are optically equivalent and therefore indistin-guishable by our methods. However, Bradbury and his collaborators (3) have used neutron scattering to eliminate models in which the mass of the fiber is concentrated in a shell at the periphery. We conclude that the structure is related to the model in Figure 1A, in which chromato-somes are arranged radially.

We next attempted to take into account the contribution of the spacer DNA. In doing so, we are obliged to perturb the chromatosome orientation shown in Figure 1A by tilting each one relative to the solenoid axis, in such a way that the total contribution to dichroism from spacer and chromatosome DNA corresponds to the experimental value. The best way to obtain information about the contribution of spacer DNA is to examine a series of chromatins from different sources in which the spacer length varies. We have therefore isolated chromatins from several such sources and measured their dichroism (Ref. 10, Table 1).

Three possible models for the path of the spacer DNA are shown in Figure 2. The first model (I) places the spacer parallel to the sole-noidal axis. Although this arrangement can be made consistent with measured dichroism by tilting the chromatosomes at angles of 40-50 degrees to the solenoid axis, it involves unreasonable assumptions about the overall geometry of the spacer. It would be necessary for spacers 80 base pairs (about 27 nm) long to connect adjacent chromatosomes that are unlikely to be separated within the fiber by more than 2 nm. To accommodate such an arrangement, it would be necessary to bend the spacer

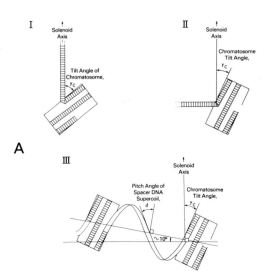

Figure 2. Possible spacer and chromatosome arrangements in the chromatin solenoid. (From Ref. 10).

abruptly. Models of the kind represented by II can also be eliminated:
at least two of the chromatins studied (sea urchin and chick erythrocyte)
have dichroisms that cannot be made compatible with such a model.

We first proposed the model of spacer DNA shown in III of Figure 2
to account for the data obtained with chicken erythrocyte chromatin (8).
In this model, the spacer DNA follows a superhelical path between
chromatosomes. Since the diameter of the solenoid and the size of the
chromatosomes are known, the distance between adjacent chromatosomes is
also known. The spacer length in any particular chromatin is fixed, and
therefore the pitch of the spacer supercoil is determined, and with it
the contribution it makes to dichroism.

We have applied this analysis to the chromatin samples listed in
Table 1. Since the total dichroism depends only upon the spacer pitch
angle and the chromatosome tilt angle, the latter can be deduced for
each chromatin sample. As shown in Table 1, the estimated tilt angles
vary slightly with spacer length, but all lie within the narrow range of
21 to 33 degrees. The packing of chromatosomes thus seems to be an
approximate invariant of the structure of the 30 nm fiber. The structure
we propose is shown in Figure 3 for the particular case of chick erythrocyte chromatin.

What are the forces that stabilize this structure? We have suggested elsewhere (8) that the amino terminal tails of the core histones,
which are not essential to the stability of individual nucleosomes,
might be implicated in the stabilization of the 30 nm fiber. Recent

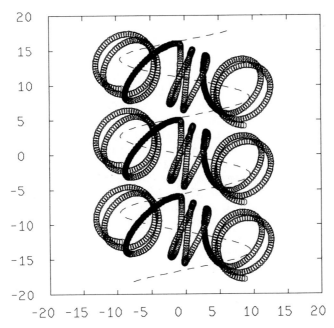

Figure 3. The path of DNA in the 30 nm solenoid from chick erythrocyte nuclei, according to the proposed model. The chromatosomes are tilted 25° from the solenoid axis. The spacer DNA segments are shaded. (See Refs. 8 and 10).

*Table 1: Chromatosome Tilt Angles, γ_c, For Various Chromatin Samples, in the Supercoiled Spacer Model			
Chromatin Source	Nucleosome Repeat Length (N bp)	Spacer DNA Length (N-166) bp	Chromatosome Tilt Angle, γ_c °
Sea Urchin Sperm	243	77	21
Chicken Erythrocyte	210	44	26
Rat Liver	196	30	30
HeLa Cells	186	20	30
CHO Cells	177	11	33

*Data from Ref. 10.

experiments by Allan <u>et al</u>. (11), which show that removal of the tails by trypsin digestion destabilizes the 30 nm fiber, support such a model.

The histone tails also contain the sites of histone acetylation, and it has occurred to a number of investigators that, if the tails are implicated in stabilizing the 30 nm fiber, then histone acetylation, which reduces the positive charge on the tails, might also destabilize the 30 nm fiber. Since high levels of histone acetylation have been correlated with transcriptional activity, one might then speculate that acetylation functions to facilitate transcription. We have made a direct test of this hypothesis by following the salt-dependent conversion of hyperacetylated chromatin from the extended 10 nm filament to the folded 30 nm fiber (12). This conversion can be monitored by measuring the dichroism, which undergoes a dramatic decrease as folding occurs. We have also made use of the change in hydrodynamic properties that accompanies folding. As shown in Figure 4, and first documented in detail by Butler and Thomas (13), a large change in sedimentation coefficient occurs during this transition. The data in Figure 4 compare

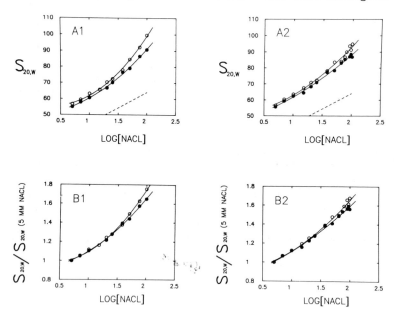

Figure 4A. Plot of $s_{20,w}$ vs. logarithm of the added NaCl concentration (millimolar) for pairs of chromatin fractions, each ~ 35 nucleosomes in length. (●) = hyperacetylated; (O) = control chromatin. All samples also contained 5 mM Tris-HCl, 1 mM Na butyrate, 0.1 mM EDTA, pH 8. Solid lines are best fit quadratics, dashed line is behavior expected for stripped chromatin (8). Figures 2A1 (left) and 2A2 (right) represent data from complete duplicate experiments, starting from different cell batches.
    B. Sedimentation data from Figure 2A, normalized to the $s_{20,w}$ measured at 5 mM NaCl, to correct for any possible DNA size differences between hyperacetylated and control chromatins. (From Ref. 11).

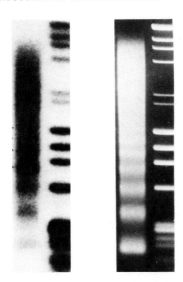

Figure 5. Micrococcal nuclease digests of erythrocyte nuclei from 14-day-old chick embryos. The DNA was purified and electrophoresed on 1.5% agarose gels, and either stained with ethidium bromide (right) or blotted to DBM paper, hybridized to a radioactive globin-specific probe, and autoradiographed (left). The probe was a 6.2 kb fragment containing the entire adult β globin gene and flanking sequences. Similar results have been obtained with a smaller probe containing only an intervening sequence of the gene.

the folding process for 30 nm fibers isolated from normal HeLa cells, with the folding of fibers obtained from cells that have been treated with sodium butyrate to induce histone hyperacetylation. We observe a small and reproducible decrease in the salt-dependent stabilization of the acetylated 30 nm fiber. However, the obvious conclusion is that hyperacetylation has remarkably little effect on this chromatin structure. This appears to rule out proposals concerning the biological effects of acetylation that envision an extensive disruption of the 30 nm fiber, although smaller changes might not be detected.

Properties of Chromatin in the Vicinity of Active Genes

Since we now have considerable information about the physical properties of the 10 nm filament and the 30 nm fiber, we are able to ask whether these properties are displayed by chromatin containing specific DNA sequences. For these studies we chose the chicken adult β globin gene. In the erythrocytes of 14-day embryos, this gene has a history of transcriptional activity and displays the sensitivity to DNase I digestion characteristic of active genes (14,15).

There is considerable evidence that the adult beta globin gene in these cells is packaged in nucleosomes. In support of this view, we

show in Figure 5 the results of an experiment in which chicken erythrocyte nuclei were digested with micrococcal nuclease to obtain a "ladder"
of DNA fragments that are multiples of the fundamental nucleosome
repeat of 210 base pairs. The right lane of Figure 5 shows the distribution of bulk DNA in the digest. The left lane shows the same material
hybridized to a globin-specific probe. The repeat is identical to that
obtained for the bulk. We conclude that a considerable fraction of
this gene is packaged in nucleosome-like structures.

We are now able to extend the analysis of the globin gene structure
to the level of the 10 nm filament and 30 nm fiber. We ask whether the
nucleoprotein containing the gene undergoes the same salt-dependent
folding into the compact solenoidal form. The experiment is carried
out exactly as in the studies summarized in Figure 4, except that
sucrose gradient sedimentation is employed. Typical data are shown in
Figure 6 for a preparation of chromatin fragments sedimented at varying
ionic strengths. In each case the movement of the bulk chromatin is
monitored by optical absorbance, while the chromatin containing the
globin gene is followed by "dot-blot" hybridization to specific globin
probes. The data show that the chromatin containing the globin gene
co-sediments with the bulk of the chromatin at every salt concentration.
Thus, within our limits of error (see Conclusion), the globin gene is
packaged in a structure of normal stability, at least with respect to
perturbation by salt.

These data suggest that even in genes capable of being transcribed,
chromatin structure is not greatly perturbed. We do not mean to imply
that this structure is maintained when the gene is actually transcribed;
it would not be surprising if the passage of RNA polymerase displaces
histones. The data presented here merely suggest that gross perturbation of chromatin structure is not associated with the potential for
transcription. Active chromatin is of course marked by certain properties that distinguish it from the bulk, notably a uniform increased
sensitivity to nucleases (15). It should be noted that the globin
chromatin fractions used in the studies shown in Figure 5 display the
full sensitivity to DNase I normally exhibited by those sequences
within erythrocyte nuclei. A second feature of transcriptionally
active genes is the presence in the 5' flanking region of sites that
are hypersensitive to nucleases (16). In the case of the adult beta
globin gene, we have shown that these sites define a region that appears
to be free of nucleosomes (17). This structural feature is correlated
with the activity of the gene: it is absent in nuclei isolated from
5-day-old embryos, in which the adult globin message is not synthesized,
as well as from nuclei isolated from other tissues.

CONCLUSION

We have discussed the structure of the 30 nm fiber, and shown how
physical methods can be extended to an examination of the properties of
chromatin containing specific, transcriptionally active genes. Within

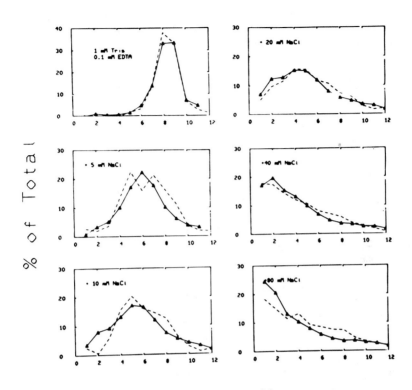

Figure 6. Sucrose gradient sedimentation of chromatin fractions in solvents of varying ionic strength. Aliquots from the gradients were used to determine absorbance (____), or to determine globin gene concentration (----). In the latter case, samples were "dot-blotted" to DBM paper, hybridized to the globin-specific probe described in Fig. 5, and autoradiographed. The abundance of globin sequence was determined by densitometry of the autoradiograms. We have also carried out experiments in which the size distributions of bulk and globin DNA have been compared in each fraction, and found to be identical (Ref. 16).

the limits of error of the methods presently available to us, the chromatin containing the globin gene appears to be packaged in a 30 nm fiber similar in properties to the bulk of chromatin. It is certainly possible that the globin chromatin fiber contains internally perturbed segments that do not affect sedimentation properties, particularly since the transcribed portion of the adult beta globin gene is only about 1 kb in length. When embedded in a segment of normal 30 nm fiber containing 6 kb or more of DNA, the perturbed structure might not be detected. We are now applying these methods to other genes with a longer coding region, in which any perturbations should have a more pronounced effect. In any case, the combination of classical physical methods with the analytical techniques made possible by the availability of gene-specific probes should eventually make possible a detailed description of chromatin structure as it affects biological function.

REFERENCES

1. McGhee, J.D. and Felsenfeld, G.: 1980, Ann. Rev. Biochem. 49, pp. 1115-1156.
2. Simpson, R.T.: 1978, Biochemistry 17, pp. 5524-5531.
3. Suau, P., Bradbury, E.M., and Baldwin, J.P.: 1979, Eur. J. Biochem. 97, pp. 593-602.
4. Thoma, F. and Koller, T.: 1981, J. Mol. Biol. 149, pp. 709-733.
5. Campbell, A.M., Cotter, R.I., and Pardon, J.F.: 1978, Nucleic Acids Res. 5, pp. 1571-1580.
6. Rattner, J.B. and Hamkalo, B.A.: 1978, Chromosoma 69, pp. 363-372 and pp. 373-379.
7. Rattner, J.B. and Hamkalo, B.A.: 1979, J. Cell Biol. 89, pp. 453-457.
8. McGhee, J.D., Rau, D.C., Charney, E., and Felsenfeld, G.: 1980, Cell 22, pp. 87-96.
9. Fredericq, E. and Houssier, C.: 1973, "Electric Dichroism and Electric Birefringence", Oxford, Clarendon Press.
10. McGhee, J.D., Nickol, J.M., Felsenfeld, G., and Rau, D.C.: 1983, Cell 33, pp. 831-841.
11. Allan, J., Cowling, G.J., Harborne, N., Cattini, P., Craigie, R., and Gould, H.: 1981, J. Cell Biol. 90, pp. 279-288.
12. McGhee, J.D., Nickol, J.M., Felsenfeld, G., and Rau, D.C.: Manuscript in preparation.
13. Butler, P.J.G. and Thomas, J.O.: 1980, J. Mol. Biol. 140, pp. 505-529.
14. Weintraub, H. and Groudine, M.: 1976, Science 193, pp. 848-856.
15. Wood, W.I. and Felsenfeld, G.: 1982, J. Biol. Chem. 257, pp. 7730-7736.
16. Wu, C., Bingham, P.M., Livak, K.J., Holmgren, R., and Elgin, S.C.R.: 1979, Cell 16, pp. 797-806.
17. McGhee, J.D., Wood, W.I., Dolan, M., Engel, J.D., & Felsenfeld, G.: 1981, Cell 27, pp. 45-55.
18. Felsenfeld, G., McGhee, J., Rau, D.C., Wood, W., Nickol, J., and Behe, M.: 1982, "Gene Regulation", New York, Academic Press, 26, pp. 121-135.

PROTEIN-NUCLEIC ACID INTERACTION: THE AMINOACYLATION OF T-RNA AND ITS
EVOLUTIONARY ASPECT

Friedrich CRAMER and Wolfgang FREIST
Max-Planck-Institut für experimentelle Medizin
Abteilung Chemie, D-3400 Göttingen, FRG

SUMMARY: The aminoacylation of tRNA occurs in two steps. In the first
step the amino acid is selected, activated and attached, in the second
step errors are removed by proofreading. Thus an overall fidelity of
aminoacylation of 10^5 to 10^6 is achieved. Details of the enzymatic
mechanism of proofreading are discussed. It is suggested that the above
figure of accuracy reflects the overall accuracy of protein biosynthesis in all steps of transcription and translation. Kinetic experiments
allow conclusions on the nature of the binding site for the amino acid.
An explanation is given for the fact that lipophilic amino acids are
attached to the 2'-OH and hydrophilic amino acids to the 3'-OH of
tRNA. In primordial systems, when tRNA had to catalyse its own aminoacylation the terminal A could form a lipophilic binding site for 2'-
attachment.

Aminoacylation of tRNA occurs in two separate chemical steps

1. Activation of the amino acid

$$aa + ATP \rightleftharpoons aminoacyl\text{-}AMP + PP$$

2. Transfer of the activated amino acid to the appropriate tRNA

$$aminoacyl\text{-}AMP + tRNA \longrightarrow aa \sim tRNA + AMP$$

It has been found earlier, that most of the aminoacyl-tRNA synthetases
attach the amino acid to either the 2'- or the 3'-OH-group of the 3'-
terminus of their tRNA with a high degree of specificity (1). The
functional reason for this specificity could later be found in the fact,
that the nonaccepting hydroxyl is involved in a proofreading process in
which this OH-group is required for a hydrolytic editing mechanism:
tRNA^{ILE} lacking the nonaccepting 3'-OH is hundred percent misacylated
by valine (2,3).

B. Pullman and J. Jortner (eds.), Nucleic Acids: The Vectors of Life, 353–364.
© *1983 by D. Reidel Publishing Company.*

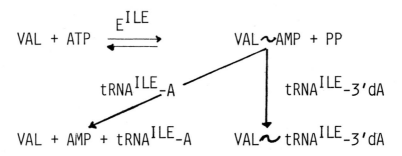

Thus, synthetases in presence of noncognate amino acids are tRNA-de-
pendant ATPases. The table of the specificities for the 2'- or 3'-site
is shown in Fig. 1.

2'	3'	2',3'	not yet assigned
ARG	ALA	TYR	ASN
ILE	GLY	CYS	ASP
LEU	HIS		GLN
MET	LYS		GLU
PHE	PRO		
TRP	SER		
VAL	THR		

Fig. 1. Specificity of aminoacylation with respect to position at terminal ribose.

As far as data are available, this specificity has been preserved in
evolution. The enzymes for tyrosine and cysteine do not exhibit a specificity. Concurrently, no proofreading is observed in these enzymes;
apparently these two amino acids show sufficiently characteristic
features in order to be distinguished from other amino acids with high
fidelity in a single step process. Thus the third column of Fig. 1 is
explained. What determines the assignment to the 2'- and 3'-position?
When looking at Fig. 1 it is obvious, that the more lipophilic amino
acids are attached to the 2'-OH and the more hydrophobic to the 3'-OH.
But why?

In this paper we would like to contribute to the answers of the
following questions:

       1. What is the mechanism of proofreading?
       2. How good is the overall fidelity of protein biosynthesis?
       3. How did this mechanism of selection evolve?

1. What is the mechanism of proofreading?

Clearly, proofreading uses the nonaccepting hydroxyl of the 3'-terminal
ribose for its mechanism. However, a direct transfer of the amino acid
to be edited to this OH-group does not seem to be necessary (4). Probably the nonaccepting OH-group of the ribose serves as a general base
for the activation of the hydrolysing water molecule. This OH-group, in
turn, may be activated by an electron-release chain in a similar manner
as in the serine hydrolases.

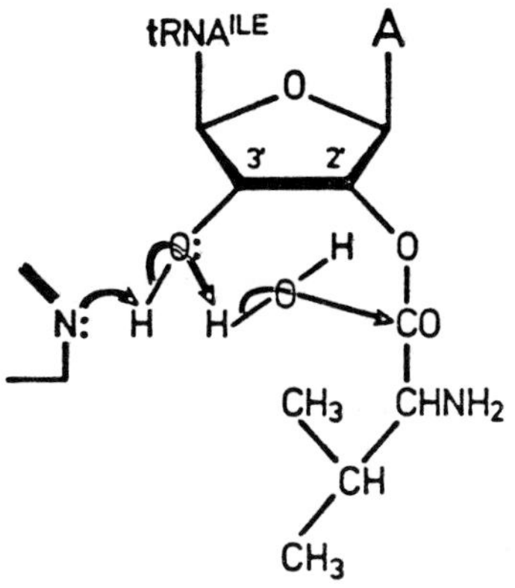

Fig. 2. Mechanism for the proofreading hydrolysis of VAL in
        VAL⁻tRNA^{ILE} (3).

It could well be that in this process the same groups are involved
which catalyse the group transfer of the amino acid to ATP and further
on to the tRNA. It is essential, however, that the two processes, transfer and proofreading, are kinetically independant. Only in this way
enhancement of fidelity is achieved. In fact, the active site of the
enzyme is used twice in two separable processes in which the fidelity
of each step does not only add to the previous but multiplies its
specificity. This is depicted in Fig. 3.

Recently we have found that in higher eucaryotic systems proofreading
can also occur at the aminoacyl-AMP-level before the amino acid is
transferred to the tRNA (5).Phe-tRNA synthetase from turkey liver uses
a pretransfer proofreading for noncognate natural amino acids and a

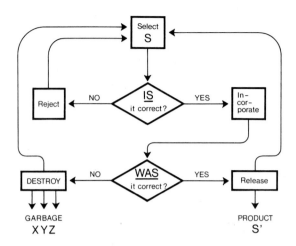

Fig. 3. Flow diagram for a two-step selection process. High
specificity is obtained by an energy consuming proofreading process in the second, kinetically independant loop of the reaction (9).

posttransfer route for synthetic analogues. Therefore it can be suggested that there is probably not a fundamental mechanistic difference between the pretransfer and posttransfer pathway. The interplay between
the active site of the enzyme, the 3'-terminal adenosine and the aminoacyladenylate may give an intermediary complex that potentially can
dissipate its energy into either pathway dependant on the structure of
the amino acid. Donor and acceptor interactions within the active site,
perhaps via a carboxylate, the 3'-OH group and a histidine residue, may
result in a correct orientation and necessary polarization, common to
both subsequent reactions. For example, in the case of tyrosine, either
the amino acid is transiently transferred to the 2'-OH group and
proofread via the influence of the 3'-OH group (E. coli, S. cerevisiae),
or the aminoacyl moiety of the aminoacyladenylate is positioned for
hydrolysis before transfer (turkey liver) (Fig. 4).

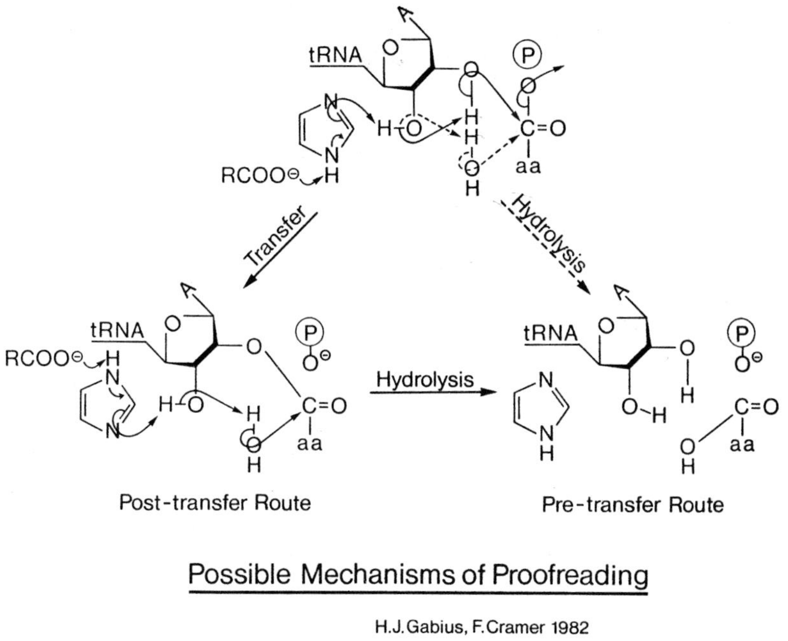

Possible Mechanisms of Proofreading

H.J.Gabius, F.Cramer 1982

Fig. 4. Possible mechanisms of proofreading according to (5).

2. How good is the overall fidelity of protein biosynthesis?

The function of any protein depends on the correct threedimensional arrangement of its functional groups; the spatial structure, in turn, is solely determined by the primary structure of the protein chain. Thus the functionality of a protein resides on the correct incorporation of each individual amino acid. Let us assume that common enzymatic reactions occur with a selectivity for correct versus incorrect substrate of 1000 : 1. For the best possible distinction between isoleucine and valine even a lower specificity of 100 : 1 has been calculated (6,7). Taking these values into account, normal proteins with a chain length of 100 to 1000 amino acids would in average have at least one error: There would be a random population of proteins with a broad functional spectrum. This is clearly an intolerable situation. Genotype and phenotype must be precisely related. It is for this reason that proofreading was invented in evolution.

How good is the classical part of the reaction and how much does the proofreading contribute? In order to measure this we have used a device to abolish proofreading. When a 3'-amino group instead of the 3'-hydroxyl group is introduced into the terminal ribose moiety, this amino function serves as a trap for the mischarged amino acid. Then, in a competition experiment the specificity without proofreading can be measured (Fig. 5) (8,12).

TRAP FOR WRONG AMINO ACID PREVENTS PROOFREADING:

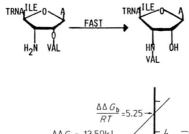

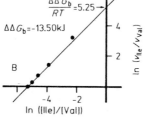

Fig. 5. Competitive aminoacylation of tRNA^{ILE}C-C-A(3'NH$_2$)
with ILE and VAL by isoleucyl-tRNA synthetase.
Determination of $\Delta\Delta G_b$ from aminoacylation of tRNA^{ILE}-
C-C-A(3'NH$_2$) under standard assay conditions with a
variable mixture of [^{14}C]isoleucine and valine. ILE
is 180 times better as substrate than VAL (12).

The primary selection between isoleucine and valine by Ile-tRNA
synthetase is 180 : 1. The contribution of the proofreading step is
measured in the following way: tRNA^{ILE}C-C-A is charged with valine as
the only amino acid. Consequently, almost all of the mischarged
VAL-tRNA^{ILE} is hydrolysed by proofreading and the corresponding amount
of ATP is consumed. The small amount of VAL-tRNA^{ILE}, which escapes the
proofreading process, is immediately bound to EFT$_u$ added in excess.
This amount can be measured on millipore filters. It turns out, that 300-
800 moles of ATP are hydrolysed before one molecule of mischarged tRNA
is formed. This is depicted in Fig. 6.

Accuracies have been measured also for other misacylation pairs

ILE/VAL (yeast)	100 000	: 1
PHE/TYR (E. coli)	370 000	: 1
PHE/LEU (E. coli)	920 000	: 1
PHE/MET (E. coli)	760 000	: 1

Thus, fidelity of aminoacylation of tRNAs is in the order of 1×10^5
to 10×10^5, a figure good enough to keep the world of the proteins in
law and order. Let us speculate for a moment about the preceding and
following steps of the translation-transscription process. Aminoacylation of tRNA is tuned to high fidelity in a sophisticated process

$$tRNA^{ILE} + VAL + ATP$$

$$\downarrow E^{ILE}$$

$$tRNA^{ILE} + VAL + AMP + PP$$
$$+ VAL - tRNA^{ILE}$$
(very little)

$$\downarrow EFTu$$

$$[VAL - tRNA^{ILE} \cdot EFTu]$$
complex, protected
$$300 - 800 : 1$$

'Classical selection 180 : 1
Proof reading   300 - 800 : 1
_____
Total fidelity ~ 100 000 : 1

Fig. 6. Total fidelity in the distinction between ILE and
VAL by the yeast enzyme.

that costs at least one extra mole of ATP. This sophistication and ex-
penditure would be useless, if the other steps were less accurate, and
vice versa a higher accuracy than 1 to 10 x 10^5 in the transscription
and in the codon-anticodon reading at the ribosome would mean an un-
necessary effort. The tendency of evolution to achieve perfection on
the one side and to act economically on the other side has reached a
compromise with this figure of accuracy, and we would like to make the
prediction that RNA-polymerase and the ribosome act at the same level
of accuracy, that is one error in 1 x 10^5 to 10 x 10^5. Detailed figures
in these systems are, however, not yet available (10,11).

3. How did the mechanism of selection of amino acids evolve?

Under special conditions tRNA^{ILE}-2'dA can be aminoacylated by the
cognate synthetase at the "unnatural" 3'-OH-position (12). The distinction between ILE and VAL which for the "natural" 2'-OH-position is about
200 : 1 (8,12) is lowered in the "unnatural" case to 6 : 1 (Fig. 7).
The specificity for the amino acid is almost lost when forcing the
system to attach the amino acid to the wrong position, suggesting that
the reactive center of the activated amino acid cannot reach the 3'-OH
when bound to its normal hydrophobic pocket.

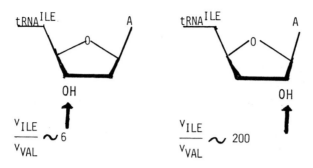

Fig. 7. Distinction between ILE and VAL at the 3'- and 2'-
        position resp. (12).

If in the aminoacylation reaction of tRNA^{ILE}-C-C-2'dA the activated
amino acids are not bound to the normal hydrophobic pocket, the question arises as to what other site they may be attached. For tyrosyl.
tRNA synthetase from B. stearothermophilus it was shown by electron
density difference maps that in complexes obtained with crystalline
enzyme and AMP or tyrosine, surprisingly these two compounds bind at
the same site, and tyrosinyl-adenylate is also bound with the tyrosyl
side chain to that same site. Obviously two hydrophobic pockets must
exist for binding of the aminoacyladenylate, one for the side chain
of the amino acid and one for the adenine moiety. For an explanation
of the 2',3'-specificity shown by aminoacyl-tRNA synthetases the following conclusions can be made. Because AMP is also bound in the amino
acid pocket it must be possible that under certain conditions the
aminoacyladenylate can also be bound rotated by about 180° with the
adenine moiety in the amino acid pocket and the aminoacyl part in the
adenine pocket.

From a molecular model of isoleucyl-adenylate and the C-A end of the
tRNA it can clearly be seen that the adenosine moiety of the adenylate
can be situated on the enzyme in a position in which it stacks to the
terminal adenosine of the tRNA if the carbonyl group of the aminoacyl
moiety is attached to the 3'-OH group (Fig. 8). If the carbonyl group
is in contact with the 2'-OH group the possibility of stacking with
the terminal adenosine moiety of the tRNA is lost; even if the adenylate is adjusted to the syn-conformation at the glycosidic bond this
would be very difficult (Fig. 8). However, the loss of stacking energy,
for wich considerable ΔH° values of 35.5 - 41.8 kJ are given (14,15),
in formation of the enzyme substrate complexes may be compensated if
the adenylate is turned by 180°. In this case a nonpolar side chain of
the amino acid could be in hydrophobic interaction with the terminal
adenine moiety and a surrounding hydrophobic binding pocket. For a
comparison of energy values in the case of the isoleucine system, only

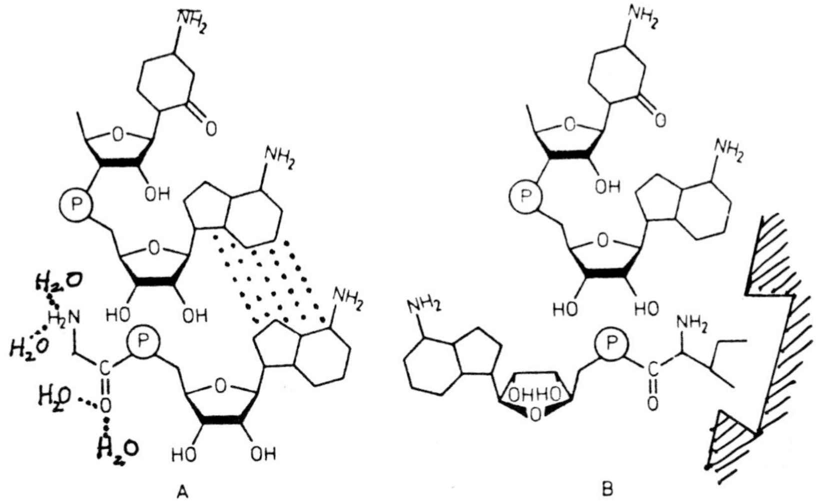

Fig. 8. The C-A end of tRNA and aminoacyl-adenylate stacked
with (A) the adenosine moiety to the terminal adenosine and the carbonyl group near 3'-OH and (B) with
the hydrophobic side chain of the amino acid near the
adenine moiety of the terminal adenosine and with
the carbonyl group next to the 2'-OH.

the Gibbs energy for an isobutyl moiety of 42.93 kJ can be mentioned
here (16,17); perhaps this shows that the amount of binding energy may
be in the same order of magnitude. In both cases the phosphorus atom
at which the aminoacyl residue is attached is in the same position.

A simple preliminary proof can be given by simulating in the aminoacylation assay the experiment done by Montheilet and Blow with crystalline tyrosyl-tRNA synthetase (13). When isoleucyl-tRNA synthetase
is preincubated with tRNA[ILE]-C-C-2'dA and ATP, a fraction of ATP molecules should be bound in the amino acid binding pocket perhaps favored
by stacking to the terminal adenosine of the tRNA. In this way ATP or
adenylate are in the suitable position for aminoacylation at the 3'-OH
group. When the aminoacylation reaction is now started by addition
of isoleucine or valine the reaction will start at once. Vice versa,
if the enzyme is preincubated with tRNA[ILE]-C-C-2'dA and isoleucine or
valine, the amino acid will be in its normal binding site which is not
suitable for aminoacylation at the 3'OH group. When the aminoacylation

reaction is started by addition of ATP, an obvious lag phase should occur because the reaction starts only when a fraction of amino acid molecules is displaced by ATP. These predicted effects were in fact observed in two experiments. When the aminoacylation of tRNA^{ILE}-C-C-2'dA is started by addition of isoleucine or valine the reaction starts practically linearly but when it is started by ATP a clear lag phase is observed (Fig. 9 A and B). When the reaction is started as usual by addition of enzyme only a small deviation from linearity can be noticed (Fig. 9 C), which may indicate a certain superposition of

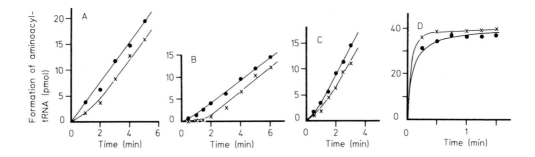

Fig. 9. Aminoacylation of tRNA^{ILE}-C-C-2'dA or tRNA^{ILE}-C-C-3'dA started by different reaction constituents.
(A) Aminoacylation of tRNA^{ILE}-C-C-2'dA with [^{14}C]iso-leucine and ATP started by addition of [^{14}C]iso-leucine (●) or ATP (X).
(B) Aminoacylation of tRNA^{ILE}-C-C-2'dA with [^{14}C]-valine and ATP started by addition of [^{14}C]valine (●) or ATP (X).
(C) Aminoacylation of tRNA^{ILE}-C-C-2'dA with [^{14}C]iso-leucine (●) or [^{14}C]valine (X) and ATP started by addition of enzyme.
(D) Aminoacylation of tRNA^{ILE}-C-C-3'dA with [^{14}C]iso-leucine and ATP under standard conditions started by addition of [^{14}C]isoleucine (●) or ATP (X). See (12).

of both effects. It must be admitted that this is only a qualitative consideration. A quantitative analysis of the lag phase will be very complex, because for a suitable special model system dealing with only one substrate the rate equation already contains coefficients which are combinations of 16 rate constants (18). If the same experiment is carried out with tRNA^{ILE}-C-C-3'dA the observations should be reversed because the aminoacylation site is changed. In experiments this effect is also observed, but due to the higher reaction velocity the differences are not so great and not so obvious as in the preceding case (Fig. 9). Nevertheless, these results are a preliminary indication that in aminoacylation at the 3'-OH group the activated isoleucine may be bound in the hydrophobic pocket which normally is the binding pocket of the adenine moiety of the aminoacyladenylate.

Taking into account the above-mentioned steric restriction seen in model building, the following generalization might be valid. Those aminoacyl-AMPs which stack their adenine moiety on the 3'-terminal adenosine of the tRNA will aminoacylate the 3'-OH of the terminal adenosine, those in which the amino acid is located next to the 3'-terminal adenosine will aminoacylate the 2'-OH. When looking at the respective specificities it is obvious that the more hydrophobic amino acids are in general esterified to the 2'-OH (see Fig. 1). Strikingly, tyrosyl-tRNA synthetase which binds AMP easily at the tyrosine binding site attaches the amino acid to the terminal 2'-OH and 3'-OH group. The terminal adenosine of tRNA seems to provide a hydrophobic pocket preselecting amino acids, perhaps an inheritance from early evolution when tRNA was the first gene, adapter and enzyme in the same molecule (Fig. 8) (19). In the primordial system when no proper aminoacylsynthetases were available and the tRNAs had to catalyse their own charging they were able to distinguish between hydrophobic and hydrophilic amino acids, attaching them to different sites.

REFERENCES

1. SPRINZL, M. and CRAMER, F. (1975). Site of aminoacylation of tRNAs from E. coli with respect to the 2'- or 3'hydroxyl group of the terminal adenosine. Proc.Natl.Acad.Sci.USA 72, pp. 3049-3053.

2. VON DER HAAR, F. and CRAMER, F. (1975). Isoleucyl-tRNA synthetase from baker's yeast: The 3'-hydroxyl group of the 3'-terminal ribose is essential for preventing misacylation of tRNA[ILE]-C-C-A with misactivated valine. FEBS Lett. 56, pp. 215-217.

3. VON DER HAAR, F. and CRAMER, F. (1976). Hydrolytic action of aminoacyl-tRNA synthetases from Baker's yeast: "Chemical proofreading" preventing acylation of tRNA[ILE] with misactivated valine. Biochemistry 15, pp. 4131-4138.

4. IGLOI, G.L., VON DER HAAR, F., and CRAMER, F. (1977). Hydrolytic action of aminoacyl-tRNA synthetase from baker's yeast: Chemical proofreading of Thr-tRNA[VAL] by valyl-tRNA synthetase studied with modified tRNA[VAL] and amino acid analogues. Biochemistry 16, pp. 1696-1702.

5. GABIUS, H.J., VON DER HAAR, F., and CRAMER, F. (1983). Evolutionary aspects of accuracy of phenylalanyl-tRNA synthetase. A comparative study with enzymes from Escherichia coli, Saccharomyces cerevisiae, Neurospora crassa, and turkey liver using phenylalanine analogues. Biochemistry, in press.

6. DEMAEYER, L.C.M. (1976). Energiebedarf der molekularen Informationsübertragung. Ber. Bunsen Ges. 80, pp. 1189-1196.

7. PAULING, L. (1958). The probability of errors in the process of
    synthesis of protein molecules. In Festschrift Arthur Stoll,
    Birkhäuser Verlag, Basel, pp. 597-602.

8. CRAMER, F., VON DER HAAR, F., and IGLOI, G.L. (1979). Mechanism
    of Aminoacyl-tRNA Synthetases: Recognition and Proofreading Pro-
    cesses. In Transfer RNA: Structure, Properties, and Recognition
    (Schimmel, P.R., Söll, D. & Abelson, J.N., eds), Cold Spring
    Harbor Laboratory, New York, pp. 267-279.

9. VON DER HAAR, F. GABIUS, H.-J., and CRAMER, F. (1981). Target
    Directed Drug Synthesis: The Aminoacyl-tRNA Synthetases as Possible
    Targets. Angew.Chemie Internat. Ed. 20, pp. 217-223.

10. REANNY, D.C. (1982). The Evolution of RNA Viruses. Ann.Rev.Micro-
    biol. 36, pp. 47-73.

11. RUUSALA, T., EHRENBERG, M., and KURLAND, C.G. (1982). Is there
    proofreading during polypeptide synthesis? EMBO J. 1, pp. 741-745.

12. FREIST, W. and CRAMER, F. (1983) Isoleucyl-tRNA Synthetase from
    Baker's Yeast. Catalytic Mechanism, 2',3'-Specificity and Fidelity
    in Aminoacylation of tRNA^{ILE} with Isoleucine and Valine Investi-
    gated with Initial-Rate kinetics using Analogs of tRNA, ATP and
    Amino acids. Eur.J.Biochem. 131, pp. 65-80.

13. MONTEILHET, C. and BLOW, D.M. (1978). Binding of Tyrosine, Adenosine
    Triphosphate and analogues to Crystalline Tyrosyl Transfer RNA Syn-
    thetase. J.Mol.Biol. 122, pp. 407-417.

14. DAVIS, R.C. and TINOCO, I. (1968). Temperature-Dependent Properties
    of Dinucleoside Phosphates (1968). Biopolymers 6, pp. 223-242.

15. LENG, M. and FELSENFELD, G. (1966). A Study of Polyadenylic Acid
    at Neutral pH. J.Mol.Biol. 15, pp. 455-466.

16. FERSHT, A. (1977). Enzyme Structure and Mechanism, W. Freeman
    and Company, Reading and San Francisco.

17. HANSCH, C. and COATS, E. (1970). α-Chymotrypsin: case study of sub-
    stituent constants and regression analysis in enzymic structure-
    activity relation. J.Pharm.Sci. 59, pp. 731-743.

18. AINSLIE, G.R., SHILL, J.P., and NEET, K.E. (1972). Transients and
    Cooperativity. J.Biol.Chem. 247, pp. 7088-7096.

19. EIGEN, M. and WINKLER-OSWATITSCH, R. (1981). Transfer-RNA: The
    Early Adaptor. Naturwissenschaften 68, pp. 217-228; 292-292.

CONFORMATION OF DNA, THE NUCLEOSOME AND CHROMATIN. HYDRATION OF DNA.

HENRYK EISENBERG
Polymer Research Department, The Weizmann Institute of
Science, Rehovot 76100, Israel

Topics reviewed in this contribution are (a) the conformational
properties and hydration of DNA in solution, (b) folding of DNA into
the compact nucleosome structure and the stability and conformational
properties of the latter, and (c) conformational transitions in the
higher order chromatin structure, as studied by light and small angle
x-ray scattering.

INTRODUCTION

The maintenance of all living matter is based on the continuing
and never ending dialogue between proteins and nucleic acids. This
report is concerned with reviewing ongoing studies in our laboratory
directed towards an understanding of some basic structural features of
the nucleoprotein complex genetic apparatus, within the broader context
of structural and functional studies currently proceeding over a wide
front (1-4).

The linearly encoded messages uniquely specifying protein structures
are translated in the ribosomal apparatus into peptide chains which then
spontaneously fold into active proteins. Thus a "higher order" folded
protein structure is necessary and sufficient for biological activity.
Proteins are active in the globular, folded, state and, compared to nucleic acids, they mostly assume moderate size and shape. New proteins
are continuously manufactured to replace "old" proteins which disappear
at the conclusion of their useful lifespan.

With nucleic acids the situation is quite different. Nucleic acids
of higher organisms encode the information for the manufacture of a
large number of proteins, many times repeated in a huge one-dimensional
array. For the successful processing of successive parts of this stag-
geringly large information storage bin, as well as for the replication
of the information it contains - for growth and transfer to succeeding
generations - it is essential that the encoded information present
itself in a conveniently accessible form. Whereas globular proteins are

B. Pullman and J. Jortner (eds.), Nucleic Acids: The Vectors of Life, 365–372.
© 1983 by D. Reidel Publishing Company.

manifested on the 40 to 100 Å scale, unfolded nucleic acids would correspondingly appear on the 10 to 100 cm scale along one dimension, while maintaining about 20 Å size along the other two dimensions. Clearly, this is an improbable situation, mechanically and topologically unrealistic, exceeding the size of a mature cell by orders of magnitude. Thus nucleic acids must be "packaged" in compact structures, while at the same time, easily available in selected regions, for replication and transcriptional "read-out". Packaging of intrinsically "stiff" DNA is achieved by interaction with selected proteins, histones for instance, and the switch from inactive, folded, to active, "unfolded", forms may be promoted by postsynthetic modifications or in interactions with further proteins, leading to subtle structural changes in delicately balanced systems. While much of this is suspected, the exact processes involved are presently unknown. Their precise evaluation presents a major challenge in understanding chromatin structure and function and discussion centered around this question will no doubt repeatedly recur at this meeting.

DNA IN SOLUTION

The basic reference conformation of DNA free in solution is the classical right-handed B double helical form, originally identified with the structure of high humidity fibers. For many years B DNA, and to a lesser extent the related right-handed A and C conformations, have domi- nated our thinking. It has now been demonstrated that DNA in such pro- tein-nucleic acid complexes as the nucleosome, for instance, is in the B form (5), though this certainly is not so for all protein DNA complexes. Single crystal X-ray analysis of protein DNA fragment complexes will eventually settle this point. Much excitement has in recent years been caused by the discovery of the left-handed unorthodox Z DNA structure (6). Quite aside from introducing a host of exciting previously unsus- pected structural variations, the tantalizing question relates to the potential role of Z DNA in relation to biological function.

DNA in solution is in strong interaction with counterions surroun- ding it and, furthermore, its conformation is sensitive to ionic strength (7,8). From polyelectrolyte theory has been derived the valuable con- cept of ion condensation (9) which has led to an extensive and suggestive, though not complete, representation of DNA-ion interactions (10). Tri- valent and higher valency ions of sign opposite to the negative charge of the DNA phosphate chains may lead to critical collapse (I avoid the term condensation, to avoid confusion with ion condensation) of the DNA coils (11,12), while maintaining some features of the original structure. High salt concentrations, for certain base compositions, may lead to the Z form, which can be obtained at physiological salt concentrations with methylated bases ,   or with some specific multivalent ions (13).

In our own work (7,8) we observe a smooth decrease in molecular dimensions of the linear form of $ColE_1$ plasmid DNA (6594 base pairs) with increasing Na^+ ion concentration between 0.007 to 4 M. From the

radii of gyration R_g determined by light scattering we calculated persistence lengths a, characteristic of the worm-like coil model. There
has been some argument with respect to the excluded volume correction
to be applied to account for coil expansion due to thermodynamic non-
ideality, in particular for low values of the ionic strength (14). At
higher values of the ionic strength little uncertainty persists leading
to a value of about 41 nm at 0.2 M $NaCl_2$, a salt concentration often
used in DNA research, asymptotically reaching about 28 nm at the limiting high values of the ionic strength. This is somewhat lower than
values obtained from hydrodynamic studies.

   To circumvent the problem inherent in correcting for thermodynamic
non-ideality of finite chains, we have extended (15) our experiments
to the study of LiDNA over a wide range of LiCl concentrations (Table 1).
It has been claimed (16), from circular dichroism and quasielastic light
scattering (17) studies, that LiDNA at high LiCl concentration, expe-
riences a critical structural condensation. We observe rather smooth
changes in LiDNA conformation up to rather high LiCl concentrations (15).
The virial coefficients A_2 (Table 1) vanish above 3 M LiCl. Thus, it
was possible to establish that the persistence length a decreases to a
limiting value of 29 nm between 3 and 5 M LiCl. This value is rather
similar to the values obtained for NaDNA in high concentrations of NaCl
(7,8) and can be related to the inherent flexibility of DNA at high
salt. The conformation of LiDNA remains essentially unchanged above and
up to 9 M LiCl, as followed by diffusion and sedimentation determinations,
precipitation occurring around 10 M LiCl concentration (18). This
agrees rather well with results from X-ray diffraction studies of DNA
fibers, immersed in various media (19).

   Another interesting aspect related to DNA structure in solution
concerns the amount of water of hydration associated with the nucleic
acid. The concept of hydration as derived from physico-chemical studies
is to a large extent operational and different values may be obtained
depending on the method used (20). We have in past years developed an
approach for determining hydration in multicomponent systems (20-22).
We evaluate hydration from a study of macromolecular buoyancy at variable solvent densities, assuming that a close shell of hydration is not
penetrated by other low molecular weight solvent components. The
approach is straightforward for non-ionic components. In the case of
charged macromolecules, such as DNA, for instance, disregard of Donnan
exclusion leads to an apparent increase in the values of hydration.
With Donnan exclusion duly considered we derive 3.7 water molecules per
nucleotide for NaDNA in NaCl and 5.9 water molecules per nucleotide for
CsDNA in CsCl (22,23) an average of 5 water molecules, with an uncertainty of one. This compares well with the value of 5 (three phosphate,
one each in the major and minor grooves) with an uncertainty of one,
estimated by Kopka *et al*. (24) from X-ray diffraction studies of a B-
DNA dodecamer. This is about half of other estimates arrived at by
various procedures (24).

Table 1. Molecular properties of linear ColE$_1$ form III DNA in solutions of LiCl[*]

	M, LiCl				
	0.2	2	3	4	5
$s^o_{20,w}$, S	15.4	16.2	17.2	17.3	17.0
$D^o_{20,w} \times 10^8$, cm^2/sec	2.07	2.36	2.50	2.60	2.60
R_g, nm	176	146	143	143	142
$A_2 \times 10^4$, ml mol/g^2	3.3	0.6	0.3	0.4	0.01
a, nm	44.3	30.0	29.1	29.1	29.1

[*] Solutions also contain phosphate buffer (4.8 meq Na$^+$) 2 mM NaEDTA, pH 7. 5 M LiCl solution contains only 4 mM NaEDTA, pH 7. From Ref. 15.

The close agreement of the result derived from the physico-chemical concept applied by us with the crystallographic evaluation of the first hydration shell transcends the satisfaction of having obtained a correct estimate in a controversial field, long before crystallographic analysis was feasible. More significance resides in the fact that we can use the method with trust in other instances, which have not been studied at this level of refinement by X-ray crystallography. I mention the unusual hydration properties of enzymes from extreme halophilic bacteria of the Dead Sea (25) or the probing of hydration or other solvent occupied spaces in the nucleosome (26) by the use of variable sized probes. More of this work is now in progress in our laboratory.

DNA IN THE NUCLEOSOME

The nucleosome core particle, a complex of 145 base pairs of DNA and an octamer of the four core histones, can be isolated as a basic repeating unit from a variety of chromatin samples of various origins (1,2). Along with many others we have investigated the chicken erythrocyte system. The nucleosome core particle is a flat cylinder of diameter 11 nm and height 5.7 nm (27). At 2 M NaCl the histones form an octamer (two each of the four core histones) but do not complex with DNA. Upon lowering NaCl concentration to 0.6 M the histone octamer decomposes to a H3,H4 histone tetramer and two H2A, H2B dimers, yet in the presence of 145 base pair long DNA fragments, nucleosome core particles are correctly reconstituted. The DNA coils around the histone core in 1.75, smoothly bent, superhelical turns. Thus, in this first stage of compaction, the radius of curvature of DNA is reduced to about 5 nm from a few tens of nm in solution. The protein nucleic acid complex is rather loose and bulky (26). An interesting aspect relates to the structure of the core histones which consist of a globular portion and unfolded lysine and arginine-rich tails at the N-amino terminal end of the peptide chains (28). It appears that the DNA core histone interaction is primarily with the globular portion of the core histones, as removal of the charged tails by trypsin action does not significantly weaken nucleosome stability .   It is conceivable that the tails may, in some yet unknown fashion, be implied with the higher order chromatin structure (29).

The stability of the nucleosome increases with lowering the NaCl concentration to 0.1 M. We now know that the conformational change accompanying this reduction in ionic strength is rather minor (26). Sedimentation and diffusion coefficients increase by about 10%, yet we have recently found that the radius of gyration R_g, determined by small angle X-ray scattering, is essentially unchanged over this range of salt concentrations (30). Considering that X-ray scattering is primarily influenced by the stronger scattering nucleic acid, we conclude that the DNA conformation is essentially unaffected and that the change in hydrodynamic properties may be due to loosening of the histone tails conformation at the higher ionic strength.

Additional support countering the concept of nucleosome core particle unfolding with increasing ionic strength derives from experiments in which the single sulfhydryl groups on the two H3 histones in the core particle have been crosslinked *in situ* without affecting the hydrodynamic properties over a range of conditions, as compared to the native particles (31). We also could show that the reversible dissociation of the nucleosome core particle into DNA and histones significantly depends, besides concentration of salt, on temperature and core particle concentration (31).

DNA IN CHROMATIN

The next levels of organization and of folding of DNA occur in chromatin structures. In the particular instance examined in our research, we isolate from the chicken erythrocyte genome stretches of chromatin containing between 14 and 42 chromatosomes (165 base pairs of DNA wrapped around the histone octamer in two superhelical turns, as well as the H1 and H5 histones) connected by 45 base pairs of linker DNA (32). In the intact nuclei the genome is organized by "packaging" DNA in a hierachy of ordered structures. Replication and transcription of active genes proceeds by mechanisms now being unravelled. The chromatin fragments at very low ionic strength assume open coil-like flexible structures (Table 2). Upon increase of NaCl concentration beyond 50 mM, or $MgCl_2$ concentration beyond 0.3mM, transition occurs to a rigid solenoidal structure (33). We have determined by the angular dependence of scattered light, with some information coming from small angle X-ray, scattering, the dimensions of the higher order structure. Assuming a solenoid with pitch 11 nm (the diameter of the nucleosome) we calculate (Table 3) a value close to six for the number n of nucleosome per helical turn. Values consistent with this finding have previously been evaluated from neutron scattering (34) and relaxation of electrically induced dichroism (35).

Table 2. Conformation of the "10 nm" chromatin coil at low
salt. (a) Corresponds to free DNA and (b) and (c) are two
models of chromatin coiling (32)

| Fraction | NaCl | R_g,exp. | N_z | (a) | | (b) | | (c) |
| | | | | L_z | $R_{g,2}$ | L_z | $R_{g,z}$ | $R_{g,z}$ |
	mM	nm		nm		nm		nm
2	5	42.6	22	1560	167	337	66.5	56.4
4	5	74.3	45	3180	245	689	103.6	80.6
5	5	81.7	53	3750	267	811	114.5	87.5
	1	102.2	"	"	"	"	"	"

Above solutions are in 1 mM Tris-HCl, 0.1 mM EDTA, pH 8.   N
and L are the number of nucleosomes and contour lengths of the
structures, subscript z signifies a z-average value.

Table 3. Calculation of number n of nucleosomes per
helical turn in the "30 nm" solenoid (32)

| | Fraction | | |
	2	4	5
R_g,nm (exp)[*]	16.6	27.8	33.5
$N_{z,z+1}$, nm	24.5	48	58
$L_{z,z+1}$, nm	47.6	91.7	113.0
n	5.7	5.8	5.6

[*] At 75 mM NaCl, 1 mM Tris-HCl, 0.1 mM EDTA, pH 8; n is the
number of nucleosomes per helical turn, the other symbols
as in Table 2; subscript z, z+1 signifies a z, z+1-average.

ACKNOWLEDGEMENT

    This work was supported by a grant from the United States-Israel
Binational Science Foundation, Jerusalem, Israel.

REFERENCES

1. McGhee, J.D. and Felsenfeld, G.: 1980, Ann.Rev.Biochem. 49,
   pp. 1115-1156.
2. Igo-Kemenes, T., Hörz, W. and Zachau, H.G.: 1982, Ann.Rev.Biochem.
   51, pp. 89-121.
3. Zimmerman, S.B.: 1982, Ann.Rev.Biochem. 51, pp. 395-427.
4. Anderson, C.F. and Record, M.T.Jr.: 1982, Ann.Rev.Phys.Chem. 33,
   pp. 191-222.
5. Cowman, M.R. and Fasman, G.D.: 1978, Proc.Natl.Acad.Sci.USA 75,
   pp. 4759-4763.
6. Wang, A.H.-J., Quigley, G.J., Kolpak, F.J., van der Marel, G.,
   van Boom,J.H. and Rich,A.:1981, Science 211, pp. 171-176.
7. Borochov, N., Eisenberg, H. and Kam, Z.: 1981, Biopolymers 20,
   pp. 231-235.
8. Kam, Z., Borochov, N. and Eisenberg, H.: 1981, Biopolymers 20,
   pp. 2671-2690.
9. Manning, G.S.: 1978, Q.Rev.Biophys. 11, pp. 179-246.
10. Record, M.T., Jr., Anderson, C.F. and Lohman, T.M.: 1978, Q.Rev.
    Biophys. 11, pp. 103-178.
11. Widom, J. and Baldwin, R.L.: 1979, J.Mol.Biol. 144,pp. 431-453.
12. Thomas, T.J. and Bloomfield, V.A.: 1983, Biopolymers 22, pp. 1097-
    1106.
13. Behe, M. and Felsenfeld, G.: 1981, Proc.Natl.Acad.Sci.USA 78,
    pp. 1619-1623.
14. Post, C.B.: 1983, Biopolymers 22, pp. 1087-1096.
15. Ausio, J., Borochov, N., Kam, Z., Reich, M., Seger, D. and
    Eisenberg, H.: 1983, in: Structure, Dynamics, Interactions and
    Evolution of Biological Macromolecules, C. Hélène (ed.),
    D. Reidel, Dordrecht, pp. 89-100.
16. Wolf, B., Berman, S. and Hanlon, S.: 1977, Biochemistry 16,
    pp. 3655-3662.
17. Parthasarathy, N., Schmitz, K.S. and Cowman, M.K.: 1980, Biopolymers
    19, pp. 1137-1151.
18. Borochov, N. and Eisenberg, H., to be published.
19. Zimmerman, S.B. and Pheiffer, B.H.: 1980, J.Mol.Biol. 142,
    pp. 315-330.
20. Eisenberg, H.: 1974, in: Basic Principles in Nucleic Acid Chemistry,
    P.O.P.Ts'o (ed.), Academic Press, New York, pp. 171-264.
21. Eisenberg, H.: 1976, Biological Macromolecules and Polyelectrolytes
    in Solution, Clarendon Press, Oxford.
22. Reisler, E., Haik, Y. and Eisenberg, H.: 1977, Biochemistry 16,
    pp. 197-203.
23. Cohen, G. and Eisenberg, H.: 1968, Biopolymers 6, pp. 1077-1100.
24. Kopka, M.L., Fratini, A.V., Drew, H.R. and Dickerson, R.E.: 1983,
    J .Mol.Biol. 163, pp. 129-146.
25. Pundak, S. and Eisenberg, H.: 1981, Eur.J.Biochem. 118, pp.463-470.
26. Eisenberg, H. and Felsenfeld, G.: 1981: J.Mol.Biol. 150, pp.537-555.
27. Finch, J.T., Lutter, L.C., Rhodes, D., Brown, R.S., Rushton, B.,
    Levitt, M. and Klug, A.: 1977, Nature 269, pp. 29-36.

28. Bohm, L., Hayashi, H., Cary, P.D., Moss, T., Crane-Robinson, C. and Bradbury, E.M.: 1977, Eur.J.Biochem. 77, pp. 487-493.

29. Allan, J., Harborne, N., Rau, D.C. and Gould, H.: 1982, J.Cell Biol. 93, pp. 285-297.

30. Reich, M.: 1982, Ph.D. Thesis, The Weizmann Institute of Science, Rehovot, Israel.

31. Ausio, J., Haik, Y., Seger, D. and Eisenberg, H.: 1983, in: Mobility and Recognition in Cell Biology, H. Sund and C. Veeger (eds.), de Gruyter, Berlin, pp. 195-211.

32. Ausio, J., Borochov, N., Seger, D. and Eisenberg, H., in preparation.

33. Finch, J.T. and Klug, A.: 1976, Proc.Natl.Acad.Sci.USA 73, pp. 1897-1901.

34. Suau, P., Bradbury , E.M. and Baldwin, J.P.: 1979, Eur.J.Biochem. 97, pp. 593-602.

35. McGhee, J.D., Rau, D.C., Charney, E. and Felsenfeld, G.: 1980, Cell 22, pp. 87-96.

36. McGhee, J.D., Nickol, J.M., Felsenfeld, G. and Rau, D.C.: 1983, Cell, in press.

NUCLEOSOMAL DNA STRUCTURE

E.N. Trifonov
Department of Polymer Research, The Weizmann Institute
of Science, Rehovot, Israel 76100

ABSTRACT

   The helical repeat of the DNA in the nucleosome is evaluated by
three independent methods: from periodical distribution of the dinucleotides along known nucleosomal DNA sequences, from topology of DNA superhelix in the nucleosome and from analysis of accessibility of DNA in the
nucleosome to DNase I, all giving similar results. All previous estimates
are critically reviewed and the average figure is obtained, 10.38 $\pm$ 0.02
base-pairs per turn, based on five independent estimates. Experimental
data available indicating that the free nucleosomal DNA is inherently
curved, are reviewed. A novel approach to the higher order structure of
chromatin is discussed which is based on steric exclusion effects imposing
specific restrictions on the relative spatial positioning of the neighboring nucleosomes.

INTRODUCTION

   Evidence is accumulating that the nucleosomes are connected by DNA
linkers of variable, rather than constant length. However, there are
certain preferred lengths so that the linker lengths distribution is
modulated by the period of about 10 base-pairs (1-4). This immediately
suggests some models of elementary chromatin filaments. If the linkers
are multiples of about 10 base-pairs (10·n) the nucleosomes can be
considered to form a parallel stack with DNA making a long continuous
superhelix (2). If their lengths fit to the 10·n + 5 series, an obvious
suggestion would be an antiparallel, zig-zag structure (1). In both these
cases the number of double-helical repeats of DNA in the core-particle
is assumed to be an integer and the relative orientation of two nucleosomes in space, therefore, is assumed to be dependent only on the length
of the linker between them (see, however, Ref. 3).

   The value of the helical repeat of the nucleosomal DNA is, however,
also of extreme importance for the relative orientation of neighboring
nucleosomes. Indeed, 145 base-pairs of DNA in the core-particle might

373

B. Pullman and J. Jortner (eds.), Nucleic Acids: The Vectors of Life, 373–385.
© *1983 by D. Reidel Publishing Company.*

correspond to some non-integer number of repeats, thus, bringing about
additional rotation of one nucleosome relative to another (3,5). In
order to calculate the angular orientation of the neighboring nucleosomes
the helical repeat of the nucleosomal DNA has to be known with high
accuracy. A bare 1% error in this value would result in about 50° angular
uncertainty ($0.01 \cdot 360^\circ \cdot 14$).

Therefore, one of our objectives is to get an estimate of the helical repeat of the nucleosomal DNA with the highest possible accuracy.

THE HELICAL REPEAT OF THE NUCLEOSOMAL DNA

Existing estimates

These estimates are listed in the Table 1. They are clearly clustered in two non-overlapping and, therefore, contradicting sets: about 10.0
and 10.35 base-pairs. The difference, 3.5%, is too large to ignore,
bearing in mind high sensitivity of relative orientations of the nucleosomes to the pitch of the nucleosomal DNA, as discussed above.

Method	The helical repeat	Reference
Topological considerations (SV40 minichromosome)	10.0	6,7
Energy calculations	10.0	6
DNA accessibility for DNase I	10.0	7
DNase I digestion (interclea-vage distance)	10.3 - 10.4	8
Modulation of nuclease sensi-tivity (beating effect)	10.33 - 10.40	9
DNA sequence periodicity	10.5 ± 0.2	10

Table 1. Helical repeat of the nucleosomal DNA (base-pairs
per turn).

Critical evaluation of corresponding data and arguments is one way
to decide which of these two estimates is wrong (11,12). However, additional estimates by means of some new, independent, approaches would be
desirable as well. In the following sections the new data relevant to
the problem are discussed.

Periodicity in the nucleosomal DNA sequences

Previously we have shown that the chromatin DNA sequences manifest characteristic periodicity of certain elements, with the period 10.3 - 10.7 bases (10,13). It was interpreted as reflection of the helical repeat of the DNA in the nucleosomes. One could argue, however, whether this periodicity indeed is related to the nucleosome structure. Direct analysis of the sequences involved in the nucleosomes would be of prime interest.

Several examples of very accurate ($\pm$ 5 base-pairs or less) location of the nucleosomes on DNA with known nucleotide sequences are reported (14-18). These are the nucleosomes reconstituted on the regulatory lac region of *E. coli* (14), on 5S rRNA gene of *X. laevis* and on certain fragment of pBR322 DNA (18) as well as nucleosomes mapped on natural chromatins: green monkey α-satellite chromatin (17) and rat satellite chromatin (15,16). Mere inspection of the corresponding nucleotide sequences does not allow any save conclusions about the periodicity in question. It becomes obvious, however, when the total occurence of various dinucleotides along the sequences is calculated. Dinucleotides AA (and, complementarily, TT) are clearly concentrated around some positions separated by about 10.4 bases or multiples of this. This is shown in Fig. 1, where the combined distribution of the dinucleotide AA along 9 nucleosomal DNA molecules (14-18) is presented. Considering the nucleosome core-particle as axially symmetrical structure, one can take for the summation both of the DNA complementary strands reading all of them, of course, from their 5'-ends. It makes a total of 18 sequences, all included in the diagram. The variation is somewhat masked by the noise components which are, probably, responsible for the absence of some peaks on their expected positions. The amplitude spectrum of the periodical components of the distribution was calculated (19) allowing a quite accurate estimation of the major period. The major maximum of the amplitude spectrum is shown in the Fig. 2. It is centered at 10.38 $\pm$ 0.03 bases. The error indicated was estimated by separate calculations for several subsets of the 18 sequences. The figure obtained is in agreement with the less accurate earlier estimate (10.3 - 10.7 bases) calculated for the chromatin DNA sequences without discriminating between the nucleosomal and linker DNA (10). It is noteworthy that the dinucleotides AA and TT appear to display the strongest periodical variation along the nucleosomal DNA, as was also found by the analysis of total chromatin DNA (13).

Our deformational anisotropy interpretation of the periodicity, 10.3 - 10.7 bases in the case of total chromatin, was recently disputed and an alternative explanation was proposed by Zhurkin (20). Namely, if in the α-helices of the proteins certain aminoacids would preferentially concentrate on one side of the α-helix, it will be reflected in the nucleotide sequence as about 10.8 base periodicity, close to the figure above. This effect, however, should be most pronounced in prokaryotic nucleotide sequences, where about 90% is occupied by the protein coding regions, and it should be very small for eukaryotic sequences, to which

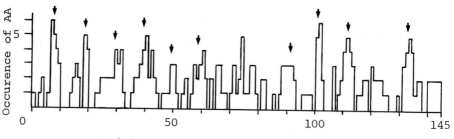

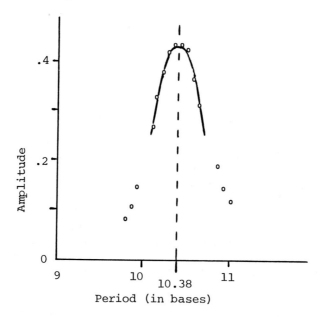

Figure 1. Distribution of the dinucleotides AA along the nucleosomal DNA sequences. Major peaks of the distribution are marked by arrows.

Figure 2. Amplitude spectrum of the AA-distribution in the range 10 - 11 bases. The smooth curve is parabolic approximation, best fit to the upmost points.

the coding regions make rather small contribution. Exactly the opposite has been actually observed (10) which substantially weakens the argument of Zhurkin (20).

DNA topology in the nucleosome

In our earlier work we suggested that the DNA in the nucleosome is smoothly wrapped around the histone octamer forming a superhelix (21,22). This was confirmed recently by neutron diffraction studies of the crystals of the nucleosome core-particles (23). The pitch of the superhelix is estimated to be 27 Å, the number of superhelical turns 1.8 (23). One could evaluate what is an additional twist around the DNA axis provided by the superhelicity of the DNA (5). For this purpose the simple formula given by Crick (24) can be used:

$$\Delta T = N \sin\alpha \cdot 360°$$

where ΔT is the total additional twist (in°), N - the number of the superhelical turns and α - ascending angle of the superhelix. Putting the known parameters of the core-particle in the formula we obtain $\Delta T = 66°$, which is equivalent to the change of the helical repeat equal to 0.14 ± 0.02 base-pairs per turn. The error indicated combines uncertainties of the superhelix parameters and of the rise per residue (3.3 ± 0.1Å).

The pitch of the unconstrained B-DNA in solution is estimated to be 10.55 ± 0.05 base-pairs per turn (25,26). If the DNA is wound in the nucleosome in the left-handed superhelix as in naturally occuring superhelical DNA, its helical repeat should then be $10.55 - 0.14 = 10.41$ (± 0.07) base-pairs. This is another independent estimate of the pitch of the nucleosomal DNA.

Correction of the Klug-Lutter estimate (7)

An interesting approach to the problem was proposed by Klug and Lutter (7). The idea is that directions of nuclease attack at any point of the DNA molecule in the nucleosome should depend on the general architecture of the nucleosome. In particular, the adjacent superhelical turn of the DNA should provide some protection. The parts of the nucleosomal DNA facing the cleft between two superhelical turns should be inaccessible for the nucleases. Similarly, the histones might provide additional limitations. The region of the permissible angles of nuclease attack can be roughly estimated as well as the pitch of DNA which would conform to these angular limitations, using experimentally determined cleavage positions (27).

Klug and Lutter compared two possible repeats of the nucleosomal DNA, 10.0 and 10.5 base-pairs per turn. It was shown that the former one is more sound in terms of the angular limitations. Unfortunately, the only known alternative for the pitch, about 10.35 base-pairs per turn (9,12) was not compared with the results.

We repeated the calculations of Klug and Lutter for the whole range of the values between 10.0 and 10.5 base-pairs (28), using the same formula (7) and the same experimental data (27). The region 10.3 - 10.4 base-pairs per turn was found to be sterically acceptable as well (Fig.3), even with more severe angular limitations than proposed by Klug and Lutter. Moreover, the pitch corresponding to the narrowest distribution of the angles of nuclease attack, thus, least interfering with the angular limitations, whatever they are, has been found equal to 10.38 ± 0.05 base-pairs per turn (28), rather than 10.0 base-pairs thought by Klug

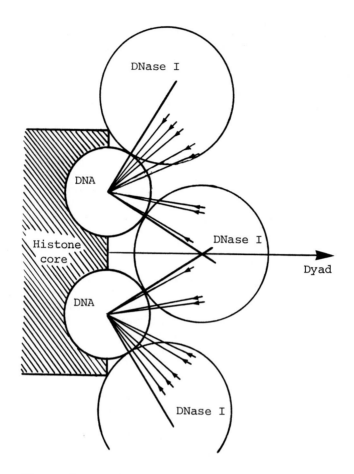

Figure 3. Angular disposition of DNase I cutting sites calculated for DNA helical repeat 10.375 base-pairs per turn. The
diagram corresponds to the cross-sections of the nucleosome
core-particle perpendicular to the DNA axes. Relative sizes of
the histone core and DNA correspond to estimates from neutron
diffraction data (23). DNA is shown half buried in the protein.

and Lutter to be the best estimate.

There are still two more estimates (Table 1) which have to be clari-
fied before we come to the already emerging conclusion about correct
value of the helical repeat of the DNA in the nucleosome. We now turn
to the

Estimate based on the topology of the SV40 minichromosome (6,7)

The SV40 minichromosome is known to contain about 25 nucleosomes
(29). On the other hand, its DNA purified from the histones has about 24
superhelical turns (30). It appears, therefore, that the nucleosome has

one superhelical turn of DNA rather than about 1.8 turns one would expect
from the known sizes of the nucleosome. To solve this paradox an additional twist of the nucleosomal DNA had to be introduced (6,7) resulting
in the value 10.0 base-pairs for the pitch of the DNA in the nucleosome.
As we have argued earlier (11), this estimate is based on the assumption
that the path of DNA in the minichromosome is ideally superhelical which
is not necessarily the case.

The ideally superhelical DNA path in the chromatin would also mean
that the linker lengths have to be multiples of about 10 base-pairs (2).
It is important, therefore, to find out whether the distribution of the
linker lengths in the SV40 minichromosome meets this expectation.

We have recently developed a sequence-directed mapping procedure
which locates the nucleosomes along the chromatin DNA nucleotide sequences (4,31). It has been successfully tested by using the experimental
mapping data available (14-18) and can be used, therefore, for quite
plausible predictions. The map for the SV40 minichromosome has been calculated (32) providing both the positions of the nucleosomes and the lengths
of the linkers. The distribution of the lengths obtained is shown in
Fig. 4. It demonstrates that the lengths multiple of the repeat (left and
right ends of the diagram) are not preferred.

Thus, the DNA path in the SV40 minichromosome does not appear to be
a simple superhelix. It will take some time before its topology will be
quantified. The numbers of the nucleosomes and superhelical turns might
then well become consistent without involving any additional twist around
the nucleosomal DNA axis. The estimate (10.0 base-pairs per turn) provided
by the topological considerations (6,7) is, therefore, based on the wrong
assumption and should not be taken into account.

Estimate from energy calculations (6)

According to these calculations the helical repeat of the free B-DNA
is 10.6 base-pairs per turn, which is close to the latest experimental
estimates (25,26) while that of the DNA smoothly bent to imitate its
structure in the nucleosome is close to 10.0 base-pairs per turn. This
latter estimate in our opinion is doubtful because of the following reasons. First, the pitch of the smoothly bent molecule was calculated for
the oversimplified model in which the influence of the histones bound to
the DNA was not taken into account. Secondly, the potential functions
used for the calculations (6) are known only as approximations, and the
computational error ($\pm$ 1%) is apparently underestimated. Qualitatively,
however, this calculation indicates that the pitch of the DNA in the
nucleosome should be smaller than for free DNA in solution which, indeed,
seems to be the case (see above, "DNA topology..." section).

On the beating effect (9)

Our first evaluation of the helical repeat of DNA in the nucleosome
was based on the experimental data on positional modulation of DNA sensi-

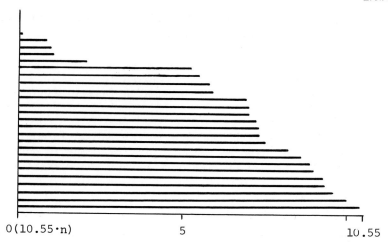

0(10.55·n)   5   10.55

Figure 4. Linker lengths distribution in the SV40 minichromosome (32). Integer numbers of repeats (10.55 base-pairs) are
subtracted from every linker and only the fractional parts
are shown.

tivity to nucleases (see references in 9). Assuming that the DNA molecule
is partially buried in the idealized cylindrical histone core, and that
its helical repeat is non-integer, we speculated that the probability of
nuclease cutting of the DNA should be modulated - the beating effect
caused by the non-integer repeat (9). The helical repeat calculated on
these assumptions is 10.33 - 10.40 base-pairs. This approach was criti-
sized (7) for unrealistically narrow angular window for the nuclease
attack, apparently required for the modulation effect to be observed.
This invites some clarification.

The angular range of accessibility, rather narrow indeed (18°) was
chosen for the purpose of illustration only. What actually matters is
the angular distance from any particular direction of the attack to the
optimal one. This optimal direction, in principle, always exists when
some anisotropy in DNA accessibility is introduced. Even if DNA is not
buried in the protein, its protection by the contacting surface from one
side should produce, theoretically, slight modulation of the cutting
probability, as soon as the helical repeat of the DNA is noninteger.

In the experiments of Rhodes and Klug with the DNA extracted from
the nucleosomes and bound to the mica surface (33), the weak modulation
one could expect was indeed documented. The probability of DNase I cutting at some sites, e.g. at sites 3, 8 and 11, was lower than for their
corresponding neighboring sites (33). According to our calculations (9)
this pattern, though not exactly symmetrical, fits best to the helical
repeats in the ranges 10.33 - 10.40 and 10.60 - 10.67 base-pairs. Both
are fairly close to the helical repeat obtained by direct counting of
the bands in the gel, in the same experiments - 10.6 ± 0.1 base-pairs
(33). More accurate measurements are required, however, to obtain more

reliable estimate of the DNA helical repeat on the mica surface, based on the orientational protection mechanism.

What we are left with

Table 2 summarizes the estimates of the helical repeat of DNA in the nucleosome including the ones obtained in this work and excluding the questionable ones as discussed above. The rough estimate (10.5 base-pairs) obtained from the previous analysis of the periodicities in the total chromatin DNA (10) is replaced by the latest, more accurate figure (10.38) for the periodicity in the nucleosomal DNA sequences.

Method	The helical repeat	Reference
DNase I digestion (inter-cleavage distance)	10.35 ± 0.05	8
Modulation of nuclease sensi-tivity (beating effect)	10.33 - 10.40	9
Periodicity in the nucleoso-mal DNA sequences	10.38 ± 0.03	19, this work
DNA accessibility for DNase I	10.38 ± 0.05	28, this work
Nucleosomal DNA topology	10.41 ± 0.07	5, this work

Table 2. Helical repeat of the nucleosomal DNA (base-pairs per turn), updated.

The average of these estimates which are obtained by five independent approaches is 10.38 (± 0.02) base-pairs per turn. It is noteworthy that this figure has to be considered as the average helical repeat of the nucleosomal DNA. There might be some local variations of the pitch along the molecule due to peculiarities of the nucleosome structure hitherto unknown, or individual sequence-dependent variations (34).

CURVED DNA

The concept of sequence-dependent deformational anisotropy and, pos-sibly, permanent slight curving of certain regions of the DNA molecule was first formulated in connection with the 10.5 or so base periodicity discovered in the chromatin DNA sequences (10,13). This idea stems from the natural assumption that certain combinations of adjacent base-pairs in the DNA molecule are not exactly parallel, thus, inclining the DNA axis in a certain direction. This inclination appears to be sequence-specific that is different for 16 possible combinations of the base-pairs, as has been illustrated by a characteristic repeating pattern of the dinucleotides found in the chromatin DNA sequence (13). The pattern can be used for location of presumed unidirectionally curved pieces of DNA, which are found to belong to the nucleosomes (4,31).

In principle, the dinucleotide periodicity observed in the nucleo-somal DNA could be of some different nature, not necessarily related to the curving of the DNA. It would be important to obtain some independent evidence in support of this hypothesis. Some experiments have been recently reported strongly indicating that, indeed, DNA molecules are curved (35,36). Results of electric linear dichroism experiments and rotational relaxation time measurements imply that DNA molecule in solu-tion has a non-linear tertiary equilibrium structure, rather than being just an isotropic, semi-rigid, molecule statistically folded in a Gaussian coil (35). Moreover, when the DNA was extracted from the nucleo-somes, the end-to-end distance for these 145 base-pair long fragments was found to be less than their contour length and the deficit observed could not be attributed to the DNA flexibility only. In other words, nucleosomal DNA appears to be inherently curved or according to Lee and Charney (35) - helically coiled. Similar evidence is provided by elec-tric dichroism studies of MboI restriction fragment of *Leishmania tarentolae* kinetoplast DNA (36). It appears to have a smaller overall size than other molecules of the same length (425 base-pairs) being pre-sumably additionally compacted by the inherent curving of the DNA axis. Interestingly, the nucleotide sequence of this anomalous fragment reveals strong periodicity of the dinucleotides AA with the period about 10.5 bases. Marini *et al.* (36) hold that the molecule is curved though in a rather irregular complex way. One more unusual property of this DNA fragment apparently related to its inherent curvature - its anomalous electrophoretic behavior. Depending on porosity of the gel, it migrates either slower or faster than other molecules of similar size.

Another dramatic example of anomalous migration of DNA in electro-phoretic gel is the Hinf restriction fragment of att P site of bacterio-phage lambda. In the gel conditions chosen this 317 base-pair fragment migrates slower as if it was of length 345 base-pairs. Deletion of only one base-pair at about 1/3 of its length does not result in some acce-leration as one would expect, but rather further slows it down as much as would be usually caused by an extra 15 base-pairs. This obviously sequence-dependent anomaly, in our opinion, can be explained only by sequence-dependent local curving of this DNA.

One could propose a direct experiment to demonstrate that the free nucleosomal DNA is curved. If the molecule is curved, then it has "sides", so that after being put down on some flat surface, it would contact with the surface by either of the sides, with equal probability. DNaseI digestion of the 5'-end labeled nucleosomal DNA bound to the sur-face should produce in this case two periodical series of fragments separated by exactly half-period (180° difference in the orientation of the molecules relative to the surface). Thus, periodicity of about 5 bases should be observed.

As a matter of fact, this experiment has already been done and the five base periodical digestion pattern was indeed observed (33), but this was ascribed by the authors to a special role which the ends of the molecule presumably play in the binding.

The free nucleosomal DNA, we think, is not necessarily an ideally curved arc-like molecule. Its axis might have rather complicated sequence-dependent trajectory, which can be considered as an arc only in the first approximation. Physically, however, it is indifferent since for the energy of smooth bending of the DNA in the nucleosome, partially saved by this curving, this first approximation arc matters most. Some small portions of the nucleosomal DNA might be curved in the wrong direction, but this has to be compensated by other parts all together making the uneven arc. If these wrongly oriented portions would be allowed to rotate freely around their axes, say, by introducing nicks in the molecules, one would expect them to reorient properly. The implication is that the distribution of nuclease cuts along any particular DNA sequence involved in the nucleosome might be rather complex due to this reorientation of some fragments during digestion. Such a complex pattern was recently observed indeed (39).

How large is the actual curvature of the nucleosomal DNA? Unfortunately, there is no answer to this question for the time being. The wedge-angles between different adjacent base-pairs have been recently estimated by energy calculations (40). It was found that due to the non-parallelness of certain combinations of the base-pairs, the DNA axis could be locally inclined toward the grooves by as much as 18°. In the perpendicular direction, the wedge-angle could approach 10° (for the AA·TT combination). These figures, however, should be considered only as order of magnitude estimates, due to uncertainties of the energy calculations.

In conclusion of this section, we believe that the unidirectonal curving of the free nucleosomal DNA is both plausible and fruitful hypothesis.

A NEW ANGLE ON THE HIGHER ORDER STRUCTURE

We now turn back to the higher order structure of the chromatin. First, let us discuss what would be a possible structure of the nucleosome dimers. In principle, the higher order structure could be reconstructed then sequentially, as soon as the relative spatial positioning is known for all neighboring nucleosomes. Only two basic dinucleosome conformations have been discussed so far: parallel and antiparallel dimers (1-3). In both cases only specific lengths of the linkers between the nucleosomes are considered, differing by integer numbers of the DNA helical repeats: "10·n" and "10·n+5" series. We do not see, however, any reasons why some intermediate orientations should not be allowed. Generally, two neighboring nucleosomes could be in many different relative orientations, all the way from the parallel to antiparallel modes, depending on the length of the linker between them. Indeed, the linker lengths obtained from the nucleosome map calculated for the SV40 DNA sequence (32) are rather uniformly distributed (Fig.4), thus, excluding both parallel and antiparallel dimers as predominant conformations. The conclusion suggests itself that all linker lengths might be possible. This cannot be true, however, since of all geometrically possible relative orientations, the ones leading to overlapping of the nucleosomes in space have to be excluded. Indeed, if e.g. the parallel dimer of the nucleosomes is taken with 10 base-pair linker between them (about 34Å), then insertion of, say, 2 base-pairs in the linker would result in about a 70° relative rotation of the nucleosomes and their penetration into each other which is, of course, impossible.

Taking known geometrical parameters of the nucleosome core particle (23) and assuming the linkers to be rigid and straight (which, of course, can be debated) we derived the sterically forbidden or at least avoided linker lengths (5): 1 to 6, 11 to 16 and 22 to 25 base-pairs, all the rest being, in principle, allowed. About the same linker lengths were found to be avoided by analyzing the calculated nucleosome maps for a large collection of the chromatin DNA sequences available (4).

As mentioned previously, the relative orientation of the nucleosomes depends as well on the pitch of the nucleosomal DNA chosen for the steric exclusion calculations. The combined estimate of the pitch given above (10.38±0.02 base-pairs per turn) is sufficiently accurate to predict the relative orientations of the nucleosomes in space with uncertainty of only about 10°.

Thus, we are in a position for the first time to make plausible computer assisted predictions on the higher order structure of any chromatin of interest, provided the nucleotide sequence of the chromatin DNA is known. One example of this kind is the green monkey α-satellite chromatin. According to our calculations (5) this is a compact, regular zig-zag structure in which every nucleosome is in tight contact with 4 other nucleosomes.

REFERENCES

1. Lohr, D., and van Holde, K.E.: 1979, Proc.Natl.Acad.Sci.USA 76, pp. 6326-6330.
2. Karpov, V.L., Bavykin, S.G., Preobrazhenskaya, O.V., Belyavsky, A.V., and Mirzabekov, A.D.: 1982, Nucl.Acids Res. 10, pp.4321-4337.
3. Strauss, F., and Prunell, A.: 1983, EMBO J. 2, pp.51-56.
4. Mengeritsky, G., and Trifonov, E.N.: submitted.
5. Ulanovsky, L., and Trifonov, E.N.: submitted.
6. Levitt, M.: 1978, Proc.Natl.Acad.Sci.USA 75, pp.640-644.
7. Klug, A., and Lutter, L.C.: 1981, Nucl.Acids Res. 9, pp.4267-4283.

8.   Prunell, A., Kornberg, R., Lutter, L., Klug, A., Levitt, M., and Crick, F.: 1979, Science 204, pp. 855-858.
9.   Trifonov, E. and Bettecken, T.: 1979, Biochemistry 18, pp.454-456.
10. Trifonov, E.N., and Sussman, J.L.: 1980, Proc.Natl.Acad.Sci.USA 77, pp. 3816-3820.
11. Trifonov, E.N.: 1981, in "Structural Aspects of Recognition and Assembly in Biol.Macromolecules" (Ed. M. Balaban, J. Sussman, W. Traub and A. Yonath, Balaban ISI, Rehovot-Philadelphia, pp. 711-720.
12. Trifonov, E.N.: 1981, in "International Cell Biology 1980-1981" (Ed. H. Schweiger, Springer-Verlag), pp. 128-138.
13. Trifonov, E.N.: 1980, Nucl.Acids Res. 8, pp. 4041-4053.
14. Chao, M.V., Gralla, J., and Martinson, G.: 1970, Biochemistry 18, pp. 1068-1074.
15. Igo-Kemenes, T., Omori, A., and Zachau, H.G.: 1980, Nucl.Acids Res. 8, pp. 5377-5390.
16. Igo-Kemenes, T., and Seligman, H.: personal communication.
17. Wu, K.C., and Varshavsky, A.: submitted.
18. Gottesfeld, J.: submitted.
19. Wartenfeld, R., Mengeritsky, G., and Trifonov, E.N.: in preparation.
20. Zhurkin, V.B.: 1981, Nucl.Acids Res. 9, pp. 1963-1971.
21. Sussman, J.L., and Trifonov, E.N.: 1978, Proc.Natl.Acad.Sci.USA 75, pp. 103-107.
22. Trifonov, E.N.: 1978, Nucl. Acids Res. 5, pp. 1371-1380.
23. Bentley, G.A., Finch, J.T., and Lewit-Bentley, A.: 1981, J.Mol.Biol. 145, pp. 771-784.
24. Crick, F.H.C.: 1976, Proc.Natl.Acad.Sci.USA 73, pp. 2639-2643.
25. Peck, L.J., and Wang, J.C.: 1981, Nature 292, pp. 375-378.
26. Strauss, F., Gaillard, C., and Prunell, A.: 1981, Eur.J.Bioch. 118, pp. 215-222.
27. Lutter, L.: 1979, Nucl.Acids Res. 6, pp. 41-56.
28. Trifonov, E.N.: submitted.
29. Müller, U., Zentgraf, H., Eicken, I., and Keller, W.: 1978, Science 201, pp. 406-415.
30. Shishido, K.: 1980, FEBS Lett. 111, pp. 333-336.
31. Trifonov, E.N.: 1983: Cold Spring Harb.Symp.Quant.Biol. 47, in press
32. Mengeritsky, G., and Trifonov, E.N.: in preparation.
33. Rhodes, D., and Klug, A.: 1980, Nature 286, pp. 573-578.
34. Kabsch, W., Sander, C., and Trifonov, E.N.: 1982, Nucl.Acids Res. 10, pp. 1097-1104.
35. Lee, C.-H., and Charney, E.: 1982, J.Mol.Biol. 161, pp. 289-304.
36. Marini, J.C., Levene, S.D., Crothers, D.M., and Englund, P.T.: 1982, Proc.Natl.Acad.Sci.USA 79, pp. 7664-7668.
37. Ross, W., Shulman, M., and Landy, A.: 1982, J.Mol.Biol. 156, pp. 505-522.
38. Ross, W., and Landy, A.: 1982, J.Mol.Biol. 156, pp. 523-529.
39. Simpson, R.T., and Stafford, D.W.: 1983, Proc.Natl.Acad.Sci.USA 80, pp. 51-55.
40. Ulyanov, N.B., Zhurkin, V.B., and Ivanov, V.I.: 1982, Stud.bioph. 87, pp. 99-100.

SEQUENCE AND SECONDARY STRUCTURE CONSERVATION IN RIBOSOMAL RNAs IN THE COURSE OF EVOLUTION

J.P. EBEL, C. BRANLANT, P. CARBON, B. EHRESMANN, C. EHRESMANN, A. KROL and P. STIEGLER
Laboratoire de Biochimie,
Institut de Biologie Moléculaire et Cellulaire,
15, rue René Descartes - 67000 STRASBOURG (FRANCE)

The ribosomal RNAs play a fundamental role in ribosome structure and function. For a long time, only the structural function was considered : they represent the backbone around which the ribosomal proteins are assembled in the course of ribosome assembly. But more recently it appeared that they may also play a functional role, as several regions within the RNAs have been found to interact directly with the macromolecules involved in protein synthesis.

Whereas the nucleotide sequence of the small E.coli 5S RNA has been established in 1968 in Sanger's group (1), the primary structures of the two large E.coli 16S and 23S RNAs were only determined ten years later between 1978 and 1980 on the corresponding genes in Noller's group (2,3) and directly on the RNAs in our own laboratory (4,5,6).
Completion of the nucleotide sequences of these RNAs has allowed investigation of the secondary structure of the nucleotide chains. It is obvious that the structural and functional role of the ribosomal RNAs can only be understood if the folding of these RNAs within the ribosome is known.
In this report, we will concentrate on two aspects : i) the determination of the secondary structure of the two large **E.coli** ribosomal RNAs, 16S and 23S RNAs ; ii) the conservation of primary and secondary structure features of these large ribosomal RNAs in the course of evolution, and its relation with ribosomal RNA function. This paper describes our recent results in these fields.

I. CRITERIA USED TO BUILD SECONDARY STRUCTURE MODELS FOR THE E.COLI RIBOSOMAL RNAs

The total number of the theoretical base pairings and their topological combinations are so immense in the case of large RNAs like 16S or 23S ribosomal RNAs that restriction must be introduced. This was achieved by integrating in computer programs all available experimental data on the topography of these RNAs within the ribosomal subunits. The criteria used to build the secondary structure models of

B. Pullman and J. Jortner (eds.), Nucleic Acids: The Vectors of Life, 387–401.
© *1983 by D. Reidel Publishing Company.*

E.coli 16S and 23S RNAs we proposed (7,8,9) were the following :

1. Accessibility to ribonucleases

- of the free 16S and 23S RNAs in solution ;
- of these RNAs within reconstituted complexes between the RNA and one or several ribosomal proteins ;
- of the RNAs within either the intact 30S or 50S subunits or partially unfolded subunits or within 70S ribosomes.

Two types of ribonucleases were used :
- single strand specific ribonucleases like T1, pancreatic or S1 ribonucleases
- a double strand specific ribonuclease extracted from **Naja oxiana** cobra venom.

These ribonucleases were very useful as secondary structure probes for an accurate mapping of residues located in either single stranded or helical regions. In addition they provided information on the most exposed RNA regions, thus giving rise to a precise picture of the RNA surface topography.

The accessibility data used for the building of our secondary structure models are referenced in (8) and (9).

2. Accessibility to chemical reagents

To complete these enzymatic data, we also used results from Noller's group on the reactivity towards kethoxal modification which modifies guanylic residues in single stranded regions (10,11). These accessibility studies were performed in either active or inactive subunits or in 70S ribosomes.

3. Experimental evidence for long range RNA-RNA interactions

Specific long range RNA-RNA interactions could be detected by mild ribonuclease hydrolysis of ribosomal RNA-protein complexes. This was achieved by isolation of the digested complexes under non-denaturing conditions preserving the RNA-RNA interactions, followed by a deproteinization of the complex and by a selective dissociation of the interacting RNA fragments (12). The analysis of these fragments showed the presence of complementary sequences widely separated in the primary structure, which were good candidates for RNA-RNA long distance interactions.

4. Comparative sequence analysis

Several complete small and large ribosomal subunit RNA sequences of prokaryotic, eukaryotic or organellar ribosomes are now available (see next paragraph), as well as numerous partial sequences (listed in (8), (9)). They were used for comparative sequence analysis.

Examination of these sequences showed that a large number of secondary structure elements proposed for the secondary structure of·

E.coli 16S and 23S RNAs are conserved in the corresponding RNAs of the other species : the replacement of one base in one strand of a helical region is compensated by the presence of a complementary base in the opposite strand, the secondary structure being thus maintained. The finding of such compensatory base changes in the structure of other ribosomal RNAs can be considered as a strong evidence for the reality of many helical structures in the E.coli rRNAs.

5. Computer assistance

Computer analysis was used to determine the base paired regions, taking into account the classical thermodynamic data. The total number of potential base pairings was restricted by integration of the accessibility data into the computer program. In addition, computer assistance was used to characterize in the various ribosomal RNAs sequence homologies and secondary structure conservation through compensatory base changes. The programs used are described in (8) and (9).

II. THE SECONDARY STRUCTURE OF 16S RIBOSOMAL RNA

Figure 1 shows the secondary structure we propose for the 1542 nucleotides long E.coli 16S rRNA. All enzymatic and chemical accessibility data have been fully described elsewhere (8). The base paired regions which are supported by phylogenetic evidence, e.g. conserved through compensated base changes, are indicated by bars. It must be noted that other helices are preserved by strict nucleotide sequence conservation. The secondary structure is characterized by the following features :
- almost 50 per cent of the nucleotides are base paired, thus forming 60 double helical regions (numbered 1 to 37 and A,B,C, in figure 1). 49 of these helices are conserved in other small subunit rRNA molecules through coordinate base changes.
- the folding of the 16S RNA chain is mainly directed by 8 long range interactions, e.g. base pairing between sequences widely separated in the primary structure. The following helices represent such long range base pairings : A, 1, 2, 15, 22, 23, 24 and 26. They have a structural key-role since they delineate 4 distinct structural domains, denoted I to IV in Fig. 1. These domains have also been defined by topographical studies (8).

The 16S rRNA folding presented in Fig. 1 has been up-dated by using new phylogenetic evidence from recently available rRNA sequences (see section IV). This led to the detection of new base-paired regions denoted A, B and C in Fig. 1 and to the extension of helix 2. The existence of these helical regions is mainly supported by compensatory base changes occurring in the chloroplastic rRNA sequences. Only a few secondary structure motifs are still tentative : part of helix 4b, helices 6, 21a, 28b, 29 and 30. In these motifs, insufficiency of experimental evidence or even contradictory sets of data do not allow any definitive conclusion. Some other base pairing possibilities may

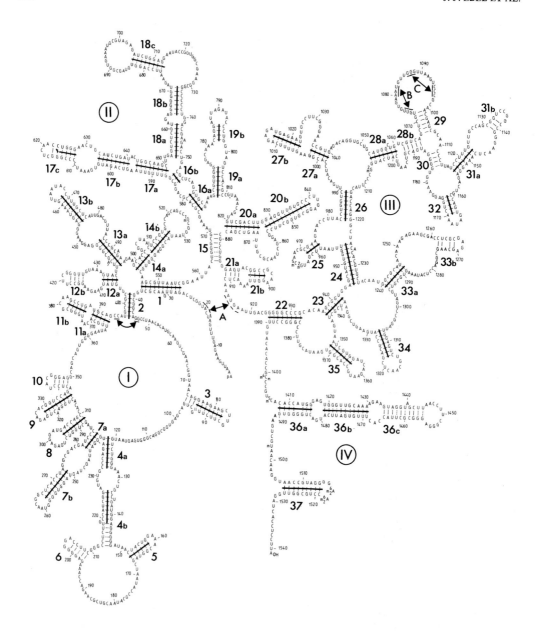

Figure 1 : SECONDARY STRUCTURE OF THE E.COLI 16S RNA

All accessibility data and the characteristic features of this folding have been fully described elsewhere (8,40). Helices supported by phylogenetic evidence are indicated by heavy lines. The various structural motifs re numbered for easy reference. New base pairings (A,B,C and extension of helix 2) are included in this up-dated scheme.

also exist between RNA regions that are left single stranded but here too, experimental evidence is not yet available for their detection.

The secondary structure we propose for the **E.coli** 16S rRNA (8) is in close agreement with the folding suggested by Woese et al (14). It also shares large similarities with a revised version of the folding proposed independently by Glotz and Brimacombe (15,16). However, several differences essentially based on accessibility and phylogenetic data can be noted between our folding and the two other ones (8). This may be accounted for by possible conformation changes, as already suggested by Herr et al (10) and Brimacombe (16).

III. THE SECONDARY STRUCTURE OF 23S RIBOSOMAL RNA

Figure 2 shows the characteristics of the secondary structure model (in a schematic representation) we propose for the 2902 nucleotides long **E.coli** 23S RNA (9). This model has been built using the same criteria as for 16S RNA. The various single and double stranded regions of 23S RNA are supported by accessibility data to various ribonucleases and to kethoxal. For a large number of helices phylogenetic evidence is brought. The folding of 23S RNA is, as for 16S RNA, directed by several long range RNA-RNA interactions, which delineate 7 distinct structural domains. Interestingly the 5'- and 3'-extremities (1-8/2984-2902) are base paired.

Here too, a limited number of secondary structure motifs are merely tentative. This is particularly true for domain IV in the central region of 23S RNA, where several alternative base pairings are possible.

Recently two other secondary structure foldings have been proposed (17,18). There is a large agreement between these two foldings and ours. However Noller's structure (18) displays some differences in domain IV and in particular an additional long distance interaction between a sequence of domain IV and one of domain II. Brimacombe's folding (17) has taken into account information brought by ultraviolet-irradiation-induced intramolecular RNA-RNA crosslinks. This folding differs from ours by small differences in the base pairing in all the domains.

IV. SEQUENCE AND SECONDARY STRUCTURE CONSERVATION BETWEEN E.COLI 16S AND 23S AND THE RIBOSOMAL RNAs FROM DIFFERENT SOURCES : EVIDENCE FOR A COMMON STRUCTURAL ORGANIZATION

We have investigated the extent of both nucleotide sequence and secondary structure conservation which might exist between the **E.coli** 16S and 23S RNAs and the following complete nucleotide sequences of small and large ribosomal subunit RNAs, covering diverse types of prokaryotes, eukaryotes and organelles :

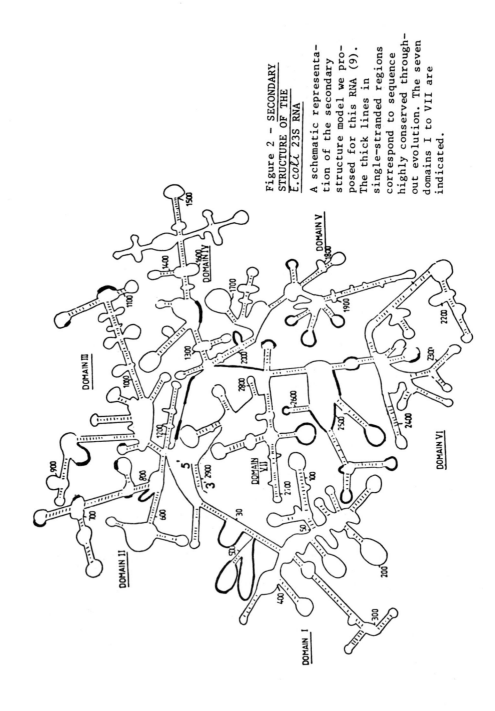

Figure 2 – SECONDARY STRUCTURE OF THE *E.coli* 23S RNA

A schematic representation of the secondary structure model we proposed for this RNA (9). The thick lines in single-stranded regions correspond to sequence highly conserved throughout evolution. The seven domains I to VII are indicated.

- small subunit rRNA sequences :
  * prokaryotic RNAs :
      Proteus vulgaris 16S RNA (19)
  * eukaryotic cytoplasmic RNAs :
      Yeast 18S RNA (20,21)
      Aspergillus nidulans 15S RNA (30)
      Xenopus laevis 18S RNA (22)
  * chloroplastic RNAs :
      Zea mays 16S RNA (23)
      Tobacco 16S RNA (24)
      Euglena gracilis 16S RNA (25)
      Chlamydomonas reinhardii 16S RNA (26)
  * mitochondrial RNAs :
      Yeast 15S RNA (27)
      Mouse 12S RNA (28)
      Rat 12S RNA (31)
      Bovine 12S RNA (32)
      Human placenta 12S RNA (29)

- large subunit rRNA sequences :
  * eukaryotic cytoplasmic RNAs :
      Yeast 26S RNA (33)
  * chloroplastic RNAs :
      Zea mays 23S RNA (34)
      Tobacco 23S RNA (35)
  * mitochondrial RNAs ;
      Yeast 21S RNA (36)
      Aspergillus nidulans 21S RNA (37)
      Paramecium primaurelia 21S RNA (38)
      Mouse 16S RNA (28)
      Rat 16S RNA (39)
      Human placenta 16S RNA (29)

All the RNA molecules examined could be folded into secondary structure schemes that brought to light a remarkable preservation of many structural motifs as well as strong sequence conservations when compared to the **E.coli** molecule (9,40).

The homology is particularly striking in the case of the prokaryotic **Proteus vulgaris** 16S RNA, where 93 per cent of sequence homology is found compared to the **E.coli** 16S RNA. Remarkable homology is also found between the chloroplastic 16S and 23S RNAs and the prokaryotic **E.coli** 16S and 23S RNAs. In this case, the fact that nearly superimposable secondary structures can be deduced for the chloroplastic rRNAs provides an even more convincing proof for the prokaryotic nature of chloroplastic ribosomes than their 70-85 per cent sequence homology. In the case of the eukaryotic Yeast and **Xenopus laevis** rRNAs, despite their larger size and an overall lower sequence homology, similar secondary structure schemes can be drawn, where many motifs are conserved. The mitochondrial ribosomal RNAs were

of particular interest since great variability is observed in both
their size and base composition : the mammalian mitochondrial rRNAs
have significant shorter chain length than their bacterial
counterpart, and a rather low G + C content. However, here too,
remarkable sequence homology and secondary structure conservation were
observed (for a detailed study see (9) and (40)).

Another interesting observation is that in all the RNAs studied
the homologous regions are interrupted by domains varying in both
length and nucleotide sequence. Furthermore the various folding
schemes proposed highlight a striking constancy in size and secondary
structure of several domains. This is illustrated in both the small
and large subunit ribosomal RNAs.

1. Sequence and secondary structure in small subunit in small ribosomal RNAs

Figure 3 presents a further refinement of the common basic
organization of the small subunit rRNAs first described in reference
40. This scheme shows the base paired elements which appear to be
common to all the ribosomal RNA molecules studied (see above). It can
be seen that the principal long distance RNA-RNA interactions which
delineate, in the **E.coli** 16S RNA, the four well defined domains are
conserved, suggesting a common basic organization. But there are also
variable domains (A-G) where extensive divergence both in base
composition and in chain length is observed. These variable domains
are all located in finite areas in each structural domain.

The RNA sequences analyzed here not only exhibit a common basic
structural organization but also display single nucleotide residues or
sequences which are strictly conserved at equivalent positions. These
residues are good candidates for being invariant nucleotides. It must
be emphasized that two (out of the 6) mitochondrial RNA sequences
diverge greatly from the **E.coli** RNA sequence, as far as potential
invariant or semi-invariant nucleotides are concerned. Therefore, if
the yeast and **Aspergillus** mitochondrial 15S RNA sequences are not
considered for the search of potential invariants, the pattern of
nucleotide residues or sequences which are conserved at equivalent
positions is even more impressive than the one presented in Fig. 3.

In the common structural scheme most of the conserved nucleotides
(85%) appear to be base paired. They are distributed in each
structural domain but are particularly clustered in the single
stranded sequences flanking motif 36 near the 3'-extremity in domain
IV. They are also concentrated in the 3'-part of domain II in the
center of the molecule. Invariant nucleotides may be involved in
tertiary structure interactions as they are in tRNA molecules. Another
possible role could be their active participation in ribosome
function. Indeed, there is strong evidence that both the central and
3'-terminal domains contain sequences that are exposed on the surface
of the subunit and located at the interface of the two subunits in the
E.coli ribosome (9,41,42), some of which making contact with the large

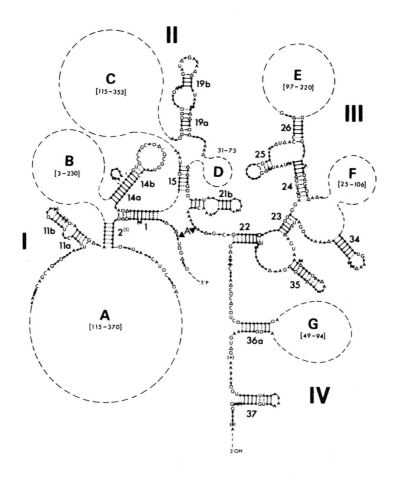

Figure 3 : COMMON BASIC STRUCTURAL ORGANIZATION OF THE SMALL
          RIBOSOMAL SUBUNITS RNAs.

Only secondary structure motifs which are common to all the RNAs
studied are shown. The helices are numbered according to Fig. 1.
Nucleotides are symbolized by filled circles and base pairs by bars.
Dashed bars indicate that base pairing between the relevant
nucleotides is not possible in every case. The four structural
domains (I-IV) are shown. Variable domains (A-G) are symbolized by a
dashed line, their size in nucleotide residues varying between the
two extreme values indicated in parentheses. Invariant nucleotides,
found in homologous positions are symbolized (□) for purines and
(△) for pyrimidines. In some cases, a single residue was inserted or
deleted and these are indicated in parentheses.

subunit (10). Furthermore, the highly conserved region in the 3'-domain of the **E.coli** 16S RNA and yeast 18S RNA between motifs 22 and 36a appears to be in contact with the anticodon of the tRNA when it is bound at the ribosomal P site as demonstrated by UV crosslinking studies (43).

2. Sequence and secondary structure conservation in large subunit ribosomal RNAs

Figure 4 represents the secondary structure elements conserved in all large rRNAs studied. The conserved nucleotide sequences are also indicated. As is the case for the RNA of the small subunit, the long distance RNA-RNA interactions which delineate the various domains are conserved.

In the model proposed by Noller et al (18), domains II and III are considered as a single domain since they are closed by a base-pairing between the sequence 579-585 and 1255-1261. Whereas the latter sequence is single stranded in our model, the former is base paired with the sequence 805-812 and delineates domain II. Interestingly, the two alternative base pairings are conserved throughout evolution, so that they might correspond to two different conformational states of the large rRNA.

It should be noticed that in the chloroplast ribosomes from higher plants, the counterpart of the 3'-terminal region of bacterial 23S RNA is a small additional RNA species : the 4.5S rRNA (44). In the same way, the cytoplasmic 5.8S rRNA is the counterpart of the 5'-terminal region of bacterial 23S RNA (Fig. 5)(33). The genes for 5.8S and 26S rRNAs are separated by a stretch of DNA which is transcribed but eliminated during the maturation process, leading to two separated species 5.8S and 26S rRNAs. Such an observation suggests that in the course of evolution an insertion has occurred in the gene coding for the large rRNA. The inserted DNA is transcribed but eliminated by processing enzymes. The same explanation may be given for the existence of the chloroplastic 4.5S RNA.

Several conclusions can be drawn from the comparison of the secondary structure of the large rRNAs studied up to now. Domains I, IV and VII and part of domain II are poorly evolutionarily conserved at the level of both primary and secondary structures. These domains are extremely reduced in size in mammalian mitochondrial rRNAs, whereas they contain large insertions in yeast cytoplasmic rRNA. The three other domains and especially domains V and VI are highly conserved throughout evolution. In these two domains long single stranded structures have very conserved primary structures.

Interestingly, the variable domains I, part of II, IV and VII correspond to regions we found to be very resistant to ribonuclease attack within the 50S subunit. On the contrary the conserved domains III, V and VI, correspond to regions highly accessible at the surface

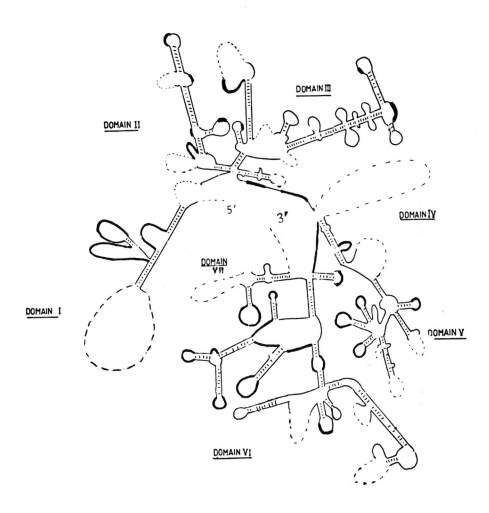

Figure 4 : CONSERVED SECONDARY STRUCTURE MOTIFS IN THE LARGE RIBOSOMAL RNA

A schematic representation of the secondary structure motifs which are present in all the large rRNAs we have studied. The broken lines correspond to variable regions. The thick lines to single stranded segments whose primary structure is highly conserved throughout evolution.

of the subunit. Furthermore, almost all the post-transcriptionally modified nucleotides are located in the conserved domains (9). These observations strongly suggest that the variable domains have essentially a structural role whereas the conserved ones are implicated in the ribosome function.

A functional role of domain V and VI is strongly suggested by other experimental data : i) domain VI contains the binding site for protein L1 (45) and a segment found associated with 5S RNA in the 50S subunit (9) ; both protein L1 and 5S RNA have been suggested to be involved in the binding of tRNA to ribosomes (46,47) ; ii) studies of mitochondrial mutants resistant to chloramphenicol or erythromycin revealed that single-point mutations of nucleotides in domain VI (2447, 2504 and 2058) are sufficient to induce resistance to the antibiotic (48,49) ; iii) a puromycin derivative bound to the ribosomal A site has been crosslinked to 23S RNA and the crosslinked 23S RNA T1 digestion product is likely the sequence 2553-2557 (50). Altogether these observations strongly suggest that a least part of domain VI belongs to the ribosomal A site. Interestingly all intron sequences observed up to now in ribosomal rDNA lie within the sequence coding either for domains V or VI (9). Finally, a hint for a functional role of domain III is given by the fact that binding of thiostrepton to the segment (1052-1112) in this domain completely inhibits ribosome function (51).

CONCLUSION

The most striking conclusion of this study is the remarkable preservation of many primary and secondary structural motifs in ribosomal RNAs in the course of evolution. Therefore extensive phylogenetic variability of the ribosomal RNAs appears to be permitted, but seems to be restricted to distinct parts of the molecule. Whereas the conserved structural features seem to be related to important structural and functional properties, the variable domains might correspond to the intrinsic peculiarities of each species. In this way, the chloroplastic RNAs appear to be more closely related to their bacterial counterpart than the mitochondrial RNAs. The homology between **E.coli** and yeast cytoplasmic ribosomal RNAs is even greater than that of **E.coli** with the various mitochondrial RNAs.

9S and 12S kinetoplast RNAs from **Trypanosoma brucei**, which have some characteristics of ribosomal RNAs, have been described (52). In these RNAs only a few sequence conservations and possible secondary structure homologies can be detected. But the unusual reduction in size and the complete lack of certain highly conserved domains in ribosomal RNAs bring some doubt about the fact that these are functional ribosomal RNAs.

Phylogenetic variability may also be related with changes in the mechanism of protein synthesis. For instance, as far as the mechanism of initiation is concerned, both cytoplasmic and mitochondrial small subunit RNAs lack the mRNA binding Shine and Dalgarno sequence (53) which is found near the 3'-extremity of both bacterial and chloroplastic RNAs.

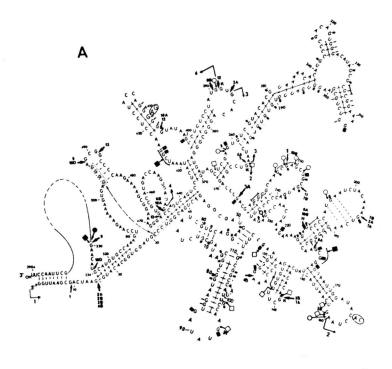

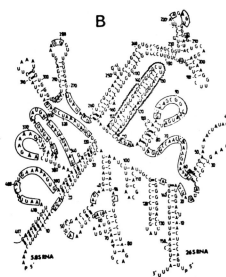

Figure 5 - DOMAIN I OF *E.coli* 23S RNA (A) AND ITS COUNTERPART IN YEAST CYTOPLASMIC RIBOSOMES (B).

In yeast cytoplasmic ribosomes domain I is constituted by 5.8S RNA and the 417 nucleotides at the 5'-end of 26S RNA. Nucleotides in single-stranded regions which are conserved as compared to *E.coli* 23S RNA are boxed with full lines.
Conserved base pairs are indicated by thick bars, and additional sequences in yeast RNA which are absent in *E.coli* RNA are denoted by ⌇⌇⌇⌇ .

REFERENCES

1. Brownlee, G.G., Sanger, F. and Barrel, B.G. (1968) J. Mol. Biol. 34, 379-412.
2. Brosius, J., Palmer, M.L., Kennedy, P.J. and Noller, H.F. (1978) Proc. Natl. Acad. Sci USA 75, 4801-48805.
3. Brosius, J., Dull, T.J. and Noller, H.F. (1980) Proc. Natl. Acad. Sci. USA 77, 201-204.
4. Carbon, P., Ehresmann, C., Ehresmann, B. and Ebel, J.P. (1978) FEBS Lett. 94, 152-156.
5. Carbon, P., Ehresmann, C., Ehresmann, B. and Ebel, J.P. (1979) Eur. J. Biochem. 100, 399-410.
6. Branlant, C., Krol, A., Machatt, M.A. and Ebel, J.P. (1979) FEBS Lett. 197, 177-181.
7. Stiegler, P., Carbon, P. Zuker, M., Ebel, J.P. and Ehresmann, C. (1980) C.R. Acad. Sci. Paris 291, 937-940.
8. Stiegler, P., Carbon, P. Zuker, M., Ebel, J.P. and Ehresmann, C. (1981) Nucleic Acids Res. 9, 2153-2172.
9. Branlant, C., Krol, A., Machatt, M.A., Pouyet, J., Ebel, J.P., Edwards, K. and Kossel, H. (1981) Nucleic Acids Res. 9, 4303-4324.
10. Herr, W., Chapman, N.M. and Noller, H.F. (1979) J. Mol. Biol. 130, 433-449.
11. Herr, W. and Noller, H. (1979) Cell 18, 55-60.
12. Ehresmann, C. Stiegler, P., Carbon, P., Ungewickell, E. and Garrett, R.A. (1980) Eur. J. Biochem. 103, 439-446.
13. Woese, C.R., Magrum, L.J., Gupta, R., Siegel, R.B., Stahl, D.A., Kop, J., Crawford, N., Brosius, J., Guttel, R., Hogan, J.J. and Noller, H.F. (1980) Nucleic Acids, Res. 8, 2275-2293.
14. Noller, H.F. and Woese, C.R. (1981) Science 212, 403-411.
15. Glotz, C. and Brimacombe, R. (1980) Nucleic Acids Res. 8, 2377-2395
16. Brimacombe, R. (1980) Biochem. Int. 1, 162-171
17. Glotz, C., Zwieb, C., Brimacombe, R., Edwards, K. and Kossel, H. (1981) Nucleic Acids Res. 9, 3287-3306.
18. Noller, H., Kop, J., Wheaton, V., Brosius, J., Gutell, R., Kopylov, A., Dohme, F., Herr, W., Stahl, D., Gupta, R. and Woese, C. (1981) Nucleic Acids Res. 9, 6167-6189.
19. Carbon, P., Ebel, J.P. and Ehresmann, C. (1981) Nucleic Acids Res. 9, 2325-2333.
20. Rubtsov, P.M., Musakhanov, M.M., Zakharyev, V.M., Krayev, A.S., Skryabin, K.G. and Bayev, A.A. (1980) Nucleic Acids Res. 8, 5779- 5794.
21. Mankin, A.S., Kopylov, A.M. and Bogdanov, A.A. (1981) FEBS Lett. 134, 11-14.
22. Salim, M. and Maden, E.H. (1981) Nature 291, 205-208.
23. Schwarz, Zs. and Kossel, H. (1980) Nature 283, 739-742.
24. Tohdoh, N. and Sugiara, M. (1982) Gene, 17, 213-218.
25. Graf, L., Roux, E., Stutz, E. and Kossel, H. (1982) Nucleic Acids Res. 10, 6369-6381.
26. Dron, M., Rahire, M. and Rochaix, J.D. (1982) Nucleic Acids Res.

10, 7609-7620

27. Sor, F. and Fukuhara, H. (1980) C.R. Acad. Sci. Paris, Ser. D. 291, 933-936.

28. Van Etten, R.A., Walberg, M.W. and Clayton, D.A. (1980) Cell 22, 157-170

29. Eperon, I.L., Anderson, S. and Nierlich, D.P. (1980) Nature 286, 460-467.

30. Kochel, H.G. and Kuntzel, H. (1981) Nucleic Acids Res. 9, 5689-5696.

31. Kobayashi, M., Seki, T., Yaginuma, K. and Koike, K. (1981) Gene 16, 297-307

32. Anderson, S., De Bruijn, M.H.L., Coulson, A.R., Eperon, I.C., Sanger, F. and Young, I.G. (1982) J. Mol. Biol. 156, 683-717.

33. Veldman, G.M., Klootwijk, J., de Regt, C.H.F., Planta, R.J., Branlant, C., Krol, A. and Ebel, J.P. (1981) Nucleic Acids Res., 6935-6952.

34. Edwards, K. and Kossel, H. (1981) Nucleic Acids Res. 9, 2853-2869.

35. Takaiwa, F. and Sugiura, M., (1982) Eur. J. Biochem., 124, 13-19.

36. Sor, F. and Fukuhara, H. (1983) Nucleic Acids Res. 11, 339-348.

37. Netzker, R., Kochel, H., Basak, N. and Kuntzel, H. (1981) Nucleic Acids Res. 10, 4784-4801.

38. Seilhamer, J. and Cummings, D. (1981) Nucleic Acids Res. 9, 6391- 6406.

39. Saccone, C., Cantatou, P., Gadeletta, G., Gallerani, R., Lanave, C. and Pepe, G. (1981) Nucleic Acids Res. 9, 4139-4148.

40. Stiegler, P., Carbon, P, Ebel, J.P. and Ehresmann, C. (1981) Eur. J. Biochem. 120, 487-495.

41. Chapman, N.M. and Noller, H.F. (1977) J. Mol. Biol. 109, 131-149.

42. Vassilenko, S.K., Carbon, P., Ebel, J.P. and Ehresmann, C. (1981) J. Mol. Biol. 152, 699-721.

43. Ofengand, J., Liou, R., Kohut, J., Schwartz, R. and Zimmermann, R.A. (1979) Biochemistry, 18, 4322-4332.

44. Machatt, M., Ebel, J.P. and Branlant, C. (1981) Nucleic Acids Res. 9, 1533-1549.

45. Branlant, C., Sriwidada, J., Krol, A., Ebel, J.P., Sloof, P., and Garrett, R. (1975) FEBS Lett. 52, 195-201.

46. Kazemi, M. (1975) Eur. J. Biochem. 58, 501-510.

47. Erdmann, V.A. (1976) Prog. Nucleic Acids Res. Mol. Biol. 18, 45-90.

48. Dujon, B. (1980) Cell, 185-197.

49. Sor, F. and Fukuhara, H. (1982) Nucleic Acids Res. 10, 6571-6577

50. Greenwell, P., Harris, R. nd Symons, R. (1974) Eur. J. Biochem. 49, 539-554.

51. Thomson, J., Schmidt, F. and Cundliffe, E. J. Biol. Chem. (1982) 257, 7915-7917.

52. Eperon, I.C., Janssen, J.W.G., Hoeijmakers, J.H.G. and Borst, P. (1983) Nucleic Acids Res. 11, 105-125.

53. Shine, J. and Dalgarno, L. (1974) Proc. Natl. Acad. Sci. USA 71, 1342-1346.

THE VERSATILE TRANSFER RNA MOLECULE : CRYSTALLOGRAPHY OF YEAST tRNA^{Asp}

D. Moras, A.C. Dock, P. Dumas, E. Westhof, P. Romby and R. Giegé
Institut de Biologie Moléculaire et Cellulaire du CNRS
15, rue René Descartes, 67084 Strasbourg Cedex (France)

ABSTRACT

The molecular structure of yeast tRNA^{Asp}, a short extra-loop tRNA, was solved at 3 Å resolution in two closely related crystal forms. The resulting model confirms the folding originally found in tRNA^{Phe} : major differences concern the conformations of the loops and the relative positioning of the acceptor and anticodon stems which are more open, conferring to the tRNA^{Asp} molecule its boomerang like shape. Crystal packing involves self-complementary GUC anticodon interactions thus making the structure a tempting model of a tRNA interacting with mRNA on the ribosome. The chemical stability of the tRNA in the crystalline state and in solution is compared using end-labelling and rapid sequencing gel methodologies. Partial splitting of the ribose-phosphate backbone in the anticodon loop was observed in solution but not in the crystal where the splitting is more pronounced in the D-loop.

1. FUNCTIONAL STATES OF tRNAs

Among the various functions of transfer ribonucleic acids (tRNAs) in living cells, their best undertood role is their participation to ribosome-mediated protein synthesis. In that process their function is to carry aminoacids to the ribosomes, to decode the messenger RNAs and help to incorporate the correct aminoacid into the protein sequence. These functions lead tRNAs to many interactions with different proteins and nucleic acids. With aminoacyl-tRNA synthetases, enzymes which attach the correct aminoacid to the 3'-end of their cognate tRNAs, the molecular recognition must be highly specific. The same is true for the decoding of the genetic code at the messenger RNA level. However, with the elongation factor, which carries the aminoacylated tRNAs to the ribosome and also with the factors allowing the peptide bond synthesis on the ribosome, the common partners imply common features. From the example of protein synthesis alone it is clear that these adaptor molecules must adapt themselves to a large variety of situations. One

403

B. Pullman and J. Jortner (eds.), Nucleic Acids: The Vectors of Life, 403–414.
© 1983 by D. Reidel Publishing Company.

way to solve the problem of adaptability is to vary the three dimensional structure via conformational changes induced and/or controlled by the reaction. This conformational flexibility is in agreement with our present knowledge of the protein-synthesizing machinery (1-3). Many experimental evidences, although indirect, have suggested the existence of conformational changes in solution (4-7).

Our three-dimensional structural knowledge on tRNAs was up to now essentially based on the pioneering work on yeast tRNA^Phe which brought a general frame for all tRNAs, the so-called L-shaped structure (8-11). But specific questions raised by versatile functioning of these molecule were not answered. The structural studies on yeast tRNA^Asp bring some light on these points.

2. PRIMARY STRUCTURE AND CRYSTALLISATION OF tRNA^Asp

tRNA^Asp was extracted from brewer's yeast Saccharomyces cerevisiae by countercurrent distribution followed by chromatography on benzoylated DEAE-cellulose (12). Its nucleotide sequence, shown on **Figure 1**, together with that of tRNA^Phe presents some features of importance for our purpose (13,14). For instance, it contains a high

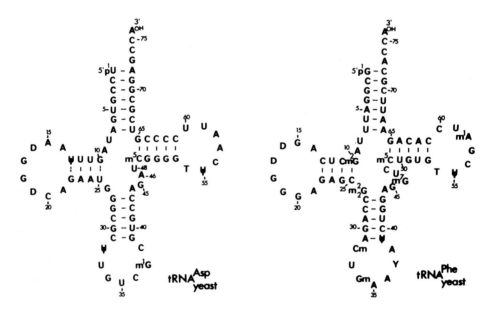

Figure 1 : The nucleotide sequence of yeast tRNA^Asp and tRNA^Phe. For convenience the numbering system of the nucleotides is that of yeast tRNA^Phe ; in the 75-nucleotide long tRNA^Asp, position 47 in the variable loop has been omitted. Non-classical Watson-Crick base pairs are indicated by broken lines.

number of G-C base pairs, except in the D-stem where two G-U(Ψ) base
pairs are present. The variable loop is made of four nucleotides versus
five in tRNA^Phe (for convenience of comparisons we kept the same
numbering, assuming a deletion at position 47). The D-loop has the same
length as that of tRNA^Phe but the two conserved G, which are crucial
for D-T loop tertiary interactions, are at positions 17 and 18 instead
of 18 and 19, thus making and ß regions of the D loop quite
symmetrical. Last but not least, the anticodon GUC presents the
peculiarity to be self-complementary, with a slight mismatch at the
uridine position. This feature was first noted by Grosjean and al. (15)
who showed the existence of a significant interaction in solution and
suggested it to be a tempting model to study tRNA-mRNA recognition.

     Crystals of tRNA^Asp were obtained using vapour diffusion
techniques (16). Like most tRNAs, tRNA^Asp displays a high
crystallographic polymorphism. The best ordered crystals are obtained
using ammonium sulfate (61 % of saturation) as the precipitant agent.
These crystals are suitable for high resolution structure determination
the best of them giving significant diffraction spots to 2.5 Å
resolution, whereas so far crystals grown in low ionic strength and/or
in presence of alcohol appear to be less ordered. In fact two
interconvertible and non isomorphous forms are obtained. The transition
between forms A and B is temperature dependent but it can also be
induced around 20% by pH changes or the addition of some heavy atoms
derivatives. It was actually almost impossible to selectively produce
one form or the other so that two independent structure determinations
had to be carried out.

3. CRYSTAL STRUCTURE OF YEAST tRNA^Asp

     Two crystal structures of tRNA^Asp have been solved by multiple
isomorphous replacement (MIR) X-ray analysis (17). In the structure
determination the heavy atom markers were gadolinium and samarium used
as sulfate salts, gold (Au(en)$_2$Cl$_3$) and mercury (HgCl$_2$ or ClHgC$_6$H$_4$NH$_2$)
which is poorly isomorphous and sits on a two-fold axis between the two
uridines of the related anticodons. Lanthanides bind to the same unique
site in the sharp turn between residues 8 and 12. This site is similar
to the one found in tRNA^Phe but it can be already stressed that the
other lanthanide binding sites present in tRNA^Phe in the anticodon and
D-loop are not present in tRNA^Asp. This fact was a first indication of
at least subtle conformational differences between the two molecules
in these loop regions.

     The ribose-phosphate backbone of the molecule could be traced from
a 4.5 Å resolution electron density map and the first Kendrew model was
build using a 3.5 Å resolution map. A diagrammatic representation of
the backbone is given in **Figure 2**. The overall folding of the chain is
similar to that found in tRNA^Phe, where it gives the L-shaped
structure. In the case of tRNA^Asp the two branches forming the L are
open by more than 10° conferring to this tRNA a boomerang-like shape.

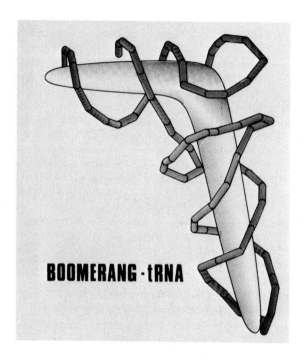

BOOMERANG · tRNA

Figure 2 : Diagrammatic representation of the ribose-backbone
of tRNA^Asp showing the boomerang-like shape of the molecule.

        The structure of one form, the lower temperature one, has
been refined in reciprocal space using the restrained least-squares
method of Hendrickson and Konnert and in real space with the graphic
modelling program FRODO. Both programs were adapted for nucleic acid
handling (18,19). The refinement was done restraining the covalent bond
lengths and bond angles,   the planarity of the bases, the chirality of
the sugar atoms,   the sugar puckers as well as the Van der Waals
contacts and hydrogen bond distances (intra and inter-molecular).   The
folding of the polynucleotide chain is then mainly achieved through
correlated rotations around the P05' and C4'-C5' bonds, maintaining the
C5'-05' rotation trans.   The other phosphodiester bond P-03' rarely
deviates from a gauche minus orientation (300°),   with the C3'-03'
torsion around 210°.   Extension of the chain or a sharp turn is
sometimes accomplished through a change of pucker from C3'-endo to
C2'-endo, with the accompanying changes in the glycosyl and C3-03'
torsion angles.

        Several structural aspects have been precised like the presence of
tertiary interactions or GU and GΨ base pairs ;   in particular the
stability of the D stem which is built up with GU, GΨ and two AU pairs
is clearly established. All these structural features also exist in
solution as shown by NMR studies (20). To summarize, two views of both
tRNA^Asp and tRNA^Phe, side by side, are displayed on Figure 3. These

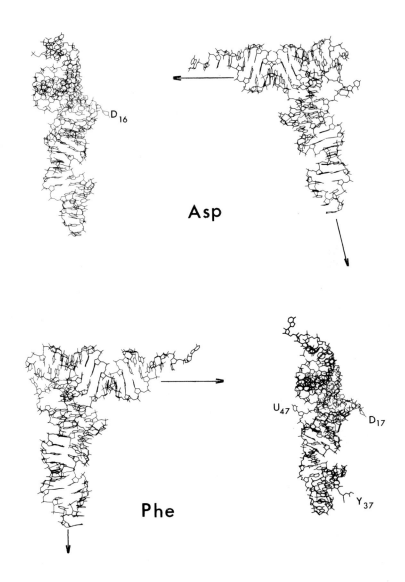

Figure 3 : Two views of the three-dimensional structures of yeast tRNA^{Asp} (top) and tRNA^{Phe} (bottom) at 90° of eachother. The coordinates of tRNA^{Asp} correspond to the refined low temperature form ; the R-factor for these data is presently 24,5% at 3 Å resolution. The CCA-end part however is not yet fully defined and was set in a standard helical conformation. The coordinates of yeast tRNA^{Phe} are those of the orthorhombic crystal form (21).

pictures show some of the main differences between the two molecules, like the opening of the L, but also stress the overall similarity of the two molecules. One can notice the similar external positioning of D16 in tRNA^{Asp} and D17 in tRNA^{Phe}.

4. PACKING CONSIDERATIONS

The structure of yeast tRNA^{Phe} and tRNA^{Asp} were both solved in two different crystal forms. For tRNA^{Phe} although the crystalline environments are quite different, the two resulting molecular structures are identical within experimental errors. Intermolecular contacts within the crystals are essentially non specific and involve stacking energy and electrostatic interactions. For tRNA^{Asp} the situation is almost opposite ; intermolecular contacts involve some specific interactions and despite a similar crystal packing the two crystal forms exhibit structural changes, especially at the D-loop level. **Figure 4** shows a dimeric association through the anticodon.

4.1 Anticodon-anticodon

In the orthorhombic crystal lattice (space group C222$_1$) tRNA^{Asp} molecules are associated through a two-fold symmetry axis parallel to the crystallographic **b** direction by anticodon-anticodon interactions. **Figure 5** represents the local conformation. The triplets of symmetrically related molecules form complementary hydrogen bonded base

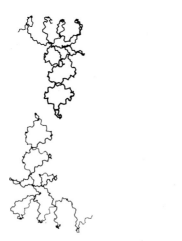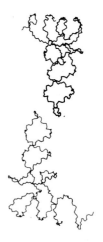

Figure 4 : Dimeric association of tRNA^{Asp} in the orthorhombic C222$_1$ crystal (stereo view). The figure shows the backbone of two tRNA^{Asp} molecules associated through GUC anticodon interactions.

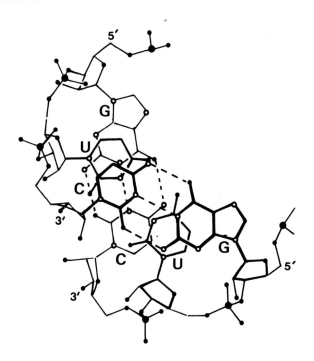

Figure 5 : Anticodon-anticodon association in the crystals of
tRNA^{Asp}. The exact nature of the interaction between the two
uridines residues cannot be determined.

pairs, arranged in a normal helical conformation. This small helix is
stabilized by stacking of the modified base m1G37 on both sides. This
packing confers a great stability to the dimeric structure and explains
the good quality of the electron density map in the anticodon regions.
A contact between the anticodon loops of two tRNAs also exists in the
orthorhombic form of yeast tRNA^{Phe}. In that case, however, the G$_m$AA
anticodons cannot be base paired and they are arranged in a stacked
conformation.

4.2 Contacts of acceptor end with D and T-loop

Another key point of the packing of tRNA^{Asp} in its orthorhombic
crystal form is the contact induced by the C mode. That translation
brings the CCA-end of one molecule in the close vicinity of both the D
and T-loop of another one. Such a contact would not be permitted if the
position of the D-loop in tRNA^{Asp} would be similar to that in
tRNA^{Phe}, since the D-loop and the CCA-end of the two neighbouring
molecules would then occupy the same physical space. This packing
explains a distorted conformation of the CCA-end, a single stranded
portion. The two last nucleotides have a weak electron density on the
map.

5. EVIDENCES FOR CONFORMATIONAL CHANGES

5.1 Crystallographic evidences

A direct evidence of molecular flexibility is brought for tRNA^{Asp} which crystallizes into two non-isomorphous and interconvertible crystal forms. Evidences from the two MIR maps show that conformational differences between the two structures are found in the D-loop region. This loop is thus mobile and in both situations different from the tRNA^{Phe}. The difference in sequence alone might account for most of the changes since the α and ß regions of the D-loop have different lenghts. Whether the alternative conformations of the D-loop in tRNA^{Asp} are correlated with different functional states or are due to particular experimental conditions remains an open question which cannot be answered using solely the crystallographic approach.

An indirect evidence for conformational change is given by the opening of the L when compared to tRNA^{Phe}. As a result the distance between phosphate 35 in the anticodon loop and phosphate 73 in the acceptor stem increases by more than 5 Å. Phosphate 73 was chosen in order to avoid bias due to conformational changes of the single stranded CCA-end induced by packing effects only. Since it is likely that the distance between anticodon and acceptor end is identical for all tRNAs on the ribosome, it is then probable that the observed variation is a consequence of a conformational change.

5.2 Non-enzymatic hydrolysis

Another way to detect the flexibility of some parts of a tRNA molecule is to look for chemical instability of that molecule. The rationale of this approach is based on the observation that RNA molecules are intrinsically unstable under certain conditions. The presence of the free 2'-hydroxyl group on the ribose moieties allows a cyclization with the neighbouring phosphates and a subsequent splitting of the ribose-phosphate chain. This mechanism is used by several ribonucleases to split RNA ; alkaline hydrolysis of RNA also is supposed to react this way (22). In the absence of ribonucleases and at a neutral pH it is clear that the extent of splitting must be close to zero, otherwise RNAs could not exist. If some chain splitting nevertheless occurs at some specific localizations in the RNA sequence, it implies the existence of a particular flexibility at those positions of the structure. This flexibility helps to overcome the energy barrier and favours the hydrolytic reaction by allowing the cyclization to occur more easily. Even so splits arise at very low extent and very sensitive methods are necessary to detect them.

The principle of the method we used to study the chemical stability of tRNA^{Asp} in the crystalline state and in solution is given in **Figure 6**. Crystals of tRNA were washed several times with mother liquor in order to remove uncrystallized material and quickly dissolved

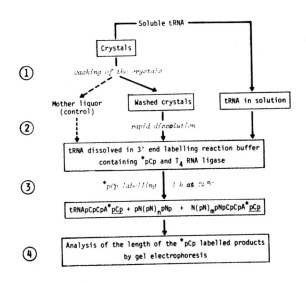

Figure 6. Scheme for identifying splits in tRNA in the crystalline state and in solution (see the text for details)

in an incubation mixture containing ^{32}P labelled pCp and T4 RNA ligase (23). In such a way the dissolved tRNA molecules were labelled at their 3'-terminus with the radioactive pCp. The labelled material was immediately deposited on a polyacrylamide slab gel in the presence of urea and the putative splitting products were resolved by high voltage electrophoresis (24,25). The tRNA in solution originating from the same batch than the crystalline tRNA, as well as non-crystallized tRNA remaining in the mother liquor were treated in the same way and electrophoresed on the same gel. The radioactive bands were visualized by autoradiography as displayed on **Figure 7**. Different degradation patterns were obtained whether the tRNA originated from solution or from the crystal. The assignements of the radioactive bands were done in a complementary experiment by sequence determination of the fragments eluted from the gel. For tRNA^{Asp} in solution splits occur in the anticodon loop of tRNAAsp in solution. In the crystals, however, practically no splitting is observed in that region ; in that case it is found in the D-loop which conversely is not split in the solution tRNA. It might be noted that the radioactivity of the splitting products represents about 5 to 8% of the total radioactivity deposited on the gel, which means that up to 8% of the tRNA molecules possess an internal break in their ribose-phosphate backbone.

What is the interpretation of this experiment ? First it illustrates the assumption made before concerning a local flexibility of the tRNA allowing at certain positions the splitting of the molecule. This local flexibility, however, depends upon the overall conformation of the tRNA. In the crystalline state of tRNA^{Asp} two molecules interact through their GUC anticodons ; this confers a structural rigidity to this loop and prevents the splitting to occur.

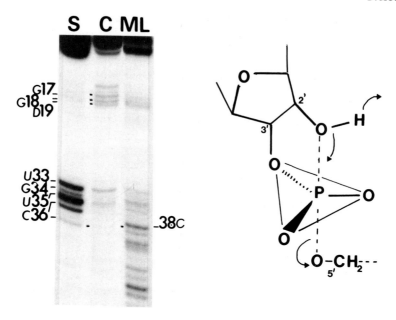

Figure 7 : Autoradiogram of a 15% acrylamide gel of tRNA^{Asp} labelled at its 3' end by pCp. The tRNA originates from solution (S), the crystals (C) or from the mother liquor of crystallization (non-crystallized tRNA) (ML). The assignment of the bands was done in complementary experiments not shown. The chain-scission mechanism is indicated.

In solution, however, only a minor part of the molecules forms dimers (15), so that the anticodon loop remains rather flexible, a situation which favours the splitting. Concerning the D-loop its reactivity in the crystalline state is in agreement with the structural results. The atoms of this loop exhibit high temperature factors from residus 18 to 20, a clear indication of internal molecular motion in this part. This part of the molecule is in the region most affected by local variations in the two crystal forms. A partial or total disruption of the G18·C56 tertiary interaction in one form is probable at the present state of refinement.

6. CONCLUSION

After yeast tRNA^{Phe}, yeast tRNA^{Asp} is the second elongator tRNA for which a three-dimensional structure at high resolution has been solved. The folding of tRNA^{Asp} although not identical, is similar to that originally found for tRNA^{Phe} (4-7). This general folding scheme is the consequence of the existence of invariant or semi-invariant residues in all tRNAs. However, in addition to the other known tRNA structures, that of tRNA^{Asp} enlightens us about the conformational state of a tRNA on the ribosome. This is due to the particular properties of the GUC anticodon of tRNA^{Asp}. In the crystals one GUC triplet form one tRNA molecule mimicks a codon of mRNA and interacts with the GUC anticodon of a second molecule. This interaction might act as a signal and trigger conformational changes elsewhere. The open structure of the tRNA might be a result of such a mechanism. Influence on the D-, T-loop association is strongly suggested by hydrolysis experiments. Packing influences however cannot be excluded, they would give the necessary extra energy.

Acknowledgments : This research was supported by grants from the Centre National de la Recherche Scientifique (CNRS), the Ministère de la Recherche et de l'Industrie and the Université Louis Pasteur, Strasbourg. We thank the European Molecular Biology Laboratory, Heidelberg, for computer facilities and Dr.H. Boshardt (EMBL) for help in the use of the Evans and Sutherland graphic systems. We acknowledge Dr. J.C. Thierry for computational help and fruitful discussions and Professor J.P. Ebel for constant support and encouragments.

References

1. Schimmel, P.R., Söll, D. and Abelson, J.N.: 1979, in Transfer RNA: Structure, Properties and Recognition (Cold Spring Harbor Monogr. Ser. 9A, New York).
2. Clark, B.F.C.: 1979, in Ribosomes: Structure, Function and Genetics, pp. 413-444 (University Park Press, Baltimore).
3. Ofengand, J.: 1982, in Protein Biosynthesis in Eukaryotes, R. Pérez-Bercoff ed., Plenum Publishing Corporation, pp.1-67.
4. Schwarz, U., Manzel, H.M. and Gassen, H.G.: 1976, Biochemistry, 15, pp. 2484-2490.
5. Ehrlich, R., Lefèvre, J.L. and Remy, P.: 1980, Eur. J. Biochem., 103, pp. 145-153.
6. Farber, N. and Cantor, C.R.: 1980, Proc. Natl. Acad. Sci., USA, 77, pp. 5135-5139.
7. Robertson, J.M. and Wintermeyer, W.: J. Mol. Biol., 151, pp.57-79.
8. Quigley, G.J., Wang, A., Seeman, N.C., Suddath, P.L., Rich, A., Sussmann, J.L. and Kim, S.H.: 1975, Proc. Natl. Acad. Sci. USA, 73, pp. 4866-4870.
9. Jack, A., Ladner, J.E. and Klug, A.: 1976, J. Mol. Biol., 108, pp.619-649.
10.Stout, C.D., Mizuno, H., Rao, S.T., Swaminathan P., Rubin, J., Brennan, T. and Sundaralingam, M.: 1978, Acta Cryst., B.34, 1529-1544.
11.Sussman, J.L., Holdbrook, S.R., Warrant, R.W., Church, G.M. and Kim, S.H.: 1978, J. Mol. Biol., 123, pp. 607-630.
12.Keith, G., Gangloff, J. and Dirheimer G.: 1971, Biochimie, 53, pp. 123-125.
13.Gangloff, J., Keith, G., Ebel, J.P. and Dirheimer, G.: 1971, Nature New Biol., 230, pp. 125-127.

14.RajBhandary, U.L. and Chang, S.H.: 1968, J. Biol. Chem., 243, pp. 598-608.

15.Grosjean, H., De Henau, S. and Crothers, D.M.: 1978, Proc. Natl. Acad. Sci. USA, 75, 610-614.

16.Giegé, R., Moras, D. and Thierry, J.C.: 1977, J. Mol. Biol., 115, pp. 91-96.

17.Moras, D., Comarmond, M.B., Fischer, J., Weiss, R., Thierry, J.C., Ebel, J.P. and Giegé, R.: 1980, Nature, 288, pp. 669-674.

18.Konnert, J.H. and Hendrickson, W.A.: 1980, Acta Cryst. A36, pp. 344-350.

19.Jones, T.A.: 1978, J. Appl. Cryst. 11, pp. 268-272.

20.Roy, S. and Redfield, A.: 1981, Nucleic Acids, Res., 9, 7073- 7083.

21.Quigley, G.J., Seeman, N.C., Wang, A.H.J., Suddath, F.L. and Rich, A.: 1975, Nucl. Acids Res., 2, pp. 2329-2339.

22.Brown, D.M.: 1974, in Basic Principles in Nucleic Acid Chemitry, vol.II, P.O.P Ts'o ed., Academic Press, New-York, pp. 1-90.

23.Bruce, A.G. and Uhlenbeck, O.C.: 1978, Nucl. Acids Res., 5, pp. 3665-3677.

24.Sanger, F. and Coulson, A.R.: 1978, FEBS Letters, 87, pp. 107-110.

25.Peattie, D.A.: 1979, Proc. Natl. Acad. Sci. USA, 76, 1760-1764.

SOLUTION CONFORMATION OF tRNAs : CORRELATION WITH CRYSTAL STRUCTURES.

R. Giegé, P. Romby, C. Florentz, J.P. Ebel,
P. Dumas, E. Westhof and D. Moras.
Institut de Biologie Moléculaire et Cellulaire du CNRS,
15, rue R.Descartes, 67084 Strasbourg Cedex (France).

Abstract : A comparative study of the solution conformation of yeast tRNA^{Asp} and tRNA^{Phe} was undertaken with chemical reagents as structural probes. The accessibility of phosphates was assayed with ethylnitrosourea, that of N7 positions in guanines and adenines with dimethylsulfate and that of the N3 position in cytosines with dimethylsulfate. Statistical modifications were done on end-labelled tRNAs subsequently split at the modified positions and analyzed using sequencing gel methodologies. The results were correlated with the crystal conformations of both tRNAs which were analyzed with a graphic modelling program. Three particular positions are discussed. In all tRNAs so far studied, including tRNA^{Asp} and tRNA^{Phe}, phosphate 60 is protected against alkylation, even in the isolated 3'half-molecule of tRNA^{Asp}. This indicates a common T-loop conformation in tRNA which is correlated with the constant presence of residue C61. A similar conformation is found in turnip yellow mosaic virus tRNA-like structure. Adenosine 21 and guanosine 45, two common residues in tRNA^{Asp} and tRNA^{Phe}, exhibit different reactivities against diethylpyrocarbonate and dimethylsulfate. A21 is reactive in tRNA^{Asp} and protected in tRNA^{Phe} ; for G45 the situation is inversed. As to A21, the different reactivity is correlated with the extra-loop structures in the two tRNAs ; in the case of G45 the results are explained by a different stacking of A9 between G45 and residue 46.

1. SEARCH FOR SPECIFIC LOCAL CONFORMATIONS IN tRNAs

After the determination of several hundred sequences of transfer ribonucleic acids (tRNAs) (1) and the X-ray crystallographic studies on a few of them (2-9) it can be stated at present that the general structural organization of these molecules is definitively established. The sequences of tRNAs (except for some mitochondrial species and tRNA-like structures of viruses 1,10) can all be folded in the classical cloverleaf model and the presence of conserved or semi-conserved residues can account for their folding in a L-shaped three-dimensional structure since these residues are involved in the so-

415

B. Pullman and J. Jortner (eds.), Nucleic Acids: The Vectors of Life, 415–426.
© 1983 by D. Reidel Publishing Company.

called "tertiary interactions". This structural frame agrees well with the functional idea that tRNA molecules must possess similar conformational features, for instance the ribosome likely recognizes similar gross features in all tRNA species. On the other hand the crystallographic studies also revealed striking conformational differences between tRNAs, for instance variable angles between the two helical domains forming the L-shaped structure. Such large structural differences between tRNAs might be related to different functional states of the molecules. In the case of tRNA^{Asp} which is packed as anticodon-anticodon dimers in the crystal lattice, the X-ray structure might correspond to the conformational state of tRNA interacting with its codon (9) ; the tRNA^{Phe} structure would be that of a free tRNA (2-5). For the discrimination between the different tRNA species, for instance by the aminoacyl-tRNA synthetases, it seems reasonable however to think that more subtle structural features are involved, whatever the overall shape of the molecule in that functional state. Such microstructures would be created by the local environment of the various residues in tRNA and might exist under different states of tRNA ; they could represent recognition sites or at least participate in the building up of specific conformations recognized by specific enzymes.

The three-dimensional structures of two tRNAs are known in great details ; first that of yeast tRNA^{Phe} (2-5) and now that of yeast tRNA^{Asp} (9,11). It becomes thus possible to compare them and it is expected to find local conformational similarities as well as differences. The relevance of such comparisons, however, can only be established when solution studies are correlated with the crystallographic data. We have therefore started a systematic biochemical screening of the solution structures of yeast tRNA^{Asp} and tRNA^{Phe} with a concomitant structural analysis on the refined structures of both molecules. The aim of such studies is to find structural rules which could explain conformations of tRNAs or more generally of RNAs for which no crystallographic data are available. An example is given with the tRNA-like structure of turnip yellow mosaic virus RNA, a molecule which is efficiently aminoacylated by valyl-tRNA synthetase (12).

2. EXPERIMENTAL APPROACHES

Solution structures of tRNAs have already been approached by a variety of physical and biochemical techniques (for a review see 13). Here we have used new biochemical methodologies which permit to probe the accessibility of the tRNA molecules to chemical reagents specific for different functional groups of the nucleic acid. Enzymatic probes can also be used (e.g.14-16), but in this study we have given the preference to the chemical reagents, because of their small size compared to the bulky enzymes. This allows to probe discrete conformational features in the tRNA. The principle of the different methods derives from the chemical sequencing methodologies of nucleic

acids (17,18) and lies in a statistical and low yield modification of
them at each potential target, in such a way that each tRNA chain
undergoes less than one modification. Two experimental conditions are
usually chosen ; one which maintain the native structure of the tRNA
and another one where the tRNA is denatured in such a way that all
potential targets become accessible and can react with the probes. The
tRNA molecules labelled at their 3' or 5' end with radioactive [^{32}P]ATP
are then specifically split at the modified positions and the resulting
end-labelled oligonucleotides are analyzed by high voltage
electrophoresis on sequencing gels followed by autoradiography. The
assignment of the bands is done by comparing their migrations with
ladders obtained after limited T$_1$ RNase digestion or alkaline
hydrolysis. In such a way it becomes possible to probe the entire tRNA
molecule in one experiment.

Three chemical reagents, ethylnitrosourea (ENU),
dimethylsulfate (DMS) and diethylpyrocarbonate (DEPC) were used. All
three are potent carcinogenic agents which react at different sites of
nucleic acids (19,20) but with a pronounced preference for those
indicated in **Figure 1.** The modification reactions labilize the
ribophosphate backbone so that it can be split easily at the modified
positions according to the scheme shown in **Figure 1.**

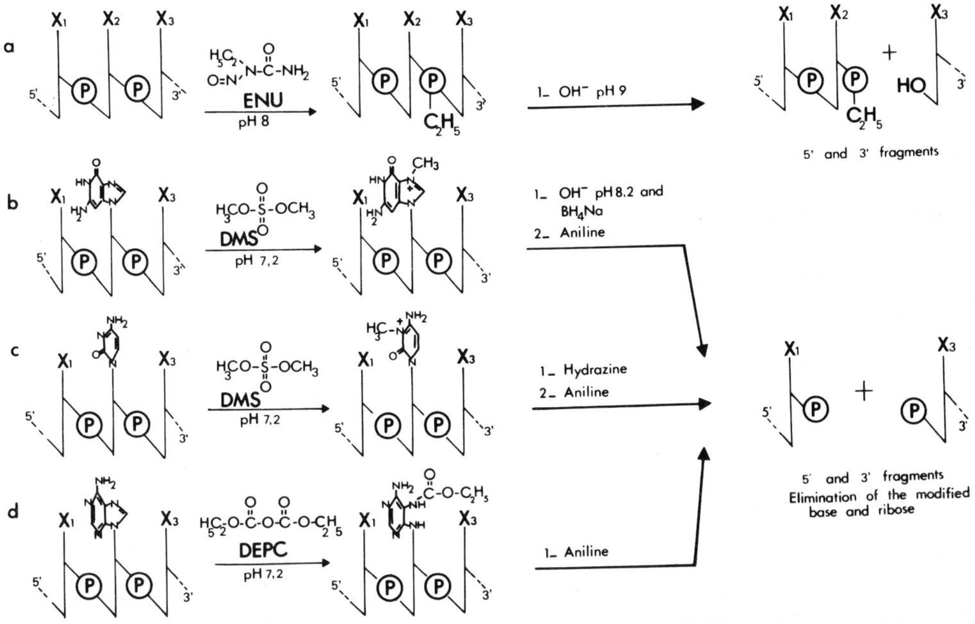

Figure 1. Principle of the chemical modification reaction of
ribonucleic acids with ENU (a), DMS (b,c) and DEPC (d) and
of strand-scission at the modified positions. The
ribophosphate backbone is schematized ; (P) represents a
phosphate group ; X$_n$ are the bases. Experimental details
for the ENU reaction are in (21), and for the base
alkylations in (24).

ENU probes the accessibility of phosphates(21). The reactivity per residue is particularly low : 0.1 to 0.3 ethyl group incorporated per phosphate. The splitting of the ribophosphate chain at the modified positions is done at alkaline pH; 3'and 5' fragments are liberated, the latter one carrying the ethyl groups. The method was established with yeast tRNA^{Phe} and it was shown that the results could be interpreted in terms of accessibilities to the reagent of the phosphates in the tertiary structure of the tRNA. This was best illustrated in a comparison between the chemical phosphate reactivities in native yeast tRNA^{Phe} and the calculated accessible areas of the anionic oxygens of the phosphates for Na$^+$ ions and for water (21-23). This comparison gave a quasi-perfect identity between the phosphodiester reactivity toward ENU and the steric accessibility of the phosphate groups.

DMS and DEPC probe the accessible bases and were introduced by Peattie and Gilbert who established the method with yeast tRNA^{Phe} (24). The relevance of the approach was supported by accessibility and potential computations of the base environments (25). DMS and DEPC react at the N7 position of G and A ; for A the modification is particularly sensitive to stacking. DMS also alkylates the N3 position of C ; this position can be involved in Watson-Crick pairings thus making this modification a powerful probe for secondary or tertiary interactions. The extent of base alkylation is about 10 fold that of phosphate modification by ENU, a favourable fact from an experimental point of view. However, the presence of naturally occuring N7 methylated purines in tRNA is a disadvantage since it could be responsible for the splitting of all molecules at these positions. Therefore the conditions for chain-scission are chosen so that the splitting is only partial.

The structural environment of the studied positions (either accessible or inaccessible to the reagents) were analyzed on a graphic interacting display system (MPS from Evans and Sutherland at EMBL, Heidelberg, Germany) using the graphic modelling program FRODO (26) adapted for nucleic acids.

3. ACCESSIBLE AND BURIED RESIDUES IN tRNAs

Chemical modification experiments were done in parallel on tRNA^{Asp} and tRNA^{Phe} under identical experimental conditions. **Figure 2** gives an example of a DEPC experiment in which the accessibility of the N7 position of adenines has been probed. The appearance of a band indicates that this particular adenine is accessible under the given experimental conditions. This is for instance the case for the constant residue A21 in the D-loop of tRNA^{Asp} which reacts with DEPC when the tRNA is under native conditions. On the contrary other residues, for instance A62, A64 in tRNA^{Phe} and A44, A46 in tRNA^{Asp} are protected in the native tRNAs. These examples illustrate how chemical modification experiments are interpreted. In the frame of this report we do not intend to present all accessibility results obtained with tRNA^{Asp} and

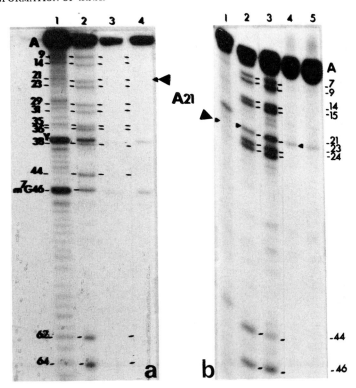

Figure 2. Adenosine accessibility experiment. Autoradiogram of 15% acrylamide gels of an alkylation experiment with DEPC on tRNA^{Phe} (a) and tRNA^{Asp} (b). (1) Formamide ladder, (2) denatured conditions : 1 mM EDTA, 50 mM sodium cacodylate buffer pH 7.2, 90°C, (3) semi-denatured conditions : 1 mM EDTA, same buffer, 37°C, (4-5) native conditions : 10 mM MgCl$_2$, same buffer, 37°C. Adenosine positions are numbered.

tRNA^{Phe}, but only to focus on 3 particular positions :P60, A21 and G45.

4. STRUCTURAL SIMILARITIES IN tRNAs : THE CONFORMATIONAL ENVIRONMENT OF PHOSPHATE 60

A comparison of the patterns of phosphate alkylation in tRNAs from yeast and E.coli (21) revealed a striking similarity essentially in the T-arm and particularly at position 60 where a total absence of alkylation can be noticed. In the three-dimensional structure of tRNA^{Phe} (2-5), as well as in that of tRNA^{Asp} (9,11), phosphate 60 is located in the hinge region where the two helical domains of tRNA join to form the characteristic L-shaped structure. A quick inspection of both molecular structures suggests a good explanation. Phosphate 60 appears to be buried in a pocket formed by A58, C60 and C61 on one side

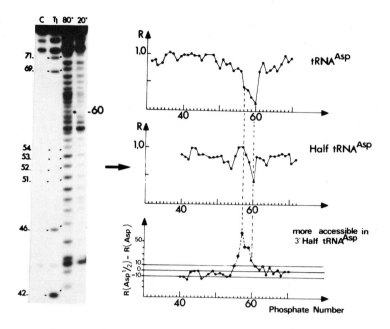

Figure 3. Comparison of phosphate reactivities toward ENU
between native and 3'half-molecule of tRNA^{Asp}. The figure
shows an autoradiogram of a 15% acrylamide gel of an alkylation experiment on the 3' half-molecule of tRNA^{Asp}
labelled at its 5' end : (C) control incubation under
conditions stabilizing the tRNA structure, (T$_1$) partial
ribonuclease T$_1$ ladder, (80°,20°) alkylations at 80°C or
20°C under conditions unfolding (80°) or stabilizing
(20°C)the half-molecule. The absence of band at position 60
is indicated. In complementarity experiments it has been
verified that the T-stem and loop exist in the half
molecule. The interpretation of data is given in plots of R
versus phosphate number for tRNA^{Asp} and half-tRNA^{Asp}
(autoradiogram for complete tRNA^{Asp} is not shown). R values
are the ratios between the intensities of the corresponding
electrophoretic bands of the alkylated folded und unfolded
molecules. A difference plot between half and complete
tRNA^{Asp} is given showing the same shielding of P60 in both
molecules, but otherwise more accessible phosphates in the
T-loop of the half- tRNA^{Asp}. This means that these
phosphates are protected by folding effects in the complete
tRNA ; this effects are lost when the T-loop is isolated
from ts structural context.

and the deep groove of the T-stem on the other side. The resulting
steric hindrance induced by the tertiary folding would thus account for
the absence of alkylation. This interpretation is best illustrated in
tRNA^{Phe} by the computation of steric inaccessibility of phosphate 60 to
sodium ions or water (22,23). In both tRNA^{Asp} and tRNA^{Phe} the existence

of an hydrogen bond between phosphate 60 and the N4 amino group of C61 and the 2'-hydroxyl group of ribose 58 is another good explanation for the lack of reactivity of this phosphate. It includes the steric inaccessibility in the hydrogen bond direction and has the additional effect of rigidifying the local conformation.

The ENU reaction on the 3'half-molecule of tRNA^{Asp} clearly shows the similar absence of reactivity of phosphate 60. **Figure 3** gives the experimental data. This experiment suggests a similar folding of the T-stem and loop. It explains the quasi-constant presence of C61 at the end of the T-stem which helps to stabilize the structure of the T-loop in a similar conformation in all tRNAs. It is interesting to mention that the tRNA-like fragment of turnip yellow mosaic virus (TYMV) RNA possesses the same local structural feature, although otherwise the secondary structure of this RNA differs from classical tRNA (16). Indeed an ENU reaction showed that two phosphates located at topologically similar positions then phosphates 59 and 60 in the T-loop of classical tRNA are also protected against alkylation. Interestingly enough this RNA possesses a C residue at the end of its pseudo T-stem. These structural data are summarized in **Figure 4.**

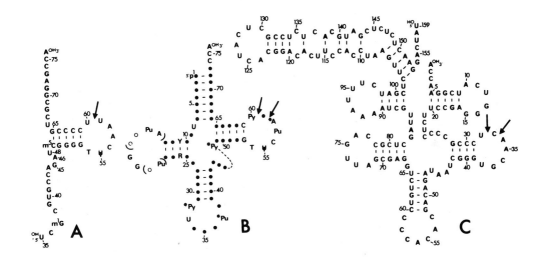

Figure 4. Comparison of the secondary structures of the 3'half tRNA^{Asp} (A), complete tRNA (B), and tRNA-like structure of TYMV RNA (16) (C), with positions of phosphates inaccessible to alkylation by ENU ; only the alkylated positions in the T-loop are indicated by arrows. In (B), tRNA is schematically represented with conserved residues in most elongator tRNAs (variable α and β region in the D-loop are between brackets ; variable residues in the extra-loop are indicated by a dashed line.

5. STRUCTURAL DIFFERENCES IN tRNA^{Asp} AND tRNA^{Phe} : THE CONFORMATIONAL ENVIRONMENT OF A21 AND G45

DMS and DEPC probe N7 positions of purines and N3 positions of cytosines. For simple structural comparison one should concentrate only on those residues located at similar positions in tRNA^{Asp} and tRNA^{Phe}. These positions are indicated in **Figure 5**. Some residues exhibit similar behaviours in both tRNAs. Other nucleosides display different chemical reactivities ; here we will discuss two of them, A21 and G45.

In tRNA^{Asp} adenosine 21 is accessible to modification at N7 by DEPC ; it is protected in tRNA^{Phe}. The explanation is linked to a different extra-loop structure in both tRNAs. tRNA^{Phe} possesses a 5 bases long extra-loop whereas tRNA^{Asp} has only 4 bases in this domain. In the three-dimensional structure of both tRNAs A21 is stacked between positions 46 and 48. Position 47 is an uridine residue in tRNA^{Phe} and bulges out the structure ; the corresponding residue is missing in tRNA^{Asp}. This confers a slightly different positioning of residues 46 and 48 in the two tRNAs as can be seen in **Figure 6.** The conformations of the extra-loops permit a perfect stacking of A21 between bases 46

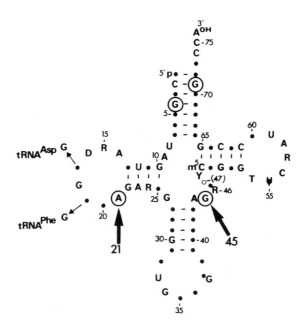

Figure 5. Composite secondary structure of yeast tRNA^{Asp} (27) and tRNA^{Phe} (28) with nucleotides common to both tRNAs. R is for common purines and Y for common pyrimidines. Those residues which exhibit a different reactivity toward DMS or DEPC are encircled. (It is recalled that U residues are not tested in this study).

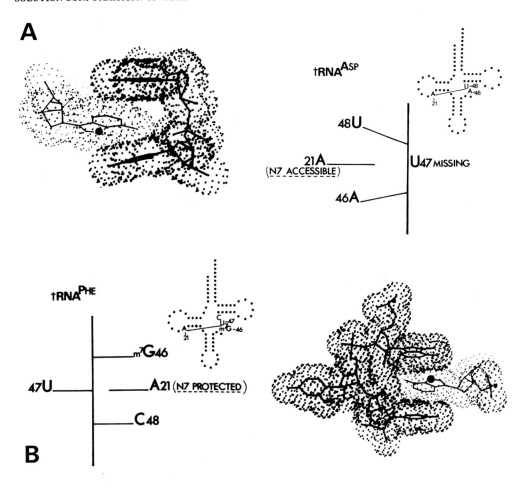

Figure 6. Graphic modelling of the structural environment of Adenosine 21 in tRNA^Asp (A) and tRNA^Phe (B) and interpretations of the display pictures. The Van der Waals spheres of atoms are shown and the N7 position of A21 is indicated by a dark dot.

and 48 in tRNA^Phe and a less good in tRNA^Asp. As a consequence the N7 position of A21 is protected in tRNA^Phe and accessible in tRNA^Asp.

As to G45 the situation is inversed. The N7 position of the guanine residue is accessible to alkylation by DMS in tRNA^Phe and protected in tRNA^Asp. The reason for that is linked to a different stacking of residue A9 between positions 45 and 46. This difference is due to the presence in tRNA^Phe of m^7G in position 46 and of A46 in tRNA^Asp. The bulky methyl group of m^7G hinders A9 to be stacked correctly in tRNA^Phe and as a consequence G45 becomes accessible to DMS. In contrast, in tRNA^Asp A9 is perfectly stacked between residues

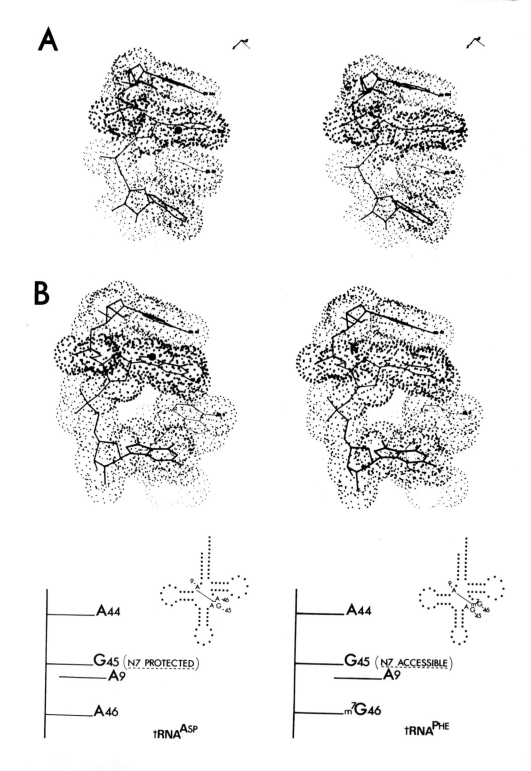

45 and 46 explaining the protection of G45. These structural differences which explain chemical reactivities are illustrated in stereo views in **Figure 7**.

6. CONCLUSIONS AND FUTURE PROSPECTS

Chemical modification studies are capable to detect discrete conformations on tRNAs. The three examples discussed before have shown that DMS and DEPC reactions are sensitive to base-stacking and that ENU alkylation can probe hydrogen bonds. These interpretations, however, could only be given after a thorough inspection of the X-ray crystal structures of tRNA with the help of graphic modelling. In good agreement with the versatile functioning of tRNAs it is shown that these molecules can possess different local conformations, even around invariant residues such as A21, but also similar local structures as illustrated by the environment around phosphate 60 which governs an intrinsic T-loop geometry. Considering these results it appears that chemical accessibility studies represent an experimental approach safe enough to reveal conformations and to predict structural rules. These rules could help to analyze the structure of tRNAs, or more generally on RNAs, for which no X-ray data are available. The use of model conformations and a modelisation with graphic systems should enable to define the stereochemistry of unknown structures. Alternatively, solution data obtained on tRNAs for which the diffraction data are poor, should allow to introduce new structural constraints in refinement calculations and improve the quality of the X-ray structure.

Acknowledgments : This research was supported by grants from the Centre national de la Recherche Scientifique (CNRS), the Ministère de la Recherche et de l'Industrie and Université Louis Pasteur, Strasbourg. We thank EMBL, Heidelberg, for computer facilities and use of graphic systems and Drs. A.C. Dock, D. Kern and J.C. Thierry for stimulating discussions. The collaboration of Dr. V.V. Vlassov at the early stages of ethylnitrosourea studies is acknowledged.

References

1. Gauss, D.H. and Sprinzl, M.: 1983, Nucleic Acids Res., 11, pp. r55-r102.
2. Quigley, G.J., Wang, A., Seeman, N.C., Suddath, F.L., Rich, A., Sussmann, J.L. and Kim, S.H.: 1975, Proc. Natl. Acad. Sci. USA, 72, pp. 4866-4870.

Figure 7. Graphic modelling of the structural environment of Guanosine 45 in tRNA^{Asp} (A) and tRNA^{Phe} (B) (stereo views) and interpretations of the display pictures. The N7 position of G45 is indicated by a dark dot.

3. Jack, A., Ladner, J.E. and Klug, A.: 1976, J. Mol. Biol., 108, pp. 619-649.

4. Stout, C.D., Mizuno, H., Rao, S.T., Swaminathan, P., Rubin, J., Brennan, T. and Sundaralingam, M.: 1978, Acta Crys. B 54, pp. 1529-1544.

5. Sussmann, J.L., Holbrook, S.R., Warrant, R.W., Church, G.M. and Kim, S.H.: 1978, J. Mol. Biol., 123, pp. 607-630.

6. Schevitz, R.W., Podjarny, A.D., Krishnamachar, N., Hugues, T.T., Sigler, P. and Sussman, J.L.: 1979, Nature, 278, pp. 188-190.

7. Wright, H.T., Manor, P.C, Beurling, K., Karpel, R.L. and Fresco, J.L. in Transfer RNA: Structure Properties and Recognition, pp. 145-160 (Cold Spring Harbor Monogr. Ser. 9A, New York, 1979).

8. Woo, N., Roe, B. and Rich, A.: 1980, Nature, 286, pp. 346-351.

9. Moras, D., Comarmond, M.B., Fischer, J., Weiss, R., Thierry, J.C., Ebel, J.P. and Giegé, R.: 1980, Nature, 286, pp. 669-674.

10. Haenni, A.L., Joshi, S. and Chapeville, F.: 1982, Progress in Nucleic Acids Research and Molecular Biology, 27, pp. 85-104.

11. Moras, D., Dock, A.C., Dumas, P., Westhof, E., Romby, P. and Giegé, R., this issue.

12. Giegé, R., Briand, J.P., Mengual, R., Ebel, J.P. and Hirth, L.: 1978, Eur. J. Biochem. 84, pp. 251-256.

13. Schimmel, P.R. and Redfield, A.G.: 1980, Ann. Rev. Biophys. Bioeng., 9, pp. 181-221.

14. Wrede, P., Woo, N.H. and Rich, A.: 1979, Proc. Natl. Acad. Sci. USA, 76, pp. 3289-3293.

15. Favorova, O.O., Fasiolo, F., Keith, G., Vassilenko, S.K. and Ebel, J.P.,: 1981, Biochemistry, 20, pp. 1006-1011.

16. Florentz, C., Briand, J.P., Romby, P., Hirth, L., Ebel, J.P. and Giegé, R.,: 1982, EMBO, J., 1, pp. 269-276.

17. Peattie, D.A.: 1979, Proc. Natl. Acad. Sci. USA, 76, pp. 1760-1764

18. Maxam, A.M. and Gilbert, W.: 1980, Methods in Enzymology, 65, pp. 499-459.

19. Leonard, N.J., Mc Donald, J.J., Henderson, R.E.L. and Reichman, M.E.: 1971, Biochemistry, 10, pp. 3335-3342.

20. Kusmierek, J.T. and Singer, B.: 1976, Biochim. Biophys. Acta, 142, pp. 536-538.

21. Vlassov, V.V., Giegé, R. and Ebel, J.P.: 1980, Eur. J. Biochem., 119, pp. 51-59.

22. Thiyagarayan, P. and Ponnuswamy, P.K.: 1979, Biopolymers, 18, pp. 2233-2247.

23. Lavery, R., Pullman, A. and Pullman, P.: 1980, Theoret. Chim. Acta (Berl.) 57, pp. 233-243.

24. Peattie, D.A. and Gilbert, W.: 1980, Proc. Natl. Acad. Sci. USA, 77, pp. 4679-4682.

25. Lavery, R., Pullman, A., Pullman, B. and de Oliveira, M.: 1980, Nucleic Acids Res., 8, pp.5095-5111.

26. Jones, T.A.: 1978, J. Appl. Cryst., 11, pp. 268-272.

27. Gangloff, J., Keith, G., Ebel, J.P. and Dirheimer, G.: 1971, Nature New Biol., 230, pp. 125-127.

28. RajBhandary, U.L. and Chang, S.H.: 1968, J. Biol. Chem., 243, pp. 598-608.

NMR STUDIES OF THE STRUCTURE OF YEAST tRNA[Phe] IN SOLUTION AND OF ITS COMPLEX WITH THE ELONGATION FACTOR Tu FROM B. STEAROTHERMOPHILUS

C.W. Hilbers, A. Heerschap, J.A.L.I. Walters and
C.A.G. Haasnoot, Department of Biophysical Chemistry,
University of Nijmegen, Toernooiveld, 6525 ED Nijmegen,
The Netherlands.

Introduction

In the last decade an enormous increase of the application of nuclear magnetic resonance spectroscopy towards the solution of problems concerning the structure and function of biomacromolecules is witnessed (1). The method could not be used to its full potential, because systematic methods for the complete interpretation of the NMR spectra of molecules as complicated as biomacromolecules were lacking. However, during the last three years there has been substantial progress in this area. On the one hand this is due to the development of two dimensional Fourier Transform (2D-FT) NMR methods (2,3) and their application to structural studies of proteins (4,5,6). On the other hand the development of NMR instruments with increased resolution and sensitivity has contributed to this progress, e.g. 500 MHz instruments have become commercially available.

In NMR studies of transfer RNA's the resonances of the imino protons involved in hydrogen bonds in secondary and tertiary basepairs have proven to be most useful. Although the application of 2D-FT NMR techniques to this type of molecules is still in its infancy (7,8), substantial progress has been made in the characterization of the imino proton spectrum by the systematic application of the 1D-nuclear Overhauser effect (NOE). In the field of tRNA research the use of NOE's was pioneered by Redfield and his coworkers (9,10,11). Subsequently, the imino proton spectrum of *yeast* tRNA[Phe] has been interpreted independently by Redfield's group (12) and in our own laboratory (13,14). Reid and his coworkers have made substantial progress in the interpretation of the spectra of *E.coli* tRNA[Ile] and tRNA[Val]₁ (15,16).

In this paper some of the most important features of the interpretation of the *yeast* tRNA[Phe] imino proton spectrum will be considered. A sequential assignment procedure permits the identification of all imino proton resonances in the tRNA spectrum. This allows a detailed comparison of the solution structure and the X-ray diffraction structure of the molecule. Subsequently we study the influence on the structure of *yeast* tRNA[Phe] of a partner of tRNA in protein synthesis, namely the elongation factor Tu from *Bacillus stearothermophilus*.

B. Pullman and J. Jortner (eds.), Nucleic Acids: The Vectors of Life, 427–441.
© 1983 by D. Reidel Publishing Company.

The nuclear Overhauser effect

The interpretation of the spectra of biomacromolecules has greatly advanced because of the use of the nuclear Overhauser effect. This is the reduction (or gain) in intensity of one group of resonances when a different resonance or group of resonances is saturated. The way in which such effects are observed is illustrated in Fig. 1. If one of the two resonances in a spectrum is saturated the intensity of the neighbouring resonance is reduced. In practice this reduction is presented in the form of a NOE difference spectrum, which is the difference between the normal and the perturbed spectrum. The observation of such effects for the imino protons of a tRNA presents some particular difficulties. The imino proton signals have to be measured in the presence of a water signal that is 100.000 times larger than that of the individual imino protons. These experimental conditions imply a 'dynamic range' problem of a factor of 10^5 that cannot be handled in a straightforward way by the current NMR data collection and processing equipment.

To overcome this problem one uses some kind of a semi selective pulse —the 214 pulse designed by Redfield is very popular (17)— to excite the spectrum in such a way that excitation of the water resonance is avoided as much as is possible. This cannot be prevented altogether and we have found it advantageous to combine the method with a so-called alternate delay acquisition or digital shift accumulation (7,18). This reduces the nominal intensity of the water resonance, which was diminished by the semi selective pulse already by a factor of about two hundred, by another factor of five hundred. An example of a spectrum of the imino protons of *yeast* tRNA[Phe] recorded using the combination of a semi selective time shared long pulse and the digital shift accumulation is shown in Fig. 2. The remainings of the water resonance are observable at 4.7 ppm. At high field, right of the water peak are resonances of the methyl groups of the tRNA. At low field, far to the left, are the imino proton resonances. Between 6 – 9 ppm are the resonances from the amino protons, the non-exchangeable aromatic protons and the sugar protons. An example of Overhauser effects measured in the acceptor stem of the tRNA is given in Fig. 3. The NOE difference spectrum (Fig. 3c) was obtained by subtracting the spectrum in which resonance R was saturated (Fig. 3b) from the unperturbed tRNA spectrum (Fig. 3a). The effects observed at the positions of the resonances marked C, O and U are Overhauser effects. The differences in NOE intensities arise from the difference in distance between the protons, for which the NOE's are observed, to the proton which is presaturated. Anticipating the conclusions presented below, this can be illustrated by considering the schematized structure of the acceptor stem in Fig. 3. Resonance R, which is presaturated, comes from the iminoproton of U69 of the GU pair in the acceptor stem. The imino proton nearest to U69N3H is the imino proton of residue G4 in the GU basepair. These protons are at a distance of 2.4 Å from each other. The other two NOE's are from the imino protons of the basepairs adjacent to the GU pair; they are at 3.4 Å (A5U68 basepair) and at 4.6 Å (G3C70 basepair). It is seen that the magnitude of the Overhauser effect depends on the distances between the nuclei considered. A more quantitative picture follows from theoretical considerations. To arrive at the results

a

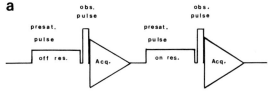

b

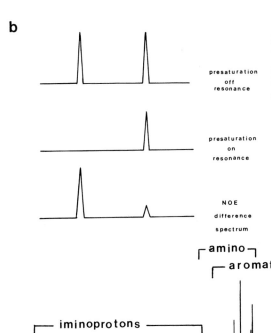

presaturation
off
resonance

presaturation
on
resonance

N O E
difference
spectrum

Fig. 1. a) Pulse sequence employed
to observe NOE's in imino proton
spectra. b) The results for a
hypothetical two spin system
without J-coupling.

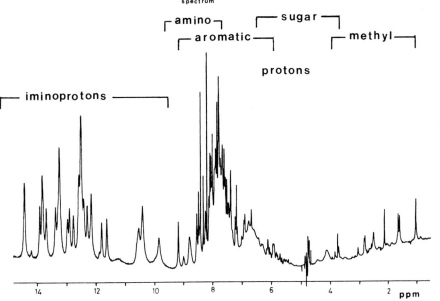

Fig. 2. A 500 MHz ^{1}H NMR spectrum recorded at 28°C from yeast tRNA^{Phe}
dissolved in H$_2$O. The buffer employed contained 110 mM Na^{+} and 5 mM Mg^{++}
ions free in solution (pH 7.0). The suppression of the waterpeak
(at 4.7 ppm) as a result of the combined application of a semi-selective
observation pulse and digital shift accumulation is demonstrated. The
spectral region in which the different types of tRNA protons resonate
are indicated.

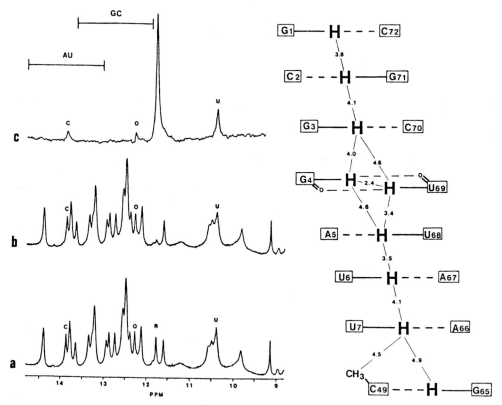

Fig. 3. Example of a NOE experiment. a) Unperturbed imino proton spec-
trum recorded at 500 MHz. b) Same spectrum with resonance R saturated.
c) NOE difference spectrum obtained after subtraction of b) from a).
The spectral regions where AU and GC pairs are resonating are indicated.
For sample conditions see Fig. 1. Also a scheme of the acceptor stem is
given. The boldface capital H's are the imino protons of the base pairs.
Average distances between the imino protons calculated from three
different crystal structures (see ref. (13)) are indicated.

described below, we followed an earlier treatment in which it was
assumed that the saturation of a spin (denoted k) is instantaneous (19).
For the complicated multispin systems we are considering, the nuclear
Overhauser effect, η_i, for spin i can be approximated by the series
expansion:

$$\eta_i = \sigma_{ik}t - \tfrac{1}{2} \sum_{j \neq i,k} \sigma_{ij}\sigma_{jk}t^2 - \tfrac{1}{2}\rho_i\sigma_{ik}t^2$$

$$+ \frac{1}{6} \sum_{l=i,j,k} \sum_{j \neq i,l} \sigma_{ij}\sigma_{jl}\sigma_{lk}t^3 + \frac{1}{6} \sum_{j \neq i,k} \rho_i\sigma_{ij}\sigma_{jk}t^3 \quad +$$

$$+ \frac{1}{6} \sum_{j=i,k} \rho_j \sigma_{ij} \sigma_{jk} t^3 + \frac{1}{6} \rho_i^2 \sigma_{ik} t^3 - \ldots\ldots\ldots \quad [1]$$

In this expression ρ_i is the spin lattice relaxation rate constant of spin i; σ_{ij} is the cross relaxation rate constant characterizing the cross relaxation between spins i and j. The first term on the r.h.s. of Eq. 1 is responsible for the first order Overhauser effect measured for spin i after presaturation of spin k. The second term on the r.h.s. is the NOE measured for spin i after presaturation of spin k, but now mediated via spin j. This is the so-called second order NOE. The third term on the r.h.s. represents the spin lattice relaxation rate of spin i; this term diminishes the NOE. For the remaining terms the physical meaning can be derived analogously.

In absence of mediated cross relaxation, i.e. of higher order NOE's, Eq. 1 can be simplified to

$$\eta_i = \frac{\sigma_{ik}}{\rho_i} (1 - e^{-\rho_i t}) \quad [2]$$

For sufficiently short presaturation pulses both Eq. [1] and [2] simplify to

$$\eta_i = \sigma_{ik} t \quad [3]$$

and the NOE becomes proportional to the length of the saturation pulse and to the cross relaxation rate constant σ_{ik}. For molecules of the size of tRNA

$$\sigma_{ik} = - \frac{\gamma^4 \hbar^2}{10 r_{ik}^6} \tau_c \quad [4]$$

where τ_c is the rotational correlation time of the molecule and r_{ik} the distance between spin i and k. The other symbols have their usual meaning. Hence under favourable conditions the Overhauser effect is proportional to $1/r_{ik}^6$.

The assignment of the imino proton resonances in tRNA

The distances between the imino protons of neighbouring basepairs varies between 3.5 and 5 Å. Using the Eq. 2 derived above, one then expects to measure NOE's of the order of 1 - 10% between these protons and this makes the method particularly suited for a sequential assignment of the imino proton resonances in nucleic acids. To begin the identification of the resonances in the spectrum it is convenient and sometimes necessary to have available a resonance that has been assigned unambiguously. The imino proton of the GU basepair in the acceptor stem may serve this

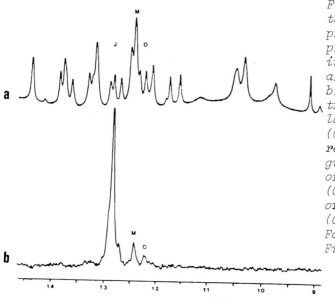

Fig. 4. Demonstration of the NOE connectivities present between the imino protons of GC basepairs in the acceptor stem.
a) Unperturbed spectrum.
b) NOE difference spectrum obtained after a long presaturation pulse (0.8 s) was given on resonance J (G1C72), which gives rise to a first order NOE at resonance M (C2G71) and a second order NOE at resonance O (G3C70).
For details see legend to Fig. 3.

purpose because in the *yeast* tRNA^{Phe} molecule these are the only imino protons which are ~2.4 Å apart from each other. Therefore these are the only candidates for Overhauser effects exceeding 20%. We have encountered this effect already when discussing Fig. 3. Preirradiation of resonance R gives rise to an NOE of about 25% on resonance U and vice versa. Such large NOE's could not be found for other pairs of imino protons and therefore resonances R and U are assigned to the two imino protons of the GU basepair (9,13). The resonances C and O arise from the imino protons of basepairs neighbouring the GU pair, i.e. A5U68 and G3C70 respectively. The NOE's at these positions have the expected magnitudes (~5%). Resonances C and O may serve as the new starting points for the sequential analysis of the spectrum. Presaturation of resonance O gives rise to the expected back NOE's on resonances R and U but also to NOE's at resonances M (and J). Hence one of resonances under M comes from C2G71, the basepair neighbouring G3C70 (resonance O). Because so many resonances are overlapping in peak M we cannot proceed from here and assign the resonance expected from the terminal basepair of the acceptor stem, i.e. G1C72. This overlap of resonances, i.e. lack of dispersion, provides an example of the limitation of the method, which can be circumvented to some extent by the application of second order Overhauser effects. When the duration of the presaturation pulse is long enough the magnetisation can be transferred from G3C70 (resonance O) via C2G71 (resonance M) to G1C73. This is a second order NOE which is brought about by the second term of the r.h.s. of Eq. 1. Preirradiation of resonance O gives rise to a second order NOE on resonance J and vice versa (Fig. 4). We therefore assign resonance J to the imino proton of basepair G1C73. At this point it is good to note that we have been careful to use second order NOE's only in the final stage of the assignment procedure, i.e. when most of the resonances were assigned, otherwise

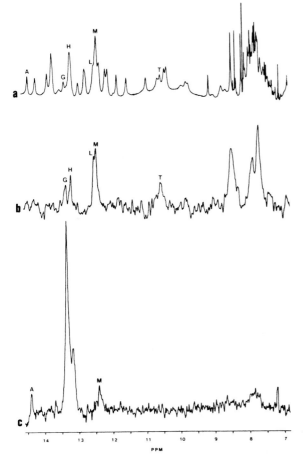

Fig. 5. Demonstration of
NOE connectivities between
protons of the T and
acceptor stem.
a) Unperturbed tRNA
   spectrum
b) NOE difference spectrum
   obtained after pre-
   saturation of the
   methyl resonances of
   m⁵C40 and m⁵C49
c) NOE difference spectrum
   observed after pre-
   irradiation of resonance
   G (U7A66). The buffer
   employed contained no
   Mg⁺⁺ and 400 mM Na⁺
   (pH=7.0).

the method may easily lead to erroneous assignments. We now return to
resonance C, which we had attributed to the imino proton of basepair
A5U68. Preirradiation of this resonance gives rise to the expected
back NOE's on the protons of the GU pair. In addition an NOE is observed for the resonance at 14.4 ppm. In a Mg⁺⁺ free solution this
resonance is split and it turns out that at these conditions the NOE
is observed at the position of resonance A (see ref. 13). We therefore
assign this resonance to the imino proton of basepair U6A67. Subsequently, using this resonance as a starting point we find that the imino
proton resonance of U7A66 corresponds with peak G (13). Then by preirradiation of peak G we find a new NOE under peak M (Fig. 5c). According to the X-ray data we expect this resonance to arise from basepair
m⁵C49G65 (Fig. 6, tRNA model). This can be demonstrated independently
by saturation the methyl resonance assigned to the methyl group of
m⁵C49. This gives rise to an Overhauser effect on resonance G (U7A66)
and resonance M (m⁵C49G65) which can only be observed when the T-stem
and the acceptor stem are stacked upon each other (Fig. 5b).
    We have discussed in some detail the assignment of the resonances in
the acceptor stem. The principle difficulties one may encounter in such

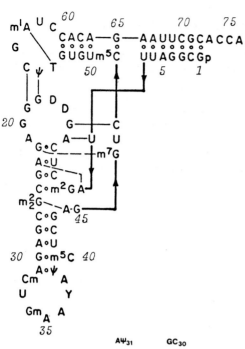

Fig. 6. Summary of the assignments of imino proton resonances of yeast tRNA^Phe. The numbering of the resonances refers to that in the scheme of the tRNA.

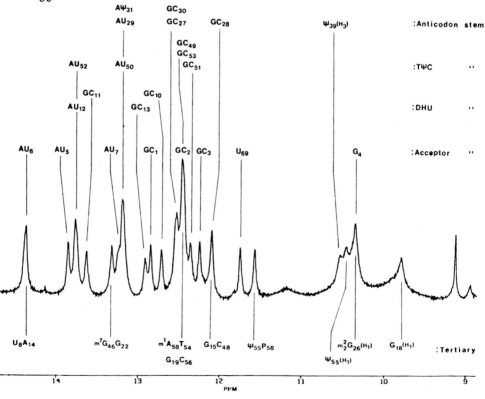

a procedure and their solution have been examined. We shall not
continue such an analysis for the rest of the spectrum; the total
assignment is summarized in Fig. 6. It is noted that all imino protons
involved in tertiary interactions have their corresponding resonances
in the spectrum except for the imino proton of U33 which in the crystal
structure hydrogen bonds to the phosphate moiety P36. This forms a
detailed demonstration that in the T-loop/D-loop region where most of
the tertiary interactions occur, the solution and crystal structure
closely resemble each other. From the Overhauser effect measurements it
can also be concluded that the acceptor stem and T-stem are stacked
upon each other as has already been mentioned above. This is also true
for the D-stem and anticodon stem which are found to be stacked in
solution with the residue m$_2^2$G26 in between.

Complex formation between EFTu•GTP and yeast tRNA^Phe

After having analyzed the imino proton spectrum of *yeast* tRNA^Phe in
such detail it is of course of interest to ask if we can go a step
further and see whether it is possible to obtain information about the
tRNA structure when it is complexed to one of the proteins tRNA

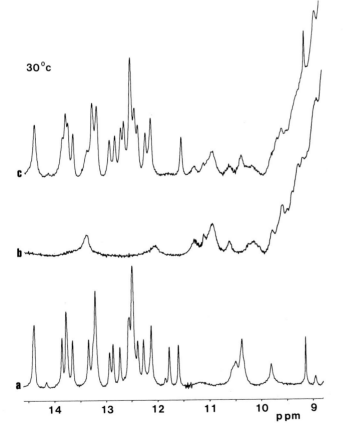

30°c

*Fig. 7. Effect of
complex formation
between yeast tRNA^Phe
and EFTu•GTP on the
imino proton spectrum
of the tRNA.*
*a) Imino proton
    spectrum of yeast
    tRNA^Phe*
b) Spectrum of EFTu•GTP
*c) Spectrum of the
    yeast tRNA^Phe –
    EFTu•GTP complex
Solution conditions are
as indicated in the
legend to Fig. 2.*

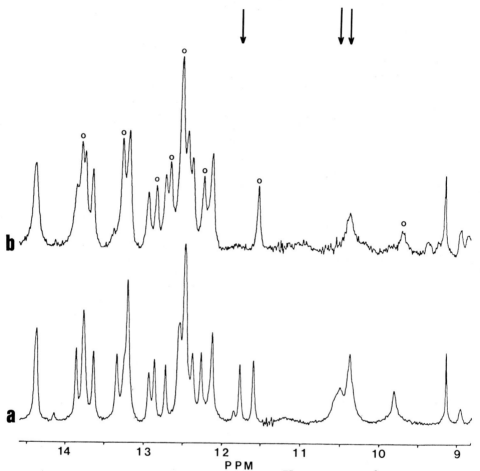

Fig. 8. Comparison of the spectra of tRNA^{Phe} and tRNA^{Phe} in the complex.
a) free tRNA^{Phe} b) spectrum obtained after subtracting the spectrum
of fig. 7b (EFTu·GTP) from the spectrum of fig. 7c (EFTu·GTP + tRNA).
The result is the spectrum of tRNA^{Phe} as it occurs in the complex.
Open circles indicate resonances of tRNA^{Phe} which shift upon complex
formation with EFTu·GTP. Arrows indicate resonances which disappear
from the spectrum upon the complex formation.

interacts with during protein synthesis.

   To this end we choose the elongation factor Tu (EFTu) from *B. stearothermophilus*. We had a preference for the protein from this thermophi-
lic bacterium instead of the corresponding protein from E.coli, because
the former is much more stable. The complex of EFTu and GTP (guanosine
triphosphate) mediates the binding of aminoacyl-tRNA to the ribosome
during the process of chain elongation in prokaryotic organisms (for
reviews see 20-23). The mechanism of the binding reaction is unknown,
but a number of speculations with regard to changes in tRNA structure

during this reaction have been made. Earlier, NMR studies have been
carried out to get some insight as to whether conformational changes
occur upon complex formation with EFTu·GTP from *E.coli* (24). From this
work it was concluded that within the available experimental accuracy
the secondary structure of aminoacylated *E.coli* tRNA[Glu] and *yeast*
tRNA[Phe] remains intact in the complex with the EFTu protein. With the
resolution available at that time it was not possible to detect the
loss of a few imino proton resonances. Moreover, the assignment of the
resonances had not yet reached the level of confidence and detail
discussed in the previous section. Contrary to the prevailing opinion,
it was found that uncharged *E.coli* tRNA[Glu] forms a complex with the
elongation factor Tu from *E.coli*. Later this was confirmed by Pingoud
et al. (25) for the complex formation of EFTu·GTP and *E.coli* tRNA[Val]
and by Bosch and his collaborators (26).

The earlier NMR experiments also showed that the resonances of the
uncharged tRNA broadened somewhat less upon complex formation with the
elongation factor than those of the aminoacylated tRNA. Therefore in
our first approach we studied the complex formation between uncharged
yeast tRNA[Phe] and EFTu·GTP. Fig. 7 shows the low field 500 MHz proton
NMR spectra of three different samples. The bottom spectrum is the
imino proton spectrum of *yeast* tRNA[Phe], which has been discussed in
detail in the previous section. The middle spectrum shows the resonances from the EFTu·GTP complex from *B. stearothermophilus* found between
15 and 9 ppm downfield from DSS. The top spectrum is from a sample in
which *yeast* tRNA[Phe] and EFTu·GTP had been added together in a 1:1 ratio.
Although there is some (ca. 30%) increase in linewidth with respect to
the bottom spectrum, the resolution is maintained remarkably well in
this spectrum. This indicates that the lifetime of the complex formed
between the tRNA and the EFTu·GTP is relatively short. Nevertheless
some significant changes in the tRNA spectrum can be observed. This is
shown in Fig. 8; the bottom spectrum is again from a tRNA solution,
while Fig. 8b presents the spectrum of *yeast* tRNA[Phe] in the complex.
The latter spectrum was obtained by substracting the EFTu spectrum
given in Fig. 7b from the spectrum of the complex given in Fig. 7c. The
most outstanding change in the spectrum of the complexed tRNA is the
disappearance of the resonance at 11.8 ppm which comes from the imino
proton of U69 involved in basepairing with G4 in the GU basepair in the
acceptor stem. Concomitantly, there is a decrease of resonance intensity at 10.4 ppm where the second imino proton of the GU pair is resonating and therefore it is likely that its resonance has disappeared also
from the spectrum in Fig. 8b. The remaining intensity at this position
then most likely arises from m_2^2G26(H1) (which also resonates at 10.4
ppm). This conclusion remains to be verified by NOE experiments.
Furthermore intensity is lost around 10.5 ppm, where the non-hydrogen
bonded imino protons Ψ39 (H3) and Ψ55 (H1) are resonating. In addition
to the disappearance of some resonances, others shift somewhat under
the influence of the EFTu protein (indicated by o in Fig. 8b). For
instance, shifts of about 0.1 ppm are observed for AU52 (T-stem), for
GC1 and GC3 (acceptor stem) and for m[7]G46G22, Ψ55P58 and G18 (H1)
(tertiary structure). Interestingly, the resonances of the D-stem are
not affected at all. Most of the resonances of the anticodon stem over-

lap with others but as far as we can see, they do not shift either, as
exemplified by the resonances of C28G42, A29U41 and A31Ψ39. It is
further noted that the resonances of imino protons involved in tertiary
interactions remain visible in the spectrum. Experiments have also been
performed at lower temperatures. These spectra are characterized by an
increased linebroadening and slight additional shifts. The hydrogen
bonds of the GU pair have the tendency of being reformed again as wit-'
nessed by the reappearance of a broad resonance at 11.8 ppm (resonance
position of U69 (H3)). Despite the diminished resolution at low temperatures, some interesting additional information about the tRNA EFTu
complex can be obtained. To this end the following experiment was performed. Resonances at 0.6 ppm in the aliphatic spectral region of the
protein were preirradiated with a long (0.8 s) powerful presaturation
pulse. Due to the strong spin diffusion in the protein, most of the
protein resonances become saturated. This saturation can be transferred
to the resonances of the tRNA in the complex provided that protons of
the tRNA and of the protein are within certain distances, i.e. ~5 Å,
and that the lifetime of the protein tRNA complex is long enough. An
example of the results of such an experiment performed at 8°C is presented in Fig. 9. In the spectrum of the protein-tRNA complex (Fig. 9a)
the linewidths have considerably increased and a good deal of the resolution seen at 30°C has disappeared. The results of the spin diffusion
experiment are shown in Fig. 9b. The broad resonances at 13.2 ppm and
at 10.8 ppm are from the EFTu protein and arise from the magnetization
transfer from the aliphatic region to these NH protons. The resonances
at 13.8 ppm are from the tRNA. The most likely candidates are A5U68 or
A62U52 from the acceptor stem or the T-stem respectively. It is these
resonances that arise from the transfer of magnetization from protein
residues to the tRNA. These experiments establish that some protein
residue is in close contact with the T-stem at the position of basepair
A62U52 or with the acceptor stem at the position of A5U68.

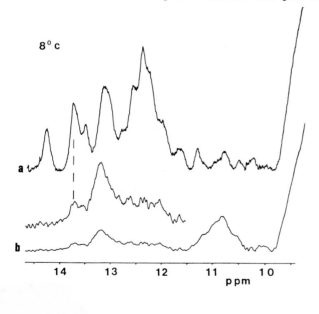

8°C

Fig. 9. Spindiffusion
experiment performed on
the complex of tRNA^Phe
and EFTu·GTP.
a) Unperturbed spectrum
    of the complex
b) Difference spectrum
    obtained after
    irradiation of EFTu·GTP
    resonances at 0.6 ppm
    during 0.8 s.
Middle trace is a blow up
of spectrum b).

a

b

14   13   12   11   10
                  ppm

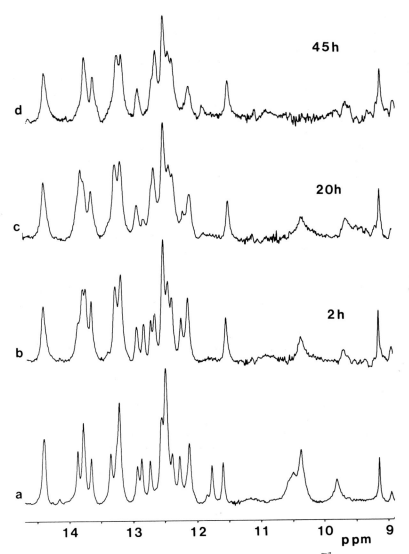

Fig. 10. 500 MHz iminoproton spectra of yeast tRNA^{Phe} complexed to EFTu·GTP as a function of time. The spectrum in (a) is of the free tRNA.

It is interesting to see that after prolonged times, spectra of the sample containing the complex show that additional resonances disappear quite selectively. In Fig. 10a the spectrum of *yeast* tRNA^{Phe} free in solution is presented and compared with the spectra of the complex obtained after 2, 20 and 45 hours. In addition to the disappearance of the GU resonance we see that resonances positioned at 13.8 ppm, 12.8, 12.5 and 12.2 ppm diminish in intensity and eventually disappear. The disappearance of these resonances is complete at 28 hours; subsequently after 45 hours no additional changes are observed in the spectrum. It is most interesting that these resonances are from basepairs A5U6, G3C7, C2G71 and G1C72 in the acceptor stem. At this point it is noted that none of these effects are observed when EFTu·GDP is mixed together with the tRNA. When EFTu·GMP-PCP is studied together with the tRNA we find that again the resonances of the GU pair are 'melted'. In a spin diffusion experiment we find for this complex transfer of magnetization from the protein to basepairs A5U68 and for A62U52 as well. However, after prolonged times the melting of the acceptor stem is not observed as a result of complex formation with this molecule. Qualitatively, these results are in accordance with the known behaviour of the elongation factor. EFTu·GDP does not complex to tRNA even at millimolar concentrations (25); the binding of EFTu·GMP-PCP to aminoacylated tRNA is weaker than that of EFTu·GTP. It is not clear yet whether these interesting phenomena of the melting of the upper part of the acceptor stem by EFTu·GTP is of physiological significance. It is noted that the EFTu from B. stearothermophilus is employed in the cell at much higher temperatures than we used in our experiments. To obtain more information on this point the experiments have to be extended to EFTu molecules from other species, to aminoacylated tRNA and other tRNA's.

A number of biochemical experiments have probed the regions of the aminoacyl-tRNA molecule which interact with the EFTu protein (27-31). There seems to be general agreement, that the protein binds at the amino acid acceptor end of the tRNA. Moreover, from nuclease digestion experiments it has been concluded that the protein binds along the acceptor-TΨC helix (28,29). The present NMR results were obtained for the complex formation of EFTu·GTP with uncharged tRNA. In so far as these results are transferable to the complex formation of EFTu·GTP and aminoacylated tRNA, they support the biochemical experiments, particularly that binding occurs at the acceptor/TΨC limb of the L-shaped molecule.

Acknowledgment

500 MHz proton NMR spectra were recorded at the Dutch National 500/200 hf NMR facility Nijmegen, which is sponsored by the Netherlands Foundation for Chemical Research (SON) with financial aid from the Netherlands Organization for the Advancement of Pure Research (ZWO). We wish to thank Ing. P.A.W. van Dael for keeping the instrument in excellent condition. It is a pleasure to acknowledge stimulating discussion with Prof. L. Bosch and his coworkers. N. Kersten-Piepenbrock is thanked for preparation of the manuscript. We thank P. Wouters for skillful technical assistance.

References

1. Jardetzky, O. and Roberts, G.C.K., NMR in Molecular Biology (1981) Academic Press, New York.
2. Aue, W.P., Bartholdi, E. and Ernst, R.R. (1976) J. Chem. Phys. 64, 2229-2246.

3. Macura, S. and Ernst, R.R. (1980) Mol. Phys. 41, 95-117.
4. Wüthrich, K., Wider, G., Wagner, G. and Brown, W. (1982) J. Mol. Biol. 155, 311-319.
5. Billeter, M., Braun, W. and Wüthrich, K. (1982) J. Mol. Biol. 155, 321-346.
6. Wagner, G. and Wüthrich, K. (1982) J. Mol. Biol. 155, 347-366.
7. Haasnoot, C.A.G. and Hilbers, C.W. (1983) Biopolymers (in the press).
8. Haasnoot, C.A.G., Heerschap, A. and Hilbers, C.W. (submitted).
9. Johnston, P.D. and Redfield, A.G. (1978) Nucl. Acids Res. 5, 3913-3927.
10. Tropp, J. and Redfield, A.G. (1981) Biochemistry 20, 2133-2140.
11. Roy, S. and Redfield, A.G. (1981) Nucl. Acids Res. 9, 7073-7083.
12. Roy, S. and Redfield, A.G. (1983) Biochemistry 22, 1386-1390.
13. Heerschap, A., Haasnoot, C.A.G. and Hilbers, C.W. (1982) Nucl. Acids Res. 10, 6981-7000.
14. Heerschap, A., Haasnoot, C.A.G. and Hilbers, C.W. (submitted).
15. Hare, D.R. and Reid, B.R. (1982) Biochemistry 21, 1835-1842.
16. Hare, D.R. and Reid, B.R. (1982) Biochemistry 21, 5129-5135.
17. Redfield, A.G. and Kunz, S.D. (1979) in NMR and Biochemistry, (Opella, S.J. and Lu, P., Eds.) pp 225-239, Marcel Dekker, New York.
18. Roth, K., Kimber, B.J. and Feeney, J. (1980) J. Magn. Res. 41, 302-309.
19. Dobson, C.M., Olejniczak, E.T., Poulsen, F.M. and Ratcliffe, R.G. (1982) J. Magn. Res. 48, 97-110.
20. Lucas-Lenard, J. and Lipmann, F. (1971) Ann. Rev. Biochem. 40, 409-448.
21. Lucas-Lenard, J. and Beres, L. (1974) in *The Enzymes* (Boyer, P. Ed.) Vol. 10, pp 53-86. Academic Press, New York.
22. Miller, D.L. and Weissbach, H. (1977) in Molecular Mechanisms of Protein Biosynthesis (Weissbach, H. and Pestka, S., Eds.) pp 323-373, Academic Press, New York.
23. Bosch, L., Kraal, B., Van der Meide, P.H., Duisterwinkel, F.J. and Van Noort, J.M. in Progr. in Nucleic Acid Res. and Mol. Biol.(1983) (Coh, W.E., Ed.) Academic Press, New York.
24. Shulman, R.G., Hilbers, C.W. and Miller, D.L. (1974) J. Mol. Biol. 90, 601-607.
25. Pingoud, A., Block, W., Wittinghofer, A., Wolf, H. and Fisher, E. (1982) J. Biol. Chem. 257, 11261-11267.
26. Van Noort, J.M., Duisterwinkel, F.J., Jonák, J., Sedláček, J., Kraal, B. and Bosch, L. (1982) EMBO J. 1, 1199-1205.
27. Jekowsky, E., Schimmel, P.R. and Miller, D.L. (1977) J. Mol. Biol. 114, 451-458.
28. Boutorin, A.S., Clark, B.F.C., Ebel, J.P., Kruse, T.A., Petersen, H.U., Remy, P. and Vassilenko, S. (1981) J. Mol. Biol. 152, 593-608.
29. Wikman, F.P., Siboska, S.E., Petersen, H.U. and Clark, B.F.C. (1982) EMBO J. 1, 10-95-1100.
30. Douthwaite, S., Garrett, R. and Wagner, R. (1983) Eur. J. Biochem. 131, 261-269.
31. Riehl, N., Giegé, R., Ebel, J.P. and Ehresmann, B. (1983) FEBS Letters 154, 42-46.

ELECTRONIC INTERACTIONS IN RNA CONSTITUENTS AND THEIR EFFECTS ON HELICAL CONFORMATIONS

Peter M. Kaiser
Institute of Biochemistry, University of Muenster
Present address: Wyeth-Pharma GmbH, Schleebrüggen-
kamp 15, D-4400 Muenster, FRG

SUMMARY

In order to elucidate further details of RNA conformations we have studied the geometry of base stacking in dinucleoside phosphates (DNP's). We used a modified MIM approximation (SCF CI MO method) to calculate UV difference spectra
and hypochromic effects of heterocyclic model systems and
CpU, UpC, CpC, ApU and UpA for a series of conformations obtained by translational and rotational operations. Results
show that only in homologoues systems Davydov splitting pro-
ducing a blue shift of transitions occur, if polarization
directions are parallel. In all other cases hypochromicity
can be observed. By fitting the hypochromicity of the DNP's
to experimental spectra we obtained distinct geometries for
different DNP sequences.

INTRODUCTION

Vertical base-base stacking interaction in RNA molecules
is a predominant stabilizing factor for the helical structure
Base stacking in connection with electronic interactions between two heterocycles especially account for the hypochromic
effect in UV absorption. Because one may consider the DNP's
as a structural starting point for a right or left handed
helix, investigation of DNP conformation is of utmost importance. Although it is evident that one can approach to the
real structures only by using all available methods, we tried
to find out new aspects while applying quantum chemical calculations.
Hypochromicity as understood is the phenomenological
effect of lowering UV absorption intensity expressed in molar
absorption, oscillator strength, or simply in %. Many efforts
have been undertaken to calculate hypochromicity by quantum
chemical approximations (1-13).

B. Pullman and J. Jortner (eds.), Nucleic Acids: The Vectors of Life, 443–455.
© 1983 by D. Reidel Publishing Company.

In the literature the view is predominant that there is a linear correlation between the magnitude of hypochromicity and the amount of molecular overlap (14).

But if one goes back to the past one can consider that the hypochromic effect often was explained by the exciton theory or the degenerate splitting of the longest wavelength band, respectively. For example, Schneider and Harris (15) calculated hypochromicity in a two-dimensional box model with the aid of the time-dependent Hartree-theory. They described their results as "Davydov-splitting of the degenerate exciton band of 259 and 211 nm peaks" (15).

Davydov splitting is a well-known effect in solid state physics. It was described by A.S.Davydov in 1948 as part of the theory of molecular excitons during the excitation of a crystal (16,17). It has been demonstrated that Davydov-splitting results only in a band shift. In order to demonstrate that Davydov-splitting does not produce a hypochromic effect we reinvestigated basically the effects on the geometrical arrangements of two heterocycles in respect to the excitation energies and oscillator strengths.

This was done with the semiempirical SCF method using "molecules in molecule" (MIM) method first introduced by Longuet-Higgins and Murrell in 1955 (18). This method was applied in the calculations of electronic properties through the fragmentation of aromatic systems (19-22).

At first we investigated the dependency of MIM results on the molecular geometry in very simple heterocyclic chromophors, i.e., pyridine and pyrimidine. Then we extended the calculations to the more complicated bases occuring in nucleic acids such as uracil, cytosine and adenine. We calculated hypochromic and hyperchromic effects during the excitation of CpU, UpC, CpC, ApU and UpA.

METHODS OF CALCULATION

Calculations of the electronic spectra of single chromophors were carried out according to the Pariser-Parr-Pople (PPP) approximation using Klessinger's version of a SCF CI MO program (23).

Interaction spectra of two stacking chromophors were calculated according to the MIM method. Program was written by Klessinger and Gerding (24). UV-spectra of a pi-system R-S can be determined through the calculation of the excited R and S electronic states. The lowest excited states of R-S can be described by a transition from an occupied into an unoccu-

pied MO in the same fragment. Thus, only local excitations
were taken into account, charge transfer transitions being
excluded because of their vanishing contribution to the
hypochromic effect. The interelectronic Coulomb repulsion
integrals $\chi_{\mu\nu}$ between each atom of the R and S system were
calculated by the Mataga-Nishimoto equation (25).

UV spectra were plotted with a PL/1 program (W.A.Visin-
tainer, unpublished) assuming a Gaussian curve for each
transition and an empirical halfwidth of 0.55 eV. The hypochromicity calculations were determined by the equation

$$\%h = [1-f(\lambda)_D/f(\lambda)_M] \times 100 \qquad (1)$$

where $f(\lambda)_M$ is the sum of oscillator strengths of the unperturbed R and S systems and $f(\lambda)_D$ is the oscillator strength
of an interacting dimer at a certain wavelength.

Programs were calculated in the computer center of the
University of Muenster using a IBM/3032 computer. Drawing of
spectra were produced by a BENSON 1232 plotter.

Model heterocycles

Calculations were performed in respect to varying geometries
in the R-S system. The geometrical model is shown in Fig. 1.
A subroutine of the MIM program allows to perform rotations
and translations of one chromophor in respect to the other
chromophor in all spatial directions.

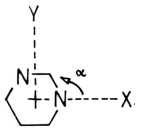

Fig. 1 Geometrical model for translations and rotations of the model heterocycles. Z axis is perpendicular to the paper plane, distance is fixed at
3.4 Å. Coordinate origin is set by a dummy atom in
the center of the six membered ring.

Calculations were performed with four CI's assuming a standard geometry (bond lengths 1.40 Å, bond angles 120°).

Nucleic acid heterocycles

Geometrical model was quite similar as in Fig. 1:

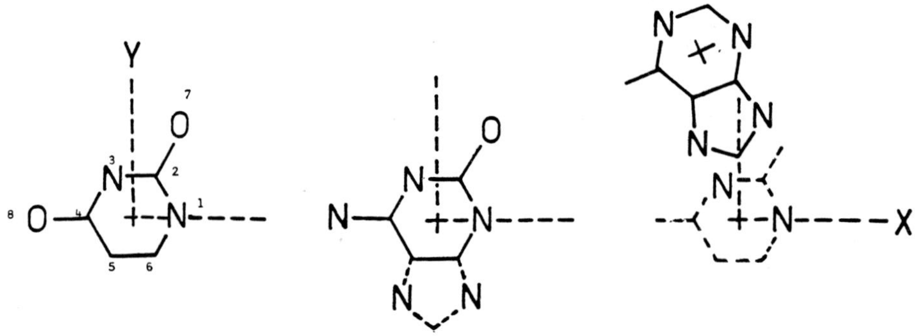

Fig. 2 Geometrical model for translational and rotational operations of nucleic acid base systems (see Fig. 1). Left: pyrimidine sequence, middle: purine/pyrimidine sequence, right: pyrimidine/ purine sequence.

Thus, one can translate and/or rotate one base while fixing the other to positions, realistically mimicing base stacking in a RNA helix. Bond lengths and angles were taken from X-ray data (26) and heteroatomparameters such as Coulomb and resonance integrals were fitted to the experimental spectra (27,28).

RESULTS

MIM calculations were performed for a series of spatial arrangements. The non-stationary molecule was translated along the x-axis (N1-C4), along the y-axis and to rotate around it's center point (α degrees).

Model heterocycles

Dependency of excitation energies and oscillator strengths in a pyrimidine/pyrimidine system in respect to α are shown in Fig. 3 and 4. Calculations were performed at intervals of 30°.

Davydov-splitting occurs if the heterocycles are exactly superimposed ($\alpha = 0°$) and when $\alpha = 180°$. Real hypochromicity in bands I and III and hyperchromicity in bands II and IV occurred at 90° and 270° because at these angles the curves for λ and f intersect. These two different effects are demonstrated in the simulated UV-difference-spectra (Fig. 5 and 6).

In the case of the heterologue pyrimidine/pyridine system energies of the first transitions in each heterocycle

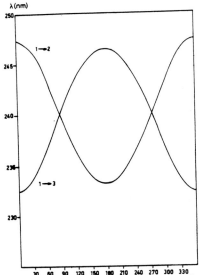

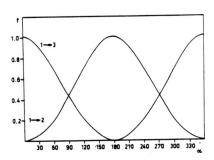

Fig. 3 Davydov-splitting of the longest wavelength transition in the pyrimidine/pyrimidine system dependent from α

Fig. 4 Oscillator strength f vs. α of the same system as in Fig. 3

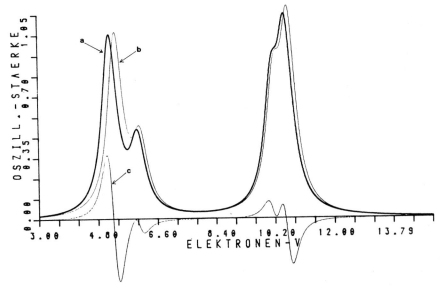

Fig. 5 Computer constructed UV spectra of the pyrimidine/pyrimidine system in terms of oscillator strength vs. excitation energy in eV. Black: sum spectrum, red: MIM interaction spectrum, blue: difference between black and red. Geometry: $\alpha=0°$, X=Y=0, Z=3.4 Å

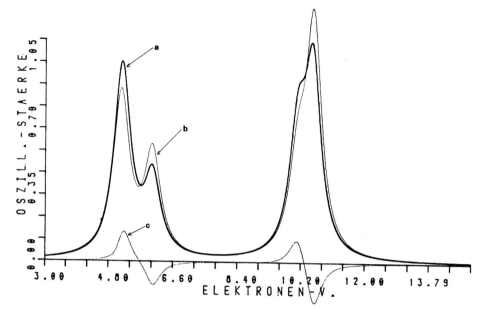

Fig. 6 Computer constructed UV spectra of the same
system as in Fig. 5. Same conditions as in Fig. 5,
except $\alpha = 90°$

were separated by 37 nm. Periodically oscillation occurs in
the curves of λ and f vs.α without crossing of the lines
(Fig. 7 and 8).

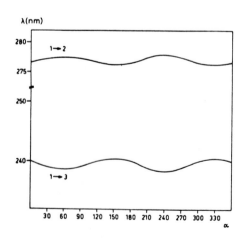

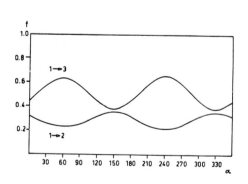

Fig. 7 Dependency of λ
on α for the pyrimidine/
pyridine system

Fig. 8 Dependency of f on
α for the same system as
in Fig. 7

The oscillation of the transitional energy has a very small amplitude, whereas the oscillation of f is stronger. Maximum separation of the f-values occurred at 60° and 240°, minimum at 150° and 330°. This corresponds to the polarizational directions of the transitions of the single heterocycles: at 60° and 240° they are perpendicular and at 150° and 330° they are parallel.

The meaning of theses effects are again demonstrated by simulated UV- and difference spectra (Fig. 9 and 10).

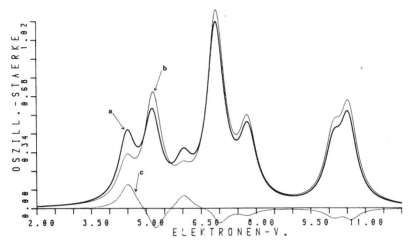

Fig. 9 Computer constructed UV spectra of the pyrimidine/pyridine system (f vs. energy). Same conditions as in Fig. 5, except α = 60°

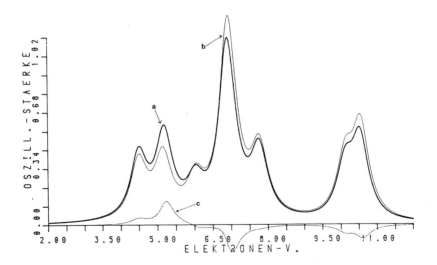

Fig. 10 Same as in Fig. 9, except α = 150°

Only hypochromic and hyperchromic effects occur in the bands (bands 1 to 8).

The effects of translations on the x-axis and the y-axis should be presented. Fig. 11 shows the vanishing of the Davydov splitting in a homologue system when one chromophor moves along the x-axis. At the distance of 4.0 Å a curious crossing point occurs at which a reason could not be elucidated.

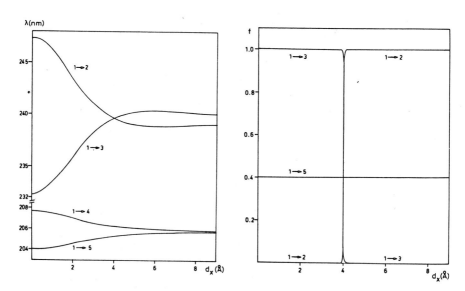

Fig. 11 Dependency of Davydov splitting on translational operation of the pyrimidine/pyrimidine system. One base moves on the x-axis. Left: energy, right: oscillator strength vs. distance between the center points.

In the case of the heterologeous pyrimidine/pyridine system the excitation energy only shifts slightly. Moving on the x-axis produces an increase of the second transition due to pyrimidine (Fig. 12). On the y-axis the same occur with the first transition due to pyridine. At least in both cases f tends to correspond to the single chromophor.

Nucleic acid heterocycles

From these results it is expected that homologue DNP's like CpC, UpU etc. exhibit strong Davydov splitting and therefore extensive difference spectra, if the chromophors are superimposed. That wouldn't correspond to a real stacking geometry either in solution or in solid state. Many efforts has been undertaken in the past to determine the stacking

part in % of a population of DNP conformations in solution
(29,30).

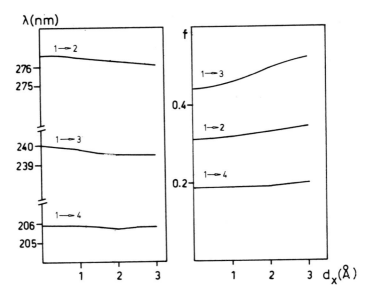

Fig. 12 Dependency of hypochromicity on translational
operation in the pyrimidine/pyridine system. Same
conditions as in Fig. 11

    Calculations carried out in this article intrinsically
presume 100 % stacking. This can be realized only in a solid
state, so one can do solid state spectroscopy.

    The publication of UV-difference spectra of all 16
natural DNP's extrapolated to 100 % stacking from thermodyna-
mical parameters brought us a large step forward (31). Thus
we decided to fit calculated spectra to the corrected experimental % h spectra published by Frechet et al. (31). These
% h spectra were corrected by us in the range 280 nm to
eliminate band broadening and other effects produced mainly
by n-π* transitions.

    From a great variety - nearly 150 - of calculated und
plotted difference spectra of distinct geometries were
chosen with the best fit (Fig. 13 and 14). Using this method
we were able to select stacking geometries from a qualitative
comparison of spectral areas under the curve. (Table 1)

    The effect of hypochromicity in the first two bands and
hyperchromicity in the shorter wavelength region is graphically shown for CpU in Fig. 13 and for ApU in Fig. 14. The
computer plots are composed of the sum of the monomer ab-
sorptions from PPP calculations, the interaction spectrum

DNP	x(Å)	y(Å)	α(degree)	$\%h_{calc.}$[a]	$\%h_{exp.}$[b]
CpU	0	2	30	32.0	25
UpC	0	2	10	11.5	18
CpC	0	3	240	30.8	26
ApU	0	1	10	12.2	20
UpA	-1	4	30	6.2	9

Table 1 Fitted geometries of DNP's and %h values.
a: Values for the longest wavelength absorption, b:
data from Frechet et al.(31), corrected

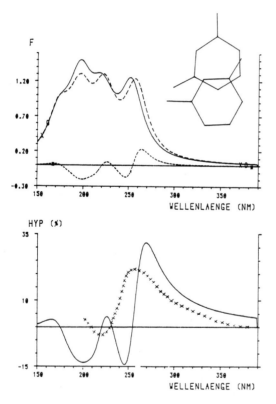

Fig. 13 UV spectra and geometry of CpU after fitting
to the extrapolated %h
spectrum (xxxx). ---- sum
spectrum, ——— interaction
(MIM) spectrum, ... difference spectrum.
Below: %h spectrum

from MIM calculations, the difference between them and the
hypochromicity of hyperchromicity curve, %h vs. wavelength.
Additionally the experimental spectra (31) after correction
are also shown.

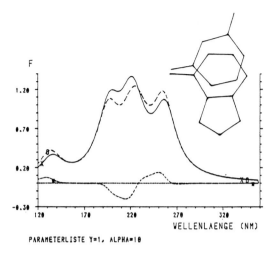

Fig. 14 UV spectra and
geometry of ApU. Same
conditions as in Fig. 13.
Other spectra of DNP's
from Table 1 are not shown
because it wasn't enough
room.

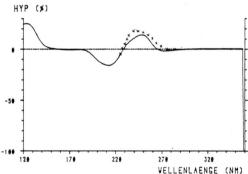

DISCUSSION

Returning to the original purpose of this article, what
is hypochromicity and what is Davydov splitting in the field
of nucleic acid constituents?

Results show that Davydov splitting only occur in homo-
logeous stacking systems with polarization directions of the
transitions nearly parallel. If they are perpendicular
(α = 90°) hypochromicity should result.

Rotational factors account for the qualitative shape of
UV-difference spectra of the stacking systems. Translating
operations correlate with a linear decrease of the band shift
(generally blue shift) or the magnitude of hypochromicity
or hyperchromicity, respectively.

In all the heterologeous systems it is only possible to
produce hypochromicity or hyperchromicity, but this does not
mean a direct correlation between hypochromicity and base

stacking or the overlapping of the chromophors. Thus, CpC shows a completely unoverlapped structure (α = 240°) yet the %h value is in the range of 30% (Table 1).

Moreover, as the DNP's CpU and UpC or ApU and UpA show there is a sequence depending effect of %h values and the geometry in each individual DNP. This may correspond to a sequence depending helix angle θ in the macrostructure of a nucleic acid helix (32). The macrostructure should therefore not be equated to the poly(A-U) and poly(G-C) models extrapolated by Rosenberg et al. (33).

Recently this view was confirmed by the findung of variations in the backbone conformation of a crystalline hexamer in the Z-DNA conformation (34) and a dodecamer in the B-DNA conformation (35).

ACKNOWLEDGMENTS

The author wish to thank Professor M.Klessinger und Dr.B.Gerding, Muenster, for their kindness and help in preparing the PPP and MIM programs for his use. Many thanks to Mr. W.Luft and Mr. W.A.Visintainer for their help in developing the plot programs.

REFERENCES

1)Bolton,H.C.,and Weiss, J.J.:1962, Nature 195, pp. 666-668
2)Nesbet,R.K.:1963-64, Mol.Phys. 7, pp. 211-221
3)Nesbet,R.K.:1964,Biopolymers Symp.No.1, John Wiley&Sons, New York, pp. 129-139
4)Ladik,J.,and Appel,K.:1966, Theor.Chim.Acta 4, pp.132-144
5)Ladik,J.,and Sundaram, K.:1969, J.Mol.Spectr. 29, 146-151
6)Sundaram,K.,and Ladik,J.:1972, Physiol.Chem.Phys. 4, pp. 483-491
7)Ladik,J.:1973, in P.-O.Löwdin (Ed.) Adv.Quant.Chem. Vol. 7, Academic Press, New York, pp. 297-445
8)Danilov,V.I.,and Pechenaya,V.I.:1974,Studia Biophys. 44, pp. 33-49
9)Danilov,V.I.:1974, Studia Biophys. 46, pp. 115-129
10)Volkov,S.N.,and Danilov,V.I.:1976, FEBS Letters 65, pp. 8-10
11)Balcerski,J.S.,and Pysh,E.S.:1976, Biopolymers 15, pp. 1873-1875
12)Volkov,S.N.:1979, Biofizika 24, pp. 408-412
13)Danilov,V.I.,Pechenaya,V.I.,and Zheltovsky,N.V.: 1980, Int.J.Quant.Chem. XVII, pp. 307-320
14)Bush,A.:1974, in P.O.Ts'o (Ed.) Basic Principles in

Nucleic Acid Chemistry, Vol. II, Academic Press, New York, pp. 91-169

15) Schneider,A.S.,and Harris,R.A.:1969, J.Chem.Phys. 50, pp.5204-5215

16) Davydov,A.S.:1971, Theory of Molecular Excitons, Plenum Press, New York/London

17) Murrell,J.N.:1967, Elektronenspektren Organischer Moleküle, Bibliographisches Institut, Mannheim

18) Longuet-Higgins,H.C.,and Murrell,J.N.: 1955, Proc.Phys. Soc. A 68, pp. 601-612

19) Simonetta,M.,Gamba,A.,and Rusconi,E.:1970, in E.D.Bergmann and B.Pullman (Eds.) Quantum Aspects of Heterocyclic Compounds in Chemistry and Biochemistry, The Jerusalem Symposia on Quantum Chemistry and Biochemistry, Vol. II, Jerusalem, pp. 213-225

20) Martin,A.,and Kiss,A.I.:1976, Acta Chim.Acad.Hung. 91, pp. 105-117

21) Martin,A.,and Ray,N.K.:1978, Indian J.Chem. Sect.B 16, pp. 517-518

22) Fabian,J.,and Scholz,M.:1981, Theor.Chim.Acta(Berl.) 59, pp. 117-125

23) Klessinger,M.:1974, Program No. 1 of the Library of Theoretical Organic Chemistry, University of Muenster, FRG

24) Klessinger,M.,and Gerding,B.:1979, Program Library of Theoretical Organic Chemistry , University of Muenster, FRG

25) Mataga,U.,and Nishimoto,K.:1957, Z.Phys.Chem. NF 12, pp. 335-338; 13, pp. 140-157

26) Voet,D.,and Rich,A.:1970, Progr.Nucl.Acid Res.Mol.Biol. 10, pp. 183-265

27) Gumbinger,H.G.,and Kaiser,P.M.:1980, Int.J.Quant.Chem. XVIII, pp. 439-448

28) Kaiser,P.M.:1981, Nucleic Acids Res.Symp.Ser. 9, pp. 1-5

29) Lee,C.-H.,Ezra,F.S.,Kondo,N.S.,Sarma,R.H.,and Danyluk, S.S.:1976 Biochemistry 15, pp. 3627-3639

30) Lee,C.-H.,Charney,E.,and Tinoco,I.:1979, Biochemistry 18, pp. 5636-5641

31) Frechet,D.,Ehrlich,R.,Remy,P.,and Gabarro-Arpa,J.: 1979, Nucleic Acids Res. 7, pp. 1981-2001

32) Kaiser,P.M.:1981, GDCh, 19. Hauptversammlung, Hamburg, 14.-18. Sept. 1981, Verlag Chemie, Weinheim/Deerfield Beach/Basel, p. 177

33) Rosenberg,J.M.,Seeman,N.C.,Day,R.O.,and Rich,A.:1976, Biochem.Biophys.Res.Commun. 69, pp.979-987

34) Wang,A.H.-J.,Quigley,G.J.,Kolpak,F.J.,van der Marel,G., van Boom,J.H.,and Rich,A.:1981, Science 211, pp. 171-176

35) Drew,H.R.,Wing,R.M.,Takano,T.,Broka,C.,Tanaka,S.,Itakura, K.,and Dickerson,R.E.:1981, Proc.Natl.Acad.Sci.USA 78, pp. 2179-2183

Correlated Internal Motions in Left-Handed DNA

Double Helices: Z-B Transition Driven by Drug Binding*

Goutam Gupta and Ramsawamy H. Sarma
Institute of Biomolecular Stereodynamics and
Center for Biological Macromolecules
State University of New York at Albany
Albany, NY 12222

From molecular model building and fiber diffraction studies, it was earlier shown[1] that Z-DNA really represents a family of left-handed zig-zag helices (LZ helices for short). LZ1 and LZ2 helices were designated as two extremes. Although these two types of helices are different in overall topology, they share the following common properties: (i) a dinucleotide repeat with guanine in (C3'-endo, syn) conformation while cytosine in (C2'-endo, anti) conformation, (ii) intrastrand stacking in the GpC sequence and interstrand stacking in the CpG sequence, (iii) oxygen atoms in the successive sugar residues in a given strand pointing in the opposite directions while the oxygen atoms of the sugars across a base-pair pointing in the same direction. The essential difference between the LZ1 & LZ2 helices is in the P-O torsions of the GpC fragment. In the LZ1 helix the P-O torsions in the GpC fragment are g^-,t while in the LZ2 helix they are g^+,g^+.

*Publication No. 4 from National Foundation for Cancer Research
State University of New York at Albany

B. Pullman and J. Jortner (eds.), Nucleic Acids: The Vectors of Life, 457–470.
© 1983 by D. Reidel Publishing Company.

Thus, by correlated internal motions around the P-O bonds in the GpC fragment it is possible to switch from LZ1 to LZ2 helix (and vice-versa) without ever disturbing the base-pairing. This (LZ1 to LZ2) is a local order to order transition achieved by correlated internal motions and hence there is little or no energy barrier. However, the effect of correlated internal motions in the LZ helices can be more dramatic than that. In this paper, we demonstrate that such motions in the LZ helices can also generate breathing modes resulting in the sequence specific drug intercalation.

Mission

Ethidium is known to bind specifically in the CpG sequence of poly(dGdC) while actinomycin D (Act D) binds in the GpC sequence. Both the drugs have been shown[2,3] to bind to poly(dGdC) in the Z-form and cause a Z to B transition. Hence, we asked ourselves the question: Can Ethidium (Act D) find an initial binding site in the CpG (GpC) sequence of poly(dGdC) such that it can finally cause the observed Z to B transition? If so, what is the precise stereochemical nature of the drug binding and the drug induced Z to B transition?

Left-Handed Intercalated Z-DNA

Figures 1a,b show the dinucleotide repeat of the LZ1 helix and the resulting duplex. It was found when two torsion angles ω' and ψ of the cytosine residue were changed from g^+ to t conformation, a nearest neighbour exclusion model of intercalation was obtained (See Table 1). The intercalated model has the following conformational features. It is a left-handed duplex with a dinucleotide repeat as shown in Figures 1c,d; all the guanine residues have (C3'-endo, syn) conformation while the cytosine residues have (C2'-endo, anti) conformation. The site of intercalation is at the CpG sequence and the GpC sequence retains almost the same kind of stacking arrangement as in the parent LZ1 helix. Thus all the properties of the LZ1 helix are preserved in the corresponding intercalated model except for the

Table 1. Torsion angles of the models described in the text.

Torsion Angles

Models	Sequence	5' χ	ψ'	ϕ'	ω'	ω	ϕ	ψ	3' ψ'	χ
LZ1	GpC	235	96	248	314	218	213	55	120	20
	CpG	20	120	256	93	86	169	154	96	235
LZ1*	GpC	265	71	233	280	196	187	138	131	-15
	CpG	-15	131	273	147	83	185	160	71	265
LZ2	GpC	235	76	187	100	100	141	75	156	17
	CpG	17	156	277	79	62	179	171	76	235
LZ2*	GpC	230	98	176	73	94	135	140	134	-20
	CpG	-20	134	280	132	100	135	138	98	230
Left-handed B-DNA	GpC	-15	138	284	167	52	218	177	135	-15
	CpG	15	135	250	220	275	130	30	138	15
Left-handed B-DNA*	GpC	-5	120	285	171	54	220	174	133	-10
	CpG	-10	133	233	220	273	158	31	120	-5

*Indicates the intercalated version.

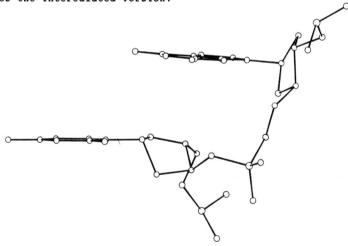

Figure 1a. The dinucleotide repeat of the LZ1 helix. The torsion angles are as indicated in Table 1. The antiparallel chain is generated by a two-fold operation around dyad between the bases. Successive dinucleotide units are generated by a rotation e = -60° and translation, h = 7.5Å.

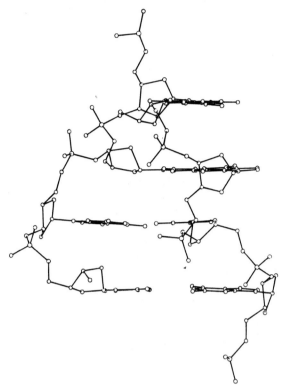

Figure 1b.   The arrangement of four base-pairs in the LZ1 helix.

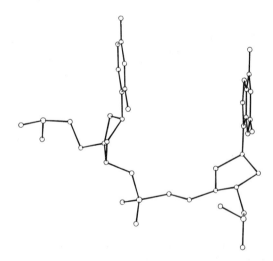

Figure 1c.   The dinucleotide repeat of the intercalated LZ1 helix.   The
antiparallel chain is generated in the same way as said in 1a.   Successive
dinucleotide units are generated by t = -60° and h = 10.2Å.

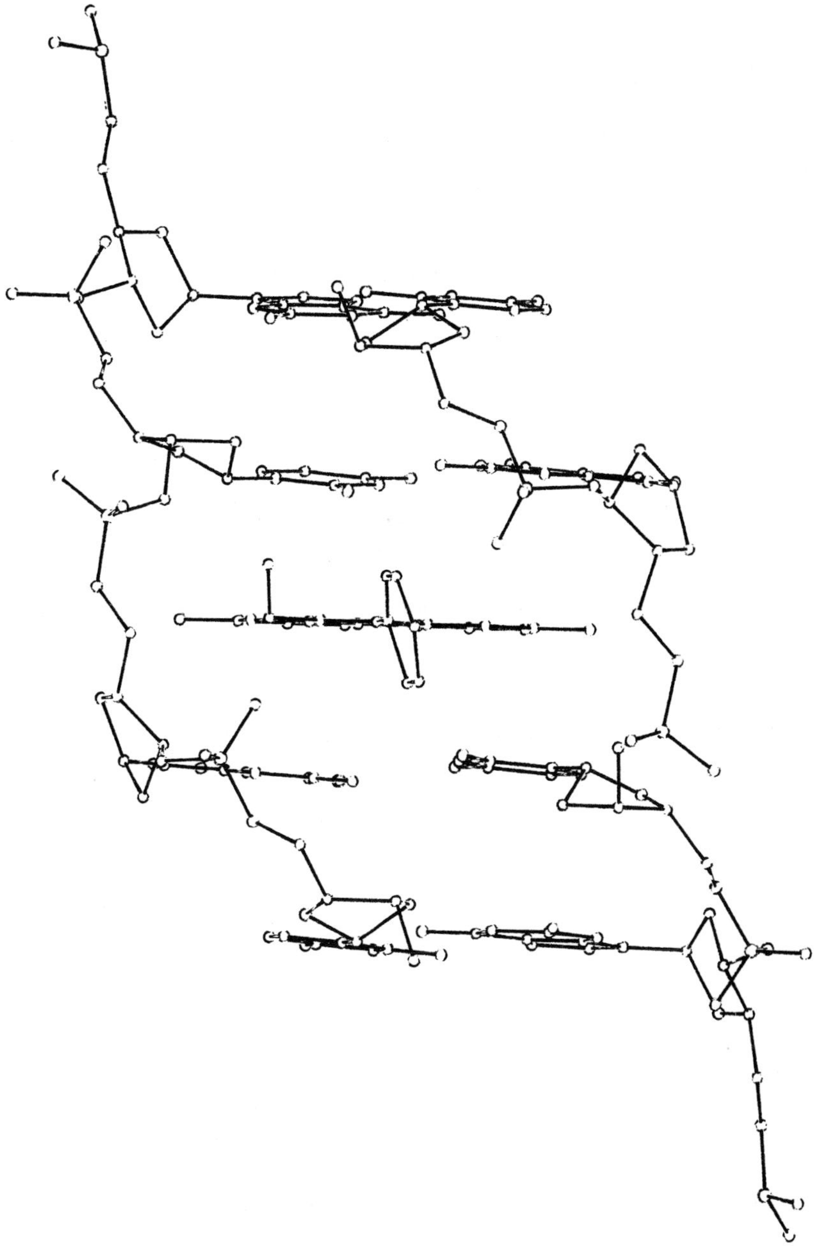

Figure 1d. The nearest neighbour exclusion model of intercalation with CpG as the site of intercalation Ethedium being the intercalator.

fact that there is a potential site of intercalation in the latter. However, it may be pointed out that the orientations of the sugar oxygen atoms are somewhat different in the intercalated model from that in the parent LZ1 helix (compare Figures 1b and d). In a similar fashion as shown in Figures 2 a-d, by changing ω' and ψ of the cytosine residue, it was possible to obtain a nearest neighbour exclusion model of intercalation starting from a LZ2 helix (see Table 1). However in this case the GpC sequence is the site of intercalation. Thus, our results clearly establish that correlated internal motions can generate sequence specific intercalation sites in the LZ helices i.e., at the CpG sequence of the LZ1 helix and the GpC sequence of the LZ2 helix.

Stereochemistry of the site specific drug binding in Left Handed Z-DNA

Our main interest being in the intercalative mode of binding, we would be concentrating on Ethidium binding to the intercalating LZ1 helix while mentioning that the phenoxazone ring of Act D can also be snuggly fitted in the GpC sequence of the intercalating LZ2 helix.

The position of Ethidium in the CpG pocket was chosen in the following manner. The approximate two-fold axis in the phenanthridine ring of Ethidium was made coincident with the dyad axis of DNA with the phenyl and ethyl group of the drug facing the minor groove. Then three rigid body rotations plus three translations were performed on the drug molecule to choose a position which gave appreciable geometric overlap between the drug and the bases but no steric compression. Figure 3 shows the intercalation geometry of Ethidium in the CpG sequence. Note that the pattern of intercalation is different from that observed in the single crystal of Ethidium-IodoCpG complex[4], though there is an approximate two-fold symmetry in the interaction of DNA and the drug in the minor groove, phenanithridine ring mainly overlaps on cytosines and hardly on guanines (exactly opposite to what is found in the single crystal). The intercalated helix being slim, the drug is quite tightly fitted

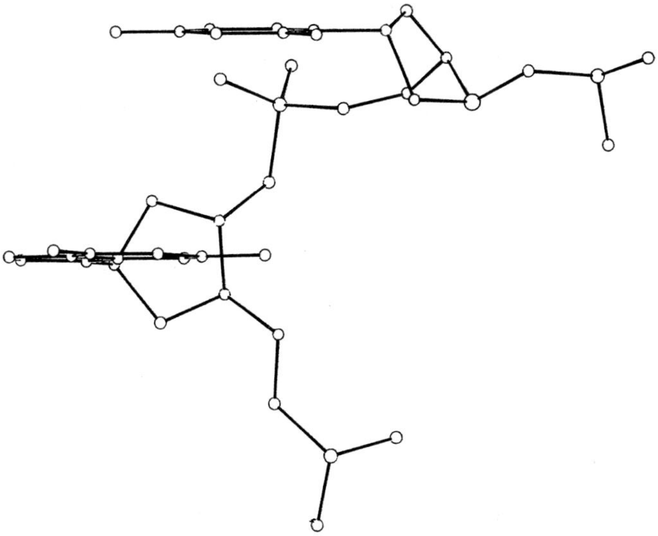

Figure 2a.   The dinucleotide repeat of the LZ2 helix with t = -60°, h=7.5Å.

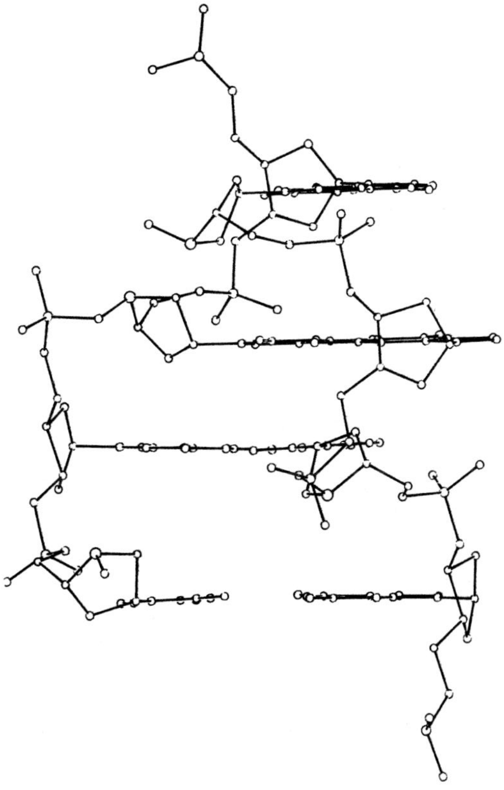

Figure 2b.   The arrangement of four base pairs in the LZ2 helix.

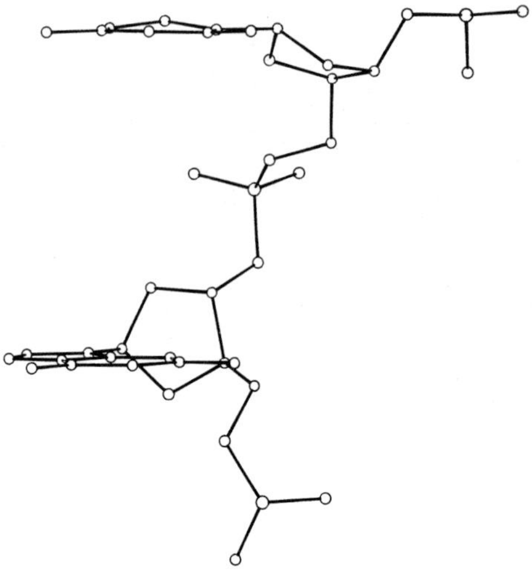

Figure 2c. The dinucleotide repeat of the intercalated LZ2 helix with t=-60°, h=10.2Å.

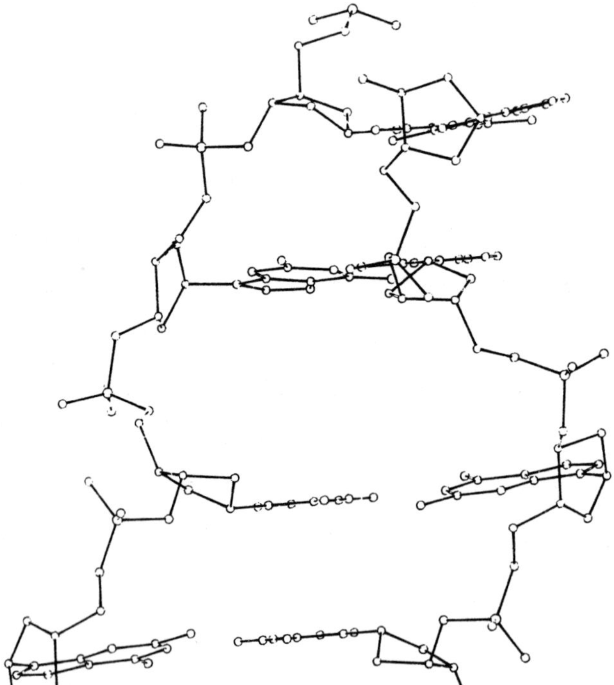

Figure 2d. The nearest neighbour exclusion model of the LZ2 helix with GpC as the site of the intercalation.

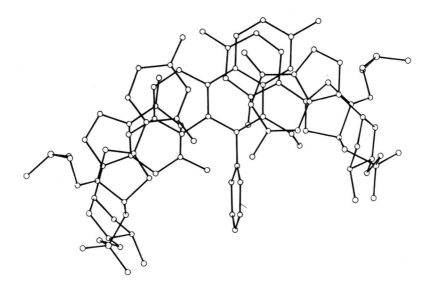

Figure 3.   Geometry of overlap of Ethedium in the CpG sequence of the
intercalated LZ1 helix.

with the phenyl ring placed symmetrically between the amino groups of two

guanines and the ethyl group close to one of the cystosine sugars.   It may

also be noted that there is hardly any change in the winding angle of CpG

sequence.   This is probably due to the fact that the winding angle in CpG

sequence of the LZ1 helix is already very small.   Hence, Ethidium binding at

the CpG sequence of the intercalating LZ1 helix provides an example of

non-disruptive yet tight mode of binding which may in turn serve as the

nucleation of Z to B transition.

Z to B Transition: possibility of a left-handed intercalated B-DNA as an

intermediate

For Z to B transition to occur, guanine residues from the edge of the

helix surface in the LZ1 helix should come to the centre and should turn from

(C3'-endo, syn) to (C2'-endo, anti) conformation, along with the change of

handedness (assuming that the B-DNA is right-handed).   Thus in this process of

drug induced transition, it is not unlikely that a left-handed intercalated

B-DNA appears as an intermediate provided that the latter can hold the drug
with binding energy close to that of the right handed intercalated B-DNA. We
discovered that such a possibility does really exist.

Figures 4a,b show the dinucleotide repeat and the resulting intercalated
B-DNA duplex. Note, both guanine and cytosine have (C2'-endo, anti)
conformation and hence, such a model can offer site of intercalation either at
the CpG or the GpC sequence. But what is more important to note is that such
a model also demonstrates the presence of correlated internal motions in DNA.
The dinucleotide repeat of Figure 4a is obtained by giving minor changes to
that of Figure 4c which leads to a regular left handed B-DNA as shown in
Figure 4d with uniform 3.4$\overset{\circ}{A}$ of base separation[5]. While in our earlier
examples we had large changes on two torsion angles; in the present case the

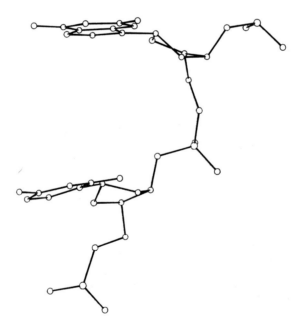

Figure 4a. The dinucleotide repeat of the left-handed intercalated B-DNA
with t = -52° and h = $10.2\overset{\circ}{A}$. Note that, there is a net unwinding of 20°
per dinucleotide repeat.

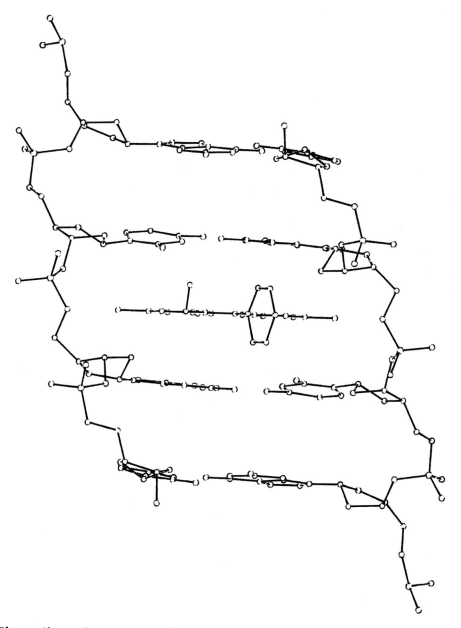

Figure 4b. The nearest neighbour exclusion model of the left-handed B-DNA. Here CpG is chosen as the site of the intercalation.

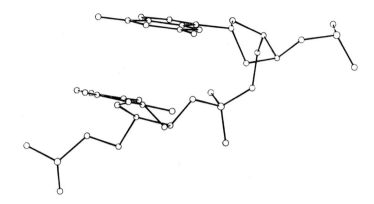

Figure 4c. The dinucleotide repeat of a regular left-handed B–DNA with t = –72° and h = 6.8Å.

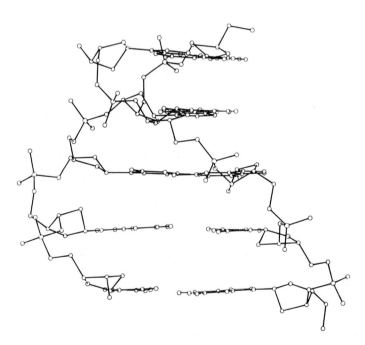

Figure 4d. The arrangement of five base pairs in the left-handed B–DNA with a dinucleotide repeat.

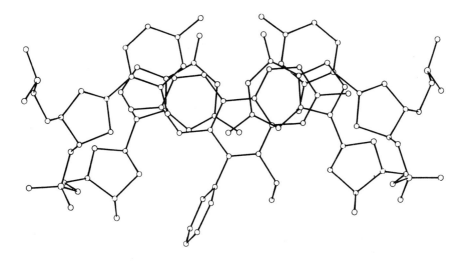

Figure 5. Geometry of overlap of Ethedium in the CpG sequence of the intercalated left-handed B-DNA.

changes are evenly distributed all over the repeat; however, the end product is still a left-handed intercalated B-DNA model from a regular one. This we consider the most striking example of correlated internal motions in DNA.

The final intercalation geometry was obtained in the same manner as described earlier. From Figure 5, the symmetric nature of interaction is very evident. In the present case, the bases being close to the helix centre the helix has got a larger dimension than the intercalated LZ1 helix; as a result the amino groups of two guanines are moved away from the phenyl ring (compare Figures 3 and 5). The overlap geometry of the phenanthridine ring in this model is very similar to that in the Ethidium-IodoCpG complex[4], i.e. overlap is only on guanines and hardly on cytosines. The close resemblance in the intercalation geometry should suggest comparable energies for Ethidium binding to right and left-handed intercalated B-DNA models.

Conclusion

Thus, we provide a clear evidence that a sequence specific intercalator like Ethidium can find a site in the LZ1 helix. The intercalation geometry, though suggests a tight binding, the entry of the drug through the minor groove is stereochemically restricted which poses a kinetic barrier. However, such a barrier disappears when the intercalated LZ1 helix gets converted into the left-handed intercalated B-form wherein the intercalation site is wide open in the minor groove and the binding is equally strong. And it is quite reasonable to expect such a gradual opening of the minor groove and in fact, this defines the essential morphological change in the drug induced Z to B transition.

References

1. Sasisekharan, V., Bansal, M., Brahmachari, S., K., & Gupta, G., in Biomolecular Stereodynamics, Volume 1, Ed., Sarma, R.H., Adenine Press, N.Y., p. 123-149 (1981).

2. Van de Sande, J.H., & Jovin, T.M., EMBO Journal, $\underline{1}$, 115-120 (1982).

3. Pohl, F.M., Jovin, T.M., Baehr, W., & Holbrook, J.J., Proc. Natl. Acad. Sci, USA, $\underline{69}$, 3805-3809 (1972).

4. Jain, S.C., Tsai, C.C., & Sobell, H.M., J. Mol. Biol., $\underline{114}$, 317-331 (1977).

5. Sasisekharan, V. (unpublished results).

The Binding of the Carcinostatic Drugs Cis-[Pt(NH$_3$)$_2$Cl$_2$] and [Ru(NH$_3$)$_5$Cl]Cl$_2$ with Yeast Transfer RNA^{Phe}. Absence of Binding to Watson-Crick Base Pairs.

J.R. Rubin, M. Sabat and M. Sundaralingam
University of Wisconsin-Madison, College of Agricultural and
Life Sciences, Department of Biochemistry, Madison, WI
53706 USA

Cis-diammedichloroplatinum(II) is a potent anticancer agent and is widely used in clinical work (1). On the other hand, the trans-diammine-dichloroplatinum(II) is inactive (2). A search for related platinum group metal containing oncostatic drugs has led to the discovery of the pentaammineruthenium(III) complex which also inhibits cellular DNA synthesis in a manner (3) similar to cis-[Pt(NH$_3$)$_2$Cl$_2$]. Solution studies have shown that the guanine bases of DNA are preferentially attacked by the Pt drug at low drug to DNA ratios (4). The lack of oncostatic activity of the trans-Pt has led to the suggestion that the cis-Pt exerts its activity by forming intrastrand crosslinks between adjacent guanines on the DNA chain (5). This mode of complexation has received some support from solution studies of polynucleotides (6) and single-crystal X-ray studies of cis-Pt complexes with nucleic acid constituents (7-9). A competing model for the intrastrand crosslinked model is the bidentate coordinaton of the cis-Pt to the N(7) and keto O(6) atoms of guanines. Neither complexation mode is possible for trans-Pt. We have carried out a crystallographic investigation of the interaction of both the cis-Pt, and the trans-Pt as well as the [Ru(NH$_3$)$_5$]$^{3+}$ ion with yeast phenylalanine transfer ribonucleic acid (tRNA^{Phe}) to provide further insights on the nature of the interaction of these metal complexes. tRNA^{Phe} is a good system since its structure is known and crystals suitable for X-ray diffraction studies can be obtained with little difficulty. The main drawback, however, is that tRNA is in some ways different from DNA and therefore the results obtained should be extrapolated to DNA with caution.

METHODS

The details of the crystallographic studies including the metal-tRNA complex preparation are described elsewhere (10,11). The metal binding sites were determined from difference Fourier maps calculated using coefficients $||F_M| - |F_{Nat}||\alpha_{calc}$, where $|F_M|$ and $|F_{Nat}|$ are the observed structure factor amplitudes for the appropriate metal-tRNA complex and the native-tRNA crystals respectively. α_{calc}

471

B. Pullman and J. Jortner (eds.), Nucleic Acids: The Vectors of Life, 471–477.
© 1983 by D. Reidel Publishing Company.

is the calculated phases from the native tRNA coordinates (12). In all cases the work was done at low resolution (4-5.5Å). This required that we assume the square planar geometry for the Pt and octahedral geometry for Ru obtained from model studies on metal-nucleoside complexes.

RESULTS

Cis-Pt Binding to tRNA^{Phe}

Contrary to earlier results (13-15) cis-Pt binds to tRNA when crystals are soaked in a solution of the reagent for about 9 days. Crystals soaked for a shorter period (4 days) showed no significant binding. Cis-Pt is found to bind at the guanine residues G15 and G18. In both cases the Pt is involved in a monodentate ligation to the base N(7) site. The complex is stabilized by interligand hydrogen bonding between the Pt ligands (NH$_3$/H$_2$O) and the exocyclic oxygen O(6) of the guanine as well as to neighboring phosphates and bases (figure 1).

Trans-Pt Binding to tRNA^{Phe}

Trans-Pt binds differently to tRNA. Crystals soaked for 23 days showed the strongest binding at the N(1) site of A73. The complex is stabilized by interligand hydrogen bonding to N(3) and O(2) of C74 and O(6) of G1 on the 5' strand (figure 2). The other trans-Pt binding sites involve monodentate ligation to N(7) sites of the guanines G^{m}34, G18 and G43 in that order of strength (Pt occupancy). G18 binding is common to trans- and cis-Pt and the coordination is also similar for both complexes. However, trans-Pt does not exhibit binding to G15.

Ru(NH$_3$)$_5$$^{3+}$ Binding to tRNA^{Phe}

Two studies were done on ruthenium binding to tRNA. In one a crystal was soaked for 25 days (Ruphe-25) in a saturated solution of [Ru(NH$_3$)$_5$Cl]Cl$_2$ while in the other the crystal was soaked for 58 days (Ruphe-58). In Ruphe-25 two binding sites were observed. Site I is located in the deep groove of the amino acid stem (figure 3). There is no direct coordination of the ruthenium to the tRNA bases at this site, but rather the [Ru(NH$_3$)$_5$(H$_2$O)]$^{3+}$ complex is hydrogen bonded through ammonia ligands to the G4-U69 base pair as well as the adjacent bases G3 and U68. Site 2 involves covalent binding to the N(7) atom of G15 in the dihydrouridine loop. The difference density maps indicate that binding here causes disruption of the G15-C48 tertiary base pair. The complex is stabilized by hydrogen bonds between ammonia ligands and the keto oxygen atom O(6) of G15, and phosphate oxygens on P14 and P15.

Ruphe-58 showed three ruthenium binding sites. It is interesting that the deep groove binding site at the G4-U69 pair observed in Ruphe-25 is no longer found. The binding of ruthenium to N(7) of G15 is observed with shifts in the position of the base, similar to those seen before. In addition to the G15 site, ruthenium is also directly bound to the N(7) position of the residues G1 (figure 4) and G18.

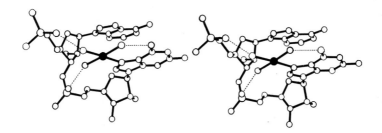

Figure 1   A steroscopic view of the <u>cis</u>-Pt binding site at G15.
           <u>Cis</u>-Pt (solid ball) is directy coordinated to N(7) of G15
           and forms interligand hydrogen bonds (dashed lines) to O(6)
           of G15 and phosphate oxygens of P14 and P15. tRNA^Phe atomic
           coordinates are from reference 18.

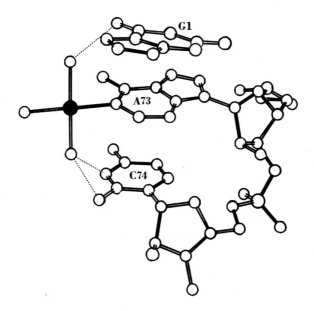

Figure 2   A plot of the <u>trans</u>-Pt binding site at A73. <u>trans</u>-Pt (solid
           ball) is directly coordinated to N(1) of A73 and forms inter-
           ligand hydrogen bonds (dashed lines) to O(6) of G1 and N(3)
           and O(2) of C74.

DISCUSSION:

Cis-Pt binds strongly at two sites in orthorhombic tRNA[Phe] crystals, viz. G15 and G18. In both cases direct coordination of Pt to the N(7) position of guanine is observed. The Pt coordination to guanine is monodentate. The previously proposed bidentate coordination to N(7) and O(6) of guanine is not observed (2). The Pt-guanine complexes are stabilized by interligand hydrogen bonding to O(6) of the base. The guanines G15 and G18 are both located in tertiary base paired regions of tRNA and neither is involved in standard Watson-Crick base pairing. Cis-Pt does not form intrastrand crosslinks between adjacent guanines in tRNA[Phe] although there are three regions in the molecule (G3-G4, G18-G19-G20 and G42-G43) which contain consecutive guanine sequences.

The trans-Pt binds at a novel site at the N(1) position of adenine (A73). In addition trans-Pt binds at the N(7) position of guanines G18, G[m]34 and G43. Although the binding site at G18 is common to cis-Pt, binding at G[m]34 and G43 are unique to trans-Pt. In monoclinic crystals of tRNA[Phe], trans-Pt showed a single binding site at G[m]34 but showed no binding to A73, G18 or G43 (13,15). The minor trans-Pt binding site at G43 is the only instance observed for platinum binding in a double-helix although it should be noted that G43 is located at the terminus of the anticodon stem.

In general, it appears that cis- and trans-Pt complexes do not bind to Watson-Crick base pairs. Model studies of cis-Pt complexes of nucleosides (7-9) indicate that the square planar platinum complex is twisted at an angle of about 90° to the base plane. The twist alleviates the steric clash between the Pt-ligand and the guanine keto oxygen atom O(6). The Pt-N(7) coordination is usually accompanied by an interligand hydrogen bond between a Pt-ligand and the O(6) atom. In this preferred coordination geometry, it is seen that binding to N(7) in the deep groove would produce unfavorable Van der Waals contacts between the Pt ligands and the base stacked on the 5' side. This steric clash is particularly severe in the compact A-RNA and A-DNA type double helical structures, but is less severe in the broad groove of the more extended B-DNA structure. Thus, the binding would require a distortion or denaturation of the DNA helix. On the other hand, in the Z-DNA conformation the N(7) position of guanines are more exposed and would be expected to bind Pt more readily. Thus, monodentate coordination of Pt to Z-DNA will be favored over B-DNA and A-DNA.

The binding of the hexacoordinated ruthenium to N(7) of G15 and G18 is similar to that observed for the square planar cis-diammedichloroplatinum(II) complex with tRNA[Phe] (11). The similarity in monodentate binding properties of these two metal complexes may underlie their carcinostatic properties. Both ruthenium and platinum are observed to bind to the N(7) site of specific guanines in the nonhelical regions of the tRNA. The binding to these guanine bases in tRNA[Phe] is related to the fact that the bases are exposed and occur in a region of the molecule where negative charge density due to phosphates is high as shown by theoretical studies (16). The same region

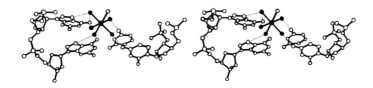

Figure 3   A stereoscopic view of the $[Ru(NH_3)_5H_2O]^{3+}$ binding site at
the G4-U69 base pair. The complex is hydrogen bonded to
N(7) and O(6) of G3 and G4 as well as to O(4) oxygens of U68
and U69 as indicated by dashed lines.

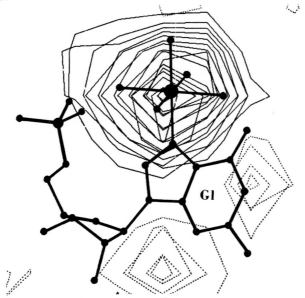

Figure 4   A composite Ru-native difference electron density map in
the vicinity of residue G1 showing the binding of
$[Ru(NH_3)_5]^{3+}$ to N(7) of G1.

(adjacent to residue G15) has been previously observed to bind the organic cationic drug ethidium (17). The complex is stabilized by interligand hydrogen bonding between the ammonia ligands and the O(6) atom of guanine and to neighboring phosphates and bases. This hydrogen bonding contributes to the binding specificity of these carcinostatic drugs for guanine bases.

It should be emphasized that the present study was done on a tRNA molecule and generalization of these results for other nucleic acids should be made with caution. Nevertheless, one may consider possible consequences of the findings presented here for construction of models for nucleic acid-metal interactions and understanding the mechanism of the carcinostatic behavior of these metal complexes. In general we find that the Pt and Ru complex ions show a reluctance to bind to bases in the Watson-Crick double helical stems. The binding of Ru to G1 at the end of the acceptor helix and trans-Pt to G43 at the end of the anticodon helix are exceptions.

The lack of covalent binding of the drug in double helical regions may be a consequence of diminished exposure of the guanine N(7) sites in the deep groove of the A-RNA helix. The steric obstruction of the binding of the metal complex seems to be mainly due to the base stacked on the 5'-side of the guanine. Therefore, the A-RNA type right-handed helix would display the steric hindrance mentioned above and the accessibility for metal binding to the helix would be limited unless appropriate exposure of the bases occurs from local helical distortions.

On the other hand, a different situation may be present in B-DNA, which is more extended than A-DNA, where the N(7) atom of the guanine is relatively more exposed in the major groove although there is still some steric obstruction. It will be expected that denatured or distorted regions of B-DNA may serve as better binding sites for the metal complexes. However, in the left-handed Z-DNA it can be seen that the availability of the guanine binding site N(7) is even more pronounced. This arises from the fact that only the interleaving G-C base pairs are twisted with respect to each other and thus the steric conflict between the metal complexes and the 5'-stacked base is minimized (10,11). On the basis of these steric arguments it would be expected that regions of Z-DNA or other types of distorted B-DNA, which expose the guanine bases, would preferentially bind metal complexes. The present study has shown that the Pt and Ru carcinostatic drugs bind by monodentate ligation to N(7) sites of guanine in non-helical regions. Intrastrand crosslinks between adjacent guanines are not formed. It will be of interest to see if these binding patterns are also observed in the metal-DNA complexes and to what extend such binding is involved in the mechanism of action of these drugs.

ACKNOWLEDGMENTS

We gratefully thank Dr. B. Rosenberg and Dr. D. T. Thompson for the generous gift of the Pt complexes and the National Institutes of Health for supporting this research through Grants GM-18455 and GM-7378.

We also acknowledge continued support by the College of Agricultural and Life Sciences of the University.

REFERENCES

1. Rosenberg, B., Van Camp, L., Trosko, J.E. and Mansour, V.H. (1969) Nature (London) 222, 385-386.
2. Rosenberg, B. (1978) Biochimie, 60, 893-899.
3. Kelman, A.D., Clarke, M.J., Edmonds, S.D. and Peresie, H.J. (1977) in Proceedings of the Third International Symposium on Platinum Coordination Complexes in Cancer Chemotherapy, Wadley Institute of Molecular Medicine, P. 1, 274-288.
4. Stone, P.J., Kelman, A.D. and Sinex, F.M. (1974) Nature (London) 251, 736-737.
5. Kelman, A.D. and Peresie, H.J. (1979) Cancer Treat. Rep. 63, 1445-1452.
6. Kelman, A.D. and Buchbinder, M. (1978) Biochimie, 60, 893-899.
7. Goodgame, D.M.L., Jeeves, I., Phillips, F.L. and Skapski, A.C. (1975) Biochim. Biophys. Acta, 378, 153-157.
8. Kistenmacher, T.J., Chiang, C.C., Chalipoyil, P. and Marzilli, L.G. (1979) J. Am. Chem. Soc., 101, 1143-1148.
9. Gellert, R.W. and Bau, R. (1975) J. Am. Chem. Soc., 97, 7379-7380.
10. Rubin, J.R. and Sundaralingam, M. (1983) Nucl. Acids Res. (submitted).
11. Rubin, J.R., Sabat, M. and Sundaralingam, M. (1983) Nucleic Acids Res. (submitted).
12. Sussman, J.L. and Kim, S.H. (1976) Biochim. Biophys. Res. Comm. 68, 89-96.
13. Stout, C.D., Mizuno, H., Rao, S.T., Swaminathan, P., Rubin, J., Brennan, T. and Sundaralingam, M. (1978) Acta Cryst. B34, 1529-1544.
14. Teeter, M.M., Quigley, G.J. and Rich, A. (1980) in Nucleic Acid-Metal Ion Interactions, Spiro, T.G. Ed., pp. 146-177, John Wiley & Sons, New York.
15. Jack, A., Ladner, J.E., Rhodes, D., Brown, R.S. and Klug, A. (1977) J. Mol. Biol. 111, 315-328.
16. Lavery, R., Pullman, A. and Corbin, S. (1981) in Proceedings of the Second SUNY Conversation in the Discipline Biomolecular Stereodynamics, Sarma, R.H., Ed., Vol. 1, pp 185-193, Adenine Press, New York.
17. Liebman, M., Rubin, J. and Sundaralingam, M. (1977) Proc. Natl. Acad. Sci. USA, 74, 4821-4825.
18. Holbrook, S.R., Sussman, J.L., Warrant, R.W., Church, G.M. and Kim, S.H. (1977) Nucleic Acids Res., 4, 2811-2820.

STRUCTURES AND MECHANISMS OF MISPAIRING IN A HELICAL ENVIRONMENT INCLUDING SEQUENCE EFFECT

Robert Rein and Masayuki Shibata
Department of Experimental Pathology and Biophysics and
National Foundation for Cancer Research Laboratory, Roswell
Park Memorial Institute, Buffalo, NY 14263

ABSTRACT

The structure and energetics of mutagenic mispairs in a helical environment are studied. All the energy components involved in substituting a normal pair with a mispair are calculated. Both alternative types of mispair, the wobble and the tautomeric one are treated. The study is comprehensive involving some of the possible mispairs which can be involved in known spontaneous transitions, transversions and 2-aminopurine (APur) induced mutations.

The results of the calculation according to the wobble pair and passive polymerase model are in good agreement with the following experimental results: Order and magnitude of observed incorporation frequencies of mispairs are $G \cdot T > A \cdot G > A \cdot C$ with the range of 10^{-5} to 10^{-8}. For the base analogue 2-aminopurine induced mutation, the results obtained by using a single hydrogen bonded APur·C model reproduced all experimentally observed trends including dTTP and dCTP competition on APur site and the increase in misincorporation of dCTP when nearest neighbor base pair A·T is replaced by G·C pair.

These results provide strong support to the conclusion that base substitution type mutation proceeds through a wobble mechanism and not a tautomeric one.

INTRODUCTION

In the recent years researchers in several scientific disciplines have focused on the elucidation of the mechanism responsible for the high degree of accuracy in DNA copying. A combination of biochemical, genetic and kinetic experiments have led to the characterization of the rate laws and reaction pathways of phage replication (1). Frequencies of the misinsertion step at the template site have been described in both in vivo and in vitro systems, for spontaneous (2,3) as well as base analogue induced mutations (4). The general view which has

B. Pullman and J. Jortner (eds.), Nucleic Acids: The Vectors of Life, 479–494.
© 1983 by D. Reidel Publishing Company.

emerged from these studies (for a review see ref. 5) is that copying fidelity involves at least three different molecular discrimination steps for reducing error rates. The first of these steps is at the template base and it is assumed to be governed by the free energy difference between correct and incorrect base pairing. The second and third steps are comprised of proofreading and post replicative repair.

The primary concern of this paper is examination of the mechanism of the first of these steps, i.e. that of the competition of a non-complementary base with a complementary one for hydrogen bonding with the template base following the primer terminus.

Previous studies have generally assumed one of the following three mechanisms for insertion of a noncomplementary base at the template site: a) mispairs involving minor tautomers (6,7), b.) wobble type mispairs (8,9) and c.) mispairs involving both tautomers and syn isomers (10). We will in this paper analyze the structure, energetics and sequence effect of the various mispairs and examine the results in light of experimentally measured misinsertion frequencies. Our overall aim is to discriminate between the various mispairing mechanisms based on these comparisons.

STRUCTURE OF MISPAIRS   LEADING TO MISINSERTION

In considering the molecular theory of point mutations a central questions one would like to clarify is the nature of base mispairs which are incorporated in DNA during replication and the relation between the mutational frequencies and the relative stabilities of the various mismatched base pairs. The theoretical approaches to these questions are simplified by looking on the misinsertion rate rather than mutational frequencies, since, in this case, a passive polymerase model and editing free mechanism is applicable (1). In view of this mechanism the energetics controlling the rate of misinsertion is the difference in the free energy of formation of a mispair and of the corresponding complementary pair. If the structure of all the pairs involved are known, this quantity is calculable by theoretical methods according to equation (1).

$$\text{Misinsertion Frequency} = \frac{[Bm]^O}{[Bc]^O} \ \text{EXP} \ \frac{-(\Delta G_m^O - \Delta G_c^O)}{RT} \tag{1}$$

where $[Bc]^O$ and $[Bm]^O$ are the complementary and noncomplementary dNTP (deoxy nucleotide triphosphate) concentration and ΔG_c^O and ΔG_m^O are the respective free energies of formation of the complementary and noncomplementary base pairs. The free energy difference between a mispair and complementary pair for the three mechanisms is given by equations 2a - 2c.

A.) Tautomerization Mechanism

$$\Delta G_m^O - \Delta G_c^O = \{ \Delta G(m) Int. - \Delta G(c) Int.\} + \Delta G^O \text{ Tautomerization} \quad (2a)$$

B.) Wobble Mechanism

$$\Delta G_m^O - \Delta G_c^O = \{ \Delta G(m) Int. - \Delta G(c) Int.\}$$

$$+ \Delta G^O \text{ Backbone deformation} \quad (2b)$$

C.) Topal-Fresco Mechanism involving Gimino-enol and <u>Asyn</u> Pair

$$\Delta G_m^O - \Delta G_c^O = \{ \Delta G(m) Int. - \Delta G(c) Int.\} + \Delta G^O \text{ Tautomerization}$$

$$+ \Delta G^O \text{ Backbone deformation} \quad (2c)$$

The quantities entering the calculation in expressions 2a-2c are:

1. The tautomerization energies of the respective minor tautomer involved in the mispairing schemes.

2. The calculation of the backbone adjustment energies of the respective wobble mispairs. This later presuposes the knowledge of the structure of the minihelix incorporating the wobble pairs.

3. Interaction energy includes the respective hydrogen bonding, base stacking and base, sugar, phosphate interaction energies.

In a previous paper (12) we have presented a compilation and a critical discussion of tautomeric equilibrium constants for mutationally relevant tautomers of the four bases, as well a compilation of the molecular orbital theory calculation of the tautomerization energies. In Table 1 we present what we consider as the best values of these constants based on the compilation.

Recently, we have studied the structure and the energetic of the wobble pairs (11,12). In short the model building procedure involved the following steps:

1.) Two dimensional optimization of the hydrogen bonding scheme between the wobble partners (13).

2.) Positioning of the optimized wobble pair in a DNA segment by our model building program GEOMOL (14).

3.) Closure of the sugar phosphate backbone using Millers AGNAS algorithm (15).

4.) Energy optimization of the whole structure using program REFINE (11).

Table 1: Comparison of Tautomeric Equilibrium
in Solution (Kcal/mol).

| Bases | Theory | | Experiment |
	Δ_E^o a)	ΔG^o b)	ΔG^o c)
C → C*	7.17	11.06	6.4
A → A*	7.55	5.81	6.3
G → G*	3.89	10.93	---
G → G*(TF)	17.15	22.13	---
T → T*	4.39	5.15	4.8

a) Intrinsic tautomerization energies by the MINDO/3 method.
b) Theoretical tautomerization energies in solution with
   solvent effect calculated by the reaction field method.
c) Averaged observed tautomerization energies.

Table 2: Deviations From B-DNA Structure After Optimized
Incorporation of Mispairs[a]

| | G·T | | Asyn·G | | A·C | |
	Strand 1	Strand 2	Strand 1	Strand 2	Strand 1	Strand 2
ΔR	0.010	0.007	0.009	0.010	0.000	0.009
$\Delta\Theta$	1.291	1.352	1.213	1.372	1.478	1.588
$\Delta\Psi$	1.909	2.771	1.235	4.342	4.982	6.457
$\Delta\lambda_1$	-0.01		-0.27		0.72	
$\Delta\lambda_2$	0.03		0.11		0.59	
$\Delta\sigma$	16.85	-16.25	16.25	0.58	-13.09	22.0
ΔP	0.9	-4.45	1.76	6.12	-26.29	-11.57
Sugar Pucker	C3'exo	C3'exo	C3'exo	C3'exo	C2'endo	C3'exo

a)All deviations calculated with respect to B-DNA geometry.
$\Delta R, \Delta\Theta, \Delta\Psi$ = Root mean square deviations in bond length, bond angle
        and fixed dihedral angle (excluding backbone angles)
        respectively.
$\Delta\lambda_1$ =Deviation in distance between C1' on sugars of opposite strand.
$\Delta\lambda_2$ =Deviation in distance between N(1 or 9) on bases of
        opposite strands.
$\Delta\sigma$ =Deviation in C1'-N-N angle for opposite strands.
ΔP =Deviation in pseudo rotation.

The detailed methodology was given in the previous paper (12) but it is worthwhile to note that the parameter set employed in the calculation of interaction energy was carefully examined to reproduce magnitudes, base sequence dependent spreads and trends of high resolution melting experiments (16).

The structural characteristics of the mispairs relative to B-DNA (17) are summarized in Table 2.

Energetics of mispairs involving normal and rare tautomers are shown in Table 3.

Table 3: Energetics of Mispairs Involving Normal and
Rare Tautomers [a]

Type of Pair	M2·M2' [b]	Def. [c]	Tau. [d]	Int. [e]	Total [f]
Wobble	G·T	3.08	-----	0.67	3.75
	Asyn·G	3.47	-----	3.38	6.85
	A·C	4.20	-----	4.32	8.52
Tautomer [g]	G·T*	0.06	5.51	-2.77	2.44
	G*·T	0.0	10.93	-3.14	7.79
	A·C*	0.0	11.06	2.87	13.93
	A*·C	0.06	5.81	4.09	9.96
	Asyn·G**	3.47	22.13	3.57	29.17

a) All energies are in Kcal/mol and the optimized B-DNA
   with A·T pair is taken as reference.
b) Base pairs in mini-helix G1·C1'/M2·M2'/C3·G3'
c) Differences in deformation energies.
d) Tautomerization energies.
e) Differences in interaction energies.
f) Differences in total energies.
g) The asterisk indicates the rare tautomer.

Tables 2 and 3 indicate that the wobble pair can be well accommodated into the double helical environment with very small deviations from the reference B-DNA structure.

2-AMINOPURINE INDUCED MUTATIONS AND EFFECT OF STACKING ENVIRONMENT

Another interesting system for studying mispairing mechanisms is the one involving the base analogue 2-aminopurine, which is known to strongly increase AT$\rightarrow$GC transitions. It is believed that 2-

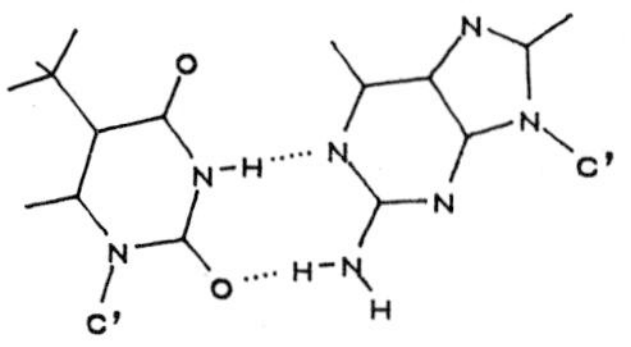

T·APur (Normal)

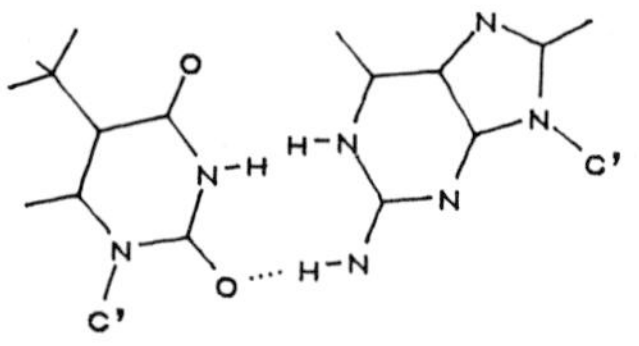

T·APur* (Tautomer)

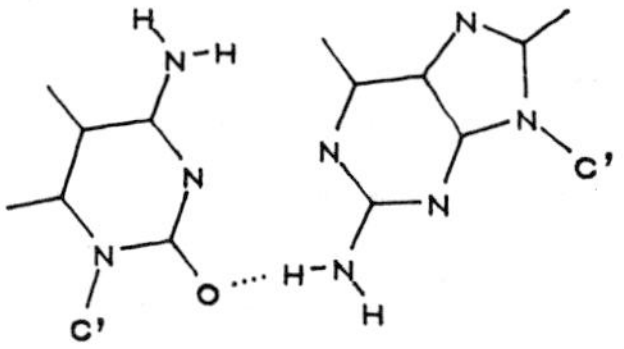

C·APur (Normal)

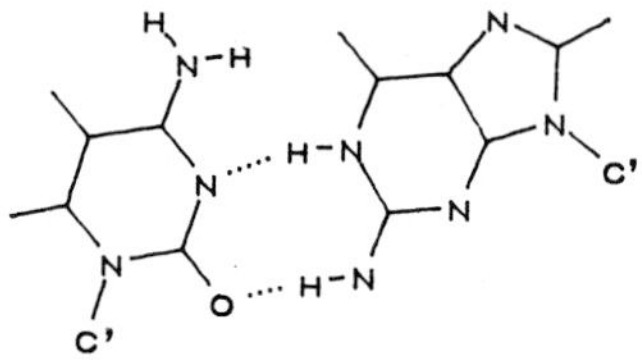

C·APur* (Tautomer)

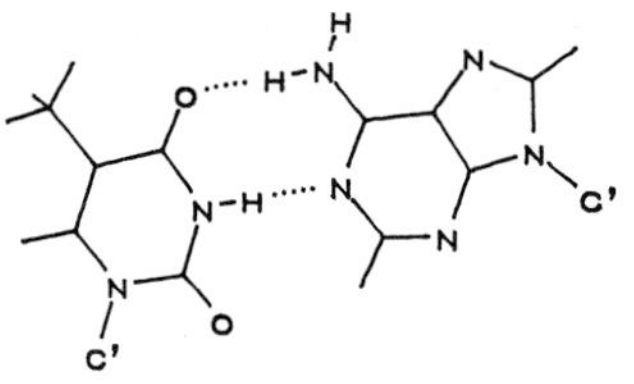

T·A (Normal)

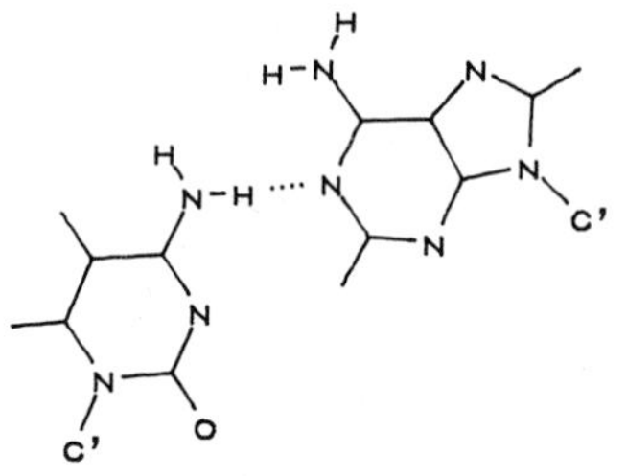

C·A (Wobble)

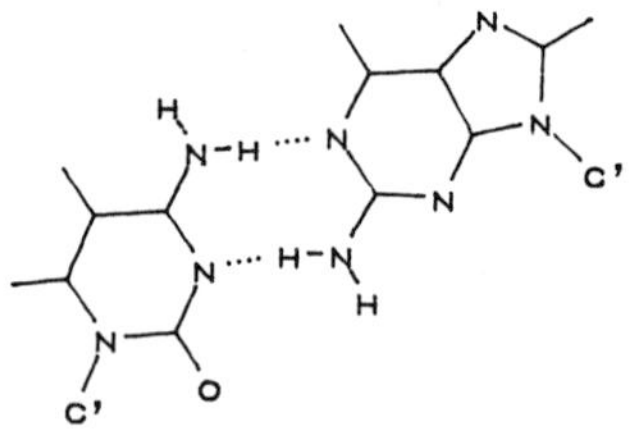

C·APur (Wobble)

Figure I.   Hydrogen bonding schemes in evaluating 2-aminopurine
induced mutation

aminopurine enhance mutation by its ability to base pair both to thymine and cytosine and in this respect it is less selective than adenine (18-20). This system has been studied extensively by both experimental (for a review see ref. 21) and theoretical (22,23,24) methods.

The competition for the 2-aminopurine site in the template between thymine and cytosine has been measured in vitro using synthetic poly nucleotide templates (25). Frequency of incorporation of 2-aminopurine, i.e. competition between 2-aminopurine and adenine for thymine at the template site as well as between 2-aminopurine and guanine for a cytosine template has been obtained in the T4 system (26). The first of these reactions has also been studied in·vitro systems (28). Two of these papers also contain information on sequence effect, i.e. the effect of stacking on the misinsertion frequencies. We will analyze this system in two stages. The first stage, relaying essentially on our previous calculations (25), where we have considered the system which is independent of stacking environment and focusing on the energetics of the alternative mispairing mechanism as well on the energetics of populating the mutationally relevant tautomer in the tautomeric mechanism. In the second stage, we reconsider the system according to a minihelical model explicity including stacking and backbone deformation energies. For the details of the interaction calculation used in the first stage we refer the reader to the original paper (25). The computation methods emphasized in the second stage based on the melting data optimized potential function have already been described or referred to in previous sections.

The possible hydrogen bonding schemes between 2-aminopurine and thymine and cytosine respectively are presented in Figure 1 as well as hydrogen bonding schemes for A·T and A·C pairs. It should be recognized that these schemes with the exception of wobble pairs, do not imply a significant deviation from the double helical geometry. Hence, the stereochemical factor is not likely to have a significant influence on base selection for both normal and tautomeric mechanisms. We will consider the wobble mechanism which involves backbone adjustment at the second stage. This leaves the option for two major alternatives for the mechanism of erroneous coupling.

i) 2-aminopurine in its normal tautomeric form can lead to mispairing with cytosine by a single hydrogen bond (C·APur (Normal)) .

ii) 2-aminopurine in its rare form can make a pair with cytosine (C·APur* (Tautomer)) in which two hydrogen bonds are involved.

The rest of the paper is concerned with the calculation of the energetics of these two alternative pathways. The objective being to describe the pairing scheme responsible for mutations.

As discussed previously (12), the semi-empirical MINDO/3 method gives the best agreement with experimentally observed tautomeric

equilibrium results. Therefore we have recalculated the
tautomerization energy of the 2-aminopurine system with this method
instead of the previously employed MINDO/2 method. The results are
shown in Table 4.

Table 4: Theoretical Tautomerization Energy

APur (Normal) $\underset{\leftarrow}{\rightarrow}$ APur* (Rare form)	
ΔE^O in vacuo	7.0 Kcal/mol
Difference in solvent effect	-0.4 Kcal/mol
ΔG^O in solution	6.6 Kcal/mol

Table 5: Calculated Energies of Interaction for Some

Normal and Abnormal Base Pairs of DNA(Kcal/mol)

Base Pair	Energy	Base Pair	Energy
A·T	-8.04	G·C	-23.54
APur·T	-5.37	APur·C	-2.75
APur*·T	+2.97	Apur*·C	-3.31

Values taken from ref. 25.

The hydrogen bonding energies for normal and tautomeric pairs were
previously calculated by the IEHT multipole interaction scheme. This
method was shown to reproduce experimental as well as ab initio MO
hydrogen bonding energies (for a review see Ref. 31). We will
reproduce the previous results in Table 5. The order of strength of
the hydrogen bonded normal base pairs agree with that proposed by
Goodman and coworkers (27) which are based upon differential hydrogen-
bonding models. Based on the above hydrogen bonding energies and the
MINDO/3 tautomerization energy, the relative probabilities for the
various base pairs are shown in Table 6 and the energy differences
characterizing competitions are shown in Table 7.

Table 6: Relative Probabilities for Various Base Pairs

Base Pair	Energy (Kcal/mol)	Boltzman Factor $Exp(-\Delta G/RT)$	Relative Probability Normalized to;	
			AT	GC
G · C	-23.53	9.19×10^{16}	----	1.0
A · T	- 8.04	6.25×10^{5}	1.0	6.81×10^{-12}
APur · T	- 5.37	7.44×10^{3}	1.19×10^{-2}	8.09×10^{-14}
APur · C	- 2.76	9.77×10	1.56×10^{-4}	1.06×10^{-15}
APur*·C	3.29	4.25×10^{-3}	6.79×10^{-8}	4.62×10^{-20}
APur*·T	9.57	1.26×10^{-7}	2.02×10^{-19}	1.37×10^{-24}

Table 7: Comparison of Theoretical and Experimental Energy

Difference Characterizing the Competitions (Kcal/mol)

Competition	Experiment	Theory	
		Normal	Tautomer
APur $\nwarrow^{C}_{T}$	1.8	2.61	6.28
T $\nwarrow^{APur}_{A}$	1.1	2.67	17.61

From these tables it is obvious that the normal pairing mechanism is much preferred to the tautomeric mechanism. The same conclusion was obtained by Goodman and coworkers (27) by intuitive differential H-bonding models. It also appears that the predicted incorporation ratio based only on H-bond energies are in reasonable agreement with the range of the experimental numbers for these reactions. The data also confirms the experimentally observed asymmetry in 2-aminopurine induced transitions favoring the AT→GC pathway over GC→AT. In order to examine how are the additional factors such as stacking and backbone deformation effecting the conclusions, we next consider a minihelical model accounting for these effects in relation to the experiments (25).

The measurement of the incorporation ratio between cytosine and thymine in direct competition for adenine or 2-aminopurine sites was carried out by Watanabe and Goodman using mammalian DNA polymerase with two different neighboring base sequence templates described below (25).

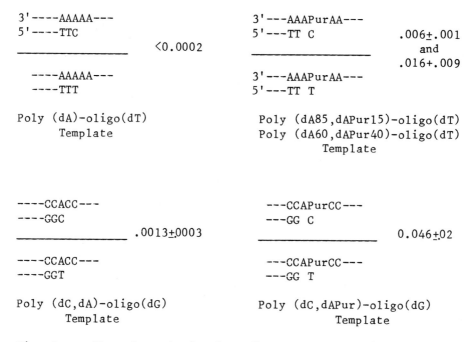

```
3'----AAAAA---                          3'---AAAPurAA---
5'----TTC                               5'---TT C              .006+.001
_____  <0.0002            _____       and
                                                              .016+.009
   ----AAAAA---                         3'---AAAPurAA---
   ----TTT                              5'---TT T

Poly (dA)-oligo(dT)                     Poly (dA85,dAPur15)-oligo(dT)
      Template                          Poly (dA60,dAPur40)-oligo(dT)
                                                   Template


----CCACC---                            ---CCAPurCC---
----GGC                                 ---GG C
_____ .0013±.0003          _____  0.046±.02

----CCACC---                            ---CCAPurCC---
----GGT                                 ---GG T

Poly (dC,dA)-oligo(dG)                  Poly (dC,dAPur)-oligo(dG)
       Template                                Template
```

Fig. 2   The schematic drawing of the system examined and
                                                                $\frac{dCMP}{dTMP}$
           experimentally observed incorporation ratios

To evaluate the stability of a mispair in the double helical environment, we have constructed the following system to mimic a template-primer complex as shown in Figure 3.

To obtain the stability of the A·C pair, we used the same geometry as the A·C wobble pair as discussed in the previous section. For the APur·C system, we have considered the wobble pair with two hydrogen bonds (APur·C(Wobble)) in addition to the normal and tautomeric pairs described at the first stage.

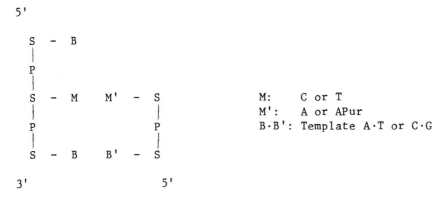

Figure 3. Schematic representation of the model consiered for the computations.

The calculation was performed in the same manner as in previous section, except that the time consuming optimization step was avoided by using the previously optimized geometry. The free energy difference was calculated according to equations 2a-2c and the results are shown in the following table.

Table 8: Comparison of Theoretical and Experimental dCTP and dTTP Competition on Different Template

Template Site Base	Neighbor Base Pair	Experiment Incorporation Ratio $\frac{dCMP}{dTMP}$ $(\times 10^{-2})$	Corresponding Energy Difference (Kcal/mol)	Theory Difference in Free Energy of Complex Formation(Kcal/mol) Normal Model	Tautomer Model	Wobble Model
A	A·T	<.02	>5.13	----	----	8.72
A	C·G	0.13±.03	4.00±14	----	----	8.42
APur	A·T	1.6±.9 ~0.6±.1	2.49±.38 ~3.08±.10	0.61	7.94	8.03
APur	C·G	4.6±2	1.85±.28	0.03	7.39	7.98

The nearest neighbor base sequence effect can be obtained by comparing dCTP and dTTP competition for A as well as APur with different templates. In both cases, it is known that the dCTP misincorporation increases when the nearest neighbor is a G·C pair as compared to a A·T pair. This effect is shown in the following table.

Table 9:   A·T and G·C Nearest Neighbor Base Pair Effect on dCTP
           Misincorporation over dTTP

| Template Site Base | Experimental | | | Theoretically obtained Difference in Energy (Kcal/mol) | |
	Increase in dCTP Incorporation [fold]	Corresponding Energy Difference [Kcal/mol]		Normal Model	Wobble Model
Adenine	6.5	−1.13		---	−0.3
APur	2.9	− .64			
	~ 7.7	~ −1.23		−.58	−0.25

DISCUSSION

       Both theory and experiment indicate that the tautomerization
energy values characterizing the mutationally interesting minor
tautomer population are in the range of 5-7 Kcal/mol with the exception
of the imino-enol tautomer of guanine proposed by Topal and Fresco
(10). The value for this tautomer is 17 Kcal/mol and is obtained by
the same methodology as the other minor tautomers (range of 3.9 to 7.5
Kcal/mol). This confirms unambiguously that this tautomeric form cannot
play any role in spontaneous mutations and thus transversion type
mutations cannot be explained by any tautomeric mechanism. This leaves
the wobble type mechanism as the only reliable alternative for
transversion.

       In the modelling of the three dimensional structure of the
mispairs in a helical environment, it is striking to find that the
incorporation of the mispairs including the purine-purine pair Asyn·G
can occur with relatively little deformation of the overall helical
structure of DNA (Table 2) This point is further confirmed from an
energetic point of view by the relatively modest energy requirement for
wobble pairs in a helical environment, which are of the order of 4 to 8
Kcal/mol above the energy for forming the corresponding Watson Crick
structures (Table 3). These results by themselves definitively show
the preference for the wobble mechanism of mispairing leading to
transversion.

       Regarding spontaneous transitions, we have compared the G·T, A·C,
wobble pairs versus the G·T*, G*·T, A·C* and A*·C tautomeric pairs.
The result obtained in these cases with the single exception of the

G·T* pair show that the wobble pairs are preferred. However, looking on the structure of the G·T wobble pair, it should be emphasized that this is the structure requiring the least deformation of the backbone of the three wobble pairs. In fact, this pair has been observed in a helical environment by direct experimental measurement (29,30). Using these energies the insertion frequencies according to equations 1 and 2a-2c, have been calculated and are presented together with experimentally measured insertion frequencies in Table 10.

Table 10:   Calculated and Experimental Misinsertion Frequencies by Mispairs

| Mutation Mispair | Misinsertion Frequencies | | | |
| | Calculated* | | Experimental | |
	Tautomeric Pair	Wobble Pair	Ref.3	Ref.2
AT→GC G·T	1.6×10^{-2} (G·T*) 1.9×10^{-6} (G*·T)	1.8×10^{-3}	6.8×10^{-6}	8.3×10^{-5}
AT→CG A·G	4.1×10^{-22} (Asyn·G**)	9.5×10^{-6}	5.6×10^{-7}	1.1×10^{-5}
AT→GC A·C	6.1×10^{-11} (A·C*) 5.0×10^{-8} (A*·C)	5.7×10^{-7}	9.0×10^{-6}	2.2×10^{-5}

*The total contributions of base mispairing and base stacking of the mini-helix G1·C1'/M2·M2'/C3·G3' agreed best with the values reported by Sinha and Haimes (1981), that is, G·T > A·G > T·T > A·C.

The calculated frequencies with the wobble pairs reproduced a set of the experimentally observed trend by Sinha and Haimes (3) for the three mispairs. Furthermore, the quantitative aspects of the comparison are also reasonable. Contrary to this, the frequencies calculated from the tautomeric mispair energies show an agreement only within the purine-pyrimidine pair. The good agreement between observed and predicted insertion frequencies based on the wobble mispairs appears as very strong evidence confirming both the validity of the passive polymerase model for insertion as well for the nature and structure of the wobble mispairs presented in this paper.

Our conclusion of the unlikely role of minor tautomers in the mechanism of mispairing is further reconfirmed in the 2-aminopurine system. In fact, both in the base pair and in the minihelix model, the tautomeric mechanism is disfavored by approximately 7 Kcal/mol which implies a population of about four orders of magnitude less than the corresponding mispair in the mechanism involving only normal tautomers.

Among the two possible pairing schemes between the normal tautomer of 2-aminopurine with cytosine, the singly H-bonded pair is the preferred one. This seem to be the consequence of the cost of the wobble energy required to fit the double hydrogen bonded structure in the minihelix. This energy is of the order of 4 Kcal/mol which is more than the difference in the hydrogen bonding energy between the two hydrogen bonding schemes.

The results presented in Tables 8 and 9 show that the normal mispairs account for all the experimentally observed trends both for C and T competition for APur and A, as well as for the competition of APur and A for T. It is rewarding to observe that the model as well accounts for the observed sequence effects on these competitions;that is in the C and T competition for A and APur, there is an enhancement of the incorporation of C over G·C template relative to the one over the A·T template.

In conclusion, we consider the results presented in this paper as very strong supporting evidence for the notion that mispairing is responsible for error incorporation at the template terminus, involving normal bases and not minor tautomers.

From this analysis it would also appear that the methods used have reached a point of refinement that they are capable in giving a satisfactory description of both the structure and energetics of the intermediates considered in this paper. We attribute this primarily to the reliability of the MINDO/3 method in the tautomerization energy calculations on the one side and the other, the use of the empirical function calibrated to reproduce the thermodynamic values of DNA melting. It seems also that the force field method used in the backbone deformation calculation is well adapted to the rest of the calculations.

ACKNOWLEDGEMENT

We would like to thank Dr. Giorgio Bolis for a computer program to draw Figure 1 and Mr. Thomas Kieber-Emmons for carefully reading the manuscript. We are also indebted to Mrs. Lynette Nelson for her skillful assistance in the preparation of the manuscript. This work was supported by the National Foundation for Cancer Research and by NASA Grant NSG 7305.

REFERENCES

1. Goodman, M.F., Hopkins, R.L., Watanabe, S.M., Clayton, L.K. and Guidotti, S.: 1980, in "Mechanistic Studies of DNA Replication and Genetic Recombination", B. Alberts and C.F. Fox, Eds., Academic Press, NY, pp. 685 - 705.

2. Fersht, A.R., Knill-Jones, J.W. and Tsui, W.-C.: 1982, J. Mol. Biol. 156, pp. 37-51.

3. Sinha, N.K. and Haimes, M.D.: 1981, J. Biol. Chem. 256, pp. 10671-10683.

4. Watanabe, S.M. and Goodman, M.F.: 1982, Proc. Natl. Acad. Sci. USA 79, pp. 6429-6433.

5. Loeb, L.A. and Kunkel, T.A.: 1982, Ann. Rev. Biochem. 52, pp. 429-457.

6. Watson, J.D. and Crick, F.H.C.: 1953, Nature 171, pp. 964-967.

7. Lowdin, P.-O.: 1965, Advs. Quantum Chem. 2, pp. 213-360.

8. Rein, R., Coeckelenbergh, Y. and Egan, J.T.: 1975, Int. J. Quantum Chem. QBS 2, pp. 145-153.

9. Garduno, R., Rein, R., Egan, J.T., Coeckelenbergh, Y. and MacElroy, R.D.: 1977, Int. J. Quantum Chem. QBS 4, pp. 197-204.

10. Topal, M.D. and Fresco, J.R.: 1976, Nature 263, pp. 285-289.

11. Kothekar, V., Bolis, G. and Rein, R.: 1983, Int. J. Quantum Chem. (in press).

12. Rein, R., Shibata, M., Garduno, R. and Kieber-Emmons, T.: 1983; in "Structure and Dynamics of Nucleic Acids", E. Clementi and R.H. Sarma, Eds., Adenine Press, NY, pp. 269-288.

13. Poltev, V.I. and Bruskov, V.I.: 1977, Mol. Biol. (USSR) 11, pp. 509-516.

14. Rein, R., Nir, S., Haydock, K. and MacElroy, R.D.: 1978, Proc. Indian Acad. Sci. 87A, pp. 95-113.

15. Miller, K.J.: 1979, Biopolymers 18, pp. 959-980.

16. Shibata, M., Kieber-Emmons, T. and Rein, R.: 1983, Int. J. Quantum Chem. (in press).

17. Arnott, S., Smith, P.J. and Chandrasekaran, R.: 1976 in "CRC Handbook of Biochemistry and Molecular Biology Vol. 2", pp.411-422.

18. Freese, E.: 1959, Proc. Natl. Acad. Sci. USA. 45, pp. 622-633

19. Freese, E.: 1959, J. Mol. Biol. 1, pp. 87-105.

20. Champe, S.P. and Benzer, S.: 1962, Proc. Natl. Acad. Sci. USA 48, pp. 532-546.

21. Ronen, A.: 1980, Mut. Res. 75, pp. 1-47.

22. Pullman, B. and Pullman, A.: 1962, Biochim. Biophys. Acta 64, pp. 403-405.

23. Danilov, V.I., Kruglyak, Yu.A., Kupriyevich, V.A. and Shramko, O.V: 1967, Biophys. (USSR) 12, pp. 840-844.

24. Rein, R. and Garduno, R.: 1976, in "Quantum Science: Methods and Structure" J.-L Calais, O. Goscinski, J. Linderberg, and Y. Ohrn, Eds., Plenum Press, NY, pp. 549-560.

25. Watanabe, S.M. and Goodman, M.F.: 1981, Proc. Natl. Acad. Sci. USA 78, pp. 2864-2868.

26. Goodman, M.F., Hopkins, R. and Gore, W.C.: 1977, Proc. Natl. Acad. Sci. USA 74, pp. 4806-4810.

27. Goodman, M.F., Watanabe, S.M. and Branscomb, E.W.: 1982, in "Molecular and Cellular Mechanisms of Mutagenesis" J.F. Lemontt and W.M. Generoso, Eds., Plenum Press, NY, pp. 213-229.

28. Pless, R.C., Vevitt, L.M. and Bessman, M.J.: 1981, Biochem. 20, pp. 6235-6244.

29. Ackermann, Th., Gramlich, V., Klump, H., Knable, Th., Schmid, E.D., Seliger, H. and Stulz, J.: 1979, Biophys. Chem. 10, pp. 231-238.

30. Early, T.A., Olmsted, J., Kearns, D.R. and Lezius, A.G.: 1978, Nucleic Acids Res. 5, pp. 1955-1970.

31. Rein, R.: 1978, in "Intermolecular Interactions: From Diatomics to Biopolymers", B. Pullman, Ed., John Wiley and Sons, NY, pp. 308-362.

DRUG NUCLEIC ACID INTERACTIONS: PROFLAVINE INTERCALATION INTO DOUBLE-HELICAL FRAGMENTS OF RNA AND DNA

S. Neidle, R. Kuroda, A. Aggarwal and S.A. Islam
Cancer Research Campaign Biomolecular Structure Research Group,
Department of Biophysics, King's College, London   WC2B 5RL

ABSTRACT

The results of crystallographic studies on proflavine complexed with three different dinucleoside phosphate systems, are discussed in terms of backbone and sugar conformations. Extent of drug-base-pair stacking and base-pair mutual orientations are compared and contrasted.

INTRODUCTION

   Proflavine (2,7-diaminoacridine, Fig. 1) is among the most studied of nucleic acid intercalating agents, as well as being structurally one of the simplest. Its lack of significant side-chains or groupings means that in principle the whole of the molecule can be involved in intercalation inbetween adjacent base pairs, as envisaged in the original Lerman concept [1]. Structurally more complex intercalators such as actinomycin D or daunomycin have important additional interactions with DNA, which often introduce base or sequence specificity, as well as inevitably making the binding processes in solution more complex to analyse.

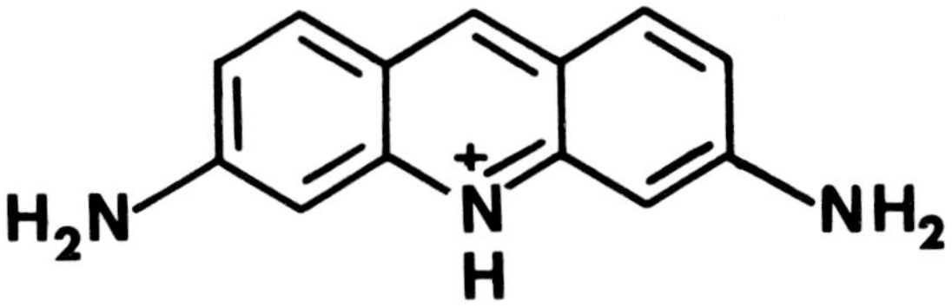

Figure 1.   The proflavine molecule.

495

B. Pullman and J. Jortner (eds.), Nucleic Acids: The Vectors of Life, 495–509.
© 1983 by D. Reidel Publishing Company.

   The physico-chemical data on the intercalative interactions
of proflavine with double-stranded polynucleotides can be summarised
as :
(i)   intercalation occurs from the major groove direction, with
        the exocyclic amino nitrogen atoms directed out from this
        groove [2].
(ii)   pyrimidine-3',5'-purine sites such as -CpG- are preferred
        [3-5].
(iii)   the drug is aligned approximately parallel to the long axis
        of the base pairs, and thus all its non-exchangeable protons
        are more or less shielded from the environment [3]. This
        indicates the overall correctness of the Lerman intercalation
        model in respect of drug-base overlap involving all four
        bases at the intercalation site, rather than asymmetric
        overlap with bases on just one polynucleotide chain as suggested
        in other models [6].
(iv)   Unwinding of closed-circular DNA [7] is 17° per drug molecule.

   Both crystallographic and model-building studies have been
extensively used to study proflavine-nucleic acid interactions,
in attempts to further clarify and understand these issues at the
molecular level. This paper concentrates on such approaches, with
particular reference to studies from our laboratory, which have
explored questions of nucleotide conformation and sequence variability.

TABLE 1

Proflavine-dinucleoside monophosphate (XpY) Crystal Structures				
XpY	Structural motif of XpY	Number of water molecules per XpY	Torsion angle esd (°)	Reference
CpG	Anti-parallel duplex	11.5	1	[7,8]
d(CpG)	Anti-parallel duplex	13	3	[9]
5-iodo-CpG	Anti-parallel duplex	7.5	8	[10]
CpA + UpG	Anti-parallel duplex	12.5	8	[11]
CpA	Parallel duplex	9	4	[12,13]
ApA	Opened-out	16.5	1.5	[14,15]

   The crystal structures of proflavine complexes with six different
dinucleoside monophosphates (the simplest systems that may in principle

display at least some polynucleotide structural features), have
been solved (Table 1). Those that do not involve anti-parallel
Watson-Crick base-paired strands are of unknown relevance to intercalation into double-stranded nucleic acids, and accordingly will
not be further discussed here.

In general, the four proflavine complexes with anti-parallel
base-paired strands all show the same gross structure, with the
drug oriented in an approximately similar manner within the binding
sites, as in the Lerman prediction. However, there are a number
of more detailed features which do distinguish one structure from
another. The 5-iodo-CpG and CpG complexes do not appear to be
significantly distinct, and thus the former is not discussed here
in view of its lesser accuracy.

THE CRYSTAL STRUCTURES

The self-complementary proflavine-CpG complex, illustrated
in Fig. 2, utilises a crystallographic two-fold axis passing through
the central ring carbon and nitrogen atoms of the drug, which is
therefore symmetrically intercalated. Additional stabilisation
of the drug is provided by hydrogen bonds (3.00Å long) between
its exocyclic amino groups and phosphate oxygen atoms of the ribo-
dinucleoside phosphate backbone. Stacking with cytosine and guanine
bases is roughly equal.

The d(CpG) complex (Fig. 3), by contrast, has the drug asymmetri-
cally intercalated, and the two dinucleoside are no longer totally
identical conformationally (see below). There are no hydrogen
bonds between the phosphate oxygens and the exocyclic nitrogen
atoms of the drug; as if to compensate for this, the overlap between
base pairs (especially with the guanines) and the drug appears
to be appreciably greater than for the ribo CpG complex, and so
it is buried deeper in the intercalation site.

The third proflavine complex recently analysed in our laboratory
[11] is the first single-crystal intercalation structure to have
non-self-complementary strands, and thus affords the opportunity
of examining the structural effects of this non-equivalence in
the drug's surroundings. The strands are again anti-parallel,
and the intercalated proflavine is flanked by a C.....G and an
A.....U base pair. Fig. 4 shows that the drug is oriented pseudo-
symmetrically, with a similarity to the CpG structure. There is
a weak hydrogen bond (of length 3.15Å), to a phosphate oxygen of
the UpG strand alone. The nitrogen-oxygen distance to the CpA
strand is 3.84Å. The extent of drug-base stacking is intermediate
between that of the CpG and the d(CpG) structures. It is unsurprising
that the base pairs in this structure have slightly different dimensions, as measured by C1'.....C1' separation (Table 2) and that the
U.....A one is slightly smaller than C.....G. The ethidium complexes
with 5-iodo-CpG [16] and 5-iodo-UpA [17] illustrate this well, with an

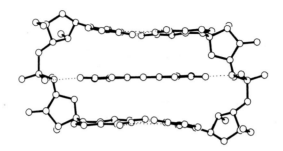

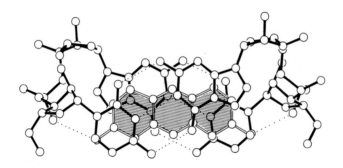

Figure 2.   Two views of the proflavine-CpG complex, with drug-
            base overlap shown shaded and hydrogen bonds with dotted
            lines.

average Cl'.....Cl' distance of 10.83Å in the former and 10.58Å
in the latter. However the variability in inter-strand phosphorus-
phosphorus distances for the three proflavine structures (Table
2) is much greater than these differences. The larger P.....P
distances reflect the diminished drug-backbone hydrogen-bonding,
compared to the CpG complex, and may be a consequence of the
difference in dinucleoside sequence.

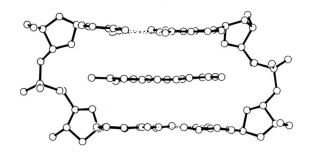

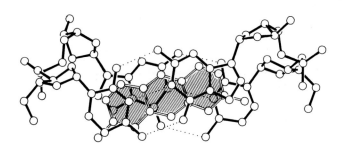

Figure 3.   The proflavine-d(CpG) complex.

UNWINDING

    These three proflavine complexes do not have identical base-
turn angles (defined as the angle subtended by interstrand C1'-
C1' vectors). The 32° angle for the CpG structure is almost identical
to that expected for an unintercalated dinucleoside with A-RNA
geometry (and a 32.7° helical twist angle between the base pairs).
Note that since all intercalated dinucleosides studied to date
do not have helical symmetry (see below) it is inappropriate to
use the term helical twist for the interglycosidic vector angle
whereas 'base turn' [18] is more accurate. The d(CpG) and CpA-
UpG complexes have 17° and 16° base turn angles respectively.
There is a good correlation between base-turn angle and P.....P
distance, which is perhaps unsurprising on the basis of simple
steric considerations, since a smaller base turn angle brings bases

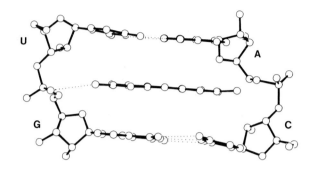

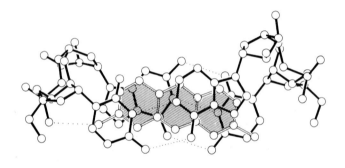

Figure 4.   The proflavine-CpA-UpG complex.

on the same strand closer together and thus will force the intervening backbone in an outwards direction (Fig. 5).

      There has been considerable discussion [18-21] as to whether
the base turn angles in dinucleoside complexes directly relate
to the unwinding angles observed in closed-circular supercoiled
DNA [22].   That for proflavine is 17°, relative to 26° for ethidium,
which might be considered to imply a resulting helical twist angle
at the intercalation site of 19° for B-DNA and 16° for A-DNA with
bound proflavine.   However, there are several factors which make
such a view unlikely.   Firstly the unwinding angle for DNA observed
in solution represents the total effect produced by an intercalator
at its binding site, taken over all affected base pairs around
a site, as well as being averaged over all sites.   It would indeed
be most surprising if the change in polynucleotide helical twist
angle was confined to the two base pairs surrounding the drug.

TABLE 2

Inter-strand distances (in Å) and base-turn angles (in °) in

proflavine dinucleoside (XpY) complexes

XpY	Distance C1'....C1'	Distance P.....P	Base turn
CpG	10.70	15.91	32
d(CpG)	10.62	16.73	17
	10.59		
CpA }	10.63	16.56	16
UpG }	10.54		

The daunomycin complex with the hexamer d(CpGpTpApCpG) demonstrates this point, with a 36° angle at the intercalation site itself and unwinding of about 8° at the adjacent residues [23]. Secondly, the crystallographic analyses of uncomplexed short oligonucleotides have shown that helical twist angles can vary over a wide range [24-26] in a sequence-dependent manner, with pyrimidine-3',5'-purine steps having lower values than others. These are the preferred sequences for proflavine intercalation [3] and so not only does it seem unlikely that they would have precisely 36° twist angles in a B-type helix, but moreover it would be expected that -CpG- and -CpA-steps would not have identical helical twist angles. Thus, one can only conclude that the observed base turn angles in the intercalated dinucleoside structures are related to sequence, environment and drug type, rather than being an intrinsic property imparted by the drug alone. Until structural information is available on oligonucleotide systems containing intercalated sites remote from end-effects, conclusions on these (and indeed other) aspects of drug-nucleic acid structure and conformation, remain at the speculative level.

Sugar Puckers around the Intercalation Site

The first two dinucleoside intercalation structures to be established were with ethidium [16,17]. Both revealed a pattern of alternating C3'endo-3',5'-C2'endo sugar pucker on each strand. This was considered to be a fundamental feature of intercalated structures. Analogous mixed sugar puckers in intercalated DNA have been suggested as being responsible for the opening-up of base pairs to 6.8Å [20]. However, this generalisation was invalidated

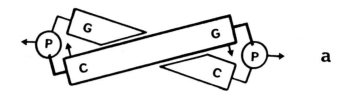

Figure 5.   Schematic representation of (a) a complex with a 33°
             base turn angle, and (b) one with a 16° angle. The
             arrows indicate directions in which phosphate groups
             and bases might move in order to pass from (a) to (b).

by the finding that the proflavine-CpG complex has all sugars displaying
C3'endo pucker (Table 3). The special features of this structure,
particularly the hydrogen-bonding between the drug's amino groups
and the phosphate oxygen atoms could play a role in stabilising
the C3'endo pucker at the 3' end of each strand. It has been suggested
[27] on the basis of molecular mechanics calculations that the
alternating C3'endo-3',5'-C2'endo pattern is inherently more stable,
but that factors such as the drug-dinucleoside hydrogen bonds can
change this. These gas-phase calculations in effect imply that
environmental factors, not least dinucleoside-water interactions,
may well be of dominent importance. This prediction is indeed borne
out by experiment.

      The proflavine-d(CpG) structure, which does not have inter-
molecular drug-dinucleoside hydrogen-bonding, displays a relatively
complex pattern of sugar conformations. Two closely-related crystal
forms have been analysed [28]. The more ordered has strand 1

TABLE 3

Sugar puckers in proflavine-dinucleoside (XpY) complexes

XpY	Pseudorotation parameter (in°)	Pucker description
CpG: 5' end	3	C3' endo
3' end	12	C3' endo
d(CpG): strand 1, 5' end	22	C3' endo
3' end	40	C3' endo
strand 2, 5' end	20	C3' endo
3' end	171	C2' endo
CpA+UpG: CpA, 5' end	1	C3' endo
3' end	153	C2' endo
UpG, 5' end	23	C3' endo
3' end	13	C3' endo

with both sugars in a C3'endo pucker, and strand 2 with the alternating C3'endo-3',5'-C2'endo pattern (Table 3). The second form
has the 3' end (i.e. guanosine) sugar on strand 1, disordered between
50% populations of C3'endo and C2'endo pucker; strand 2 has the
same pattern of puckers in both crystal forms. This difference
between ordered and disordered sugars is ascribable to a thermal
disruption of the otherwise ordered water structure surrounding
the latter so that one crucial water molecule, which forms a hydrogen
bond with the guanosine O3' atom in the ordered structure, is absent
in the disordered one. Thus, this sugar may be thought of as no
longer being constrained by its environment to take up a particular
conformation.

The proflavine-CpA-UpG complex also shows a pattern of sugar pucker
asymmetry on one strand compared to another (Table 3). It is notable
that the UpG strand with C3'endo-3',5'-C3'endo pucker, is the one
with a hydrogen bond to the drug molecule, lending further support
to the view that this interaction stabilises C3'endo pucker at
the 3' end of intercalated dinucleosides.

These structures suggest that the 5' end C3'endo sugar pucker
in intercalated ribo- and deoxy-dinucleosides is invariant, whereas
the 3' end sugar is flexible, and adopts the conformation most
appropriate to its surroundings in a crystal. It seems likely
that this disparity between the flexibilities of the two ends,
extends to the conformations in solution.

BACKBONE CONFORMATIONS

The high degree of correspondance between the backbone conformation angles, in all intercalated dinucleoside structures, has been fully analysed elsewhere [18,21,29]. Table 4 shows that the proflavine complexes follow these findings, with all five dinucleoside strands having essentially the same conformation.

TABLE 4

Backbone and glycosidic conformational angles (in°) in pro-flavine dinucleoside complexes. The convention used has the angle about the P-05' bond designated as α, and the glycosidic angle χ is defined with respect to the atoms O1'-C1'-N9-C4 for purines and O1'-C1'-N1-C2 for pyrimidines.

XpY	$\chi(5')$	α	β	γ	$\delta(5')$	$\delta(3')$	ϵ	ζ	$\chi(3')$
CpG	-172	-73	-126	53	75	79	-156	-68	-93
d(CpG):									
strand 1	-162	-70	-141	46	84	85	-150	-70	-108
strand 2	-165	-73	-142	73	79	149	-157	-60	-76
CpA	-172	-80	-132	65	74	96	-156	-58	-91
UpG	-165	-62	-141	50	84	100	-156	-83	-95

The $\delta(3')$ value for the UpG strand is unreliable on account of disorder in the 3' sugar, which could not be fully resolved in the crystal structure.

The backbone angles mostly adopt values neighbouring those of A-RNA or the very similar A'-DNA, with α and ζ being gauche⁻, β between trans and gauche⁻, γ gauche⁺ and ϵ trans. The major changes from the polymer values are in β and $\chi(3')$ with the former having ∿30-40° higher and the latter ∿60-70° higher values in the intercalated structures. The resulting asymmetry in 3' and 5' glycosidic angles means that these dinucleosides do not show simple helical symmetry. Fig. 6 shows a $\beta : \chi(3')$ plot for uncomplexed and complexed dinucleosides. The larger scatter in values for the latter suggests that a somewhat greater degree of conformational flexibility can be induced by drug binding. Since alterations in only β and $\chi(3')$ are sufficient to alter an A-type ribo- or an A'-type deoxy-ribodinucleoside to one with intercalation geometry, we assume that the pathway between the two sets of values in Fig. 6 is a linear one. Further extrapolation of this plot to higher values increases the base pair separation beyond the ∿6.8Å in the complexes, to over 8Å, although at such large separations the base pair hydrogen-bonding becomes increasingly distorted.

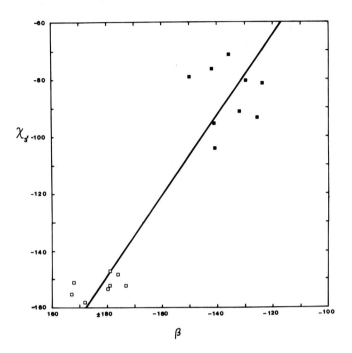

Figure 6.   The glycosidic $\chi(3')$ angle plotted against the β (about
             $05'-C5'$) torsion angle for complexed (shaded squares)
             and uncomplexed dinucleoside (open squares).

     The flexibility around the phosphodiester linkage, although
not large, is again greater than in uncomplexed dinucleosides (Fig.
7), presumably because of the varying effects of the different
bound drugs. In general, these angles have a $\sim20°$ range of softness
in intercalation complexes. There is a measure of anti-correlation
in these angles α and ζ within the two strands of any one complex.
This is best shown in the CpA-UpG case, where the UpG strand has
the smaller α value, compensated by a larger ζ angle.

     The flexibility of the 3'end sugar pucker (measured by the
$\delta(3')$ angle), can be correlated with the glycosidic angle at this
end (Fig. 8). As has been noted for the dodecanucleotide structure
[26], these correlated values cluster around particular puckers,
with the lower χ and δ angles corresponding to C3'endo, and the
higher ones to C2'endo puckers.

CONCLUSIONS

     We now have a considerable body of structural information

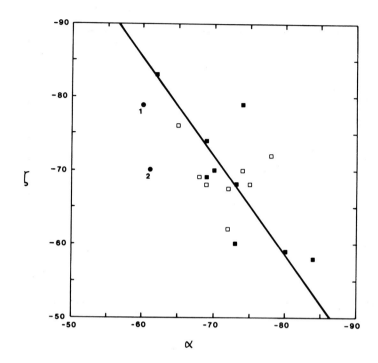

Figure 7.   Plot of α(P-05') against ζ(03'-P) torsion angles for
             complexed (shaded squares) and uncomplexed dinucleosides
             (open squares).   Points 1 and 2 are for A-RNA and
             A'-DNA respectively.

on proflavine intercalation into dinucleosides, and some beginning
of rationalisations for these observations. However, as yet, one
can only speculate as to how even this simple drug intercalates
into longer sequences; it is nonetheless clear that knowledge of
just one structure will not be sufficient to describe the effects
of sequence variability at and around the intercalation site, which
may well have significant biological consequences.

ACKNOWLEDGEMENTS

     We are grateful to Helen Berman for her collaboration on many
of the topics discussed here, and for fruitful discussions. The
Cancer Research Campaign is thanked for support and a Career
Development Award (to SN) and the Science and Engineering Research
Council for a studentship (to AA).

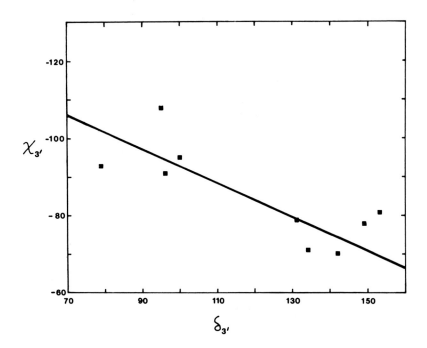

Figure 8.   Plot of 3'end glycosidic angle against $\delta(3')(C4'-C3')$
             torsion angle, for intercalated dinucleosides.

REFERENCES

[1]   Lerman, L.S: 1961, J. Mol. Biol. 3, pp. 18-30.

[2]   Li, H.J. and Crothers, D.M.: 1969, J. Mol. Biol. 39, pp. 461- 477.

[3]   Patel, D.J. and Canuel, L.L.: 1977, Proc. Natl. Acad. Sci.
      USA 74, pp. 2624-2628.

[4]   Broyde, S. and Hingerty, B.: 1979, Biopolymers 18, pp. 2905- 2910.

[5]   Kollman, P.A., Weiner, P.K. and Dearing, A.: 1981, Ann. N.Y.
      Acad. Sci. 367, pp. 250-268.

[6]   Pritchard, N.J., Blake, A. and Peacocke, A.R.: 1966, Nature
      212, pp. 1360-1361.

[7]   Neidle, S., Achari, A., Taylor, G.L., Berman, H.M., Carrell, H.L., Glusker, J.P. and Stallings, W.C.: 1977, Nature 269, pp. 304-307.

[8]   Berman, H.M., Stallings, W., Carrell, H.L., Glusker, J.P., Neidle, S., Taylor, G. and Achari, A.: 1979, Biopolymers 18, pp. 2405-2429.

[9]   Shieh, H.-S., Berman, H.M., Dabrow, M. and Neidle, S.: 1980, Nucleic Acids Res. 8, pp. 85-97.

[10] Reddy, B.S., Seshadri, T.P., Sakore, T.D. and Sobell, H.M.: 1979, J. Mol. Biol. 138, pp. 787-812.

[11] Agggarwal, A., Islam, S.A., Kuroda, R. and Neidle, S.: 1983, J. Mol. Biol. submitted for publication.

[12] Westhof, E. and Sundaralingam, M.: 1980, Proc. Natl. Acad. Sci. USA 77, pp. 1852-1956.

[13] Westhof, E., Rao, S.T. and Sundaralingam, M.: 1980, J. Mol. Biol. 142, pp. 331-361.

[14] Neidle, S., Taylor, G., Sanderson, M., Shieh, H.-S. and Berman, H.M.: 1978, Nucleic Acids Res. 2, pp. 4417-4422.

[15] Shieh, H.-S., Berman, H.M., Neidle, S., Taylor, G. and Sanderson, M.: 1982, Acta. Crystallogr. B38, pp. 523-531.

[16] Jain, S.C., Tsai, C.-C. and Sobell, H.M.: 1977, J. Mol. Biol. 114, pp. 317-331.

[17] Tsai, C.-C., Jain, S.C. and Sobell, H.M.: 1977, J. Mol. Biol. 114, pp. 333-365.

[18] Berman, H.M., Neidle, S. and Stodola, R.K.: 1978, Proc. Natl. Acad. Sci. USA 75, pp. 828-832.

[19] Tsai, C.-C., Jain, S.C. and Sobell, H.M.: 1975, Proc. Natl. Acad. Sci. USA 72, pp. 628-632.

[20] Sobell, H.M., Tsai, C.-C., Jain, S.-C. and Gilbert, S.G.: 1977, J. Mol. Biol. 114, pp. 333-365.

[21] Berman, H.M. and Neidle, S.: 1980, in : Nucleic Acid Geometry and Dynamics (Sarma, R.H. ed.) Pergamon Press, New York, pp. 367-382.

[22] Gale, E.F., Cundcliffe, E., Reynolds, P.E., Richmond, M.H. and Waring, M.J.: 1982, The Molecular Basis of Antibiotic Action, 2nd ed., John Wiley, London.

[23] Quigley, G.J., Wang, A.H.-J., Ughetto, G., van der Marel, G., Van Boom, J.H. and Rich, A.: 1980, Proc. Natl. Acad. Sci. USA 77, pp. 7204-7208.

[24] Fratini, A.V., Kopka, M.L., Drew, H.R. and Dickerson, R.E.: 1982, J. Biol. Chem. 257, pp. 14686-14707.

[25] Shakked, Z., Rabinovich, D., Kennard, O., Cruse, W.B.T., Salisbury, S.A. and Viswamitra, M.A.: 1983, J. Mol. Biol. in press.

[26] Dickerson, R.E. and Drew, H.R.: 1981, J. Mol. Biol. 149, pp. 761-786.

[27] Dearing, A., Weiner, P. and Kollman, P.A.: 1981, Nucleic Acids Res. 9, pp. 1483-1497.

[28] Neidle, S., Berman, H.M. and Shieh, H.S.: 1980, Nature 288, pp. 129-133.

[29] Neidle, S. and Berman, H.M.: 1982, in : Molecular Structure and Biological Activity (Griffin, J.F. and Duax, W.L. eds.) Elsevier, New York, pp. 287-301.

CONSEQUENCES OF SUBSTITUTING 2NH$_2$A FOR A IN SYNTHETIC DNA'S

Frank B. Howard and H. Todd Miles
Laboratory of Molecular Biology, NIADDK, National Institutes
of Health, Bethesda, MD 20205

Abstract: Chemical and spectroscopic consequences of replacing A with
2NH$_2$A have been examined in a variety of synthetic DNA's. This
substitution, which permits formation of a third hydrogen bond in AT
pairs, increases the stability of these pairs. The T$_m$ elevation,
however, is much smaller in the deoxy (ΔT$_m$ 12-15°) than in the ribo
series (ΔT$_m$ 27-33°). Sequence effects appear to be small. CD spectra
of all helices having the 2NH$_2$A substitution have a relatively strong
extremum at 286 to 298 nm. This band is positive for B-form helices
(deoxy-deoxy pairs in low salt) and negative for A-form helices (ribo-
ribo and deoxy-ribo pairs). These results are consistent with the
unusual CD spectrum of S-2L DNA (Kirnos et al.). This natural DNA has
all A's replaced by 2NH$_2$A and positive CD bands at 290 nm and 265 nm.
We assign the band at ~290 nm in these helices to the B$_{2u}$ transition
of 2NH$_2$A, displaced to longer wavelength by exciton splitting, and
suggest that it is relatively unperturbed by transitions of other
bases. Alternating (d2NH$_2$A-dT)$_n$ undergoes a cooperative transition to
an altered conformation in the presence of 4M NaCl or 2 x 10^{-4}M
hexammine cobalt. CD, IR, and ^{31}P NMR experiments reveal similarities
to the behavior of (dG-dC)$_n$ as well as some differences. The results
are consistent with a Z conformation for the high salt form but do not
establish it. The alternating polymers (d2NH$_2$A-dC)$_n$·(dG-dT)$_n$ and
(d2NH$_2$A-dC)$_n$·(dI-dT)$_n$ were also observed with CD. The former did not
undergo a discrete transition in high salt. The latter did undergo a
transition, but the structural nature of the change is not clear.

INTRODUCTION

An amino group in the purine 2-position alters the hydrogen bonding potential of the common nucleic acid bases while maintaining the
usual base pairing specificity. In the case of G in DNA, RNA, and
synthetic polynucleotides this property and some of its consequences
have long been familiar. Substitution of the 2-amino group of A has
been studied extensively in the ribopolynucleotide series [1-3], but
there has been relatively little information about the deoxy series.

B. Pullman and J. Jortner (eds.), Nucleic Acids: The Vectors of Life, 511–520.
© 1983 by D. Reidel Publishing Company.

While 2NH$_2$A is not known to occur in RNA, a novel phage DNA in which all A residues are replaced by 2NH$_2$A has been reported by Kirnos et al. [4-5]. The usual AT, GC equivalence was observed. We have recently reported properties of the homopolymer pair [6] (d2NH$_2$A)$_n$·(dT)$_n$ and present here studies of other properties and of more complex sequences.

RESULTS

We consider first the effect of 2NH$_2$dA substitution on helix transition temperatures. In the ribo series, T_m's of 2NH$_2$A·U and 2NH$_2$A·T helices are elevated 33° and 27°, respectively, above the A·U and A·T values. If these results are used to estimate the expected thermal stability of S-2L DNA [4,5] (68.7% GC), ΔT_m values, 10.3° and 8.5, respectively, are obtained. The observed value of 3.6° is markedly lower than these estimates from the ribo series. A much closer agreement is obtained when the deoxyhomopolymer pair (d2NH$_2$A)$_n$·(dT)$_n$ is used for reference: an expected elevation of 3.7°. The question remains, how satisfactory is a homopolymer pair as a model for the natural DNA?

Table 1.   Thermal Transitions[*]

Polymer	T_m(0.1 M Na$^+$)	dlog[Na$^+$]/dT
a) (d2NH$_2$A)$_n$·(dT)$_n$	77°	16.2° ± 0.6°
b) (d2NH$_2$A-dT)$_n$·(d2NH$_2$A-dT)$_n$	75.8°	13.3° ± 0.5°
c) (dA)$_n$·(dT)$_n$	65°	17.1° ± 0.6°
d) (dA-dT)$_n$·(dA-dT)$_n$	61°	21°
e) (d2NH$_2$A-dC)$_n$·(dG-dT)$_n$	95°	14.3° ± 0.3°
f) (d2NH$_2$A-dC)$_n$·(dI-dT)$_n$	66°	14.3° ± 0.7°
g) (dA-dC)$_n$·(dG-dT)$_n$	89°	16.2° ± 0.4°
h) (dA-dC)$_n$·(dI-dT)$_n$	56.5°	15.2° ± 0.8°
i) (dG)$_n$·(dC)$_n$	(100°)	18°
j) (dG-dC)$_n$·(dG-dC)$_n$	(108.5°)	13.5°
k) (dI)$_n$·(dC)$_n$	43°	16°
l) (dI-dC)$_n$·(dI-dC)$_n$	54°	15.5°

[*]Data for d,i,j,k,l from ref. [29]. All other data this work. Values in parentheses extrapolated.

wavelength maximum greatly intensifies and shifts to 287 nm, and the
first minimum intensifies and shifts from 242 to 261 nm. We attribute
the 287 nm maximum primarily to contributions from the B$_{2u}$ transitions
of 2NH$_2$A and G. It appears that the magnitude of the 2NH$_2$A contribution is greater.

The final substitution in this group is that of I for G to produce
(d2NH$_2$A-dC)$_n$·(dI-dT)$_n$. The first maximum is shifted to longer wavelength and decreased in intensity, evidently as a result of removing
the contribution of the B$_{2u}$ transition of G from the high wavelength
region. The B$_{2u}$ transition of 2NH$_2$A is now evidently the only significant one in this region, and the first pair of extrema is consistent
with exciton splitting of this transition: crossover occurs at the UV
maximum of 2-aminoadenosine, and the extrema at 268 and 292 nm are
equally displaced from the crossover. The intense band at 220 nm in
this helix and in that (d2NH$_2$A-dC)$_n$·(dG-dT)$_n$ are attributed primarily
to the E$_{1ua}$ transition of 2NH$_2$A at 214 nm (a negative band is observed
at 205 nm), as in the homopolymer duplexes discussed above. Other
transitions of the remaining bases presumably make contributions in the
middle wavelength region, though overlap, and possibly cancellation,
make these difficult to identify.

We next examine the effect of a 2NH$_2$ substitution in A on conversion of B-form DNA to an alternative conformation in high salt. Single
crystal studies of dC-dG oligomers have revealed a new left-handed Z
helix [17,18], and (dG-dC)$_n$ [19] and (dG-dMeC)$_n$ [20] are believed to
have very similar conformations in solutions of high ionic strength.
The CD spectra of (d2NH$_2$A-dT)$_n$ in 0.1 Na$^+$ (curve 1) and in 4 M Na$^+$
(curve 2) are shown in Fig. 4. Like the deoxyhomopolymer pair in Fig.
2, the low salt form of the alternating polymer has a first positive
extremum at 293 nm which we assign to exciton splitting of the B$_{2u}$
transition of 2NH$_2$A. The second component of the splitting contributes
to the minimum at 270 nm, as does, presumably, the B$_{2u}$ transition of T.
The shoulder at 255 nm may arise from the B$_{1u}$ transition and that at
220 nm from the E$_{1u}$ transition of 2NH$_2$A. In higher salt the positive
first band disappears and a much more intense negative band appears at
277 nm. This change is similar in some respects, though not all, to
that observed in (dG-dC)$_n$ by Pohl and Jovin [19]. The change is quite
cooperative in [Na$^+$] with a midpoint at 1.5 M (Fig. 4). Similar CD
changes are observed with hexammine cobalt (cf. [20]). The dependence
here is also cooperative, with the midpoint at 4 x 10^{-5} M hexammine
cobalt, a value similar to that observed for (dG-dC)$_n$ by Behe and
Felsenfeld. Overlays of spectra at differing salt concentrations have
isodichroic points at 267 nm, suggesting that only two species are
present during the transition.

Infrared spectra (D$_2$O solution) of (d2NH$_2$A-dT)$_n$ and (dG-dC)$_n$
exhibit similar changes in the carbonyl region when the salt concentration is raised to 4 M NaCl (Fig. 5). (d2NH$_2$A-dT)$_n$, like most double
helices containing U or T [21], has two carbonyl bands above 1650 cm^{-1}:
a strong shoulder at ~ 1686 and a peak at 1670 cm^{-1}. A T ring vibration

To examine the matter of possible sequence effects on T_m, we have
prepared polynucleotides containing $2NH_2A$ residues and compare their
transition temperatures with appropriate reference polymers in Table 1.

From these data, observed ΔT_m values for $2NH_2A$ of 12° (a-c), 15°
(b-d), 12° (2[e-g]), and 19° (2[f-h]) are found. When expressed as
expected elevation in T_m for DNA of 68.7% GC content, these values
correspond to 3.7°, 4.7°, 3.8°, and 5.8°. It is also interesting to
compare the effect of the $2NH_2$ group of G with that of $2NH_2A$ (Table 1).
T_m elevations of 57° (i-k), 54.5° (j-l), 58° (2[e-f]), and 65° (2[g-h])
are found for G, or values 3 to 4 times larger than those for $2NH_2A$.
ΔT_m for $2NH_2A$ is much smaller in the deoxy than in the ribo series, as
is the value for G ([6] and Fig. 1). This difference may be a consequence of A-form geometry in the ribo and B-form in the deoxy series.
Values for RNA-DNA hybrids are the same as those for RNA, in agreement
with this suggestion.

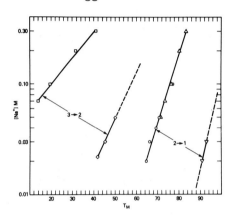
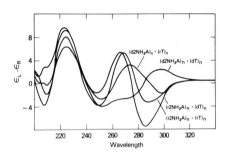

Fig. 1 (left). Phase diagrams for the helices $(d2NH_2A)_n \cdot (dT)_n$ and
$(r2NH_2A)_n \cdot (rT)_n$. The two curves on the left give salt dependence for
the 3 → 2 transitions and those on the right for the 2 → 1 transitions.
The deoxy polymers are the lower melting members of each pair.

Fig. 2 (right). CD spectra of four homopolymer pairs formed by
$(d2NH_2A)_n$, $(r2NH_2A)_n$, $(dT)_n$, and $(rT)_n$ in 0.1 M Na^+, pH 7.5.

CD spectra of the four $2NH_2A \cdot T$ homopolymer duplexes are shown in
Fig. 2. Significant features of the spectra are the long wavelength
first extrema (286-298 nm), which are unusual in being negative in
three of the four cases. The first two bands are conservative in the
four cases, with the crossover occurring at or near the mean of the
wavelengths of these bands. The positive and negative lobes are
roughly of equal area. We assign the longest wavelength bands to the
B_{2u} transition of $2NH_2A$, displaced from the UV maximum at ~ 280 nm by

exciton splitting [7]. The second lobe of this splitting presumably
contributes to the second band (269 - 275 nm), as does, we assume, the
B$_{2u}$ transition of T. The B$_{1u}$ transition of 2NH$_2$A (~ 257 nm) evidently
produces only weak CD bands in both (r2NH$_2$A)$_n$ [2] and (d2NH$_2$A)$_n$ and
appears to do so also in the spectra of Fig. 2. We suggest the two
bands at ~ 224 and ~ 205 nm result from exciton splitting of the 2NH$_2$A
transition at ~ 214 nm (for reasons supporting these and later assignments, cf. references [2,8]).

The only positive first extremum in Fig. 2 is that of (d2NH$_2$A)$_n$·
(dT)$_n$, and the conservative pair of bands in this spectrum is a mirror
image of those of (d2NH$_2$A)$_n$·(rT)$_n$. We suggest that 2NH$_2$A in B-form
helices contributes a positive first extremum, and note that the
anomalous CD spectrum of S-2L DNA [4,5] has a positive first band at
290 nm as well as a second well-resolved positive band at 265 nm.
Since RNA-DNA hybrids as well as RNA duplexes are considered to have
A-form structures [9-11], it appears that the same transition in
A-form helices leads to a negative first extremum (cf. Fig. 2 and other
examples in references [2,8,12]). The example in Fig. 2 appears to be
the first instance in which a change from B to A-form geometry leads to
simple inversions of the long wavelength bands, without change of band
shape or wavelength.

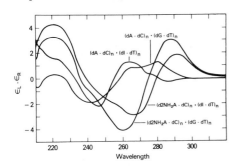

Fig. 3. CD spectra of (dA-dC)$_n$·
(dG-dT)$_n$, (dA-dC)$_n$·(dI-dT)$_n$,
(d2NH$_2$A-dC)$_n$·(dG-dT)$_n$, and
(d2NH$_2$A-dC)$_n$·(dI-dT)$_n$ in 0.1 M Na$^+$,
pH 7, 0.002 M cacodylate buffer.

In examining synthetic DNA's containing four bases, we begin with
(dA-dC)$_n$·(dG-dT)$_n$ [13,14] and modify it by replacing A with 2NH$_2$A and
G with I (Fig. 3). This helix has two positive long wavelength bands,
not well resolved, at 280 and 264 nm. Replacement of G with I results
in loss of the 280 nm maximum, suggesting that this band in the
original polymer has a major contribution from the B$_{2u}$ transition of G
(cf. ref [2]). We note that this polymer has a weak negative first
extremum at 284 nm ($\Delta\varepsilon$, - 0.5), an unusual feature in B-form DNA. A
similar negative extremum was observed in (dI-dC)$_n$·(dI-dC)$_n$ [15]
(λ_{min} 285 nm, $\Delta\varepsilon$, - 0.66). The origin of the band is unclear, though
it may arise from a B$_{2u}$ transition of I(the NH$_4^+$-poly(rI) complex has a
positive band at 284 nm, $\Delta\varepsilon$, +2 [16]), or from interactions of other
transitions. Replacement of A in the original polymer by 2NH$_2$A causes
large changes in the appearance of the CD spectrum (Fig. 3). The long

occurs at ~ 1646 cm^{-1}. In 4 M NaCl the higher frequency band disappears, possibly shifting under the envelope of the remaining intense band at 1668 cm^{-1}. A shoulder is present at ~ 1642 cm^{-1}. (dG-dC)$_n$ has two carbonyl bands at 1686 and 1656 cm$_{-1}$ in 0.1 M NaCl. The 1686 cm^{-1} band either disappears or decreases greatly in intensity in high salt (Fig. 5). A strong band is present at 1668 cm^{-1} and a resolved band at moderate intensity at 1638 cm^{-1}. The marked similarity in response of the carbonyl vibrations in the two helices to high salt suggests there may be a common structural basis for this pattern of change.

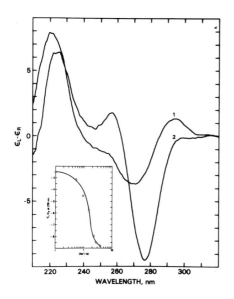

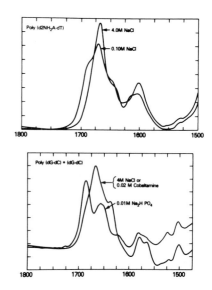

Fig. 4 (left). CD spectra of (d2NH$_2$A-dT)$_n$ in 0.1 M NaCl (curve 1) and in 4 M NaCl (curve 2). Inset, cooperative dependence of $\Delta\varepsilon$ on log [Na$^+$] is observed.

Fig. 5. (right). Infrared spectra in D$_2$O solution of (d2NH$_2$A-dT)$_n$ (above) and (dG-dC)$_n$ (below) in low salt and high salt, as indicated.

In a collaborative study with C. Chen and J. S. Cohen [22], ^{31}P-NMR spectra of (d2NH$_2$A-dT)$_n$ were observed with different cations as a function of salt concentration. A doublet (δ-4.1, -4.4 ppm upfield

from TMS) observed in low salt presumably indicates a B-DNA structure
with a dinucleotide repeat (cf. [23-25]). With increasing [Na⁺], both
peaks move progressively downfield (final values -3.3 and -4.0 ppm),
the peak separation increasing in a cooperative manner (Fig. 6).
NMR and CD data can be scaled to the same curve, showing that both
methods monitor the same transition (Fig. 6). Presence of only two
peaks at intermediate salt concentrations indicates fast exchange
between conformations.

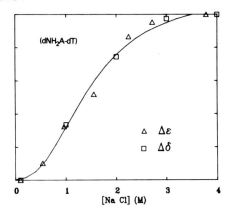

Fig. 6. Cooperative dependence of ^{31}P peak separation ($\Delta\delta$) is
coincident with that of CD peak magnitude ($\Delta\varepsilon$) when plotted on a
linear [Na⁺] scale.

The cooperative salt dependence of $(d2NH_2A-dT)_n$ resembles those
seen with $(dG-dC)_n$ [19] and $(dG-dMeC)_n$ [20,26] and differs from the
linear salt dependence of $(dA-dT)_n$ [22]. The pattern of infrared
spectral changes in high salt is also similar to that of $(dG-dC)_n$.
The magnitude of peak separation in 4 M NaCl (0.65 ppm) is intermediate
between those of $(dA-dT)_n$ (0.35 ppm) and $(dG-dC)_n$ (1.5 ppm) or
$(dG-dMeC)_n$ (1.3 ppm). As noted above, fast exchange is observed with
$(d2NH_2A-dT)_n$ and slow exchange for the two GC polymers. The available
evidence appears to be consistent with a Z conformation for the high
salt form, though clearly it does not establish this structure. If the
polymer is not in the Z-form, the work of Zimmerman and Pheiffer [27]
showing a B rather than a C diffraction pattern for DNA fibers in high
concentrations of most salts (cf. [28]) suggests some variant of the B-
form as the most likely alternative.

We consider next the salt dependence of the alternating polymers
containing four bases. $(d2NH_2A-dC)_n \cdot (dI-dT)_n$ exhibits large changes in
magnitude on going from 0.1 to 5.2 M Na (Fig. 7), though there are no
sign inversions and no significant wavelength shifts. The conformation
is clearly altered by high salt concentration, and an isodichroic point
at 246 nm suggests only two structures are present. There is currently

no reliable way, however, to deduce the high salt conformation from the CD spectrum and no compelling reason to conclude that it is the Z-form. A similar set of changes with $[Na^+]$ is observed with $(dA-dC)_n \cdot (dI-dT)_n$, again suggesting a second conformation but not identifying it. Of the polynucleotides in Table 1 which contain four bases, only one, $(d2NH_2A-dC)_n \cdot (dG-dT)_n$, permits a clear answer to the question of a Z conformation in high salt, and the answer is negative. Though there is some perturbation by 5 to 6 M NaCl in the 250 to 260 nm region (Fig. 8), changes elsewhere are small and do not at all resemble those observed by Pohl and Jovin [19]. Changes in 1 M Mg^{++} and 3×10^{-3} cobalt hexammine are even smaller than those in 6 M NaCl. This polymer evidently does not have a Z-form under these conditions of high salt concentration.

From these observations at high salt, it appears that $2NH_2A$ may facilitate a B $\rightarrow$ Z conversion in $(d2NH_2A-dT)_n$, but negative or ambiguous results with the other alternating polymers suggest that any effect is not decisive.

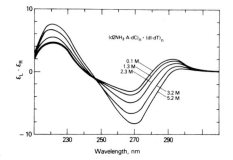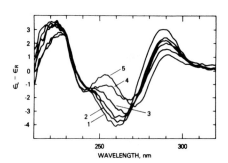

Fig. 7. (left). CD spectra of $(d2NH_2A-dC)_n \cdot (dI-dT)_n$ in 0.1 M NaCl and 5 M NaCl.

Fig. 8. (right). CD spectra of $(d2NH_2A-dC)_n \cdot (dG-dT)_n$ as a function of NaCl concentration; 1, 0.1 M; 2, 3.5 M; 3, 4.6 M; 4, 5.2 M; 5, 5.9 M. Cacodylate buffer, 0.002, pH 7.

CONCLUSIONS

Replacement of A with $2NH_2A$ in synthetic DNA's (and at least one naturally occurring DNA) increases the transition temperature, as one would anticipate from addition of a third hydrogen bond. This elevation, however, is much smaller than in the ribo series and is only a third to a quarter of that resulting from replacement of I with G in the deoxy series.

Helices containing 2NH$_2$A have prominent first extrema at 286 to 296 nm, which we attribute to exciton splitting of the B$_{2u}$ transition of 2NH$_2$A. This band is positive in the DNA series and negative in the RNA series. In hybrid helices, which have A-form geometry, the long wavelength band is negative, suggesting a change from B to A conformation is responsible for a sign inversion of this band.

The alternating copolymer (d2NH$_2$A-dT)$_n$ undergoes a cooperative transition to a new helical conformation when the salt concentration is increased. CD and IR spectral changes resemble those observed with (dG-dC)$_n$. ^{31}P NMR changes show both similarities to and differences from G·C polymers believed to exist in the Z-form. The results are consistent with a Z conformation of (d2NH$_2$A-dT)$_n$ in high salt but do not establish it. (d2NH$_2$A-dC)·(dI-dT) is converted to a new conformation in high salt, as judged by CD changes, but the structure cannot be identified at present. The CD spectra of (d2NH$_2$A-dC)$_n$·(dG-dT)$_n$ changes less with [Na$^+$] than other polymers reported here, and evidently is not converted to the Z-form.

REFERENCES

1. Howard, F.B. & Miles, H.T.: 1966, J. Biol. Chem. 241, pp. 4293-4295.

2. Howard, F.B., Frazier, J. & Miles, H.T.: 1976, Biochemistry 17, pp. 3783-3795.

3. Muraoka, M., Miles, H.T. & Howard, F.B.: 1980, Biochemistry 19, pp. 2429-2439.

4. Kirnos, M.D., Khudyakov, I.Y., Alexandrushkina, N.I. & Vanyushin, B.F.: 1977, Nature 270, pp. 369-370.

5. Khudyakov, I.Y., Kirnos, M.D., Alexandrushkina, N.I. & Vanyushin, B.F.: 1978, Virology 88, pp. 8-18.

6. Howard, F.B. & Miles, H.T.: 1983, Biopolymers 22, pp. 597-600.

7. Tinoco, I.: 1964, J. Am. Chem. Soc. 86, pp. 297-298.

8. Mathelier, H.D., Howard, F.B. & Miles, H.T.: 1979, Biopolymers 18, pp. 709-722.

9. Milman, G., Langridge, R. & Chamberlin, M.J.: 1967, Proc. Natl. Acad. Sci. USA 57, pp. 1804-1810.

10. O'Brien, E.J. & MacEwan, A.W.: 1970, J. Mol. Biol. 48, pp. 243-261.

11. Arnott, S.: 1970, Prog. Biophys. Mol. Biol. 21, pp. 267-319.

12. Ikeda, K., Frazier, J. & Miles, H.T.: 1970, J. Mol. Biol. 54, pp. 59-84.

13. Wells, R.D., Ohtsuka, E., & Khorana, H.G: 1965, J. Mol. Biol. 14, pp. 221-237.

14. Gray, D.M. & Ratliff, R.L.: 1975, Biopolymers 14, pp. 487-498.

15. Mitsui, Y., Langridge, R., Grant, R.C., Kodama, M., Wells, R.D., Shortle, B.E. & Cantor, C.K.: 1970, Nature 228, pp. 1166-1169.

16. Howard, F.B. & Miles, H.T.: 1982, Biopolymers 21, pp. 147-157.

17. Wang, A.H., Quigley, J.J., Kolpak, F.J., Crawford, J.L., van Boom, J.H., van der Marel, G. & Rich, A.: 1979, Nature 282, pp. 680-686.

18. Drew, H., Takano, T., Tanaka, S., Itakura, K. & Dickerson, R.E.: 1980, Nature 286, pp. 567-573.

19. Pohl, F.M. & Jovin, T.M.: 1972, J. Mol. Biol. 67, pp. 375-396.

20. Behe, M. & Felsenfeld, G.: 1981, Proc. Nat. Acad. Sci. USA 78, pp. 1619-1623.

21. Miles, H.T.: 1971, Procedures Nuc. Acid Res. 2, pp. 205-232.

22. Cohen, J.S., Chen, C-w., & Knop, R.H.: 1983, These Proceedings.

23. Shindo, H., Simpson, R.T. & Cohen, J.S.: 1979, J. Biol. Chem. 254, pp. 8125-8128.

24. Patel, D.J., Canuel, L.L. & Pohl, F.M.: 1979, Proc. Natl. Acad. Sci. USA 76, pp. 2408-2511.

25. Klug, A., Jack, A., Viswamitra, M.A., Kennard, O., Shakked, Z. & Steitz, T.A.: 1979, J. Mol. Biol. 131, pp. 669-680.

26. Chen, C-w., Cohen, J.S. & Behe, M.: 1983, Biochemistry 22, in press.

27. Zimmerman, S.B. & Pheiffer, B.H.: 1980, J. Mol. Biol. 142, pp. 315-330.

28. Rhodes, N.J., Mahendrasingam, A., Pigram, W.J., Fuller, W., Brahms, J., Vergne, J. & Warren, R.A.J.: 1982, Nature 296, pp. 267-269.

29. Szybalski, E.H. & Szybalski, W.: 1975 Handbook of Biochemistry and Molecular Biology (3rd ed.) Vol. I, pp. 575-588.

NONIONIC OLIGONUCLEOTIDE ANALOGS AS NEW TOOLS FOR STUDIES ON THE STRUCTURE AND FUNCTION OF NUCLEIC ACIDS INSIDE LIVING CELLS

Paul S. Miller, Cheryl H. Agris, Kathleen R. Blake, Akira Murakami, Sharon A. Spitz, Parameswara M. Reddy, and Paul O.P. Ts'o
Division of Biophysics, School of Hygiene and Public Health, The Johns Hopkins University, Baltimore, Maryland 21205 USA

ABSTRACT

    Two types of nonionic oligonucleotide analogs, deoxyribonucleotide alkyl phosphotriesters and deoxyribooligonucleoside methylphosphonates, have been synthesized to serve as selective inhibitors of cellular nucleic acid function. The backbones of these analogs are resistant to nuclease hydrolysis and the analogs are taken up by mammalian cells and certain bacterial cells in culture. Sequence specific analogs inhibit tRNA aminoacylation and translation of mRNA in both mammalian and bacterial cell-free systems in a specific manner as a result of oligomer binding to complementary sequences of the target nucleic acid. These analogs also inhibit cellular protein synthesis and growth of living cells. Selective inhibition of bacterial versus mammalian cell growth is observed with a methylphosphonate oligomer complementary to the Shine-Dalgarno sequence of 16S rRNA. Methylphosphonate complementary to the 5'-end of U_1 RNA and to the donor splice site of SV40 large T antigen pre-mRNA inhibit T-antigen production in SV40-infected cells.

INTRODUCTION

    Studies on nucleic acid analogs possessing modified internucleoside linkages have made important contributions to our understanding of nucleic acid conformation and have provided materials for a variety of biochemical and biological studies (1-12). We have studied two types of nonionic oligonucleotides, oligonucleotide alkylphosphotriesters and oligodeoxyribonucleoside methylphosphonates, whose structures are shown in Figure 1. The 3'-5' linked internucleotide bonds of these analogs closely resemble the size and geometry of the nucleic acid phosphodiester bond. However, since the sugar-phosphate backbones of these analogs are electroneutral, the analogs have unique physical and biological properties. These properties include (1) their ability to form stable hydrogen-bonded complexes with complementary polynucleotides; (2) their resistance to hydrolysis by nucleases; and (3) their ability to be taken up intact by mammalian and certain bacterial cells.

521

B. Pullman and J. Jortner (eds.), Nucleic Acids: The Vectors of Life, 521–535.
© *1983 by D. Reidel Publishing Company.*

Figure 1

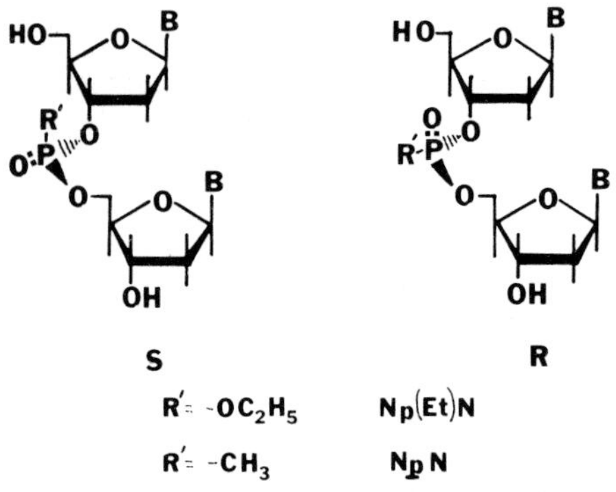

S   R

R'= -OC$_2$H$_5$ Np(Et)N

R'= -CH$_3$ NpN

The properties of nonionic oligonucleotides suggest they could specifically bind to single-stranded regions of cellular nucleic acids. As a consequence of binding, the analogs may inhibit the function or expression of cellular or viral nucleic acids in a selective manner. We have tested this possibility by examining the effects of sequence-specific analogs on aminoacylation of tRNA, translation of mRNA and splicing of pre-mRNA both in the test tube and in living cells. The results of our experiments suggest nonionic oligonucleotides may indeed be designed to specifically control nucleic acid function. We will first briefly describe the physical properties of these analogs and then describe experiments designed to examine the biochemical and biological properties of these molecules.

PHYSICAL PROPERTIES

Dideoxyribonucleoside methyl and ethyl phosphotriester [dNp(R)N] and methylphosphonate [dNpN] dimers occur as a pair of diastereoisomers which differ in their configuration about the phosphorous atom (Figure 1). The effects of the phosphotriester group configuration on dimer conformation were studied by NMR (13). The detailed conformations of methylphosphonate diastereoisomers were studied by circular dichroism and by NMR (14,15). The absolute configuration of d-ApT has been determined by X-ray crystallography (16) while that of d-ApA has been assigned by NMR nuclear Overhauser enhancement experiments (15).

As shown by CD and ^{1}H NMR, dinucleoside methylphosphonates adopt stacked conformations in aqueous solution similar to those of dinucleoside monophosphates. The conformation of the sugar-phosphonate backbone as defined by the puckering of the deoxyribofuranose rings and the rotation about ψ, ϕ, and ϕ' is very similar to that of the dinucleoside monophosphates. The base stacking of d-ApA-S-isomer is slightly greater than that of the R isomer and is almost identical to that of d-ApA. These differences in conformation may result from the differences in solvation of the two isomers. Thus the hydrophobic methyl group of the S-isomer of d-ApA is located near the hydrophobic base stacking region which would tend to stabilize base stacking interactions in the dimer. In contrast, the methyl group in the R-configuration is directed away from the base stacking region and may be expected to destabilize stacking interactions.

In some cases, the configuration of the alkylphosphotriester group or the methylphosphonate group can influence interactions of nonionic oligomers with complementary polynucleotides. For example, d-ApA forms 2U:1A triple stranded complexes with polyuridylic acid (Table I).

TABLE I. Interaction of Oligonucleoside Methylphosphonates with Complementary Polynucleotides[a]

Oligomer	Tm with poly(rU) (2U:1A) (°C)	Tm with poly(dT) (2T:1A) (°C)
d-ApA R-isomer	15.4	18.7
S-isomer	19.8	18.4
d-ApApA	33.0	36.8
d-ApApApA	43.0	44.5
d-ApA	7.0	9.2
d-ApApApA	32.0	35.5

(a) 5×10^{-5} M total (nucleotide), 10 mM Tris and 10 mM $MgCl_2$, pH 7.5.

The melting temperature of the S isomer complex is approximately 4° higher than that of the R isomer complex, while both complexes have Tm's higher than that of d-ApA·poly U (14). Similar increases in Tm are seen for complexes between d-ApApA and d-ApApApA and poly U and poly(dT)(17). The sharpness of the melting curves indicates the various diastereoisomers form complexes of similar stability.

More dramatic effects of configuration are seen for oligothymidy-late ethylphosphotriesters (18) and oligothymidylate methylphospho-nates (Table II). The triester d-[Tp(Et)7]T, which consists of 2^7 diastereoisomers, forms a 1:1 complex with poly (dA), which displays a rather broad melting curve. The octamer triester does not bind to poly(dA)·poly(dT) and interacts with poly(rA) only at low temperature. Similar results were obtained for the methylphosphonate, d-(Tp)8T. These results suggest isomers of different backbone configuration form

TABLE II.   Interaction of Nonionic Oligothymidylates with
             Complementary Polynucleotides

Oligomer	Tm with poly(dA)	Tm with poly(rA)
d-[Tp(Et)]$_7$T $^{(a)}$	18° (1T:1A)	<0°
d-(T$\underline{p}$)$_8$T $^{(b)}$	22° (1T:1A)	<0°
d-T$\underline{p}$(TpT$\underline{p}$)$_4$T $^{(b)}$		
Isomer 1	33.5° (1T:1A)	19.5° (1T:1A)
Isomer 2	2° (2T:1A)	0°
d-(Tp)$_9$T $^{(b)}$	22.5 (1T:1A)	18.0° (1T:1A)

(a) 1 x 10^{-4} $\underline{M}$ total [nucleotide], 0.15 $\underline{M}$ NaCl and 0.04 $\underline{M}$
    potassium phosphate, pH 6.9.
(b) 3.5 x 10^{-5} $\underline{M}$ total [nucleotide], 0.10 $\underline{M}$ sodium cacodylate,
    pH 6.8.

complexes of unique stability with poly(dA) and poly(rA). This conclusion was confirmed by examining the alternating methylphosphonate/
phosphodiester oligothymidylate analog, d-T$\underline{p}$(TpT$\underline{p}$)$_4$T where the configuration of each methylphosphonate linkage is the same throughout the
backbone of the oligomer and is denoted as type 1 or type 2 (19). As
shown in Table II, the oligomer with type 1 configuration forms stable
complexes with both poly(dA) and poly(rA) while that with type 2 configuration forms a 2U:1A complex with poly(dA) and no complex with
poly(rA).

The Tm values of nonionic oligonucleotide/polynucleotide complexes
are not affected by changes in salt concentration. This effect results
from the reduced charge repulsion between the nonionic backbone of the
oligomer and negatively charged sugar-phosphate backbone of the polynucleotide. The lack of charge repulsion also explains the increased
stabilities of nonionic oligonucleotide/polynucleotide complexes versus
those of oligonucleotide phosphodiester/polynucleotide complexes.

SEQUENCE-SPECIFIC INHIBITION OF CELL-FREE AMINOACYLATION AND PROTEIN
SYNTHESIS BY NONIONIC OLIGONUCLEOTIDES

Sequence-specific nonionic oligonucleotides form hydrogen-bonded
complexes with the -ACCA- amino acid accepting stem and anticodon loop
regions of tRNA (17,20,21). For example, the binding constants of
tritium-labeled G$_{\underline{p}}^{\underline{m}}$(Et)G$_{\underline{p}}^{\underline{m}}$(Et)U with tRNA^{Phe} yeast, unfractionated

tRNA$_{E.coli}$ and unfractionated tRNA$_{E.coli}$ lacking the 3'-CpA terminus and d-GpGpT with tRNA$_{E.coli}$ are shown Table III.

TABLE III. Interaction of Nonionic Oligonucleotides with Transfer RNA[a]

Oligomer	Temp. (°C)	K (M^{-1}) tRNAphe yeast	K (M^{-1}) tRNA$_{E.coli}$	tRNA$_{E.coli}$-CA
G$_p^m$(Et)G$_p^m$(Et)U	0	3,100	9,300	1,600
	25	3,100	1,900	–
	37	1,700	2,000	–
G$_p^m$G$_p^m$U	0	63,500	103,000	4,000
	25	5,300	12,300	–
	37	750	1,100	–
d-GpGpT	0	–	1,000	–
	25	–	200	–
	37	–	100	–

(a) The binding constants were measured by equilibrium dialysis in 0.01 $\underline{M}$ NaCl, 10 m$\underline{M}$ MgCl$_2$, 10 m$\underline{M}$ Tris, pH 7.5.

The binding constants of the nonionic oligomers show relatively small changes over the temperature range studied, while that of the diester, G$_p^m$G$_p^m$U, dramatically diminishes with increasing temperature. This effect may be due to self-aggregation of the nonionic oligomers at low temperatures, which would result in a decreased apparent binding to the tRNA. The apparent association constants of d-GpGpT are significantly less than those of 2'-O-methylribooligonucleotide ethyl phosphotriester. This difference may reflect overall differences in the conformation of the deoxyribo- versus 2'-O-methylribo backbones of these oligomers. Removal of the 3'-CpA nucleotides from unfractionated tRNA$_{E.coli}$ by treatment with snake venom phosphodiesterase results in a dramatic reduction of the binding constants for the oligomers. This indicates the major binding site is indeed the 3'-amino acid accepting end of the tRNA. The observed residual binding may be due to binding to other complementary single-stranded regions of the unfractionated tRNA.

As shown in Table IV, sequence-specific nonionic oligonucleotides inhibit cell-free aminoacylation of tRNA (17,21,22). Oligodeoxyadenosine methylphosphonates and the parent diester, d-ApApApA selectively inhibit cell-free aminoacylation of tRNA$_{E.coli}^{Lys}$. The extent of inhibition is temperature dependent and parallels the ability of the oligomers to bind to poly(rU) (Table I). These observations and the previously demonstrated interaction of r-ApApApA with tRNA$_{E.coli}^{Lys}$ (23) suggest the inhibition is a consequence of oligomer binding to the –UUUU– anticodon loop of the tRNA. The lower extent of inhibition

observed with d-ApApGpA is consistent with this explanation, since
interaction of this oligomer with the anticodon loop would involve
formation of a less stable G·U base pair.

TABLE IV.   Effects of Nonionic Oligonucleotides on Cell-Free
            Aminoacylation of Unfractionated tRNA$_{E.coli}^{Lys}$ (a)

| | % Inhibition | | | | |
| | Phe | Leu | Lys | | |
Oligomer	0°C	0°C	0°C	22°C	37°C
d-ApA	6	0	7	–	–
d-ApApA	9	0	62	15	0
d-ApApApA	9	12	88	40	16
d-ApApGpA	12	12	35	0	–
d-GpGpT	31	5	34	9	15
G$\overset{m}{p}$(Et)G$\overset{m}{p}$(Et)U	39 [b]	–	–	–	–
d-ApApApA	0	7	71 [c]	15 [c]	–
d-GpGpT (400 μM)					

(a) Reactions were carried out in 100 mM Tris-HCl, pH 7.4,
    10 mM Mg(OAc)$_2$), 5 mM KCl, 2 mM ATP, 4 μM ^{3}H-labeled
    amino acid, 2 μM tRNA using unfractionated E.coli amino-
    acyl synthetase in the presence of 50 μM oligomer.
(b) 37°C.
(c) [oligomer] = 100 μM.

     Since the anticodon loop of tRNA$_{E.coli}^{Lys}$ forms part of the synthetase
recognition site (24,25), inhibition of aminoacylation by the methylphosphonates could result from a reduction in the affinity of the synthetase for the tRNA^{Lys}-oligonucleotide complexes.   Alternatively,
oligomer binding to the anticodon loop may induce conformational changes
in the tRNA, thus leading to a lower rate and extent of aminoacylation.
The greater inhibition by d-ApApApA versus d-ApApApA may be a consequence of greater binding of the methylphosphonate analog to the anticodon loop or to a decreased ability of the synthetase to displace the
nonionic oligonucleotide analog.

Both phenylalanine and lysine aminoacylation are inhibited by the d-G_pG_pT at 0°, while little effect is observed on leucine aminoacylation. These differences may reflect differences in the ability of the oligomer to bind to the -ACC- ends of the tRNAs. Inhibition of lysine aminoacylation by d-G_pG_pT is very temperature dependent while $G^m_p(Et)G^m_p(Et)U$ effectively inhibits phenylalanine aminoacylation even at 37^oC. This behavior parallels the ability of the oligomers to bind to tRNA (Table III).

As shown in Table V, oligodeoxyribonucleoside methylphosphonates effectively inhibit polypeptide synthesis in cell-free systems derived from E.coli and rabbit reticulocytes (17). Poly(U)-directed polyphenylalanine synthesis is inhibited by oligodeoxyadenosine analogs in both cell-free systems. The extent of inhibition reflects the stabilities of the oligomer/poly(U) complexes (Table I). Thus, d-$A_pA_pG_pA$, which forms a less stable complex with poly(U), is 4.5-fold less effective than d-$A_pA_pA_pA$. These observations suggest inhibition results from complex formation between the poly(U) message and the oligomers. It is unlikely inhibition results from non-specific interactions of the oligodeoxyadenylate analogs with protein components of the translation systems, since no inhibition of globin mRNA translation by these analogs is observed in the reticulocyte system.

TABLE V.   Effects of Oligonucleoside Methylphosphonates on Bacterial and Mammalian Cell-Free Protein Synthesis at 22^oC

| | % Inhibition | | |
| | E. coli | Rabbit Reticulocyte | |
Oligomer	Poly(U) Directed[a]	Poly(U) Directed[a]	Globin mRNA Directed (b)
d-A_pA	20	-	-
d-A_pA_pA	84	81	-
d-$A_pA_pA_pA$	100	77	0
d-$A_pA_pG_pA$	22	-	0
d-$C_pC_pA_pT$	-	-	61[c]
d-$G_pC_pA_pC_pC_pA_pT$	-	-	40[d]
d-$(T_p)_5T$	-	-	0[e]
d-$A_pA_pA_pA$	13	18	0

(a) [poly(U)] = 360 µM in U; [oligomer] = 175-200 µM in base.
(b) [oligomer] = 200 µM in base.
(c) [oligomer] = 246 µM in strand.
(d) [oligomer] = 289 µM in strand.
(e) [oligomer] = 300 µM in strand.

Although d-ApApA and the phosphodiester d-ApApApA form complexes
with poly(U) which have very similar Tm values (Table I), the methylphosphonate analog more effectively inhibits translation. This effect
may result from a decreased ability of the ribosome to displace the
nonionic methylphosphonate oligomer from the poly(U). Alternatively
the phosphodiester oligomer may be susceptible to degradation by
nucleases in the cell-free translation systems.

d-CpCpApT is complementary to the -AUGG- initiation codon region
of globin mRNA and to the anticodon region of tRNA^{his}. d-GpCpApCpCpApT
and d-(Tp)$_5$T are complementary respectively to the initiation codon
regions and poly(A) tails of rabbit α and β globin mRNA. Both
d-CpCpApT and d-GpCpApCpCpApT effectively inhibit incorporation of
of [^{3}H]-leucine into globin, while d-(Tp)$_5$T has no effect on translation.
The greater inhibition by d-CpCpApT could be due to oligomer binding to
a number of complementary sequences along the coding region of the
globin mRNA as well as to the anticodon region of tRNA^{his}. The lack of
inhibition by d-(Tp)$_5$ suggests potential binding to the poly(A) tail of
globin mRNA does not affect translation and also shows the observed
inhibition is sequence specific.

Specific inhibition of bacterial protein synthesis can be affected
by disrupting the interaction between ribosomal RNA and mRNA (26).
Oligonucleoside methylphosphonates were synthesized whose base sequences are complementary to the Shine-Dalgarno sequence (-ACCUCCU-)
found at the 3'-end of bacterial 16S rRNA. This sequence is required
for binding of the 40S ribosomal subunit to bacterial mRNA. A similar
sequence is lacking in eukaryotic 18S rRNA, and ribosome binding most
likely begins by recognition of the 5'-cap site of eukaryotic mRNAs.

The interactions of d-ApGpGpApGpGp[^{3}H]-T and d-ApGpGp[^{3}H]T with
70S ribosomes were studied by equilibrium dialysis. The heptamer has a
high apparent binding constant which diminishes with increasing temperature (4.67 x 10^5 M^{-1} at 0°C; 1.72 x 10^5 M^{-1} at 22°C; 2.0 x 10^4 M^{-1} at
37°C). The tetramer has an approximately ten-fold lower binding
constant (1.44 x 10^4 M^{-1} at 22°C). As shown in Table VI, d-ApGpGpApGpGp
and d-ApGpGpApGpGpT exhibit significant inhibitory activities when MS-2
RNA is the message, but show less effect on poly(A)-directed polyphenylalanine or poly(A)-directed polylysine synthesis.

TABLE VI. Effects of Deoxyribonucleoside Methylphosphonates on Cell-
Free Translation in an E. Coli System

Oligomer	Conc. μM	Poly(U) [a] 22°C	37°C	Poly(A) [b] 22°C	37°C	MS2 RNA 22°C
d-ApGpGp	100	8	0	0	0	5
d-ApGpGpT	100	-	-	-	-	0
d-ApGpGpApGpGp	12.5	-	-	-	-	45
	25	0	0	0	0	75
	50	19	0	29	14	88
	100	39	18	80	27	-
d-ApGpGpApGpGpT	25	0	0	0	0	77

(a) 260 μM in UMP residues.
(b) 225 μM in AMP residues.

Inhibition is temperature and concentration dependent. The shorter
oligomers d-ApGpGp and d-ApGpGpT show little or no inhibitor activity,
even at high nucleotide concentrations. In contrast to their effects
on the E.coli system, neither d-ApGpGpApGpGp or d-ApGpGpApGpGpT show
appreciable inhibitory effects on translation of globin mRNA in a
cell-free reticulocyte system (at 100 μM and 22°C, 16% and 17%, respectively).

These results strongly suggest specific inhibition of MS2 RNA
translation in the E.coli cell-free system is a consequence of oligomer
binding to the Shine-Dalgarno sequence of 16S rRNA. This binding prevents the 40S ribosome from binding to the mRNA. Because the synthetic
mRNAs, poly(U) and poly(A) lack specific initiation sites, much lower
inhibition of translation by the oligomers is observed. Although the
3'-end sequences of 18S rRNA and 16S rRNA are similar, 18S rRNA
specifically lacks the –CCUCCU– sequence found in 16S rRNA. Thus, the
oligonucleoside methylphosphonates cannot form stable complexes with
the 18S rRNA of reticulocyte ribosomes.

UPTAKE OF NONIONIC OLIGONUCLEOTIDES BY LIVING CELLS

The internucleotide bonds of alkylphosphotriesters and methylphosphonate oligomers are completely resistant to hydrolysis by exo-
and endonucleases and nuclease and esterase activities found in mammalian sera (13,17,19,20,21). Oligomer analogs which have been incubated
with mammalian cells in culture are recovered completely intact from
the culture medium. Tritium-labeled oligonucleotide ethylphosphotriesters

and oligonucleoside methylphosphonates are readily taken up by mammalian cells in culture. In the case of $G_p^m(Et)G_p^m(Et)[^3H]$-U and d-$[Tp(Et)]_n[^3H]T$ (n = 1,4,6), the oligomers are rapidly taken up by transformed Syrian hamster fibroblasts (21; Miller and Jayaraman, unpublished results) and subsequently metabolized. Analysis by chromatography of the radioactivity recovered from cell lysates after a 2 hr. incubation with $G_p^m(Et)G_p^m(Et)[^3H]$-U shows 27% of the label occurs in the trinucleotide species $G_p^m(Et)G_p^m(Et)U$, $G_p^mG_p^m(Et)U$, $G_p^m(Et)G_p^mU$ and $G_p^mG_p^mU$, 28% is incorporated as uridine or cytidine in high-molecular-weight RNA, and the remainder is found in various mono- and dimeric species. These results suggest the triester is taken up intact by the cells, deethylated, and the resulting phosphodiester linkages may then be further hydrolyzed by nucleases.

The uptake of oligonucleoside methylphosphonates by transformed Syrian hamster fibroblasts is quite different from that of the oligonucleotide ethylphosphotriesters (17). The rate and extent of uptake is consistent with passive diffusion of the oligomer across the cell membrane. Thus, after 1.5 hr., the calculated intracellular concentration is ~177 μM when cells are incubated with 100 μM d-$Tp[^3H]T$. Both d-$Tp[^3H]T$ and d-$(Tp)_8[^3H]T$ are taken up at approximately the same rates and to the same extents which suggests there is no size restriction to uptake over this chain-length range.

Examination of lysates of cells exposed to the labeled methylphosphonates for 18 hrs. showed ~70% of the labeled thymidine was associated with intact oligomer while the remainder was found in thymidine triphosphate and in cellular DNA. These results suggest the methylphosphonates which are recovered intact from the culture medium are slowly degraded within the cell. This degradation may result from cleavage of the 3'-terminal $[^3H]$thymidine N-glycosyl bond with subsequent reutilization of the thymine base. The relatively long half lives of the oligodeoxyribonucleoside methylphosphonates may be of value in potential pharmacological applications of the analogs.

Uptake experiments with E.coli B cells show they are permeable to d-$Ap[^3H]T$, d-$Tp[^3H]T$, and d-$TpTp[^3H]T$, but not to d-$(Tp)_4[^3H]T$ or d-$(Tp)_8[^3H]T$. Thus, it appears analogs longer than 4 nucleotide units cannot enter the bacterial cell. This size cutoff agrees with that found by others for oligosaccharides and oligopeptides (27,28). Similar results were obtained for other wild type gram positive and gram negative bacteria such as Bacillus subtilis and Pseudomonas auerogenosa. Oligomers up to 7 nucleotides in length (e.g. d-$ApGpGpApGpGp[$ H$]T$) are taken up by a permeable mutant of E.coli, E.coli ML 308-225. The outer membrane of the cell wall of this mutant contains only small quantities of lipopolysaccharide (29) which may increase the permeability of the cell wall toward the longer oligonucleoside methylphosphonates.

CELLULAR PROTEIN SYNTHESIS AND GROWTH

Nonionic oligonucleotides which inhibit cell free aminoacylation of tRNA or cell free protein synthesis also inhibit cellular protein synthesis and growth of bacterial cells and transformed hamster and human cells in culture (17,21,26). For example, $G_p^m(Et)G_p^m(Et)U$ inhibits cellular protein synthesis in a dose-dependent manner in transformed Syrian hamster fibroblasts (up to 90% at 100 μM). During prolonged incubation, protein synthesis is inhibited for the first 4 hrs. and then resumes at approximately the same time when oligomer uptake begins to level off. Cellular RNA synthesis, however, increases slightly during the first 4 hrs. and then returns to control levels. The reversible inhibitory effects most likely occur as a result of degradation of the triester within the cell.

As shown in Table VII, $G_p^m(Et)G_p^m(Et)U$ and oligonucleoside methylphosphonates which inhibit cell-free aminoacylation and protein synthesis also inhibit growth of mammalian and bacterial cells as assayed by their effects on colony formation.

TABLE VII.   Effects of Nonionic Oligonucleotides on Colony Formation by Bacterial and Mammalian Cells in Culture

| | % Inhibition | | | |
| | E. Coli B | | BP-6[a] | HTB1080[b] |
Oligomer	50 μM	160 μM	50 μM	50 μM
d-ApApA	3	44	29	31
d-ApApApA	19	78	36	19
d-GpGpT	7	11	7	9
$G_p^m(Et)G_p^m(Et)U$	–	–	50[c]	–

(a)   BP-6 = transformed Syrian hamster fibroblasts.
(b)   HTB1080 = Human tumor cells.
(c)   [oligomer] = 25 μM.

This inhibition may occur as a result of binding of the analogs to complementary sequences on cellular tRNAs and mRNAs. The triester, $G_p^m(Et)G_p^m(Et)U$, was found to be a more effective inhibitor of BP6 colony formation than was d-GpGpT. This result is consistent with the relative inhibitory effects of these oligomers on cell-free aminoacylation (see Table IV).

Oligonucleoside methylphosphonates which are complementary to the
Shine-Dalgarno sequence of 16S rRNA inhibit protein synthesis in E.
coli ML 308-225 but not in E.coli B cells. Thus, for example,
d-ApGpGpApGpGpT inhibits protein synthesis 20-45% but has no effect on
RNA synthesis. This heptamer is taken up by E.coli ML 308-225 but not
E.coli B. d-ApGpGpT has no effect on either cellular protein or RNA
synthesis. This lack of inhibition was also observed in cell-free systems (see Table VI). d-ApGpGpApGpGpT also specifically inhibits
colony formation by E.coli ML 308-225 (see Table VIII). This analog
and d-ApGpGpT had no effect on colony formation by E.coli B and only a
small inhibitory effect on colony formation by transformed human cells.

TABLE VIII.   Effects of Deoxyribonucleoside Methylphosphonates on
              Colony Formation by Bacterial and Human Cells

Oligomer	% Inhibition		
	E. coli B[b]	E.coli ML 308-225[b]	HTB1080[c]
d-GpGpT	−	5	−
d-ApGpGpT	0	0	−
d-ApGpGpApGpGp	0	78 − 97	−
d-ApGpGpApGpGpT	0	67 − 97	10

(a) [oligomer] = 75 μM.
(b) At either 22°C or 37°C.
(c) HTB1080 = Human tumor cells at 37°C.

We have also begun to investigate the possibility of inhibiting
processing (splicing) of pre-mRNA by oligonucleoside methylphosphonates.
For example, we have prepared analogs complementary to nucleotides 5
through 10 (d-GpGpTpApApG) and 8 through 13 (d-CpCpApGpGpTp) of U_1 RNA.
These sequences encompass the region of U_1 RNA believed to be involved
in pre-mRNA splicing (30,31). We have also prepared a nonamer,
d-ApApTpApCpCpTpCpA, which is complementary to the exon/intron junction
of the donor splice site of SV40-large T-antigen pre-mRNA.

As shown in Table IX, d-CpCpApGpGpTp inhibits the growth of transformed human fibroblasts in mass culture and also inhibits colony
formation by transformed Syrian hamster fibroblasts. The greater
inhibition of colony formation by the hexamer may result from perturbation of the cells during the critical period of attachment of the
cells to the dish. The effects of d-CpCpApGpGpTp on hamster cell
protein synthesis and RNA synthesis were also examined. In these experiments, RNA synthesis was inhibited 66% in the presence of 50 μM
oligomer, while protein synthesis was inhibited 25%.

TABLE IX. Effects of d-CpCpApGpGpTp on Growth and Colony Formation by Mammalian Cells in Culture

	% Inhibition	
Oligomer Conc. (μM)	HTB1080[a] Growth	BP6[b] Colony Formation
25	–	47
50	5	78
75	–	94
100	30	–
200	53	–

(a) HTB1080 = Transformed human fibroblasts.
(b) BP6 = Transformed Syrian hamster fibroblasts.

The effects of d-ApApTpApCpCpTpCpA, d-GpGpTpApApG and d-(Tp)$_5$T on T-antigen synthesis in SV40-infected African green monkey kidney cells (BSC40) were studied. None of these oligomers (25 μM) show any cyto-toxic effects on the growth of the BSC40 cells over a three-day period. The production of T-antigen was determined by an immunofluorescent assay after BSC40 cells were infected with SV40 in the presence of oligomer for 27 hours. Table X shows both d-ApApTpApCpCpTpCpA and d-GpGpTpApApG lower the levels of T-antigen in the infected cells sufficiently to prevent its detection by the antibodies. d-(TP)$_5$T on the other hand appears to have little or no effect.

TABLE X. Effects of Oligonucleoside Methylphosphonates on SV40-Infected African Green Monkey Kidney Cells

Oligomer	Conc. μM	% Reduction of T-antigen Positive Nuclei
d-ApApTpApCpCpTpCpA	1	20
	5	30
	25	45
d-GpGpTpApApGp	1	10
	5	25
	25	30
d-(Tp)$_5$T	1	6
	5	6
	25	0

The results of our experiments suggest mRNA function may be selectively inhibited by nonionic oligonucleotides at two levels. Oligomers may be designed to inhibit translation of mRNA or alternatively processing of pre-mRNA may be prevented. In theory it should be possible to specifically inhibit the function of a single cellular or viral mRNA. Experiments are underway in our laboratory to further characterize and extend selective inhibition of nucleic acid function by oligonucleoside methylphosphonates.

ACKNOWLEDGEMENT

We gratefully acknowledge our collaboration with Dr. Thomas Kelly and Mr. Ron Wides (Department of Genetics and Molecular Biology, The Johns Hopkins University) on the SV40 experiments. This work was supported in part by the following grants from the National Institutes of Health: GM16066 (P.O.P.T.); GM 31927 (P.S.M.) and GM 25795 (P.S.M.).

REFERENCES

1. Jones, G.H., Albrecht, H.P., Damodaran, N.P. and Moffatt, J.C.: 1970, J. Am. Chem. Soc. 92,5510.
2. Jones, A.S., MacCoss, M. and Walkter, R.T.: 1973, Biochim. Biophys. Acta 294, 365.
3. Pitha, J., Pitha, P. and Stuart, E.: 1971, Biochemistry 10, 4595.
4. Letsinger, R.L., Wilkes, J.S., and Dumas, L.B.: 1976, Biochemistry 15, 2810.
5. Blob, L.N., Vengris, V.E., Pitha, P.M. and Pitha, J.: 1977, J. Med. Chem. 20, 356.
6. Mungall, W.S. and Kaiser, J.K.: 1977, J. Org. Chem. 42, 703.
7. Vosberg, H.P. and Eckstein, F.::1977, Biochemistry 16, 3633.
8. Eckstein, F.: 1979, Acts. Chem. Res. 12, 204.
9. Eckstein, F.: 1980, Trends. Biochem. Sci. 5, 157.
10. Burgers, P.M.J. and Eckstein, F.: 1979, J. Biol. Chem. 254, 6889.
11. Eckstein, F., Romaniuk, P.J. and Connolly, B.A.: 1982, Methods Enzymol., 87, 197.
12. Potter, B.V.L., Romaniuk, P.J., and Eckstein, F.: 1983, J. Biol. Chem. 258, 1758.
13. Miller, P.S., Fang, K.N., Kondo, N.S., and Ts'o, P.O.P.: 1971, J. Am. Chem. Soc. 93, 6657.
14. Miller, P.S., Yano, J., Yano, E., Carroll, C., Jayaraman, K. and Ts'o, P.O.P.: 1979, Biochemistry 18, 5134.
15. Kan, L.-S., Cheng, D.M., Miller, P.S., Yano, J. and Ts'o, P.O.P.: 1980, Biochemistry 19, 2122.
16. Chacko, K.K., Lindner, K., Saenger, W. and Miller, P.S.: 1983, Submitted for publication.
17. Miller, P.S., McParland, K.B., Jayaraman, K., and Ts'o, P.O.P.: 1981, Biochemistry 20, 1874.
18. Pless, R.C. and Ts'o, P.O.P.: 1977, Biochemistry 16, 1239.
19. Miller, P.S., Dreon, N., Pulford, S.M. and McParland, K.B.: 1980, J. Biol. Chem. 255, 9659.

20. Miller, P.S., Barrett, J.C., and Ts'o, P.O.P.: 1974, Biochemistry 13, 4887.
21. Miller, P.S., Braiterman, L.T., and Ts'o, P.O.P.: 1977, Biochemistry 16, 1988.
22. Barrett, J.C., Miller, P.S. and Ts'o, P.O.P.: 1974, Biochemistry 13, 4898.
23. Moller, A., Schwarz, U., Lipecky, R. and Gassen, H.G.: 1978, FEBS Lett. 89, 263.
24. Saneyoshi, M. and Nishimura, S.: 1971, Biochim. Biophys. Acta 246, 123.
25. Ramberg, E.S., Ishag, M., Rulf, S., Moeller, B. and Horowitz, J.: 1978, Biochemistry 17, 3978.
26. Jayaraman, K., McParland, K., Miller, P., and Ts'o, P.O.P.: 1981, Proc. Natl. Acad. Sci. USA 78, 1537.
27. Decad, G. and Nikaido, H.: 1976, J. Bacteriol. 128, 325.
28. Payne, J.W. and Gilvary, C.: 1968, J. Biol. Chem. 243, 6291.
29. Kaback, H.R.: 1971, Methods in Enzymol. 22, 99.
30. Braulant, C., Krol, A., Ebel, J.P., Lazar, E., Gallinaro, H., Jacob, M., Sri-Widada, J., and Jeanteur, P.: 1981, Nucl. Acids Res. 8, 4143.
31. Lerner, M.R., Boyle, J.A., Mount, S.M., Wolin, S.L., and Steitz, J.A.: 1980, Nature 283, 220.

SOLUTION CONFORMATIONAL CHARACTERISTICS OF CHEMICALLY
MODIFIED PURINE(β)PENTOSIDES

H.-D. Lüdemann[a], G. Knopp[a], F. Hansske[b], M.J. Robins[b]

[a] Institut für Biophysik und Physikalische Biochemie
Universität Regensburg, Postfach 397
D-8400 Regensburg, Germany

[b] Department of Chemistry, The University of Alberta
Edmonton, Alberta, Canada   T6G 2G2

ABSTRACT
The solution conformation of two series of adenosine analogues has been
studied by proton HRNMR. The first series comprises 2' and 3' deoxy
derivatives of the pentoses. In the second series the 2'3' hydroxyl
group of 9β-D-ribo-,ara-and xylo-furanosyl-adenine have been replaced
by amino groups. The furanoside ring is analyzed in the Altona-
Sundaralingam model. It is shown that mainly sterical interactions
determine the conformational preference and that the interaction
between the 5'CH$_2$OH and the 3' substituent has the dominant influence.

The motivation to synthesize analogues of the naturally occurring
nucleosides is derived from the various physiological activities exhibited by this class of compounds [1,2]. In the data presented here,
the solution conformation of various adenine(β)nucleosides modified
chemically in the sugar moiety was analyzed by NMR in order to characterize the influence of the various local substitutions.
One objective in this kind of analysis is, of course, the search for
correlations between the preferred three dimensional structure in
solution and physiological behaviour or chemical reactivity. However,
almost all nucleoside analogues studied hitherto present mixtures of
several conformers, and simple thermodynamic calculations reveal, that
the free energy differences between these forms amount at most to a
few kJ·mole^{-1}.
It appears thus beyond doubt, that any specific interaction with a
biological macromolecule or a macromolecular assembly should be strong
enough to change the population of these allowed conformers considerably.
The emphasis in this work is thus to obtain a systematic and if possible
quantitative description of the contributions of chemical modifications
at the 2' and 3' carbons to the overall solution conformation of
purine(β)nucleosides. The study includes work on modified ribosides,
arabinosides and xylosides. Since the nature of the base has a small
but until now unpredictable influence upon the sugar conformation, we
restricted our analysis to analogues with an unmodified adenine in the
base moiety.
Many of the analogues studied here have only a marginal solubility

B. Pullman and J. Jortner (eds.), Nucleic Acids: The Vectors of Life, 537–544.
© *1983 by D. Reidel Publishing Company.*

in water. The soluble analogues show in aqueous solutions at the
concentrations necessary for NMR-analysis pronounced self association,
which influences the chemical equilibria between the conformers. In
order to avoid these complications and to be able to standardize the
conditions the compounds were dissolved in liquid deuteroammonia
(c = 5mg per 0.5 cm^3). A discussion of the effects of this solvent upon
the solution conformation and details of the experimental procedures
have been given previously (3,4,5).

CONFORMATIONAL ANALYSIS

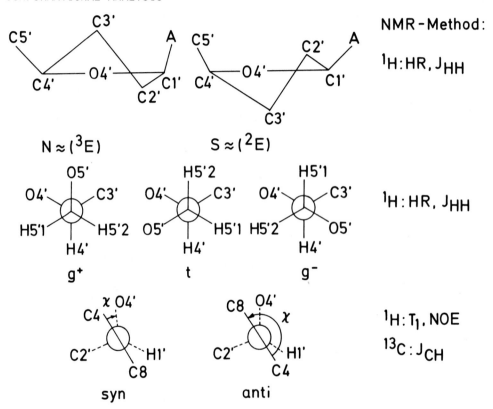

NMR-Method:

^{1}H:HR, J$_{HH}$

^{1}H:HR, J$_{HH}$

^{1}H:T$_1$, NOE
^{13}C:J$_{CH}$

Fig. 1: Schematic drawing of the three major modes of conformational
flexibility in purine(β)pentosides

The proton high resolution NMR-spectra were obtained at 100.1 MHz on a
Varian XL-100 FT spectrometer. The coupling constants and chemical
shifts were determined from the experimental spectra by application of
the computer programme LAME (QCPE no. 111).
The conformation of the sugar moiety was analyzed in the two state N↔S
model proposed by Altona and Sundaralingam (6,7). In this model the conformational equilibria are derived from the vicinal proton-proton coupling constants by application of the Karplus equation

(1) $^3J_{HH} = A \cos^2\theta_{HH} + B \cos\theta_H$ $\theta(\psi_m, P)$

(2) $^3J_{HH} = [N] \, ^3J_{HH}^N + [1-N] \, ^3J_{HH}^S$ $[N] + [S] = 1$

The constants A and B of (1) depend among others on the electronegativity of the carbon substituents. Originally they were derived for the unmodified ribosides. Recently Haasnoot et al. (8) designed a multiparameter equation which considers the relative changes in electronegativity, and is supposed to be considerably more accurate than eq. (1). Fig. 2 gives the influence of changes in the electronegativity E ($\Delta E = E_{OH} - E_R$) upon ($J_{1'2'} + J_{3'4'}$) and $J_{2'3'}$ for a variety of modified ribosides (9). It appears, that any systematic influence of ΔE upon A $\simeq \pm 0.4$ and B $\simeq 0.1$. We have thus neglected this contributions.

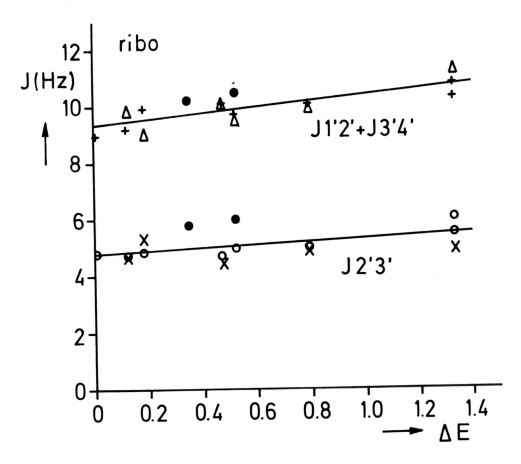

Fig. 2: $^3J(H_2)$ vs. differences in electronegativity $\Delta E \equiv E_{OH} - E_R$ of the various substituents R at 213 K in liquid ND_3.

For the arabinosides (9,10) and the xylosides (11,12,13) it was shown previously that they can be characterized with a suitable modified Altona Sundaralingam treatment. In the case of the xylosides sterical

interference between the substituent at C3' and C5' forces the N-xylose
into a ⁴₃T conformation. The A-S model demands, that in the arabinosides
J1'2' should vary little with the position of the N ⇔ S equibibrium and
J2'3' ≈ J3'4', while for the xylosides J1'2' ≈ J2'3' is expected. Fig.3
shows that this behaviour is observed in both cases, the scatter of the
data might stem in part from the neglect of electronegativity
corrections (for details see (9,13)).

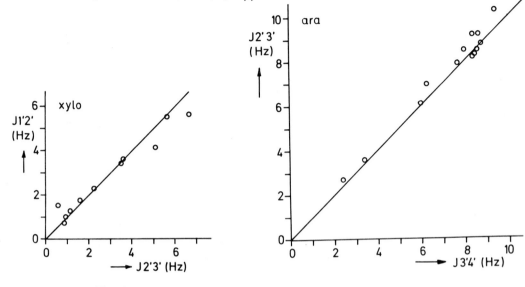

The lines in the diagrams represent slope 1

Fig. 3A: Correlation between   Fig. 3B: Correlation between
J₁'₂' and J₂'₃' in Xylo Deri-   J₂'₃' and J₃'₄' in AraA Derivatives.   vatives.

The rotamer population around the C4'-C5' bond is derived according
to Hruska et al. (14). The glycosyl torsion angle is as described
earlier (3,9,13) derived from proton-T₁ studies.

RESULTS AND DISCUSSION
Previous studies (9,13,15) have been complemented by a systematic
analysis of deoxypentosides in which the hydroxyl groups at the
various position have been replaced by hydrogens or deuterons in order
to minimize steric interference in the sugar moiety. In a second
series the hydroxyl groups have been replaced by amino groups a substitution that should enhance the steric interference of ring substituents. Table 1 and 2 compile the results of the analysis for the solutions
in ND₃ at 213 K. The complete set of coupling constants has for lack of
space not been included, when indicated it has been published before,
for the other substances it is available from the authors upon request.
In 2'3' dideoxy adenosine (2'3'didrA) only the 5'hydroxymethyl group
and the adenine can influence the position of the N ⇔ S equilibrium of
the furanoside ring. Bulky substituents of a five membered ring do have
the tendency to occupy the position where the least crowding with the

Table 1: Conformational characterisation of the various deoxyadenosine derivatives in liquid ND_3 at 213 K

Abbrev. R	2'3'didA	2'drA	3'drA	2'dxA	3'daA	rA	aA	xA	1A
R1	H	H	H	H	OD	H	OD	H	OD
R2	D	H	OD	H	H	OD	H	OD	H
R3	D	H	H	OD	H	H	H	OD	OD
R4	H	OD	H	H	H	OD	OD	H	H
ref.		5	5			5	15	13	13
[N]	.78	.36	.95	.93	.91	.43	.65	.95	.85
[g⁺]	.59	.56	.86	.09	.69	.70	.53	.07	.09
[t]/[g⁻]	.19	.21	.06	.38	.14	.20	.21	.42	.35

Table 2: Conformational description of the various amino adenosine derivatives in liquid ND_3 at 213 K

Abbrev. R	2'NH₂rA	3'NH₂rA	2'3'diNH₂rA	2'NH₂aA	3'NH₂aA	2'3'diNH₂aA	2'NH₂xA	3'NH₂xA
R1	H	H	H	ND₂	OD	ND₂	H	H
R2	ND₂	OD	ND₂	H	H	H	ND₂	OD
R3	H	H	H	H	H	H	OD	ND₂
R4	OD	ND₂	ND₂	OD	ND₂	ND₂	H	H
ref.	16	16		10	10	10	13	13
[N]	.13	.94	.48	.88	.88	1.0	.86	.30
[g⁺]	.52	.92	.72	.90	.74	.89	.11	.67
[t]/[g⁻]	.25	.07	.10	.04	.14	.00	.43	.16

substituents on neighbouring ring carbons occurs. Geometrical calculations, using bond angles and bond length from X-ray data show,
that the base has the least interference with the ring in the S-state,
while the 5'CH$_2$OD-group has minimum hindrance in the N-state. The data
for 2'3' didA show that the 5'CH$_2$OD influences the conformational
equilibrium more than the base. Introduction of a hydroxyl at the
2'2(3'drA) and 2'1(3'daA) position significantly stabilizes the N-ring.
In both cases sterical interaction between the base atoms and the 2'2
resp. 2'1 hydroxylgroup is minimized. A stabilization of the N-conformer is also the result of the introduction of a hydroxylgroup at the
3'1 (2dxA) position. In this case the distance between the 3'hydroxyl
group and the 5'CH$_2$OD is greater in the N- than in the S-state.
The 3'2 hydroxyl group does shift the equilibrium towards S again the
sterical interference between the substituents at C3' and C4' is responsible for this effect.
Addition or removal of the 2'1 or 2'2 hydroxylgroups in order to form
2'drA and 2'dxA has obviously a smaller effect upon the N ⇔ S equilibria than the respective changes at the 3'hydroxylgroups. Taken together,
these observations lead to the suggestion that the interaction between the 5'CH$_2$OD and the substituent at C3' makes the dominant contribution to the conformational equilibria of the sugar moiety.
In Table 2 analogues are compiled, in which the 2' and 3'hydroxylgroups
are replaced systematically by aminogroups. Compared to the hydroxylgroup of the unmodified sugars this introduced a larger substituent at
the sugar ring. All changes brought about by this modification can be
explained by purely sterical arguments. The bulkier substituent
attempts to occupy a pseudo-equatorial position at the furanoside ring
in order to minimize sterical interference with the atoms or groups of
atoms bound to the neighbouring ring carbon.

Table 3: Pseudoorientation of the substituents at C2' and C3' in the
N- and S-state

Position	N	S
2'1	equatorial	axial
2'2	axial	equatorial
3'1	axial	equatorial
3'2	equatorial	axial

For the ribosyl adenine derivatives (substitution pattern 2'2/3'2) the
introduction of NH$_2$ at C2' leads thus to a pronounced preference for
the S-sugar while the same substitution at C3' forces the furanoside
ring almost completely into the N-conformer. The simultaneous replacement of the 2' and 3'hydroxylgroup leaves, since the effects at C2'
and C3' counteract each other, the N ⇔ S equilibrium almost unchanged.
All modifications of arabinosyladenine (arabinose substitution pattern
2'1/3'2) stabilize according to the prediction the N-furanose the
double modification seen in 2'3'diNH$_2$aA further amplifies this trend.
The two xylosyladenines (substitution pattern 2'2/3'1) are also in
accord with this model, and shift both of the ring equilibria towards

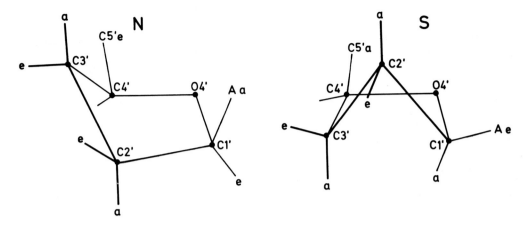

Fig. 4: "Quasiequatorial" (e) and quasiaxial (a) position of the
substituents at the furanoside ring in N and S

the S-xylose. Here again it is most clearly seen that modifications at
C3' have more impact than changes at C2'. This simple sterical concept
explains qualitatively all conformational changes observed in more
than fifty 9β-D-pentofuranosyl adenine analogues (4,5,10,13,15).
Compared to the sterical interaction, electrostatic interactions
appear to play at most a minor role in determining the sugar conformation. Stereoelectronic effects may be more significant (17).

Financial support by the Deutsche Forschungsgemeinschaft, the Fonds
der Chemischen Industrie and the Natural Sciences and Engineering
Research Council (A 5890) is gratefully acknowledged. The calculations
necessary for the evaluation of the data were performed at the Computer
Center of the Universität Regensburg.

REFERENCES
1. Chemistry, Biology and Clinical Uses of Nucleoside Analogs,
    Ann. N.Y. Acad. Sci., Vol. 225 (A. Bloch, ed.), New York 1975.
2. Conference proceeding of the 3rd Round Table on "Nucleosides,
    Nucleotides, and Biological Applications", Montpellier 1978.
3. H.-D. Lüdemann, E. Westhof, and O. Röder, Eur. J. Biochem. 1974,
    49, 143.
4. H.-D. Lüdemann, and E. Westhof, The Jerusalem Symposia on
    Quantum Chemistry and Biochemistry (B. Pullman ed.) Reidel,
    Dordrecht, 1978, 11, 41.
5. E. Westhof, H. Plach, I. Cuno, H.-D. Lüdemann, Nucleic Acid Res.,
    1977, 4, 939.
6. C. Altona and M. Sundaralingam, J. Am. Chem. Soc. 1972, 94, 8205.
7. C. Altona and M. Sundaralingam, J. Am. Chem. Soc. 1973, 95, 2333.
8. C.A.G. Haasnoot, F.A.A.M. de Leeuw and C. Altona, Tetrahedron,
    1980, 36, 2783.
9. I. Ekiel, M. Remin, E. Darzynkiewicz, and D. Shugar, Biochim.
    Biophys. Acta 1979, 562, 177.

10. G. Klimke, I. Cuno, H.-D. Lüdemann, R. Mengel, and M.J. Robins, Z. Naturforsch. 1980, 35c, 853.
11. I. Ekiel, E. Darzynkiewicz, L. Dudycz, and D. Shugar, Biochemistry, 1978, 17, 1530.
12. A. Jaworski, I. Ekiel, and D. Shugar, J. Am. Chem. Soc. 1978, 100, 4357.
13. G. Klimke, I. Cuno, H.-D. Lüdemann, R. Mengel, and M.J. Robins, Z. Naturforsch. 1980, 35c, 865.
14. F.E. Hruska, A.A. Grey, and I.C.P. Smith, J. Am. Chem. Soc. 1970, 92, 4088.
15. G. Klimke, I. Cuno, H.-D. Lüdemann, R. Mengel, and M.J. Robins, Z. Naturforsch. 1979, 34c, 1075.
16. H. Plach, E. Westhof, H.-D. Lüdemann, and R. Mengel, Eur. J. Biochem. 1977, 80, 295.
17. W.K. Olson, J. Am. Chem. Soc. 1982, 104, 278.

DEOXYRIBONUCLEOTIDE BIOSYNTHESIS: A CRITICAL PROCESS FOR LIFE

Hartmut Follmann
Fachbereich Chemie der Philipps-Universität, Biochemie,
D-3550 Marburg, Fed.Rep. Germany

SUMMARY. Deoxyribonucleotides have their own specific biochemistry which, although overlooked occasionally, is essential for understanding the DNA synthesis (S) phase of a cell. The unusual complexity of the only pathway leading to DNA precursors, enzymatic reduction of ribonucleotides, is exemplified in studies with synchronously growing green algae (Scenedesmus obliquus). The process involves flexible nucleotide-protein interactions, dithiols as reductant, transition metals (Mn, Fe, Co) for catalysis, and radical intermediates. Chemical and biochemical evidence suggests that the origin of DNA is inseparably connected with the evolution of such a catalytic process.

Deoxyribonucleic acid may be the most important vector of life, transmitting the genetic information from one cell to another, or from one generation to the next, but it is a vector utterly dependent on a scalar quantity, the intracellular deoxyribonucleotide concentration. Of course the physical existence of all biological macromolecules relies on the availability of their monomers, yet the case of deoxyribonucleotides and DNA is not a trivial one. I wish to accentuate in this contribution why this is so, and that in fact DNA could only become the essential element of life as a biosynthetic pathway to deoxyribonucleotides developed.

BIOCHEMISTRY OF DEOXYRIBONUCLEOTIDES

2-Deoxyribose, unlike ribose, is not a common intermediate of the sugar metabolism in its free, or phosphorylated form, and hence does not serve as a direct precursor molecule for DNA synthesis. All organisms, prokaryotes and eukaryotes, produce deoxyribonucleotides de novo in a specific reduction reaction in which the 2' hydroxyl group of ribonucleotides, usually quite abundant cell constituents, is replaced by hydrogen. The process is catalyzed by a unique group of enzymes, the ribonucleotide reductases (EC 1.17.4.) and requires small dicysteine polypeptides, thioredoxin or glutaredoxin, as hydrogen donors (1-3). The

545

B. Pullman and J. Jortner (eds.), Nucleic Acids: The Vectors of Life, 545–557.
© 1983 by D. Reidel Publishing Company.

reaction is summarized in Scheme I. The oxidized disulfide is re-reduced
by NADPH but this proceeds independent of ribonucleotide reduction.

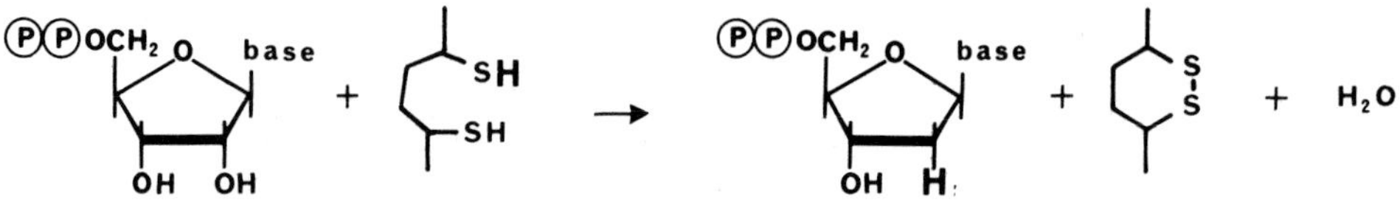

One of the four products of ribonucleotide reduction, deoxyuridylate,
is subsequently alkylated to thymidylate = 5-methyldeoxyuridylate by the
second indispensable enzyme of DNA precursor synthesis, thymidylate
synthase (EC 2.1.1.45). Furthermore some kinases of various specificity
are required for production of the deoxyribonucleoside-5'-triphosphates
which finally serve as the DNA polymerase substrates. Thus DNA depends
critically on the proper functioning of a multicomponent enzyme apparatus before the macromolecular template can be replicated (Figure 1).
Only in special cases, in mutant cells or in embryonic tissues, can
some DNA be made from reutilized ("salvaged") deoxyribonucleotides or
from small amounts of stored deoxyribose derivatives (4-6).

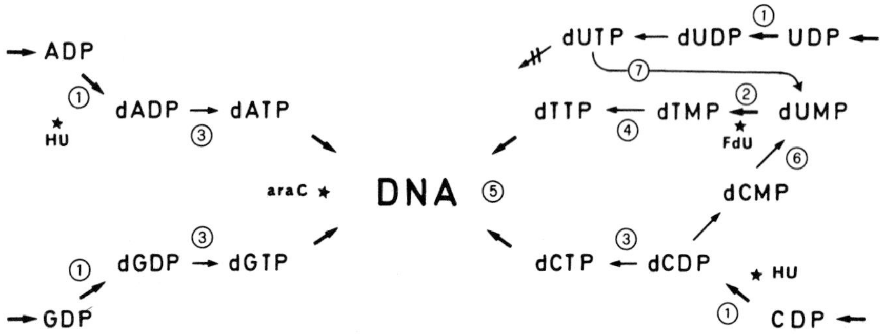

Figure 1. The main enzymes of deoxyribonucleotide synthesis. 1: Ribonucleotide reduction; 2: thymidylate synthesis; 3: nucleoside diphosphate
kinases; 4: thymidylate kinase; 5: DNA polymerase; 6: deoxycytidylate
deaminase; 7: dUTPase. Some specific inhibitors (*) are hydroxyurea (HU)
in ribonucleotide reduction, 5-fluoro-2'-deoxyuridine (FdU) in thymidylate synthesis, or arabinonucleosides (e.g., araC) in DNA replication.

CELL CYCLE DEPENDENCE

Not only are deoxyribonucleotides and the enzymes mentioned in Figure 1 very specific metabolites and catalysts, not normally involved in
other biochemical reactions, they also exhibit highly characteristic
dynamic patterns (Figure 2). Both the very small intracellular pools of
dATP, dCTP, dGTP, and dTTP, and the activities of the two main enzymes
ribonucleotide reductase and thymidylate synthase have distinct maxima

during the DNA synthesis phase (S phase) of a cell cycle but are barely detectable before and after that time. In other words, DNA precursors are produced only for, and coordinately with each round of DNA replication, suggesting tight regulation between the two processes.

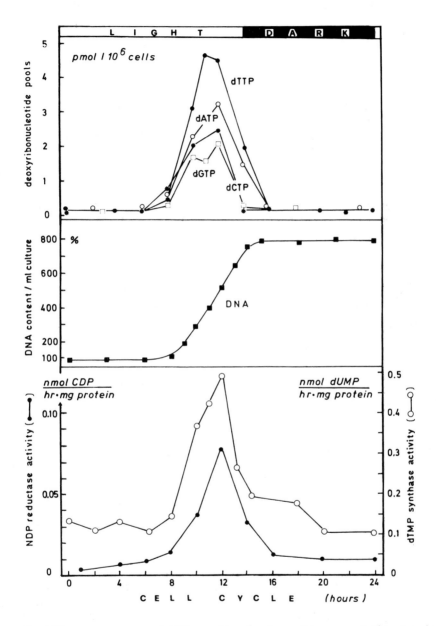

Figure 2. DNA precursor and DNA synthesis in cultures of the freshwater green algae, Scenedesmus obliquus, synchronized in a light:dark regime of 14:10 hours (7,8).

It is not at all obvious how to interpret cause and effect in this system, nor has the problem be resolved to date. It is obvious, however, that any study of the cellular or molecular regulation mechanisms, as well as of the isolated enzymes and their properties has to make use of synchronous cells or cell cultures. We have taken advantage of the capacity of unicellular green algae like Chlorella or Scenedesmus to multiply synchronously in an almost natural light-dark regime (i.e., conditions well-known in photosynthesis or protein biosynthesis research (9)) and to produce sizeable deoxyribonucleotide concentrations and enzyme activities because of the formation of eight daughter cells per cell division. The undisturbed cell cycle dependence of enzymes and products is shown in Fig.2 (7,8). I will describe experiments to probe the nature of their interrelation and to gain insight into enzyme properties and mechanisms. It may be concluded from a comparison with the work of others on yeast and animal cells (10-12) that our analysis of DNA precursor biochemistry in algae is fairly representative for the whole universal process.

The S phase peaks of ribonucleotide reductase and dTMP synthase represent de novo-protein synthesis and are entirely blocked by cyclohex-imide. Production of these two enzymes is tightly coupled, presumably on the level of gene expression. This follows from studies in which an inhibitor specific for one of them and inactive towards the other enzyme in vitro always affects both enzymes in parallel when applied in vivo. Thus hydroxyurea (NH_2-CO-NHOH), a potent metal-chelating and radical-scavenging inhibitor of iron-containing ribonucleotide reductases including the Scenedesmus enzyme also strongly reduces thymidylate synthase activity when present in the culture medium at 0.1 mM concentration. In contrast synthesis of the two enzymes is derepressed in presence of 0.1 mM 5-fluorodeoxyuridine, a strong inhibitor of cellular thymidylate formation (Figure 3). Enzyme activities rise about 30-fold over the regular peak maximum while DNA synthesis ceases due to lack of thymidylate. On the other hand thymidylate excess, exogenously produced in thymidylate-uptake mutants of yeast (Saccharomyces cerevisiae) also leads to an increase in ribonucleotide reductase activity (13).

Another special property of ribonucleotide reductases is their allosteric interaction with deoxyribonucleotides as positive or negative effectors. As there are four different products affecting the reduction of four different substrates to different degrees one usually observes very complex kinetic patterns which cannot be outlined here in detail. It appears certain that such allosteric effects measured in vitro also operate in vivo (2). An example is seen in Figure 3, right panel: Despite greatly elevated ribonucleotide reductase activity and lack of DNA synthesis only one of the deoxyribonucleotides, dATP, accumulates. This is so because dATP is an allosteric inhibitor of CDP, GDP, and UDP reduction at concentrations above 50 µM which are indeed reached intracellularly under these conditions; consequently the levels of dCTP and dGTP do not increase.

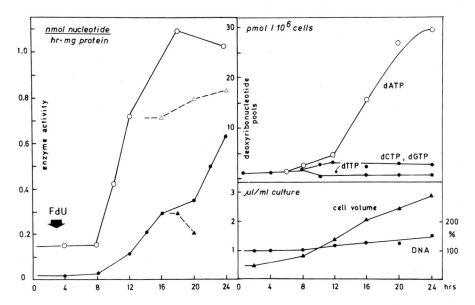

Figure 3. Effects of 20 μg/ml fluorodeoxyuridine (FdU) in cultures of
Scenedesmus obliquus during the 24 hr cell cycle (cf. Fig.2). ●:Ribo-
nucleotide reductase, o: thymidylate synthase. Algae make no DNA under
these conditions but continue to grow to abnormally large cells (lower
right).

        Accumulated evidence of this kind suggests that it is deoxyribo-
nucleotide pool imbalances (depletion, or expansion) which regulate the
enzyme peaks and DNA synthesis during a cell cycle. A specific nucleotide, or ratio of specific nucleotide concentrations, cannot yet be pin-
pointed with certainty. The dCTP concentration appears particularly
critical (14). The completion of DNA replication (or its experimental
blockage) may also be involved as a signal (15); however in Scenedesmus
DNA replication per se appears less influential as arabinucleosides which
inhibit replication have no significant effect on the enzyme activities.
In any case DNA precursor biosynthesis must be viewed as a largely self-
determined process in nucleic acid formation and functions.

ENZYMOLOGY OF RIBONUCLEOTIDE REDUCTION

It is, then, important to understand the mechanisms of enzymatic deoxyribonucleotide synthesis and the properties of the enzymes. Thymidylate
synthase has been well characterized and shown to catalyze essentially
the same, tetrahydrofolate-dependent methylation reaction in each organism. In contrast the ribonucleotide reductases have a reputation as
delicate, difficult-to-purify enzymes of great diversity, and one is far
from a complete picture. The most unusual feature for catalysts of a
central, universal biological process is that there are three different

types:
- ribonucleotide reductases which utilize deoxyadenosylcobalamin (coenzyme B12) as the catalytic component are widely distributed among many (but not all) bacteria and cyanobacteria; the eukaryotic algae, Euglena gracilis, also has this type of enzyme;
- ribonucleotide reductases of some bacteria (Escherichia coli), bacteriophages, and from all mammalian sources studied (calf thymus, rodent or human cell lines and tumors) possess non-heme-iron containing catalytic protein subunits; the iron is in a binuclear Fe(III)-O-Fe(III) protein-bound complex;
- a third type was discovered only recently in gram-positive bacteria like Brevibacterium or Micrococcus strains (16) where manganese ions serve a catalytic function; the molecular structure of these enzymes is still under investigation.

In view of this it was obviously necessary to establish the type of ribonucleotide reduction present in plant cells.

We have characterized the ribonucleotide reductase of Scenedesmus obliquus and found it to belong to the iron enzyme group, in accord with the proteins of most other eukaryotes but unlike in the algae Euglena (17). The iron dependence is seen in almost complete inhibition by hydroxyurea or o-phenanthroline and subsequent reactivation by iron ions (Figure 4).

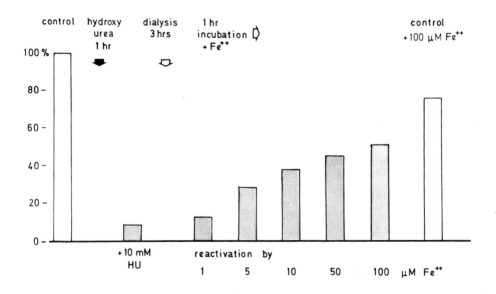

Figure 4. Inactivation and reactivation of partially purified ribonucleotide reductase from S.obliquus by hydroxyurea and Fe(II) ions,respectively. Fe(III) was less effective, but incubation with Fe(II) had to be under aerobic conditions.

A certain similarity of the algal enzyme with the proteins isolated from E.coli or calf thymus(12) is seen during its purification.Affinity chromatography on a reductase-specific adsorbent, aminohexyl-dATP-Sepharose separates a nucleotide-binding subunit U1 from an unbound, catalytic subunit U2; both are inactive separately but must be combined for restoration of enzyme activity (Figure 5). All the known iron-containing, and most likely the manganese ribonucleotide reductases dissociate readily in this way in vitro. Moreover in algal, like in mammalian reductase, the catalytic subunit is present in sub-stoichiometric amount. These proteins more resemble a coenzyme (like coenzyme B12 in the third type of reductases) than a typical enzyme subunit despite their large size.

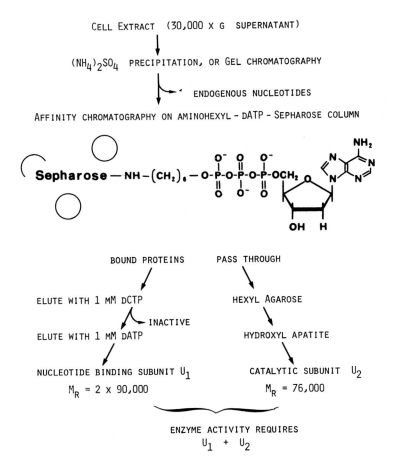

Figure 5. Purification scheme for ribonucleotide reductase from FdU-stimulated, overproducing Scenedesmus obliquus cells.

During isolation of ribonucleotide reductase from the green algae we have noted that other enzyme activities of deoxyribonucleotide and DNA synthesis, i.e. thymidylate synthase, nucleoside diphosphokinase, and DNA polymerase (cf.Fig.1), are co-purified with the reductase in many

fractionation steps. For example all these proteins possess the same
size corresponding to about 250,000 molecular weight, and similar properties on various chromatography materials. These observations are in
accord with the recently recognized organisation of a "replitase" multi-
enzyme aggregate in mammalian cells or phage-infected bacteria (18,19),
in which separate, yet structurally and functionally related enzymes are
combined in "quinary protein structure" (20). Such a complex offers great
kinetic adavantage for channeling ribonucleotides into DNA. It could be
demonstrated as a large cluster of protein ($M_r > 10^6$) in animal cells
but is obviously not stable enough to remain intact under the conditions
needed to homogenize plant cells. However it is observed that individual
Scenedesmus enzymes like ribonucleotide reductase and dTMP synthase,sepa-
rated by affinity chromatography methods, increase in activity when re-
combined, indicating some sort of mutually favorable interaction.

An interesting mechanistic detail of ribonucleotide reduction is the
involvement of free radicals, vjz. a deoxyadenosyl/cobalt(II) radical
pair in B12-dependent, and a protein-bound, Fe-coordinated tyrosyl (phen-
oxyl) radical in the iron-containing enzymes. A nucleotide radical intermediate could not be identified so far but it appears safe to predict
that the terminal transfer of hydrogen from reduced enzyme, $E(SH)_2$ to a
ribonucleotide substrate is a radical reaction (Figure 6).

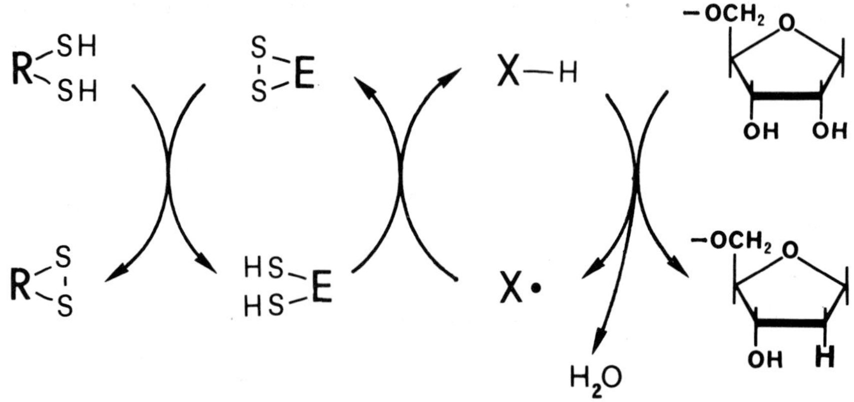

Figure 6. The complete hydrogen transfer chain from thioredoxin or glutaredoxin (R) to ribonucleotide reductase enzyme (E) and ribonucleotide
substrate involves a free radical species of coenzyme, or protein (X).

The mechanism of initial radical generation in this system is a critical
step of the entire catalysis. In the iron ribonucleotide reductases it
requires aerobic environment; at least for the eukaryotic proteins (from
Scenedesmus, or from calf thymus and mouse cells) oxygen must be continuously present while enzyme activity is lost during anaerobic storage
(21). The important consequence is that DNA replication in eukaryotes
depends on oxygen in a catalytic function not linked to respiration and
energy supply. We could recently verify this situation in anaerobically
cultured Ehrlich ascites tumor cells which are not energy-limited due to

their high glycolytic capacity but cease to make DNA after the G1 phase because of lack of deoxyribonucleotides. An exogenous supply of deoxy-cytidine, but not cytidine, enables such cells to enter S phase (22).

CONCERNING THE ORIGIN OF DNA

I have described selected aspects of the main reaction of DNA pre-cursor biosynthesis, ribonucleotide reduction, in order to deduce the bare essentials for existence of DNA itself. The advantage of DNA as a stable informational macromolecule and template for replication and transcription besides RNA is so obvious that the duality must have been an early property of protocells on the way to life. However it is very unlikely that DNA could ever be formed in abiotic reactions comparable to the ones known for production of ribonucleotides and oligoribonucleot-ides (23). Deoxyribose and its glycosides are much less stable than ri-bose derivatives in general, and have not been found in any significant amount among the products of "organic soup" experiments. Rather all che-mical arguments and the available experimental evidence agree that the only feasible pathway for formation of deoxyribonucleotides, of present-day structure, is from preformed ribonucleotides, not requiring a con-densation of deoxysugar and nucleobases or other reactions involving strong acid or base catalysts (Figure 7). This is, of course, exactly the biochemical situation outlined above, and the parallelism appears far from fortuitous.

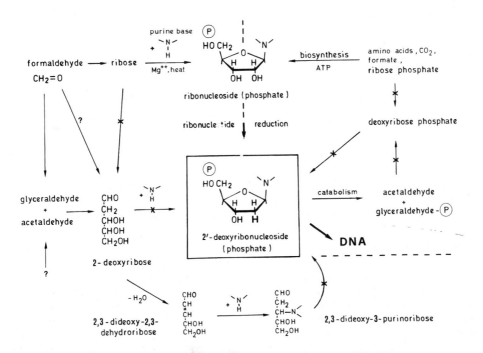

Figure 7 (see overleaf for explanation)

Two questions remain: 1. Can reduction of ribo- to deoxyribonucleotides, for which there is no good precedent in organic chemistry, be
understood under primitive conditions? 2. Given the formation of deoxyribonucleotides, how has their limited stability and quantity in a prebiotic environment to be assessed? I believe that the questions can be
reasonably answered.

With respect to the first question, it appears possible to extract
from the biochemical systems some basic chemistry necessary for reduction
of a riboside to a deoxyriboside. It should include
- dithiols (dicysteine peptides) as reductant. Favorable $\triangle G°'$ (25°C)
  values of about -70 kJ/mol are estimated for the reaction

  alcohol (-CHOH-) + 2 -SH $\rightarrow$ methylene (-CH$_2$-) + -SS- + H$_2$O

  but the unknown activation energy may be high.
- radicals for H transfer. A sugar hydroxyl is a poor leaving group and
  a very reactive species is required for its displacement.
- metal ions for generation and stabilization of such radicals in aqueous
  solution. ESR studies have shown that formation, life time, and H transfer reactions of free radicals, including phenoxy radicals of unsub-
  stituted phenols are greatly favored in the coordination field of
  transition metal complexes (24).
- a nucleotide, and not a free sugar(phosphate),as substrate which is
  locked in the more rigid ribofuranose N-glycoside structure, and
  finally
- a macromolecular carrier (polypeptide) for binding and locally concentrating all these components.

Proteins and ribonucleotides are among the most likely organic compounds believed to have existed on the primordial earth. Both have well-
known patterns of amino acid side chain/nucleobase interaction (25) for
mutual binding; nucleotide-binding enzymes are an ancient protein family.
The other components are also likely constituents of prebiotic, reducing
environments. It is thus possible without difficulty to visualize an
"ur-ribonucleotide reductase" which encompasses these structures. It may
have been one of the earliest catalysts of nucleic acid-protein aggregates or protocells.

Evidence is presented elsewhere (3,23) that all the known ribonucleotide reductases, despite their catalytic differences mentioned
above can in fact be considered to be built and to function in essentially the same way. The differences concern the type of metal complex
and mode of generating a radical species. In the B12-dependent enzymes

---

Figure 7. Deoxyribonucleotides, and hence DNA, can only be synthesized
via ribonucleotides by both biochemical and abiotic, chemical reactions.
Conversion on the sugar level is not possible (x), nor can deoxyribose
be condensed with a purine or pyrimidine base to yield the proper N-1'-
glycosidic bond. The dashed lines separate biochemistry from chemistry.

it is an organocobalt-corrin complex where the cobalt-carbon bond can be cleaved photolytically. In the iron and probably also in the manganese containing proteins oxygen is required for initial one-electron oxidation of a tyrosine residue. The first group of enzymes which today predominates in anaerobic bacteria and facultative anaerobes like cyanobacteria may therefore represent the very earliest form of catalysts (26).

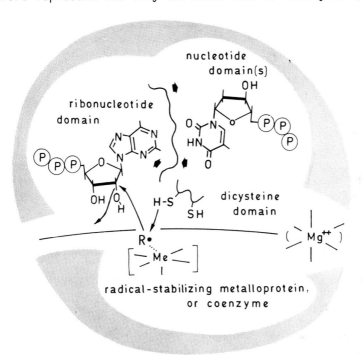

Figure 8. An "ur-reductase" is thought to combine two (or more) nucleotide binding domains, one for substrates and one for allosteric effectors, with a dicysteine center in one polypeptide chain, and a metal-coordinated radical in another polypeptide, or coenzyme. In some, but not all present-day enzymes the subunits are held together by divalent cations (Mg, or Ca).

     The chemistry of deoxyribonucleotides may also provide a clue as to why their formation is, or at least had to be so tightly coupled to DNA synthesis; such linkage is unknown in other monomer/macromolecule (RNA, or protein) relations. Early ribonucleotide reductases, having several targets for inhibition or inactivation (the SH group, the radicals) cannot have been to efficient catalysts and were probably never able to produce a constant and rich supply of DNA precursors. Moreover deoxyribonucleotides are more sensitive to degradation by hydrolysis or elimination reactions than ribonucleotides. Thus, as selective pressure operated in favor of rapid and efficient replication of the "selfish" DNA molecules (27) there had to evolve direct cooperation of the reductase and DNA polymerase enzyme proteins; newly formed monmers could then be immediately incorporated into the growing macromolecule. The specificities of ribo-

nucleotide reductases as well as of DNA polymerases to accept all four
nucleotides as substrates regardless of purine or pyrimidine base are
in accord with or even prerequisite for such cooperation. The "replitase"
complex of enzymes, and the S phase timing of deoxyribonucleotide formation (Fig.2) become understandable as result of the very simple material
constraints on DNA replication, irrespective of the great regulatory
advantage these properties also have in contemporary cells.

To summarize, I feel that the evolution of deoxyribonucleotide
(bio)synthesis marks a checkpoint on the way from protein-RNA aggregates
to "true", selfreplicating cells. Catalysis of ribonucleotide reduction
must have been the main event but chemical and biochemical arguments,
not outlined here, could also be given for the antiquity of other processes in Figure 1, in particular for thymidylate synthesis and the
simple thioredoxin polypeptides which probably developed in parallel
with it. Ribonucleotide reductases, or ribonucleotide reductase genes
are not presently available for sequencing and phylogenetic studies for
a variety of technical reasons. However several of the conclusions drawn
in this paper and predictions made about prebiotic nucleotide chemistry
can be, and will be tested experimentally in the future.

ACKNOWLEDGEMENTS

Our work on DNA precursor biosynthesis is supported by Deutsche
Forschungsgemeinschaft, Sonderforschungsbereich 103 (Zelldifferenzierung).
I thank all members of the research group for their valuable contribu-
tions to the results described here and for their continuing efforts.

REFERENCES

1. P. Reichard: Eur. J. Biochem. 3, 259-266 (1968).
2. L. Thelander and P. Reichard: Ann. Rev. Biochem. 48, 133-158 (1979).
3. M. Lammers and H. Follmann: Structure and Bonding 54 (1983)in press.
4. H.R. Woodland and R.Q.W. Pestell: Biochem. J. 127, 597-605 (1972).
5. C.K. Mathews: Exptl. Cell Res. 92, 47-56 (1975).
6. G. Schimpff, H. Müller and H. Follmann: Biochim. Biophys. Acta
    520, 70-81 (1978).
7. W. Feller, G. Schimpff-Weiland and H. Follmann: Eur. J. Biochem.
    110, 85-92 (1980).
8. B. Bachmann, R. Hofmann and H. Follmann: FEBS-Lett. 152, 247-250
    (1983).
9. P.C.L. John, C.A. Lambe, R. McGookin, B. Orr and M. J. Rollins:
    J. Cell Sci. 55, 51-67 (1982).
10. M. Lowdon and E. Vitols: Arch. Biochem. Biophys. 158, 177-184 (1973).
11. H. L. Elford, E.L. Bonner, B.H. Kerr, S.D. Hanna and M. Smulson:

IUPAC-IUB Joint Commission on Biochemical Nomenclature (JCBN)

Abbreviations and Symbols for the Description of Conformations of Polynucleotide Chains

Recommendations 1982

CONTENTS

INTRODUCTION

Several conventions and notations for polynucleotide conformation have been used by various authors [1−9]. To overcome this confusion, the Joint Commission on Biochemical Nomenclature appointed a panel of experts to review the problem and make recommendations. Their proposals, together with suggestions from the members of

Document of the IUPAC-IUB Joint Commission on Biochemical Nomenclature (JCBN) whose members are P. Karlson (chairman), H. B. F. Dixon, C. Liébecq (as chairman of the IUB Committee of Editors of Biochemical Journals), K. L. Loening, G. P. Moss, J. Reedijk, S. F. Velick and J. F. G. Vliegenthart. Comments on and suggestions for future revision of these recommendations may be sent to the secretary of JCBN, H. B. F. Dixon, University Department of Biochemistry, Tennis Court Road, Cambridge, England, CB2 1QW, or to any member. JCBN thanks the expert panel and many others consulted for their work in drafting the document, including other members of the Nomenclature Committee of IUB (H. Bielka and N. Sharon).

The members of the panel were C. Altona, S. Arnott, S. S. Danyluk, D. B. Davies (chairman), F. E. Hruska, A. Klug, H.-D. Lüdemann, B. Pullman, G. N. Ramachandran, A. Rich, W. Saenger, R. H. Sarma, M. Sundaralingam. Associate members were P. Karlson, O. Kennard, S.-H. Kim, V. Sasisekharan and H. R. Wilson. The document is based on work by a previous subcommission (chairman, F. Cramer) and by an informal committee formed at the Fifth Jerusalem Symposium on Quantum Chemistry and Biochemistry (chairman: M. Sundaralingam).

JCBN and other scientists, are presented here as recommendations that have been approved by IUPAC and IUB.

The nomenclature proposed here is consistent with that recommended for polypeptide conformation [10] as well as with recommendations for polysaccharide conformation [12] and stereochemistry of synthetic polymers [13]. The recommendations on polypeptide conformation [10] also cover general problems of specifying the conformation of biopolymers. Nomenclature of nucleic acids and symbols for their constituents follow published recommendations [14].

RECOMMENDATIONS

1. GENERAL PRINCIPLES OF NOTATION

1.1. *Chain Direction*

The atoms of the main chain are denoted in Fig. 1. The direction of progress of a polynucleotide chain is from the 5′-end to the 3′-end of the sugar residue.

Notes

a) The definition of chain direction is in accord with the definition of the nucleotide unit (see 1.2).

b) The definition of chain direction with respect to the *sugar carbon atoms* of the nucleotide unit is consistent with the alternative description of polynucleotide sequences as progressing from the 3′-end of one unit to the 5′-end of the next *through the phosphate group*, i.e. in the chain sequence $L-(3'\rightarrow5')-M-(3'\rightarrow5')-N$, etc. (written as LpMpN or $L-M-N$ for a known sequence, or L, M, N for an unknown sequence).

1.2. *Definition of a Nucleotide Unit*

A nucleotide unit is the repeating unit of a polynucleotide chain; it comprises three distinct parts: the D-ribose or 2-deoxy-D-ribose (2-deoxy-D-*erythro*-pentose) sugar ring, the

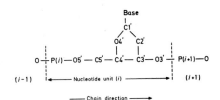

Fig. 1. *Designation of chain direction and main chain atoms of* ith *unit in a polynucleotide chain*

B. Pullman and J. Jortner (eds.), Nucleic Acids: The Vectors of Life, 559–565.
© *1983 FEBS.*

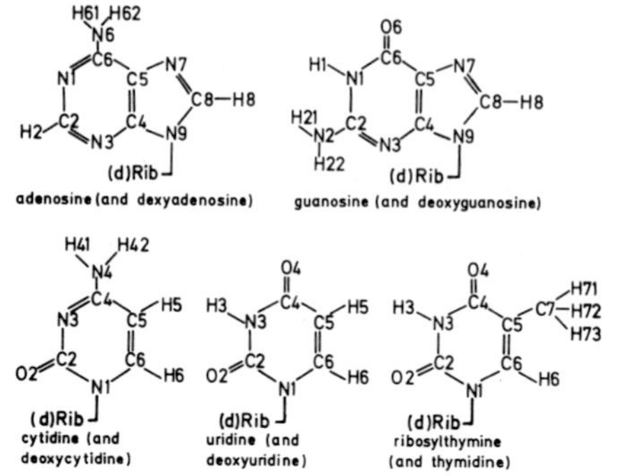

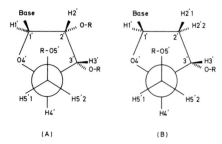

Fig. 2. *The atom numbering for the bases of common nucleosides and nucleotides.* Hydrogen atoms carry the same numbers as the heavy atoms to which they are attached. The name in parenthesis applies when the 'd' in parenthesis in the formula is present

phosphate group, and the purine or pyrimidine base. The sugar ring and the phosphate group form the backbone of the polynucleotide chain; the base ring linked to the sugar residue consitutes the side chain as shown in Fig. 1.

A nucleotide unit is defined by the sequence of atoms from the phosphorus atom at the 5′-end to the oxygen atom at the 3′-end of the pentose sugar; it includes all atoms of the sugar and base rings.

Specific units ($i, j \cdots$ or 3, 4, 5, etc.) are designated by the letter or number in brackets. The units are numbered sequentially in the chain direction, starting at the first nucleotide residue, irrespective of the presence or absence of a phosphate group at the 5′-terminal unit.

The same numbering, A(1), pU(2), pU(3) etc. would apply to the sequence ApUpUp- and pApUpUp-.

1.3. Atom Numbering

The atom numbering of the constituents of the nucleotide unit is shown in Fig. 1 − 3. The numbering scheme for the bases shown in Fig. 2 is the same as that recommended by IUPAC (Rule B-2.11 on page 58 in [15] and on page 5567 in [16]). The atoms belonging to the sugar moiety are distinguished from those of the base by the superscript prime mark on the atom number. Atoms are specified by the appropriate number after the symbol, e.g. C2, N3 (for base) and C1′, C5′, O5′ (for sugar).

Atoms of a specific unit ($i, j \cdots$ or 3, 4, 5) may be designated by the letter or number of the unit in brackets e.g. O3′(i), P(i+1) and N1(3), C2′(4).

Hydrogen atoms carry the same number as the heavy atoms to which they are attached, e.g. base ring H6 (pyrimidine) and H8 (purine) as shown in Fig. 2; sugar ring H1′, H2′ etc, as shown in Fig. 3. Where there is more than one hydrogen atom (such as at C5′ of the sugar ring), the atoms are designated numerically (e.g. H5′1 and H5′2); H5′1 and H5′2 correspond to the *pro-S* and *pro-R* positions [17], respectively (Fig. 3).

The atom-numbering scheme for the 2′-deoxyribonucleotide chain is the same as that for the ribonucleotide chain shown in Fig. 1. The numbering for the sugar ring atoms of both D-ribose and 2-deoxy-D-ribose rings is shown in Fig. 3. (Note the absence of a prime in '2-deoxy-D-ribose' in the

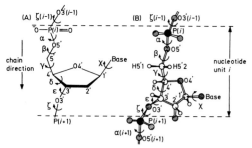

Fig. 3. *Designation of sugar ring atoms and hydrogen atoms.* (A) In β-D-nucleosides and nucleotides; (B) in their 2′-deoxy derivatives

Fig. 4. *Section of a polynucleotide backbone showing the atom numbering and the notation for torsion angles.* (A) Conventional representation; (B) absolute stereochemistry

previous sentence; C2′ of a nucleotide is C2 of its ribose residue.) The two hydrogen atoms attached to the C2′ atom of a nucleoside are denoted by H2′1 and H2′2, corresponding to the *pro-S* and *pro-R* positions respectively (Fig. 3).

The hydrogen atoms of hydroxyl groups are specified as in O5′H, O3′H and O2′H, where appropriate, whereas the hydroxyl groups are specified as OH5′, etc.

Notes

a) Designation of the sugar-ring oxygen atom by O4′ conforms with chemical nomenclature; it has been widely but inaccurately denoted by O1′ in the past.

b) Detailed atom numbering for the modified nucleotides is not considered here.

c) The numerical designation of C5′ and C2′ methylene hydrogen atoms supersedes that introduced by Davies [9].

1.4. Bonds, Bond Lengths and Interatomic Distances

Covalent bonds are denoted by a hyphen between atoms, e.g. O5′−C5′, C5′−H5′1 and C2−N3. Atoms in specified nucleotide units are indicated by putting the number of the unit in parentheses, e.g. O5′(i)−C5′(i), O3′(i)−P(i+1). Bond lengths are denoted by $b(O5′, C5′)$ or $b[O3′(i), P(i+1)]$. Use of the symbol *l* for bond length is avoided because it can be confused with the numeral 1 and because *l* is used for vibration amplitude in electron diffraction (section 1.4 of [10]).

Hydrogen bonds are denoted by a dotted line, with the donor atom being written first, if it can be specified, e.g.

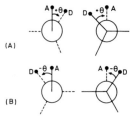

(A)

(B)

Fig. 5. *Newman projections illustrating (A) positive and (B) negative torsion angles.* (A) A clockwise turn of the bond containing the front atom about the central bond is needed for it to eclipse the bond to the back regardless of the end from which the system is viewed; hence the value of θ is positive (+ θ). (B) A counterclockwise turn of the bond containing the front atom is needed for it to eclipse the bond to the back atom regardless of the end from which the system is viewed; hence the value of θ is negative (− θ)

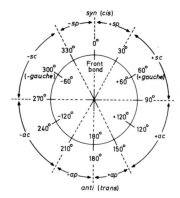

Fig. 6. *Relationship between the* syn-anti *terminology for describing conformational regions [17, 18] and the magnitude of the torsion angle* 0 − 360° *(or* 0 ± 180°*) with the front bond specifying the zero position.* sp, synperiplanar; sc, synclinal; ap, antiperiplanar; ac, anticlinal. Other descriptions of particular torsion angles are also given for comparison: 0°, *cis* (c); 60°, + *gauche* (g⁺); 180°, *trans* (t); 300°, − *gauche* (g⁻)

intramolecular O5′ · · · N3 hydrogen bonding in some purine derivatives or O5′(i) · · · N3(j) for intermolecular hydrogen bonding. The position of the hydrogen may also be indicated as in O5′−H · · · N3. Hydrogen-bonded base pairs are considered separately (see section 4.1).

Distances between non-bonded atoms are denoted by a dot, e.g. O5′(i) · O3′(j).

1.5. Bond Angles

The bond angle included between three atoms $A − B − C$ is written $\tau(A, B, C)$, which may be abbreviated to $\tau(B)$ if there is no ambiguity.

1.6. Torsion Angles [17]

If a system of four atoms $A − B − C − D$ is projected onto a plane normal to $B − C$, the angle between the projection of $A − B$ and the projection of $C − D$ is described as the *torsion angle* about bond $B − C$; this angle may also be described as the angle between the plane containing atoms A, B and C, and the plane containing atoms C and D.

The torsion angle is written in full as $\theta(A, B, C, D)$, which may be abbreviated, if there is no ambiguity, to $\theta(B, C)$. In the statement of this rule, the angle θ is used as a general angle rather than as referring to any particular bond (see Rule 2 for designation of main-chain torsion angles).

The zero-degree torsion angle $(\theta = 0°)$ is given by the conformation in which the projections of $A − B$ and $C − D$ coincide (this is also known as the eclipsed or *cis* conformation). When the sequence of atoms $A − B − C − D$ is viewed along the central bond $B − C$, a torsion angle is considered positive when the bond to the front must be rotated clockwise in order that it may eclipse the bond to the rear as shown in Fig. 5 A. When the bond to the front must be rotated counterclockwise in order to eclipse the bond at the rear, the angle is considered negative as shown in Fig. 5 B.

Angles are usually measured from 0° to 360°, but they may be expressed as − 180° to + 180° when special relationships between conformers need to be emphasized.

Illustrations of the definition of torsion angles are shown for positive and negative values of θ in Fig. 5 A and B respectively. It should be noted that a clockwise turn of the bond containing the front atom about the central bond gives a positive value of θ from whichever end the system $A − B − C − D$ is viewed in Fig. 5 A; similar considerations apply to the conformation with a negative torsion angle (Fig. 5 B).

1.7. Conformational Regions

If the precise torsion angle for a conformation is not known, it may be convenient to specify it roughly by naming a conformational region, i.e. a range in which the torsion angle lies. For this the Klyne-Prelog nomenclature [17, 18], accepted in organic chemistry, is recommended. The relationship between the terms used, ± synperiplanar (± *sp*), ± synclinal (± *sc*), ± anticlinal (± *ac*) and ± antiperiplanar (± *ap*), and the magnitudes of the torsion angles are shown in Fig. 6.

The range 0 ± 90° is denoted as *syn* and the range 180 ± 90° is denoted as *anti*.

Note

In order that conformations described by the torsion angle defined in rule 1.6 be consistent in sign and magnitude with the conformational regions (± *syn*, ± *anti*) shown in Fig. 6, it is necessary, when looking down the B→C (or C→B) bond, that the front bond A−B (or D−C) should define the zero (0°) position, and that the back bond should define the conformational region.

Examples

a) In Fig. 3 the O5′−C5′ bond makes an angle of + 60° to C4′−C3′ and 300° (− 60°) to the C4′−O4′ bond. With the O5′−C5′ bond defining the zero position these conformations correspond to the + *sc* and − *sc* regions, respectively.

b) For the sequence of atoms $A − B − C − D$ as shown in Fig. 5 A the same conformation with the torsion angle θ being positive (+ θ) is found when looking along either the B−C bond (A to the front) or the C−B bond (D to the front);

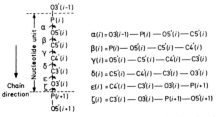

$\alpha(i) = O3'(i-1) - P(i) - O5'(i) - C5'(i)$

$\beta(i) = P(i) - O5'(i) - C5'(i) - C4'(i)$

$\gamma(i) = O5'(i) - C5'(i) - C4'(i) - C3'(i)$

$\delta(i) = C5'(i) - C4'(i) - C3'(i) - O3'(i)$

$\epsilon(i) = C4'(i) - C3'(i) - O3'(i) - P(i+1)$

$\zeta(i) = C3'(i) - O3'(i) - P(i+1) - O5'(i+1)$

Fig. 7. *Torsion angles for backbone conformations of the ith nucleotide in polynucleotide chains*

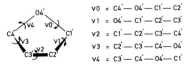

$v0 = C4' - O4' - C1' - C2'$

$v1 = O4' - C1' - C2' - C3'$

$v2 = C1' - C2' - C3' - C4'$

$v3 = C2' - C3' - C4' - O4'$

$v4 = C3' - C4' - O4' - C1'$

Fig. 8. *Torsion angles in sugar rings of β-D-nucleosides and nucleotides*

The sequence of atoms used to define each backbone torsion angle is shown in Fig. 8; e.g. $v0$ refers to the torsion angle of the sequence of atoms $C4' - O4' - C1' - C2'$, etc.

this conformation is described as $+sc$. Similarly the conformation designated by a negative value of θ $(-\theta)$, shown in Fig. 5 B, corresponds to the $-sc$ region.

c) See section 2.3, notes (a) and (c), for the example in Fig. 11.

2. THE NUCLEOTIDE UNIT

The notations used to designate the various torsion angles in the nucleotide unit are indicated in three sections: sugar-phosphate backbone chain, sugar ring and sugar-base side chain.

2.1. *Sugar-Phosphate Backbone Chain* (main chain)

The backbone of a polynucleotide chain consists of a repeating unit of six single bonds as shown in Fig. 1, viz. $P - O5', O5' - C5', C5' - C4', C4' - C3', C3' - O3'$ and $O3' - P$. The torsion angles about these bonds are denoted, respectively, by the symbols $\alpha, \beta, \gamma, \delta, \epsilon, \zeta$. The symbols $\alpha(i) - \zeta(i)$ are used to denote torsion angles of bonds within the ith nucleotide unit as shown in Fig. 4 and 7.

The sequence of main-chain atoms used to define each backbone torsion angle is shown in Fig. 7.

Note

The recommended $\alpha - \zeta$ notation differs from the ϕ, ψ, ω notation [4, 7] adopted by many workers and from an earlier $\alpha - \zeta$ notation [8].

A substantial majority of the subcommittee favoured the $\alpha - \zeta$ notation because it is convenient to remember for a backbone repeat of six bonds. The recommended $\alpha - \zeta$ notation is the second of the systems proposed by Seeman et al. [8] and was chosen because it starts at the phosphorus atom which is the first atom of the nucleotide unit, has the highest atomic number, and is the only atom of its kind in the backbone.

2.2. *Sugar Ring*

2.2.1. Endocyclic Torsion Angles

The sugar ring occupies a pivotal position in the nucleotide unit because it is part of both the backbone and the side chain. In order to provide a complete description of the ring conformation, it is necessary to specify the endocyclic torsion angles for the ring as well as the bond lengths and bond angles. The five endocyclic torsion angles for the bonds $O4' - C1', C1' - C2', C2' - C3', C3' - C4'$ and $C4' - O4'$ are denoted by the symbols, $v0, v1, v2, v3$ and $v4$, respectively.

Notes

a) The backbone torsion angle δ and the endocyclic torsion angle $v3$ both refer to rotation about the same bond, $C4' - C3'$. Both angles are needed for complete description of the main-chain and sugar-ring conformations in some studies.

b) The notation $\tau_0 - \tau_4$ previously used [4, 19–21] to represent torsion angles about the bonds in the sugar ring is superseded by the present notation (v), which is consistent with polysaccharide nomenclature [12]. The symbol τ is now used to denote a bond angle, which is consistent with polypeptide nomenclature [10].

2.2.2. Puckered Forms

Since the sugar ring is generally non-planar, its conformation may need designation. If four of its atoms lie in a plane, this plane is chosen as a reference plane, and the conformation is described as envelope (E); if they do not, the reference plane is that of the three atoms that are closest to the five-atom, least-squares plane, and the conformation is described as twist (T) [22, 23]. Atoms that lie on the side of the reference plane from which the numbering of the ring appears clockwise are written as superscripts and precede the letter $(E$ or $T)$; those on the other side are written as subscripts and follow the letter (Fig. 9). These definitions [23] mean that atoms on the same side of the plane as C5 in D-ribofuranose derivatives are written as preceding superscripts.

Notes

a) The present E and T notations for puckered forms of the sugar ring conform to those recommended for the conformational nomenclature of five and six-membered rings of monosaccharides and their derivatives [23].

b) The E/T notation has superseded the *endo/exo* description [24], in which atoms now designated by superscripts were called *endo*, and those now designated by subscripts were called *exo*. Fig. 10 shows both systems of designation. Examples:

$C3'$-*endo*/$C2'$-*exo* has become $^3T_2$

$C3'$-*endo* has become ^3E.

c) Symmetrical twist conformations, in which both atoms exhibit equal displacements with respect to the five-atom plane, are denoted by placing the superscript and subscript on the same side of the letter T, e.g. ^2_3T, ^4_3T, etc.

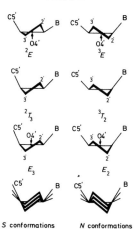

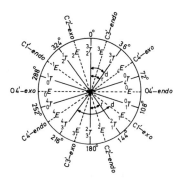

S conformations   N conformations

Fig.9. *Diagrammatic representation of sugar-ring conformations of β-D-nucleosides and their relation to the pseudorotational N-type and S-type conformers (Section 2.2.3.). The purine or pyrimidine base is represented by B. (Figure adapted from that of Saenger [25])*

Fig.10. *The pseudorotational pathway of the D-aldofuranose ring, showing the relation between phase angle of pseudorotation P (0–360°), the envelope (E) and twist (T) notations and the endo and exo notations. N-Type conformations correspond to the northern half (P = 0 ± 90°) and S-type correspond to the southern half of the pseudorotational cycle. The symbols 'r' and 'd' represent the usual range of P values for N and S conformations of ribo- (r) and 2'-deoxyribo- (d) furanose rings of β-D-nucleosides and nucleotides. (Diagram adapted from the work of Altona and Sundaralingam [26])*

2.2.3. Pseudorotational Analysis

The sugar ring conformation has also been described by Altona and Sundaralingam [26] using the concept of pseudorotation, which has been found advantageous in describing the conformational dynamics of the sugar ring [27].

Each conformation of the furanose ring can be unequivocally described by two pseudorotational parameters: the phase angle of pseudorotation, P, and the degree of pucker, ψ_m. A standard conformation ($P = 0°$) is defined with a maximally positive $C1'-C2'-C3'-C4'$ torsion angle [i.e. the symmetrical ^3_2T form], and P has value $0-360°$. Conformations in the upper or northern half of the circle ($P = 0 \pm 90°$) are denoted N and those in the southern half of the circle ($P = 180 \pm 90°$) are denoted S conformation. The relationship between P and the *endo/exo* and T/E notations is illustrated in Fig. 10. It may be seen that the symmetrical twist (T) conformations arise at even multiples of 18° of P and the symmetrical envelope (E) conformations arise at odd multiples of 18° of P.

Note. The present designation of the degree of pucker (ψ_m) differs from the original notation (τ_m) of Altona and Sundaralingam [26] in order to avoid confusion with the notation for bond angles.

2.3. N-Glycosidic Bond

The torsion angle about the N-glycosidic bond ($N-C1'$) that links the base to the sugar is denoted by the symbol χ which is the same as the notation used to denote side-chain torsion angles in polypeptides [10]. $\chi(i)$ denotes the torsion angle in the ith nucleotide unit.

The sequence of atoms chosen to define this angle is $O4'-C1'-N9-C4$ for purine and $O4'-C1'-N1-C2$ for pyrimidine derivatives. Thus when $\chi = 0°$ the $O4'-C1'$ bond is eclipsed with the $N9-C4$ bond for purine and the $N1-C2$ bond for pyrimidine derivatives. The definitions of torsion angles (section 1.6) of the N-glycosidic bond are illustrated, looking along the bond, in Fig. 11.

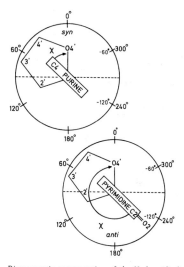

Fig.11. *Diagrammatic representation of the N-glycosidic bond torsion angle χ and the syn and anti regions for purine and pyrimidine derivatives. The purine derivative (upper left) is viewed down the N9–C1' bond and is shown in the + sc conformation. The pyrimidine derivative (lower right) is viewed down the N1–C1' bond and is shown in the − ac conformation. The sugar ring is shown as a regular pentagon*

Notes

a) The choice of bond sequence to define χ is based on accepted chemical nomenclature [13] and, at the same time, the use of the terms *syn* and *anti* to describe different conformational regions of χ for purine and pyrimidine derivatives is now consistent with accepted chemical nomenclature (Rule 1.7), viz.

$$syn, \chi = \quad 0 \pm 90°$$
$$anti, \chi = 180 \pm 90°.$$

Examples of syn and anti conformations are shown in Fig. 11. In the new convention most conformations formerly described as syn and anti remain syn and anti respectively, except the high-anti region which may be described as $-syn$-clinal $(-sc)$, Rule 1.7.

b) Many of the conventions for defining the torsion angle of a bond in nucleic acids [1–7,28,29] have been based on the sugar ring O4' atom [4,5,25] or C2' atom [3,29] in conjunction with the base ring C8(Pur)/C6(Pyr) atoms [4,5,29] or base ring C4(Pur)/C2(Pyr) atoms [3,28]. Approximate relationships between the definitions of torsion angles have been summarised by Sundaralingam [30], and these aid the comparison of conformations described in the older literature.

A substantial portion of the literature has used the nomenclature based on sugar ring O4'–C1' and base ring N9–C8-(purine) and N1–C6(pyrimidine). χ_{old} is related to the present definition χ_{new} by the relation,

$$\chi_{new} \approx \chi_{old} \pm 180°.$$

c) Following the definitions of conformational regions in Rule 1.7 and Fig. 6 the anti conformation of the pyrimidine example in Fig. 11 corresponds to the $-$anticlinal $(-ac)$ conformation and the syn conformation of the purine derivative corresponds to the $+$synclinal $(+sc)$ conformation.

2.4. Orientation of Side Groups

For the precise definition of the orientation of any pendant groups specification of the torsion angle about the exocyclic bond is necessary. The exocyclic torsion angle may be denoted by the symbol η with a locant to indicate the atom to which it refers. Examples:

Ribose Rings. The symbol $\eta2'$ may be used to denote the torsion angle about the C2'–O2' bond for the sequence of atoms C1'–C2'–O2'–X where X = H, CH₃, PO₃²⁻, etc. If no confusion is possible, the symbol η (without any additional index) may be used for the C2'–O2' bond.

Base. Torsion angles for bonds in base rings e.g. C6–N6 in adenine, C2–N2 in guanine and C4–N4 in cytosine may be specified by $\eta6$, $\eta2$ and $\eta4$, respectively.

When the groups are substituted by hydrogen atoms only (see Fig. 2), the relevant dihedral angles defined by the sequence rules are:

$$\eta61 = N1-C6-N6-H61$$
$$\eta62 = N1-C6-N6-H62$$
$$\eta21 = N1-C2-N2-H21$$
$$\eta22 = N1-C2-N2-H22$$
$$\eta41 = N3-C4-N4-H41$$
$$\eta42 = N3-C4-N4-H42.$$

The rules may be adapted for substituted base and sugar rings, such as those of minor components of tRNA. Examples:

1-Methyladenosine: use $\eta11$, $\eta12$, $\eta13$ for the C–CH₃ group conformation.

2'-O-Methyladenosine: use $\eta2'$ for rotation about the C2'–O2' bond and $\eta2'1$, $\eta2'2$, $\eta2'3$ for the OCH₃ group.

Note. Recommendations governing the description of conformations of side chains and derivatives follow the sequence rules (Rule 2) and the side-chain rules (Rule 4) of the recommendations for polypeptides [10].

3. HYDROGEN BONDS

3.1. Polarity of Hydrogen Bonds

In specifying a hydrogen bond the atom covalently linked to the hydrogen atom is mentioned first, as in X–H · · · Y. The polarity of a hydrogen bond is from the hydrogen-atom donor to the acceptor.

3.2. Geometry of Hydrogen Bonds

The hydrogen bond may be described by extension of the nomenclature of sections 3.1, 1.4, 1.5 and 1.6, so that for the hydrogen bond in the system $C(i)-X(i)-H(i) \cdots Y(k)-C(k)$ the following symbols may be used:

$$b[H(i) \cdots Y(k)] \qquad \text{or } b[H(i),Y(k)]$$
$$\tau[X(i)-H(i) \cdots Y(k)] \quad \text{or } \tau[X(i),H(i),Y(k)]$$
$$\tau[H(i) \cdots Y(k)-C(k)] \quad \text{or } \tau[H(i),Y(k),C(k)].$$

Where the positions of hydrogen atoms are not available the following may be used:

$$b[X(i),Y(k)] \text{ and } \tau[C(i),X(i),Y(k)].$$

4. HELICAL SEGMENTS

A regular helix is strictly of infinite length, with the torsion angles in a nucleotide unit the same for all units. Two or more polynucleotide chains may associate in a helical complex by hydrogen bonding between base pairs. Torsion angles for each residue may differ for different chains in the same double, triple, etc. complex.

A helical segment of a polynucleotide chain may be described in terms of torsion angles of the nucleotide units or in terms of the helix characteristics summarized in section 4.2.

4.1. Base Pairs

Base pairs with different geometries have been observed. These geometries should be denoted by the appropriate hydrogen-bonding scheme specifying both the heterocyclic base (e.g. Ade, Ura, Gua, Cyt) and the heteroatoms involved in the hydrogen bonding. In some cases it is also desirable to specify the nucleotide unit (i, j, etc.). Typical examples are: for a Watson-Crick A:U base pair:

AdeN6:O4Ura, UraN3:N1Ade
[if necessary, Ade(i)N6:O4Ura(j), etc.]

for a reversed Watson-Crick A:U base pair:

AdeN6:O2Ura, UraN3:N1Ade.

4.2. Helix Characteristics

In the description of helices or helical segments the following symbols should be used:

n = number of residues per turn
h = unit height (translation per residue along the helix axis)
$t = 360°/n$ = unit twist (angle of rotation per residue about the helix axis)
p = pitch height of helix = $n.h$.

A polynucleotide may be accurately described in terms of the polar atomic co-ordinates r_i, ϕ_i, z_i where for each atom i, r_i is the radial distance from the helix axis and ϕ_i and z_i are the angular and height differences respectively, relative to a reference point. The reference point should be either a symmetry element, as in RNA and DNA, or the C1' atom of a nucleotide if no symmetry element between polynucleotide chains is present.

REFERENCES

1. Donohue, J. & Trueblood, K. N. (1960) *J. Mol. Biol. 2*, 363 – 371.
2. Sasisekharan, V., Lakshminarayanan, A. V. & Ramachandran, G. N. (1967) in *Conformation of Biopolymers II* (Ramachandran, G. N., ed.) pp. 641 – 654, Academic Press, New York.
3. Arnott, S. & Hukins, D. W. L. (1969) *Nature (Lond.)*, *224*, 886 – 888.
4. Sundaralingam, M. (1969) *Biopolymers, 7*, 821 – 860.
5. Lakshminarayanan, A. V. & Sasisekharan, V. (1970) *Biochim. Biophys. Acta, 204*, 49 – 59.
6. Olson, W. K. & Flory, P. J. (1972) *Biopolymers, 11*, 1 – 23.
7. Sundaralingam, M., Pullman, B., Saenger, W., Sasisekharan, V. & Wilson, H. R. (1973) in *Conformations of Biological Molecules and Polymers* (Bergman, E. D. & Pullman, B., eds) pp. 815 – 820, Academic Press, New York.
8. Seeman, N. C., Rosenburg, J. M., Suddath, F. L., Kim, J. J. P. & Rich, A. (1976) *J. Mol. Biol. 104*, 142 – 143.
9. Davies, D. B. (1978) in *NMR in Molecular Biology* (Pullman, B., ed.) pp. 509 – 516, Reidel, Dordrecht.
10. IUPAC-IUB Commission on Biochemical Nomenclature (CBN) Abbreviations and symbols for the description of the conformation of polypeptide chains, Tentative rules 1969 (approved 1974), *Arch. Biochem. Biophys. 145*, 405 – 421 (1971); *Biochem. J. 121*, 577 – 585 (1971); *Biochemistry, 9*, 3471 – 3479 (1970); *Biochim. Biophys. Acta, 229*, 1 – 17 (1971); *Eur. J. Biochem. 17*, 193 – 201 (1970); *J. Biol. Chem. 245*, 6489 – 6497 (1970); *Mol. Biol.* (in Russian) *7*, 289 – 303 (1973); *Pure Appl. Chem. 40*, 291 – 308 (1974); also on pp. *94 – 102* in [11].
11. International Union of Biochemistry (1978) *Biochemical Nomenclature and Related Documents*, The Biochemical Society, London.
12. IUPAC-IUB Joint Commission on Biochemical Nomenclature (JCBN) Symbols for specifying the conformation of polysaccharide chains, Recommendations 1981, *Eur. J. Biochem. 131*, 5 – 7 (1983).
13. IUPAC Commission on Macromolecular Nomenclature (CMN), Stereochemical definitions and notations relating to polymers, *Pure Appl. Chem. 53*, 733 – 752 (1981).
14. IUPAC-IUB Commission on Biochemical Nomenclature (CBN) Abbreviations and symbols for nucleic acids, polynucleotides and their constituents, Recommendations 1970, *Arch. Biochem. Biophys. 145*, 425 – 436 (1971); *Biochem. J. 120*, 449 – 454 (1970); *Biochemistry, 9*, 4022 – 4027 (1970); *Biochim. Biophys. Acta, 247*, 1 – 12 (1971); *Eur. J. Biochem. 15*, 203 – 208 (1970) corrected *25*, 1 (1972); *Hoppe-Seyler's Z. Physiol. Chem.* (in German) *351*, 1055 – 1063 (1970); *J. Biol. Chem. 245*, 5171 – 5176 (1970); *Mol. Biol.* (in Russian) *6*, 167 – 174 (1972); *Pure Appl. Chem. 40*, 277 – 290 (1974); also on pp. *116 – 121* in [11].
15. International Union of Pure and Applied Chemistry (1979) *Nomenclature of Organic Chemistry, Sections A, B, C, D, E, F and H*, (Rigandy, J. & Klesney, S. P., eds) Pergamon Press, Oxford.
16. IUPAC Commission on the Nomenclature of Organic Chemistry (CNOC) Definitive rules for nomenclature of organic chemistry, *J. Am. Chem. Soc. 82*, 5545 – 5574 (1960).
17. IUPAC Commission on Nomenclature of Organic Chemistry (CNOC) Rules for the nomenclature of organic chemistry, Section E: Stereochemistry, Recommendations 1974, *Pure Appl. Chem. 45*, 11 – 30 (1976); also on pp. 473 – 490 in [15] and on pp. *1 – 18* in [11].
18. Klyne, W. & Prelog, V. (1960) *Experientia, 16*, 521 – 523.
19. Arnott, S. & Hukins, D. W. L. (1972) *Biochem. Biophys. Res. Commun. 47*, 1504 – 1509.
20. Pullman, B. Perahia, D. & Saran, A. (1972) *Biochim. Biophys. Acta, 269*, 1 – 14.
21. Saran, A. & Govil, G. (1971) *J. Theor. Biol. 33*, 407 – 418.
22. Hall, L. D. (1963) *Chem. Ind. (Lond.)* 950 – 951.
23. IUPAC-IUB Joint Commission on Biochemical Nomenclature (JCBN) Conformational nomenclature for five and six-membered ring forms of monosaccharides and their derivatives, Recommendations 1980, *Arch. Biochem. Biophys. 207*, 469 – 472 (1981); *Eur. J. Biochem. 111*, 295 – 298 (1980); *Pure Appl. Chem. 53*, 1901 – 1905 (1981).
24. Jardetzky, C. D. (1960) *J. Am. Chem. Soc. 82*, 229 – 233.
25. Saenger, W. (1973) *Angew. Chem. Int. Ed. Engl. 12*, 591 – 601.
26. Altona, C. & Sundaralingam, M. (1972) *J. Am. Chem. Soc. 94*, 8205 – 8212.
27. Altona, C. & Sundaralingam, M. (1973) *J. Am. Chem. Soc. 95*, 2333 – 2334.
28. Kang, S. (1971) *J. Mol. Biol. 58*, 297 – 315.
29. Saenger, W. & Scheit, K. H. (1970) *J. Mol. Biol. 50*, 153 – 169.
30. Sundaralingam, M. (1973) in *Conformations of Biological Molecules and Polymers* (Bergman, E. D. & Pullman, B., eds) pp. 417 – 455, Academic Press, New York.

INDEX OF SUBJECTS

Cancer Res. 37, 4389-4394 (1977).
12. L. Thelander, S. Eriksson and M. Akerman: J. Biol. Chem. 255, 7426-7432 (1980).
13. M. Lammers and M. Brendel, unpublished results (1983).
14. G. Bjursell and P. Reichard: J. Biol. Chem. 248, 3904-3909 (1973).
15. D. Filpula and J.A. Fuchs: J. Bacteriol. 130, 107-113 (1977); J. Bacteriol. 139, 694-696 (1979).
16. G. Schimpff-Weiland, H. Follmann and G. Auling: Biochem. Biophys. Res. Commun. 102, 1276-1282 (1981).
17. R. Hofmann, Ph. D. thesis, Philipps-Universität, Marburg, Germany (1983).
18. G.P.V. Reddy and A.B. Pardee: Proc. Natl. Acad. Sci. USA 77, 3312-3316 (1980); J. Biol. Chem. 257, 12526-12531 (1982).
19. J.R. Allen, G.P.V. Reddy, G.W. Lasser and C.K. Mathews: J. Biol. Chem. 255, 7583-7588 (1980).
20. E.H. McConkey: Proc. Natl. Acad. Sci. USA 79, 3236-3240 (1982).
21. L. Thelander, A. Gräslund and M. Thelander: Biochem. Biophys. Res. Commun. 110, 859-865 (1983).
22. M. Löffler, G. Schimpff-Weiland and H. Follmann: FEBS-Lett. in press (1983).
23. H. Follmann: Naturwissenschaften 69, 75-81 (1982).
24. A. Tkac and L. Omelka: Org. Magn. Reson. 13, 406-416 (1980); 14, 109-119 (1980).
25. C. Helene and J.-C. Maurizot: CRC Critical Rev. Biochem. 10, 213-258 (1981).
26. S.R. Dickman: J. Mol. Evol. 10, 251-260 (1977).
27. W.F. Doolittle and C. Sapienza: Nature 284, 601-603 (1980); L.E. Orgel and F.H.C. Crick: Nature 284, 604-607 (1980).

RETURN **CHEMISTRY LIBRARY**
TO ➡ 100 Hildebrand Hall 642-3753

LOAN PERIOD 1	2	3
7 DAYS		
4	5	6

ALL BOOKS MAY BE RECALLED AFTER 7 DAYS
Renewable by telephone

DUE AS STAMPED BELOW

MAY 9 1985		
JUL 2 3 198		
SEP 1 2 19		
SEP 2 3 19		
MAY 1 3 1987		

FORM NO. DD5, 3m, 12/80

UNIVERSITY OF CALIFORNIA, BERKELEY
BERKELEY, CA 94720